MATHEMATICAL METHODS IN ENGINEERING

Mathematical Methods in Engineering

Edited by

K. TAŞ
Çankaya University,
Balgat-Ankara, Turkey

J.A. TENREIRO MACHADO
Institute of Engineering of Porto
Porto, Portugal

and

D. BALEANU
Çankaya University,
Balgat-Ankara, Turkey

Springer

A C.I.P. Catalogue record for this book is available from the Library of Congress.

ISBN-10 1-4020-5677-X (HB)
ISBN-13 978-1-4020-5677-2 (HB)
ISBN-10 1-4020-5678-8 (e-book)
ISBN-13 978-1-4020-5678-9 (e-book)

Published by Springer,
P.O. Box 17, 3300 AA Dordrecht, The Netherlands.

www.springer.com

Printed on acid-free paper

Contents

Applications of wavelets

Dynamical systems and control theory

Preface

This book contains some of the contributions under five main titles that are carefully selected according to the reports of referees, presented at the International Symposium, MME06 Mathematical Methods in Engineering, held in Çankaya University, Ankara, April 27–29, 2006.

The Symposium provided a setting for discussing recent developments in Fractional Mathematics, Neutrices and Generalized Functions, Boundary Value Problems, Applications of Wavelets, Dynamical Systems and Control Theory.

The members of the organizing committee were Dumitru Baleanu, Ronald A. DeVore, J.A. Tenreiro Machado, Ali H. Nayfeh and Kenan Tas (Chairman). Lecturers of the Symposium were Om P. Agrawal, Brian Fisher, J.A. Tenreiro Machado, Francesco Mainardi, Hans J. Stetter.

The editors of this book are grateful to the President of the board of trustees of Çankaya University Sitki Alp, to the Rector Prof.Dr. Ziya Aktaş for their continuous support of the Symposium activities. We are also obliged to the TUBITAK (The Scientific and Technological Research Council of Turkey) for their co-sponsorship.

We would like to thank all the referees and other colleagues who helped in preparing this book for publication. Our thanks are also due to all participants for their contributions to the Symposium and to this book.

The Organizing Committee wishes to express their thanks to Prof.Dr. Emel Dogramaci, Dean of the Faculty of Arts and Science and to the colleagues of the Department of Mathematics and Computer Science of Çankaya University: A. Eris, F. Kumsel, M. Cenk, A. Bilgen, H. Baydar, Ozlem Defterli and A. Kabarcik who provided valuable work during the Symposium, and to our students: C. Acar, S. Arik, M.T. Dinc, K. Dogan, K.U. Aydin, R. Cebe, S. Ozsahin, B. Tasar, H. Hatipoglu, Y. Kaya, I.N. Ekim for their assistance.

Our special thanks are due to Nathalie Jacobs and Anneke Pot from Springer-Verlag for their continuous help and work in connection with this book. Finally, we would like to express our gratitude to Isil Gence, web master of the Symposium.

Ankara, August 8, 2006

Kenan Tas, J.A. Tenreiro Machado, Dumitru Baleanu
Editors

Fractional Mathematics

Fractional calculus and regularized residue of infinite dimensional space

Asada Akira

Faculty of Science, Sinsyu University, Matumoto, Japan asada-a@poporo.ne.jp

We have proposed regularization of infinite dimensional integral via fractional calculus. It is done on a Hilbert space H equipped with a Schatten class operator G. The ζ-function $\zeta(G, s)$ of G is assumed to be holomorphic at s=0. Regularization is done by using $\zeta(G, s)$. After reviewing this regularization, it is shown regularized Cauchy kernel of a Hilbert space with the determinant bundle exists if and only if $\nu = \zeta(G, 0)$ is an integer. Regularized residue on an infinite dimensional space is obtained as an application of regularized Cauchy kernel.

1 Fractional calculus and regularized infinite product

Let $\{H, G\}$ be a pair of a Hilbert space and a positive Schatten class operator G such that $\zeta(G, s) = \mathrm{tr}G^s$ is holomorphic at $s = 0$. $\zeta(G, s)$ is assumed to have its first pole at $s = d$. We also set

$$\nu = \zeta(G, 0), \quad detG = e^{\zeta'(G,0)}, \quad c = \mathrm{Res}_{s=d}\zeta(G, s).$$

We often need integrity of ν. If H is the Hilbert space of square integrable sections of a bundle E over a compact Riemannian manifold X and G is the Green operator of a positive elliptic operator D acting on the sections of E, choosing suitable mass term m and replace D by $D + mI$, ν becomes an integer. Hence integrity of ν is not restrictive for practical use (cf.[Asa04a]).

The complete ortho-normal basis $e_1, e_2, \ldots$, are taken from eigenvectors of G: $Ge_n = \mu_n e_n$, $\mu_1 \geq \mu_2 \geq \ldots > 0$. By using G, we introduce Sobolev metric $\|x\|_k$ by $\|G^{-k}x\|$. The Sobolev space constructed by H and $\|\cdot\|_k$ is denoted by W^k. The complete ortho-normal basis of W^k is given by $e_{1,k}, e_{2,k}, \ldots$, $e_{n,k} = \mu_n^k e_n$. We set $e_{\infty,k} = \sum_{n=1}^{\infty} \mu_n^{d/2} e_{n,k}$. $e_{\infty,k}$ does not belong to W^k, but belongs to W^l, $l < k$. If $k = 0$, we denote e_∞, instead of $e_{\infty,0}$.

Definition 1. *The Hilbert space $W^{k,\natural}$ is $W^k \oplus \mathcal{K}e_{\infty,k}$ with the inner product*

$$\langle e_{n,k}, e_{m,k}\rangle = \delta_{n.m}, \quad \langle e_{\infty,k}, e_n\rangle = 0, \quad \langle e_{\infty,k}, e_{\infty,k}\rangle = c. \tag{1}$$

Kenan Taş et al. (eds), Mathematical Methods in Engineering, 3–11.

Here $\mathcal{K}$ is $\mathcal{R}$ if H is a real Hilbert space, and $\mathcal{C}$ if H is a complex Hilbert space. If $k=0$, we denote $H^{\natural}$, instead of $W^{0,\natural}$. We identify W^k and $W^k \oplus 0e_{\infty,k} \subset W^{k,\natural}$. Then the above inner product on $W^{k,\natural}$ coincides with the inner product of W^k. While the inner products $\langle e_{\infty,k}, e_{n,k}\rangle$ and $\langle e_{\infty,k}, e_{\infty,k}\rangle$ come from

$$\langle e_{\infty,k}, e_{n,k}\rangle = \lim_{s\downarrow 0}(\sqrt{s}G^{s/2-k}e_{\infty,k}, \sqrt{s}G^{s/2-k}e_{n,k}),$$

where (,) is the inner product of H.

By definition, $x \in W^{k,\natural}$ is uniquely written as $x_f + te_{\infty,k}$. Hence we can write

$$x = x_f + te_{\infty,k} = \sum_{n=1}^{\infty} x_{f,n}e_{n,k} + te_{\infty,k} = \sum_{n=1}^{\infty} x_n e_{n,k}, \quad x_n = x_{f,n} + \mu_n^{d/2}t. (2)$$

Let $I_n^a f$ be the fractional integral $I_n^a f(x_n) = \dfrac{1}{\Gamma(a)}\displaystyle\int_0^{x_n} \frac{f(t)}{(x-t)^{1-a}}dt$. Then we have

$$\lim_{n\to\infty} I_1^{\mu_1^s}\cdots I_n^{\mu_n^s}1 = \prod_{n=1}^{\infty}\Gamma(1+\mu_n^s)\prod_{n=1}^{\infty}x_n^{\mu_n^s}.$$

Since

$$\log\prod_{n=1}^{\infty}\Gamma(1+\mu_n^s) = -\gamma\zeta(G,s) + \sum_{m=2}^{\infty}(-1)^m\frac{\zeta(m)}{m}\zeta(G,ms),$$

taking a path $C=C(s)$; $0\le s\le 1$ in the right half plane such that $C(0)=1$ and does not tangent to real and imaginary axes, the analytic continuation of $\prod_{n=1}^{\infty}\Gamma(1+\mu_n^s)$ to $s=0$ along C takes the value 1.

Definition 2. *Let $x = \sum_{n=1}^{\infty} x_n e_n$ be an element of $W^{k,\natural}$. Then we define regularized infinite product $:\prod_{n=1}^{\infty} x_n:$ of $x_1, x_2, \ldots$ by*

$$:\prod_{n=1}^{\infty} x_n := \prod_{n=1}^{\infty} x_n^{\mu_n^s}|_{s=0}. \tag{3}$$

Here $|_{s=0}$ means analytic continuation to $s=0$.

It is known $:\prod_n x_n:$ is linear in each variable x_n and

$$|:\prod_{n=1}^{\infty} x_n:| =: \prod_{n=1}^{\infty}|x_n|:, \quad (:\prod_{n=1}^{\infty} x_n)^m =: \prod_{n=1}^{\infty} x_n^m: . \tag{4}$$

If $x = x_f + te_{\infty,k} \in W^{k,\natural}$ and $t \neq 0$, we have

$$:\prod_{n=1}^{\infty} x_n := t^{\nu}(detG)^{k+d/2}\prod_{n-1}^{\infty}(1+\frac{\mu_n^{-(k+d/2)}x_{f,n}}{t})^{\mu_n^s}|_{s=0}.$$

Then regarding $W^l, l>k$ to be a subset of $W^k \subset W^{k,\natural}$, and $W^{1,l}$, etc., to be ℓ^1-type subset $\{\sum_n x_n e_{n,l} | \sum |x_n| < \infty\}$, etc., of W^l, etc., we have

Proposition 1. *$:\prod_n x_n:$ is a single valued function if and only if ν is an integer, and the followings hold;*

1. *If $t \neq 0$ and $x_f \in W^{1,k+d/2}$, then $:\prod_n x_n:$ exists. If $x \in W^{k+d/2}$, it exists if and only if $\sum_{n=1}^{\infty} \mu_n^{s-(k+d/2)} x_{f,n}$ is holomorphic at $s=0$.*
2. *$:\prod_n x_n:$ is analytic on $W^{k+d/2} \oplus \mathcal{C}e_{\infty,k}$.*

2 Regularized determinant

Let T be a densely defined linear operator on H. Then its regularized trace (renormalized trace) with respect to G is defined by

$$tr_G T = \mathrm{tr}(G^s T)|_{s=0},$$

[CDP02, Payc01]. For example, $\mathrm{tr}_G I = \nu$. By using regularized trace, we define

Definition 3. *If T has the logarithm $S = \log T$; $T = e^S$, then we define regularized determinant $det_G T$ of T with respect to G by*

$$det_G T = e^{\mathrm{tr}_G T} = e^{\mathrm{tr}(G^s T)}|_{s=0}. \tag{5}$$

Note 1. Since $\log T$ is not unique, $det_G T$ is not unique in general.

Example 1. If $I = I_x$; $x = (x_1, x_2, \ldots)$, is a scaling operator $I_x e_n = x_n e_n$, then $\log I_x$ is $I_{\log x}$; $\log x = (\log x_1, \log x_2, \ldots)$. Hence we have

$$det_G I_x = e^{\sum_{n=1}^{\infty} \mu_n^s \log x_n}|_{s=0} = \prod_{n=1}^{\infty} x_n^{\mu_n^s}|_{s=0} =: \prod_{n=1}^{\infty} x_n : .$$

Especially, we have

$$det_G G = det G, \quad det_G D = det D, \; G = D^{-1}, \tag{6}$$

where $det D$ is the Ray-Singer determinant of D.

Note 2. We have $det_G(I_x + N) = det_G I_x$, if N is a generalized nilpotent.

On the other hand, we have only $det_G PTP^{-1} = det_{P^{-1}GP} T$ in general. It may different from $det_G T$. For example, if G and T are

$$Ge_{2n-1} = \frac{1}{n} e_{2n-1}, Ge_{2n} = \frac{1}{n+1} e_{2n}, \; Te_{2n-1} = 2e_{2n-1}, Te_{2n} = 3e_{2n},$$

and $Pe_{2n-1} = e_{2n}$, $Pe_{2n} = e_{2n-1}$, then

$$det_G T = 2^{-1/2} 3^{-3/2} \neq det_G PTP^{-1} = 2^{-3/2} 3^{-1/2}.$$

We have $det_G T = det_G PTP^{-1}$ if $P \in GL(\infty)$, where $GL(\infty)$ is the closure of the group of invertible linear operators of the form $I + K$, K is a compact operator.

3 Regularized integral

Let W^k be a real Hilbert space, f a function on W^k which is extended to $W^{k,\natural}$ and expressed as $f = \lim_{n\to\infty} f(x_1, \ldots, x_n)$. Then the regularized integral $\int_{W^{k,\natural}} f : d^\infty x :$ is defined by

$$\int_{W^{k,\natural}} f : d^\infty x := \lim_{n\to\infty} \int_{\mathcal{R}^n} f(x_1, \ldots, x_n) d(x_1^{\mu_1^s}) \cdots d(x_n^{\mu_n^s})|_{s=0}, \tag{7}$$

[Asa04b], cf.[Asa04a, Asa04c]. Regularized integral on

$$W_+^{k,\natural} = \{\sum_n x_{f,n} e_{n,k} + t e_{\infty,k} \in W^{k,\natural} | x_{f,n} \geq 0, n = 1, 2, \ldots, t \geq 0\},$$

is similarly defined.

Regularized integral simplifies the fractional calculus $\lim_{n\to\infty} I_1^{\mu_1^s} \cdots I_n^{\mu_n^s} f|_{s=0}$. It is also interpreted as an application of the weak limit

$$\lim_{N\to\infty} \frac{\partial^N}{\partial x_1 \cdots \partial x_N} : \prod_{n=1}^{\infty} x_n := 1,$$

which is hold on suitable function space [Asa04b].

Theorem 1. *Let I_a, $a = (a_1, a_2, \ldots)$; $I_a : W^k \to W^l$ be a scaling operator, and let $I_a^a st f(x) = f(I_a x)$, f a function on W^l. Then we have*

$$\int_{W^{l,\natural}} f : d^\infty x := \int_{W^{k,\natural}} |det_G I_a|^{-1} I_a^* f : d^\infty x : . \tag{8}$$

If I_a maps $W_+^{k,\natural}$ to $W_+^{l,\natural}$, then we also have

$$\int_{W_+^{l,\natural}} f : d^\infty x := \int_{W_+^{k,\natural}} |det_G I_a|^{-1} f : d^\infty x : .$$

Example 2. To set $e^{-\pi\|x\|^2} = 0$, if $\|x\| = \infty$, we extend $e^{-\pi\|x\|^2}$ to $H^\natural$. If G is the Green operator of an elliptic operator D, we have

$$e^{-\pi(x,Dx)} = I_{\sqrt{D}}^* e^{-\pi\|x\|^2}, \quad I_{\sqrt{D}} e_n = \sqrt{\lambda_n} e_n, \ \lambda_n = \mu_n^{-1}.$$

Hence we get

$$\int_{W^{1/2,\natural}} e^{-\pi(x,Dx)} : d^\infty x := \int_{H^\natural} |det\sqrt{D}|^{-1} e^{-\|x\|^2} : d^\infty x := \frac{1}{\sqrt{detD}}.$$

This justifies physicist's calculation $\int e^{-\pi(x,Dx)} \mathcal{D}x = \frac{1}{\sqrt{detD}}$.

4 Regularized Cauchy kernel

In the rest, we assume H is a complex Hilbert space. In $W^{k,\natural}$, we set

$$\mathcal{T}_r^{\infty,k} = \{\sum_{n=1}^{\infty} z_n e_{n,k} \in W^{k,\natural} | |z_n| = \mu_n^{d/2} r\}. \tag{9}$$

If $r = 1$, we denote $\mathcal{T}^{\infty,k}$ instead of $\mathcal{T}_1^{\infty,k}$. Considering $\mathcal{C}^n$ to be $\{\sum_{j=1}^n z_j e_j\}$, we have

$$\mathcal{T}_r^{\infty,k} \cap \mathcal{C}^n = \{\sum_{j=1}^{n} z_j e_j | |z_j| = \mu_j^{k+d/2} r\}.$$

We denote this set by $\mathcal{T}_r^{n,k}$ and set $D_r^{n,k} = \{\sum_{j=1}^n z_j e_j | |z_j| \leq r\mu_j^k\}$. Here k is omitted if $k = -d/2$ and r is omitted if $r = 1$.

By the map $w = z^a$, the circle $\{z = e^{i\theta} | 0 \leq \theta < 2\pi\}$ is mapped to $\{w = e^{i\phi} | 0 \leq \phi < 2a\pi\}$. That is we have

$$\frac{(2\pi i)^{a-1}}{a} \int_{|z|=1} \frac{d(z^a)}{z^a} = (2\pi i)^a, \qquad \int_{|z|=1} dz = \int_0^{2\pi} ie^{i\theta} d\theta.$$

Hence we have

$$\lim_{n\to\infty} \int_{\mathcal{T}^n} \left(\frac{(2\pi i)^{\mu_1^s - 1}}{\mu_1^s} \frac{d(z_1^{\mu_1^s})}{z_1^{\mu_1^s}} \cdots \frac{(2\pi i)^{\mu_n^s - 1}}{\mu_1^s} \frac{d(z_n^{\mu_n^s})}{z_n^{\mu_n^s}}\right)|_{s=0} = (2\pi i)^{\nu}. \tag{10}$$

Here, $\mathcal{T}^n$ is considered to be $\{e^{\theta_1 i} | 0 \leq \theta_1 < 2\pi\} \times \cdots \times \{e^{\theta_n i} | 0 \leq \theta_n < 2\pi\}$. We set : $d^{\infty} z : |_{\mathcal{T}^n} = \prod_{n=1}^{\infty} \left(\frac{(2\pi i)^{\mu_n^s - 1}}{\mu_n^s} d(z_n^{\mu_n^s})\right)|_{s=0}$. Then by (10), we have

$$\int_{\mathcal{T}^{\infty}} \frac{: d^{\infty} z : |_{\mathcal{T}^n}}{: \prod_{n=1}^{\infty} z_n :} = (2\pi i)^{\nu}. \tag{11}$$

This formula is valid if we regard $\mathcal{T}^{\infty} = \{e^{\theta_1 i} | 0 \leq \theta_1 < 2\pi\} \times \{e^{\theta_2 i} | 0 \leq \theta_2 < 2\pi\} \times \cdots$, because $: \prod_n z_n :$ is not single valued unless ν is an integer. But if ν is an integer, we can regard $\mathcal{T}^{\infty}$ to be an ∞-dimensional torus.

On the other hand, since $d(z^a)/z^a = adz/z$, we have

$$\lim_{a\to 1} \int_{\gamma} f(z) \frac{d(z^a)}{z^a} = \int_{\gamma} f(z) \frac{dz}{z} = 2\pi i f(0),$$

if γ is a closed curve in D^1 surrounding 0 and f is holomorphic on D^1. Hence we have

Theorem 2. *If ν is an integer, f is a holomorphic function on D^{∞} and $\gamma = \gamma_1 \times \gamma_2 \times \cdots, \gamma_n$ is a closed curve in $\{z_n | |z_n| < 1\}$ surrounding 0. Then we have*

$$f(0) = \frac{1}{(2\pi i)^{\nu}} \int_{\gamma} f(z) \frac{: dz^{\infty} : |_{\mathcal{T}^{\infty}}}{: \prod_{n=1}^{\infty} z_n :}. \tag{12}$$

Here, we say a function f on D^∞ to be holomorphic, if $\frac{\partial f}{\partial \bar{z}_n} = 0$, $n = 1, 2, \ldots$. In other words, f is holomorphic if it allows Taylor expansion $f(z) = \sum_{i_1,\ldots,i_m} c_{i_1,\ldots,i_m} z_1^{i_1} \cdots z_m^{i_m}$.

Since $dz_{n,k}/z_{n,k} = dz_n/z_n$, $z_{n,k} = \mu_n^{-k} z_n$, (12) is valid if f is holomorphic on $D^{\infty,k}$ and $\gamma \subset D^{\infty,k}$. By (12), if $\gamma_n = \partial \Gamma_n$, $\Gamma = \Gamma_1 \times \Gamma_2 \times \cdots$ and $\zeta = (\zeta_1, \zeta_2, \ldots)$, $\zeta_n \in \Gamma_n$, then we have the following *Cauchy's integral expression of a holomorphic function f on $D^{\infty,k}$*

$$f(\zeta) = \frac{1}{(2\pi i)^\nu} \int_\gamma f(z) \frac{: d^\infty z_n : |_{\mathcal{T}^\infty}}{: \prod_{n=1}^\infty (z_n - \zeta_n) :}. \tag{13}$$

Note 3. If ν is an integer, $: \prod_n z_n :$ is an analytic function, but not holomorphic. For this function, we have

$$\frac{1}{(2\pi i)^\nu} \int_{\mathcal{T}^\infty} : \prod_{n=1}^\infty z_n : \frac{: dz^\infty : |_{\mathcal{T}^\infty}}{: \prod_{n=1}^\infty (z_n - c_n)} = 0, \quad |c_n| < 1, \tag{14}$$

$$\frac{1}{(2\pi i)^\nu} \int_{\mathcal{T}^\infty} : \prod_{n=1}^\infty z_n : \frac{: dz^\infty : |_{\mathcal{T}^\infty}}{: \prod_{n=1}^\infty (z_n - c_n)} = \prod_{n=1}^\infty c_n, \quad |c_n| > 1. \tag{15}$$

Therefore $: \prod_n z_n :$ behaves as if the principal part of a meromorphic function.

5 De Rham type cohomology with ∞-degree elements

In the rest of this paper, we assume ν to be an integer.

Existence of regularized Cauchy kernel implies existence regularized volume form $: dv(\mathcal{T}^\infty) :$ on $\mathcal{T}_r^{\infty,k}$. To set $z_n = r_n e^{i\theta_n}$, we may set

$$: dv(\mathcal{T}^\infty) := \prod_{n=1}^\infty i(2\pi i)^{\mu_n^s - 1} d\theta_n|_{s=0}. \tag{16}$$

We also set

$$: dv(\mathcal{T}^{\infty - \{i_1,\ldots,i_p\}}) := \prod_{n \notin \{i_1,\ldots,i_p\}} i(2\pi i)^{\mu_n^s - 1} d\theta_n|_{s=0}, \tag{17}$$

and define

$$d\theta_{j_1} \wedge \cdots \wedge d\theta_{j_q} \wedge : dv(\mathcal{T}^{\infty - \{i_1,\ldots,i_p\}}) :=$$
$$= \begin{cases} \pm : dv(\mathcal{T}^{\infty - \{k_1,\ldots,k_r\}}) : if \{j_1,\ldots,j_q\} \cup \{k_1,\ldots,k_r\} = \{i_1,\ldots,i_p\}, \\ 0 \qquad\qquad\qquad\qquad : otherwise. \end{cases}$$

The cohomology algebra $H^*(\mathcal{T}^\infty, \mathcal{C})$ of $\mathcal{T}^\infty$ is the Grassmann algebra generated by $d\theta_1, d\theta_2, \ldots$. To define Hodge $*$-operator (Poincaré duality) by

$$*(d\theta_{i_1} \wedge \cdots \wedge d\theta_{i_p}) = (-1)^{i_1+\cdots+i_p-p(p-1)/2} : dv(\mathcal{T}^{\infty-\{i_1,\ldots,i_p\}}) :, \tag{18}$$

we obtain a de Rham type cohomology algebra

$$H^{*,*}(\mathcal{T}^\infty, \mathcal{C}) = H^*(\mathcal{T}^\infty, \mathcal{C}) \oplus *\big(H^*(\mathcal{T}^\infty, \mathcal{C})\big). \tag{19}$$

Note 4. Since multiplicative structure of $H^{*,*}(\mathcal{T}^\infty, \mathcal{C})$ depends on ν, it is *not* a topological invariant.

Let $W_*^{k,\natural}$ be $\{\sum_n z_n e_n \in W^{k,\natural} | z_n \neq 0, n = 1, 2, \ldots\}$, and $W_+^{k,\natural}$ is same as in §3. Then we have

$$W_*^{k,\natural} = \mathcal{T}^{\infty,k} \times (W_*^{k,\natural} \cap W_+^{k,\natural}). \tag{20}$$

Hence we can define de Rham type cohomology with infinite degree elements $H^{*,*}(W_*^{k,\natural}, \mathcal{C})$ of $W_*^{k,\natural}$ by the same way. In this case, we denote

$$*dz_{i_1} \wedge \ldots \wedge dz_{i_p} =: dz^{\infty-\{i_1,\ldots,i_p\}} : |_{\mathcal{T}^\infty}. \tag{21}$$

Let $W_{i_1,\ldots,i_p}^{k,\natural}$ be the subspace of $W^{k,\natural}$ defined by $z_{i_1} = 0, \ldots, z_{i_p} = 0$. Then the Cauchy kernel of $W_{i_1,\ldots,i_p}^{k,\natural}$ is $\dfrac{: dz^{\infty-\{i_1,\ldots,i_p\}} : |_{\mathcal{T}^\infty}}{: \prod_{n \notin \{i_1,\ldots,i_p\}} z_n :}$, and we have

$$\begin{aligned} &H^{*,*}(W_*^{k,\natural}, \mathcal{C}) \cong H^{*,*}(\mathcal{T}^\infty, \mathcal{C}). \\ &H^{*,*}(W_*^{k,\natural}, \mathcal{C}) = H^*(W_*^{k,\natural}, \mathcal{C}) \oplus *(H^*(W_*^{k,\natural}, \mathcal{C})). \end{aligned} \tag{22}$$

$H^*(W_*^{k,\natural}, \mathcal{C})$ is isomorphic to $H^*(\mathcal{T}^\infty, \mathcal{C})$. Hence it is an ∞-dimensional Grassmann algebra.

Note 5. Since there is the regularized volume form $: d\omega :$ of the sphere $\hat{S}^\infty$ of $\hat{H}$, Hilbert space added the longitude, we can define the real coefficients de Rham type cohomology $H^{*,*}(\hat{S}^\infty, \mathcal{R})$ of $\hat{S}^\infty$ by

$$\begin{aligned} &H^{*,*}(\hat{S}^\infty, \mathcal{R}) = H^0(\hat{S}^\infty, \mathcal{R}) \oplus H^\infty(\hat{S}^\infty, \mathcal{R}), \\ &H^\infty(\hat{S}^\infty, \mathcal{R}) = *H^0(\hat{S}^\infty, \mathcal{C}) \cong \mathcal{R} : d\omega : . \end{aligned} \tag{23}$$

We conclude this section asking are there any relation between de Rham type cohomology with ∞-degree elements and entire cyclic cohomology, or stochastic de Rham complexes (cf. [Con98, Cun02, Léan03]).

6 Regularized residue

We set $W_{n,*}^{k,\natural} = \{\sum_{m>n} z_m e_m \in W^{k,\natural} | z_m \neq 0, m = n+1, \ldots\}$ and $\mathcal{C}_*^n = \{\sum_m z_m e_m \in \mathcal{C}^n | z_m \neq 0\}$. We also denote $W_n^{k,\natural}$ the subspace of $W^{k,\natural}$ defined by $z_1 = 0, \ldots, z_n = 0$. If $m \leq n$, we regard

$$\mathcal{C}^m_* \times W^{k,\natural}_n \subset W^{k,\natural}_{n-m}, \quad \mathcal{C}^m \times W^{k,\natural}_{n,*} \subset W^{k,\natural}_{n-m}.$$

Then composing the residue maps

$$res : H^p(\mathcal{C}^m_* \times W^{k,\natural}_n, \mathcal{C}) \to H^{p-1}(\mathcal{C}^{m-1} \times W^{k,\natural}_n, \mathcal{C}),$$

the composed residue map [Ler59], cf.[Asa68]

$$res^m : H^m(\mathcal{C}^m_* \times W^{k,\natural}_n, \mathcal{C}) \to H^0(W^{k,\natural}_n, \mathcal{C}) \cong \mathcal{C},$$

is obtained,and we have

$$\int_{|z_1|=\epsilon_1,\dots,|z_m|=\epsilon_m} \phi = (2\pi i)^m res^m(\phi).$$

Definition 4. *If $p \geq n$, we define the map $res^{\infty-p}$ by*

$$res^{\infty-p} * \left(\frac{: dz^{\infty-\{i_1,\dots,i_p\}} : |_{\mathcal{T}^\infty}}{: \prod_{n \notin \{i_1,\dots,i_p\}} z_n :}\right) = \frac{dz_{i_{n+1}}}{z_{i_{n+1}}} \wedge \dots \wedge \frac{dz_{i_p}}{z_{i_p}}. \tag{24}$$

Since $H^{\infty-n}(W^{k,\natural}_*, \mathcal{C}) = *H^n(W^{k,\natural}_*, \mathcal{C})$, $res^{\infty-p}$ induces the map

$$res^{\infty-p} : H^{\infty-n}(W^{k,\natural}_*, \mathcal{C}) \to H^{p-n}(\mathcal{C}^{p-n}_*, \mathcal{C}) \cong \mathcal{C},$$

and we have the following *regularized residue formula*

$$\int_{\mathcal{T}^{\infty-n,k}_r} *\phi^n = (2\pi i)^{\nu-n+p} \int_{\mathcal{T}^{p-n}} res^{\infty-p}(*\phi^n). \tag{25}$$

Here, $\mathcal{T}^{\infty-n,k}_r$ is the torus in $W^{k,\natural}_n$ defined by $|z_m| = r$, $m \geq n$ and $\mathcal{T}^{p-n}$ is the torus in $\mathcal{C}^{p-n}$ defined by $|z_j| = c_j$, $j = 1,\dots,p-n$. The integral in the right hand side is done in usual sense, but the the integral in the left hand side is the regularized integral. Cauchy's integral formula on $W^{k,\natural}$ is a consequence of this formula.

By using the map $res^{\infty-p}$, we have the following exact sequence

$$H^{\infty-p}(\mathcal{C}^{p-n} \times W^{k,\natural}_{p-n,*}, \mathcal{C}) \longrightarrow^{\iota} H^{\infty-p}(W^{k,\natural}_*, \mathcal{C}) \longrightarrow$$
$$\longrightarrow^{res^{\infty-p}} H^{p-n}(\mathcal{C}^{p-n}_*, \mathcal{C}) \longrightarrow^{\delta} H^{\infty-p+1}(\mathcal{C}^{p-n} \times W^{k,\natural}_{p-n,*}, \mathcal{C}).$$

This sequence is not embedded in long exact sequence of de Rham type cohomology groups. Because $res^{\infty-p}$ is a kind of composed residue. But we can not get $res^{\infty-p}$ composing ordinary residue maps.

Note 6. If X is an orientable ∞-dimensional smooth manifold modeled by $\hat{H}$ and Y is an orientable smooth r-dimensional submanifold of X ($r < \infty$), then if $H^{*,*}(X, \mathcal{R})$ and $H^{*,*}(X - Y, \mathcal{R})$ are defined, the *regularized residue map* $res : H^{\infty-p}(X - Y, \mathcal{R}) \to H^{r-p+1}(Y, \mathcal{R})$ may defined and we may have the following *regualrized residue exact sequence* (cf.[1])

$$\dots \longrightarrow H^{\infty-p}(X, \mathcal{R}) \longrightarrow^{\iota} H^{\infty-p}(X - Y, \mathcal{R}) \longrightarrow^{res}$$
$$\longrightarrow H^{r-p+1}(Y, \mathcal{R}) \longrightarrow^{\delta} H^{\infty-p+1}(X, \mathcal{R}) \longrightarrow \cdots.$$

References

[Asa04a] Asada, A.: Regularized Calculus; An application of zeta-regularization to infinite dimensional Geometry and Analysis. Int. J. Geom. Meth. Mod. Phys., **1**, 107-157 (2004)

[CDP02] Cardona, A., Ducourtioux, C., Paycha, S.: From tracial anomalies to anomalies in quantum field theory. Commun. Math. Phys., **242**, 31-65 (2002)

[Payc01] Paycha, S.: Renormalized trace as a looking glass into infinite dimensional geometry. Infin. Dim. Anal. Quantum Prob. Relat. Top., **4**, 221-266 (2001)

[Asa04b] Asada, A.: Zeta-regularization and calculus on infinite dimensional spaces. AIP Conference Proceedings, **729**, 71-83 (2004)

[Asa04c] Asada, A.: Regularized volume form of the sphere of a Hilbert space with the determinant bundle. Differential Geometry and Its Applications, matfyzpress, 397-409 (2005)

[Con98] Connes, A.: Entire cyclic cohomology of Banach algebra and characters of θ-summable Fredholm modules. K-theory, **1(6)**, 519-548 (1988)

[Cun02] Cuntz, J.: Cyclic Theory, Bivariant K-theory and the Chern-Connes Character. In: Cyclic Cohomology in Non-Commutative Geoemtry. EMS, Springer, Berlin (2002)

[Léan03] Léandre, R.: Stochastic algebraic de Rham complexes. Acta Appl. Math., **79**, 217-247 (2003)

[Ler59] Leray, J.: Le calcul différentiel et intégral ssur une varié té complexe. Bull. Soc. Math. France, **87**, 81-180 (1959)

[Asa68] Asada, A.: Currents and residue exact sequences. J. Fac. Sci. Shinshu Univ., **3**, 85-151 (1968)

Fractional spaces generated by the positive differential and difference operators in a Banach space

Allaberen Ashyralyev

Department of Mathematics, Fatih University, Istanbul, Turkey
aashyr@fatih.edu.tr

The structure of the fractional spaces $E_{\alpha,q}(L_q[0,1], A^x)$ generated by the positive differential operator A^x defined by the formula $A^x u = -a(x)\frac{d^2u}{dx^2} + \delta u$, with domain $D(A^x) = \{u \in C^{(2)}[0,1] : \ u(0) = u(1), u'(0) = u'(1)\}$ is investigated. It is established that for any $0 < \alpha < \frac{1}{2}$ the norms in the spaces $E_{\alpha,q}(L_q[0,1], A^x)$ and $W_q^{2\alpha}[0,1]$ are equivalent. The positivity of the differential operator A^x in $W_q^{2\alpha}[0,1](0 \le \alpha < \frac{1}{2})$ is established. The discrete analogy of these results for the positive difference operator A_h^x a second order of approximation of the differential operator A^x, defined by the formula

$$A_h^x u^h = \left\{ -a(x_k)\frac{u_{k+1} - 2u_k + u_{k-1}}{h^2} + \delta u_k \right\}_1^{M-1}, u_h = \{u_k\}_0^M, Mh = 1$$

with $u_0 = u_M$ and $-u_2 + 4u_1 - 3u_0 = u_{M-2} - 4u_{M-1} + 3u_M$ is established. In applications, the coercive inequalities for the solutions of the nonlocal boundary-value problem for two-dimensional elliptic equation and of the second order of accuracy difference schemes for the numerical solution of this problem are obtained.

1 Introduction

It is a well-known (see, e.g., [Kre66, Gri84, Fat85]) that the study of the various properties of partial differential equations is based on a positivity property of the differential operator in a Banach space. The positivity of the wider class of differential operators has been studied by many researchers (see [Sol59, Sol60, KZPS76, Ste80]). To prove stability, in a number of works (see [AS94]-[AS84] and the references given therein) difference schemes were treated as operator equations in a Banach space, and the investigation was based on the positivity property of the operator coefficient. Important progress has been made in the

Kenan Taş et al. (eds), Mathematical Methods in Engineering, 13–22.

study of positive operators from the viewpoint of the stability analysis of high order of accuracy difference schemes for partial differential equations. Application of theory of fractional spaces generated by the positive operators in a Banach space permits us to establish the stability and coercive stability of the difference schemes in various norms for partial differential equations specially when we cannot use approaches of a maximum principle and energy method. We introduce the Banach spaces $E_{\alpha,q} = E_{\alpha,q}(E, A)(0 < \alpha < 1)$, consisting of all $v \in E$ for which the following norms are finite:

$$\| v \|_{E_{\alpha,q}} = \left(\int_0^\infty \| z^\alpha A(z+A)^{-1} v \|_E^q \frac{dz}{z} \right)^{\frac{1}{q}}, 1 \le q < \infty,$$

$$\| v \|_{E_{\alpha,\infty}} = \sup_{z>0} \| z^\alpha A(z+A)^{-1} v \|_E, q = \infty.$$

The positive operator A commutes with its resolvent $(\lambda + A)^{-1}$ for all $\lambda, \lambda \in (0, \infty)$. Therefore, using the definition of the fractional spaces $E_{\alpha,q} = E_{\alpha,q}(E, A)$, we obtain

$$\| (\lambda + A)^{-1} \|_{E_{\alpha,q} \to E_{\alpha,q}} \le \| (\lambda + A)^{-1} \|_{E \to E} \tag{1}$$

for all $\alpha, \alpha \in (0, 1)$ and $q, q \in [1, \infty]$.This means that from the positivity of operator A in E it follows the positivity of this operator A in $E_{\alpha,q}$ for all $\alpha, \alpha \in (0, 1)$ and $q, q \in [1, \infty]$.

The investigation of the well-posedness of the various types of boundary value problems for parabolic and elliptic differential and difference equations is based on the positivity of elliptic differential and difference operators A in various Banach spaces E and on the structure of the fractional spaces $E_{\alpha,q}$ generated by these positive operators. Note that an excellent survey of works in the theory of fractional spaces generated by the positive multidimensional difference operators in the space and its applications to partial differential equations parabolic and elliptic types was given in the books [AS94, AS04, Ash92]. Theory and applications of positive operators in Banach spaces have been studied extensively by many researchers (see [Sob71, AS77, AS79], and [SS81]-[AY06] and the references therein). We consider the differential operator A^x defined by the formula

$$A^x u = -a(x) \frac{d^2 u}{dx^2} + \delta u, \tag{2}$$

with domain $D(A^x) = \{u \in C^{(2)}[0, 1] : u(0) = u(1), u'(0) = u'(1)\}$. Here $a(x)$ is a smooth function defined on the segment $[0, 1]$ and $a(x) \ge a > 0$, $\delta > 0$.

We introduce the Banach space $C^\beta[0, 1](0 < \beta < 1)$ of all continuous functions $\varphi(x)$ defined on $[0, 1]$ and satisfying a Holder condition and $\varphi(0) = \varphi(1)$ for which the following norm is finite:

$$||\varphi||_{C^{\beta}[0,1]} = ||\varphi||_{C[0,1]} + \sup_{0\leq x<x+\tau\leq 1} \frac{|\varphi(x+\tau)-\varphi(x)|}{\tau^{\beta}},$$

where $C[0,1]$ is the space of all continuous functions $\varphi(x)$ defined on $[0,1]$ and $\varphi(0)=\varphi(1)$ with the usual norm

$$||\varphi||_{C[0,1]} = \max_{0\leq x\leq 1} |\varphi(x)|.$$

In the paper [AK95] the following two theorems on a structure of fractional spaces $E_{\alpha}(C[0,1],A^{x})$ and on the positivity of A^{x} in $C^{2\alpha}[0,1](0<\alpha<\frac{1}{2})$ were established.

Theorem 1. *For $0<\alpha<1/2$ the norms of the spaces $E_{\alpha}(C[0,1],A^{x})$ and $C^{2\alpha}\ [0,1]$ are equivalent.*

Theorem 2. *For all $\lambda\in R_{\varphi}$, $|\lambda|\geq K_0>0$ and $0<\alpha<1/2$ the resolvent $(\lambda+A^{x})^{-1}$ is subject to the bound*

$$\left\|(\lambda+A^{x})^{-1}\right\|_{C^{2\alpha}[0,1]\to C^{2\alpha}[0,1]} \leq \frac{M(\varphi,\delta)}{\alpha(1-2\alpha)}(1+|\lambda|)^{-1},$$

where $M(\varphi,\delta)$ does not depend on λ and α.

In the papers [AK01] and [AYA05] the positive difference operators A_h^x of a first order of approximation of the differential operator A^x, defined by the formula

$$A_h^x u^h = \left\{-a(x_k)\frac{u_{k+1}-2u_k+u_{k-1}}{h^2}+\delta u_k\right\}_1^{M-1}, u_h=\{u_k\}_0^M \tag{3}$$

with $u_0=u_M$ and $u_1-u_0=u_M-u_{M-1}$ and of a second order of approximation of the differential operator A^x, defined by the formula

$$A_h^x u^h = \left\{-a(x_k)\frac{u_{k+1}-2u_k+u_{k-1}}{h^2}+\delta u_k\right\}_1^{M-1}, u_h=\{u_k\}_0^M \tag{4}$$

with $u_0=u_M$ and $-u_2+4u_1-3u_0=u_{M-2}-4u_{M-1}+3u_M$ was presented. It was proved that the spaces $E_{\alpha}(C_h,A_h^x)$ and $C_h^{2\alpha}$ coincide for any $0<\alpha<\frac{1}{2}$, and their norms are equivalent uniformly in $h, 0<h\leq h_0$. The positivity of the difference operators A_h^x in $C_h^{2\alpha}(0\leq\alpha<\frac{1}{2})$ was obtained.

In the present paper we study the structure of the fractional spaces $E_{\alpha,q}(L_q[0,1],A^x)$ generated by the positive differential operator A^x defined by the formula(2). It is established that for any $0<\alpha<\frac{1}{2}$ the norms in the spaces $E_{\alpha,q}(L_q[0,1],A^x)$ and $W_q^{2\alpha}[0,1]$ are equivalent. The positivity of the differential operator A^x in $W_q^{2\alpha}[0,1](0 \le \alpha < \frac{1}{2})$ is established. Here the Banach space $W_q^{\beta}[0,1]$ is the space of the all integrable functions $f(x)$ defined on $[0,1]$, equipped with the norm

$$\| f \|_{W_q^{\beta}[0,1]} = \{ \int_0^1 \int_0^1 \frac{|f(x) - f(x+y)|^q}{|y|^{1+\beta q}} dxdy + \| f \|_{L_q[0,1]} \}^{\frac{1}{q}},$$

$$0 < \beta < 1, 1 \le q \le \infty,$$

where $L_q[0,1]$ is the space of the all integrable functions defined on $[0,1]$, equipped with the norm

$$\| f \|_{L_q[0,1]} = \{ \int_0^1 |f(x)|^q dx \}^{\frac{1}{q}}.$$

Moreover, the discrete analogy of these results for the positive difference operator A_h^x defined by the formula (4) is investigated. It is established that the spaces $E_{\alpha,q}(L_{q,h}, A_h^x)$ and $W_{q,h}^{2\alpha}$ coincide for any $0 < \alpha < \frac{1}{2}$, and their norms are equivalent uniformly in h, $0 < h \le h_0$.The positivity of the difference operator A_h^x in $W_q^{2\alpha}[0,1]_h(0 \le \alpha < \frac{1}{2})$ is established. In applications, the coercive inequalities for the solutions of the nonlocal boundary-value problem for two-dimensional elliptic equation and of the second order of accuracy difference schemes for the numerical solution of this problem are obtained.

2 The positivity of differential operator A^x.The structure of fractional spaces $E_{\alpha,q}(L_q[0,1], A^x)$

Theorem 3. *For any $0 < \alpha < \frac{1}{2}$ the norms of the spaces $E_{\alpha,q}(L_q[0,1], A^x)$ and $W_q^{2\alpha}[0,1]$ are equivalent.*

The proof of this theorem follows the scheme of the proof of the theorem in [AK95] and it is based on the formulas

$$A^x (\lambda + A^x)^{-1} f(x) = \frac{\delta}{\lambda + \delta} f(x) + \int_0^1 J(x, s; \lambda + \delta)(f(x) - f(s))ds,$$

$$f(x) = \int_0^1 \int_0^\infty J(x, s; \lambda + t + \delta) A^x (\lambda + t + A^x)^{-1} f(s) dtds$$

for the positive differential operator A^x and on the pointwise estimates of the Green's function of the resolvent equation

$$A^x u + \lambda u = f$$

or

$$-a(x)\frac{d^2u(x)}{dx^2} + \delta u(x) + \lambda u(x) = f(x), 0 < x < 1, \tag{5}$$

$$u(0) = u(1),\ u'(0) = u'(1)$$

and its derivative.

Theorem 4. *For all* λ, $\lambda \in R_\varphi = \{\lambda : |\arg \lambda| \leq \varphi, \varphi < \pi/2\}, \alpha \in (0, \frac{1}{2})$ *and* $|\lambda| \geq K_0 > 0$ *the resolvent* $(\lambda I + A^x)^{-1}$ *is subject to the bound*

$$\left\|(\lambda I + A^x)^{-1}\right\|_{W_p^{2\alpha}[0,1] \to W_p^{2\alpha}[0,1]} \leq \frac{M(\varphi, \delta)}{\alpha(1 - 2\alpha)}(1 + |\lambda|)^{-1},$$

where $M(\varphi, \delta)$ *does not depend on* λ.

The proof of this theorem follows the scheme of the proof of the theorem in [AK95] and it is based on the estimate (1) and on the positivity of differential operator A^x in $L_p[0, 1]$. The proof of the positivity of differential operator A^x in $L_p[0, 1]$ is based on the formula

$$(\lambda + A^x)^{-1} f(x) = \int_0^1 J(x, s; \lambda + \delta) f(s) ds$$

and on the pointwise estimates for the Green's function of the resolvent equation (5) and its derivative.

Now, we consider the nonlocal boundary-value problem for two-dimensional elliptic equation

$$\begin{cases} -\frac{\partial^2 u}{\partial t^2} - a(x)\frac{\partial^2 u}{\partial x^2} + \delta u = f(t, x), 0 < t < T, 0 < x < 1, \\ u(0, x) = \varphi(x), u(T, x) = \psi(x),\ 0 \leq x \leq 1, \\ u(t, 0) = u(t, 1),\ u_x(t, 0) = u_x(t, 1),\ \ 0 \leq t \leq T, \end{cases} \tag{6}$$

where $a(x)$, $\varphi(x)$, $\psi(x)$ and $f(t, x)$ are given sufficiently smooth functions and $a(x) \geq a > 0,\ \delta > 0$ is a sufficiently large number.

Theorem 5. *For the solution of the boundary value problem (1) the following coercive inequalities are valid:*

$$\| \frac{\partial^2 u}{\partial t^2} \|_{L_p([0,T], W_q^{2\alpha}[0,1])} + \| \frac{\partial^2 u}{\partial x^2} \|_{L_p([0,T], W_q^{2\alpha}[0,1])}$$

$$\leq M(q,p,\alpha) \parallel f \parallel_{L_p([0,T],W_q^{2\alpha}[0,1])} + M(\alpha)(||\varphi||_{W_q^{2\alpha}[0,1]} + ||\psi||_{W_q^{2\alpha}[0,1]}),$$

$$1 < p, q < \infty, 0 < \alpha < \frac{1}{2}.$$

Here $M(q,p,\alpha)$ and $M(\alpha)$ are independent of $f(t,x)$, $\varphi(x)$ and $\psi(x)$.

The proof of Theorem 5 is based on the Theorem 3 on the structure of the fractional spaces $E_{\alpha,q}(L_q[0,1], A^x)$ and the Theorem 4 on the positivity of the operator A^x in $W_q^{2\alpha}[0,1]$ and on the following theorems on the structure of the fractional spaces $E_{\alpha,q}(L_q[0,1], (A^x)^{\frac{1}{2}})$[Ash92, Tri78] and on coercivity inequalities in $L_p(E_{\alpha,q})$[AS04] for the solution of the abstract boundary-value problem for differential equation

$$-v''(t) + Av(t) = f(t) \quad (0 \leq t \leq T), v(0) = v_0, v(T) = v_T \tag{7}$$

in an arbitrary Banach space E with the linear positive operator A .

Theorem 6. *The spaces $E_{\alpha,q}(L_q, A^x)$ and $E^*_{2\alpha,q}(L_q, (A^x)^{\frac{1}{2}})$ coincide for any $0 < \alpha < \frac{1}{2}$, and their norms are equivalent.*

Theorem 7. *Let $1 < p,\ q < \infty$ and $0 < \alpha < 1$.Suppose that A is the positive operator in a Banach space E. Then problem (7) is well posed in $L_p(E^*_{\alpha,q})$ and the coercivity inequality holds:*

$$||v''||_{L_p(E^*_{\alpha,q})} + ||Av||_{L_p(E^*_{\alpha,q})} \leq \frac{M(q)p^2}{\alpha(1-\alpha)(p-1)} ||f||_{L_p(E^*_{\alpha,q})}$$

$$+M(||A\varphi||_{E^*_{\alpha,q}} + ||A\psi||_{E^*_{\alpha,q}}),$$

where $M, M(q)$ do not depend on α, p, φ, ψ and $f(t)$. Here, the Banach space $E^*_{\alpha,q} = E^*_{\alpha,q}(E, A^{\frac{1}{2}})(0 < \alpha < 1, 1 < q < \infty)$ consists of those $v \in E$ for which the norm

$$\parallel v \parallel_{E^*_{\alpha,q}} = (\int_0^\infty \lambda^{1-\alpha} \parallel A^{\frac{1}{2}} exp\{-\lambda A^{\frac{1}{2}}\}v \parallel_E^q \frac{d\lambda}{\lambda})^{1/q}$$

is finite.

3 The positivity of difference operator A_h^x.The structure of fractional spaces $E_{\alpha,q}(L_q[0,1]_h, A_h^x)$

We denote $L_{q,h} = L_{q,h}[0,1]_h$ and $W^\alpha_{q,h} = W_q^\alpha[0,1]_h, 1 \leq q < \infty$ the Banach spaces of all grid functions $v^h = \{v_k\}_1^{M-1}$defined on $[0,1]_h = \{x_k = kh,\ 0 \leq k \leq M,\ Mh = 1\}$ equipped with the norms

$$\parallel \varphi^h \parallel_{L_{q,h}} = (\sum_{k=1}^{M-1} |\varphi_k|^q h)^{\frac{1}{q}},$$

$$\| \varphi^h \|_{W^{\beta}_{q,h}} = (\sum_{k=1}^{M-1} \sum_{m=1}^{M-1} \frac{|\varphi_k - \varphi_{k+m}|^q}{|mh|^{1+\beta q}} h^2 + \| \varphi^h \|_{L_{q,h}})^{\frac{1}{q}},$$

$$0 < \beta < 1, 1 \leq q < \infty.$$

Note that the Banach space $E'_{\alpha,q} = E'_{\alpha,q}(E,B)(0 < \alpha < 1)$ consists of those $v \in E$ for which the norm

$$\| v \|_{E'_{\alpha,q}} = (\int_0^\infty [z^\alpha \| B(z+B)^{-1} v \|_E]^q \frac{dz}{z})^{\frac{1}{q}}, \ 1 \leq q < \infty,$$

$$\| v \|_{E'_\alpha} = \| v \|_{E'_{\alpha,\infty}} = \sup_{\lambda > 0} \lambda^\alpha \| B(\lambda + B)^{-1} v \|_E$$

is finite.

Theorem 8. *The the spaces $E_{\alpha,q}(L_{q,h}, A_h^x)$ and $W_{q,h}^{2\alpha}$ coincide for any $0 < \alpha < \frac{1}{2}$, and their norms are equivalent uniformly in $h, 0 < h \leq h_0$.*

The proof of this theorem follows the scheme of the proof of the theorem in [AK01] and it is based on the formulas

$$A_h^x (\lambda + A_h^x)^{-1} f_k = \lambda \sum_{j=1}^{M-1} J(k,j;\lambda+\delta) [f_k - f_j] h + \frac{\delta}{\lambda+\delta} f_k, 0 \leq k \leq M,$$

$$f_k = \int_0^\infty \sum_{j=1}^{M-1} J(k,j;t+\lambda+\delta) A_h^x (t+\lambda+A_h^x)^{-1} f_j h dt, 0 \leq k \leq M$$

for the positive difference operator A_h^x and on the pointwise estimates for the Green's function of the resolvent equation

$$A_h^x u^h + \lambda u^h = f^h$$

or

$$-a_k \frac{u_{k+1} - 2u_k + u_{k-1}}{h^2} + \delta u_k + \lambda u_k = f_k, \tag{8}$$

$$a_k = a(x_k), f_k = f(x_k), x_k = kh, 1 \leq k \leq M-1,$$

$$u_0 = u_M, -u_2 + 4u_1 - 3u_0 = u_{M-2} - 4u_{M-1} + 3u_M$$

and its difference derivative.

Theorem 9. *For all λ, $\lambda \in R_\varphi = \{\lambda : |\arg \lambda| \leq \varphi, \varphi < \pi/2\}, \alpha \in (0, \frac{1}{2})$ and $|\lambda| \geq K_0 > 0$ the resolvent $(\lambda I + A_h^x)^{-1}$ is subject to the bound*

$$\left\| (\lambda I + A_h^x)^{-1} \right\|_{W_{p,h}^{2\alpha} \to W_{p,h}^{2\alpha}} \leq \frac{M(\varphi,\delta)}{\alpha(1-2\alpha)} (1+|\lambda|)^{-1},$$

where $M(\varphi,\delta)$ does not depend on λ and h.

The proof of this theorem follows the scheme of the proof of the theorem in [AK01] and it is based on the estimate (1) and on the positivity of difference operator A_h^x in $L_{p,h}$.The proof of the positivity of difference operator A_h^x in $L_{p,h}$ is based on the formula

$$(\lambda + A_h^x)^{-1} f_k = \sum_{j=1}^{M-1} J(k,j;\lambda+\delta) f_j h, 0 \le k \le M,$$

$$(\lambda + A^x)^{-1} f(x) = \int_0^1 J(x,s;\lambda+\delta) f(s) ds$$

and on the pointwise estimates for the Green's function of the resolvent equation (8) and its difference derivative.

In applications, we consider the difference scheme of the second order of accuracy

$$\begin{cases} -\frac{1}{\tau^2}(u_{k+1}^n - 2u_k^n + u_{k-1}^n) - a^n \frac{1}{h^2}(u_k^{n+1} - 2u_k^n + u_k^{n-1}) + \delta u_k^n = \varphi_k^n, \\ \varphi_k^n = f(t_k, x_n), \quad a^n = a(x_n), \quad t_k = k\tau, \quad x_n = nh, \\ 1 \le k \le N-1, 1 \le n \le M-1, N\tau = 1, Mh = 1, \\ u_0^n = \varphi^n, u_N^n = \psi^n, \varphi^n = \varphi(x_n), \quad \psi^n = \psi(x_n), x_n = nh, 0 \le n \le M, \\ u_k^0 = u_k^M, -u_k^2 + 4u_k^1 - 3u_k^0 = u_k^{M-2} - 4u_k^{M-1} + 3u_k^M, 0 \le k \le N \end{cases} \tag{9}$$

for the approximate solution of the nonlocal boundary-value problem (6).

Theorem 10. *Let τ and h be a sufficiently small numbers. For the solution of the difference problem (9) the following inequalities are valid:*

$$\| \{\tau^{-2}(u_{k+1}^h - 2u_k^h + u_{k-1}^h)\}_1^{N-1} \|_{L_{p,\tau}(W_{q,h}^{2\alpha})}$$

$$+ \| \{\{h^{-2}(u_k^{n+1} - 2u_k^n + u_k^{n-1})\}_1^{M-1}\}_1^{N-1} \|_{L_{p,\tau}(W_{q,h}^{2\alpha})}$$

$$\le M(p,q,\alpha) \| \{\varphi_k^h\}_1^{N-1} \|_{L_{p,\tau}(W_{q,h}^{2\alpha})}$$

$$+ M(p,\alpha)(\| \{h^{-2}(\varphi^{n+1} - 2\varphi^n + \varphi^{n-1})\}_1^{M-1} \|_{W_{q,h}^{2\alpha}}$$

$$+ \| \{h^{-2}(\psi^{n+1} - 2\psi^n + \psi^{n-1})\}_1^{M-1} \|_{W_{q,h}^{2\alpha}}), 1 < p,q < \infty, 0 < \alpha < \frac{1}{2},$$

where $M(p,q,\alpha)$ and $M(p,\alpha)$ do not depend on $\{\varphi_k^h\}_1^{N-1}$,φ^h, ψ^h, h and τ.

The proof of Theorem 10 is based on the Theorem 8 on the structure of the fractional spaces $E_{\alpha,q}(L_{q,h}, A_h^x)$ and the Theorem 9 on the positivity of the operator A_h^x in $W_{q,h}^{2\alpha}$ and on the following theorems on the structure of the fractional spaces $E_{\alpha,q}(L_{q,h}, (A_h^x)^{\frac{1}{2}})$ [Ash92] and on coercivity inequalities in $L_{p,\tau}(E_{\alpha,q})$ [AS04] for the solution of the second order of accuracy difference scheme

$$\begin{cases} -\frac{1}{\tau^2}(u_{k+1} - 2u_k + u_{k-1}) + Au_k = f_k, f_k = f(t_k), t_k = k\tau, \\ \qquad 1 \le k \le N-1, N\tau = 1, u_0 = \varphi, u_N = \psi \end{cases} \tag{10}$$

for the approximate solution of the boundary-value problem (7).

Theorem 11. *The spaces $E_{\alpha,q}(L_{q,h}, A_h^x)$ and $E_{2\alpha,q}(L_{q,h}, (A_h^x)^{\frac{1}{2}})$ coincide for any $0 < \alpha < \frac{1}{2}$, and their norms are equivalent uniformly in h, $0 < h \le h_0$.*

Theorem 12. *Let $1 < p,\ q < \infty$ and $0 < \alpha < 1$.Suppose that A is the positive operator in a Banach space E. Then problem (10) is well posed in $L_{p,\tau}(E_{\alpha,q})$ and the coercivity inequality holds:*

$$||\{\frac{1}{\tau^2}(u_{k+1} - 2u_k + u_{k-1})\}_1^{N-1}||_{L_{p,\tau}(E_{\alpha,q})} + ||\{Au_k\}_1^{N-1}||_{L_p(E_{\alpha,q})}$$

$$\le \frac{M(q)p^2}{\alpha(1-\alpha)(p-1)}||\{f_k\}_1^{N-1}||_{L_p(E_{\alpha,q})} + M(||A\varphi||_{E_{\alpha,q}} + ||A\psi||_{E_{\alpha,q}}),$$

where $M, M(q)$ do not depend on α, p, φ, ψ, $\{f_k\}_1^{N-1}$ and τ.

References

[Kre66] Krein, S.G.: Linear Differential Equations in a Banach Space. Nauka, Moscow (1966) (Russian); English transl.: Linear Differential Equations in Banach space, Translations of Mathematical Monographs. Vol.23, American Mathematical Society, Providence RI (1968)

[Gri84] Grisvard, P.: Elliptic Problems in Nonsmooth Domains. Patman Adv. Publ. Program, London (1984)

[Fat85] Fattorini, H.O.: Second Order Linear Differential Equations in Banach Spaces. Mathematics Studies, North-Holland (1985)

[Sol59] Solomyak, M.Z.: Analytic semigroups generated by elliptic operator in space L_p. Dokl. Acad. Nauk SSSR, **127(1)**, 37-39 (1959) (Russian)

[Sol60] Solomyak, M.Z.: Estimation of norm of the resolvent of elliptic operator in spaces L_p. Usp. Mat. Nauk, **15(6)**, 141-148 (1960) (Russian)

[KZPS76] Krasnosel'skii, M.A., Zabreiko, P.P., Pustyl'nik, E.I., Sobolevkii, P.E.: Integral Operators in Spaces of Summable Functions. Nauka, Moscow (1966) (Russian); English transl.: Integral Operators in Spaces of Summable Functions. Noordhoff, Leiden (1976).

[Ste80] Stewart, H.B.: Generation of analytic semigroups by strongly elliptic operators under general boundary conditions. Trans. Amer. Math. Soc., **259**, 299-310 (1980)

[AS94] Ashyralyev, A., Sobolevskii, P. E.: Well-Posedness of Parabolic Difference Equations. Birkhäuser Verlag, Basel Boston Berlin (1994)

[AS04] Ashyralyev, A., Sobolevskii P.E.: New Difference schemes for Partial Differential equations. Birkhäuser Verlag, Basel Boston Berlin (2004)

[Sob71] Sobolevskii, P.E.: The coercive solvability of difference equations. Dokl. Acad. Nauk SSSR, **201(5)**, 1063-1066 (1971) (Russian)

[AS77] Alibekov, Kh.A., Sobolevskii, P.E.: Stability of difference schemes for parabolic equations. Dokl. Acad. Nauk SSSR, **232(4)**, 737-740 (1977) (Russian)

[AS79] Alibekov, Kh.A., Sobolevskii, P.E.: Stability and convergence of difference schemes of a high order for parabolic differential equations. Ukrain.Mat.Zh., **31(6)**, 627-634 (1979) (Russian)

[AS84] Ashyralyev, A., Sobolevskii, P. E.: The linear operator interpolation theory and the stability of the difference-schemes. Doklady Akademii Nauk SSSR, **275(6)**, 1289-1291 (1984) (Russian)

[Ash92] Ashyralyev, A.: Method of positive operators of investigations of the high order of accuracy difference schemes for parabolic and elliptic equations. Doctor of Sciences Thesis, Ins. of Math. of Acad. Sci., Kiev (1992) (Russian)

[SS81] Smirnitskii, Yu.A., Sobolevskii, P.E.: Positivity of multidimensional difference operators in the $C-$norm. Usp. Mat. Nauk, **36(4)**, 202-203 (1981) (Russian)

[Smi82] Smirnitskii, Yu.A.: Fractional powers of elliptic difference operators. PhD Thesis, Voronezh State University, Voronezh (1983) (Russian)

[Dan89] Danelich, S.I.: Fractional powers of positive difference operators. PhD Thesis, Voronezh State University, Voronezh (1989) (Russian)

[AY98] Ashyralyev, A., Yakubov, A.: Structures of fractional spaces generating by the transport operator. In: Muradov, A.N.(ed) Modeling the Processes in Exploration of Gas Deposits and Applied Problems of Theoretical Gas Hydrodynamics. Ilym, Ashgabat (1998) (Russian)

[AY06] Ashyralyev, A., Yaz N.: On structure of fractional spaces generated by positivity operators with the nonlocal boundary conditions. In: Agarwal, R.P.(ed) Proceedings of the Conference Differential and Difference Equations and Applications. Hindawi Publishing Corporation, USA (2006)

[AK95] Ashyralyev, A., Karakaya, I.: The structure of fractional spaces generated by the positive operator. In: Ashyralyev, Ch. (ed) Abstracts of Conference of Young Scientists. Turkmen Agricultural University, Ashgabat (1995)

[AK01] Ashyralyev, A., Kendirli B.: Positivity in Holder norms of one dimensional difference operators with nonlocal boundary conditions. In: Cheshankov, B.I., Todorov, M.D.(ed) Application of Mathematics in Engineering and Economics **26**. Heron Press-Technical University of Sofia, Sofia (2001)

[AYA05] Ashyralyev, A., Yenial-Altay N.: Positivity of difference operators generated by the nonlocal boundary conditions. In: Akca, H., Boucherif, A., Covachev, V.(ed) Dynamical Systems and Applications. GBS Publishers and Distributors, India (2005)

[Tri78] Triebel, H.: Interpolation Theory, Function Spaces, Differential Operators. North-Holland Mathematical Library Vol. 18, Amsterdam (1978)

Sub-diffusion equations of fractional order and their fundamental solutions

Francesco Mainardi[1], Antonio Mura[1], Gianni Pagnini[2]
and Rudolf Gorenflo[3]

[1] University of Bologna, and INFN, Department of Physics, Via Irnerio 46, I-40126 Bologna, Italy {mainardi,mura}@bo.infn.it
[2] ENEA: National Agency for New Technologies, Energy and the Environment, Centre "E. Clementel", Via Martiri di Monte Sole 4, I-40129 Bologna, Italy gianni.pagnini@bologna.enea.it
[3] Free University of Berlin, Department of Mathematics and Informatics, Arnimallee 3, D-14195 Berlin, Germany gorenflo@mi.fu-berlin.de

The time-fractional diffusion equation is obtained by generalizing the standard diffusion equation by using a proper time-fractional derivative of order $1 - \beta$ in the Riemann-Liouville (R-L) sense or of order β in the Caputo (C) sense, with $\beta \in (0,1)$. The two forms are equivalent and the fundamental solution of the associated Cauchy problem is interpreted as a probability density of a self-similar non-Markovian stochastic process, related to a phenomenon of sub-diffusion (the variance grows in time sub-linearly). A further generalization is obtained by considering a continuous or discrete distribution of fractional time-derivatives of order less than one. Then the two forms are no longer equivalent. However, the fundamental solution still is a probability density of a non-Markovian process but one exhibiting a distribution of time-scales instead of being self-similar: it is expressed in terms of an integral of Laplace type suitable for numerical computation. We consider with some detail two cases of diffusion of distributed order: the double order and the uniformly distributed order discussing the differences between the R-L and C approaches. For these cases we analyze in detail the behaviour of the fundamental solutions (numerically computed) and of the corresponding variance (analytically computed) through the exhibition of several plots. While for the R-L and for the C cases the fundamental solutions seem not to differ too much for moderate times, the behaviour of the corresponding variance for small and large times differs in a remarkable way.

Kenan Taş et al. (eds), Mathematical Methods in Engineering, 23–55.

1 Introduction

The main physical purpose for adopting and investigating diffusion equations of fractional order to describe phenomena of *anomalous diffusion* usually met in transport processes through complex and/or disordered systems including fractal media. In this respect, in recent years interesting reviews, see *e.g.* [MK00, MK04, PSW05, Zas02], have appeared, to which (and references therein) we refer the interested reader.

All the related models of random walk turn out to be beyond the classical Brownian motion, which is known to provide the microscopic foundation of the standard diffusion, see *e.g.* [KS05, SK05]. The diffusion-like equations containing fractional derivatives in time and/or in space are usually adopted to model phenomena of anomalous transport in physics, so a detailed study of their solutions is required.

Our attention in this paper will be focused on the time-fractional diffusion equations of a single or distributed order less than 1, which are known to be models for sub-diffusive processes.

Since in the literature we find two different forms for the time-fractional derivative, namely the one in the Riemann-Liouville (R-L) sense, the other in the Caputo (C) sense, we will study the corresponding time-fractional diffusion equations separately. Specifically, we have worked out how to express their fundamental solutions in terms of an integral of Laplace type suitable for a numerical evaluation. Furthermore we have considered the time evolution of the variance for the R-L and C cases. It is known that for large times the variance characterizes the type of anomalous diffusion.

The plan of the paper is as follows.

In Section 2, after having shown the equivalence of the two forms for the time-fractional diffusion equation of a *single order*, namely the R-L form and the C form, we recall the main results for the common fundamental solution, which are obtained by applying two different strategies in inverting its Fourier-Laplace transform. Both techniques yield the fundamental solution in terms of special function of the Wright type that turns out to be self-similar through a definite space-time scaling relationship.

In Section 3 we apply the second strategy for obtaining the fundamental solutions of the time-fractional diffusion equation of *distributed order* in the R-L and C forms, assuming a general order density. We provide for these solutions a representation in terms of a Laplace-type integral of a Fox-Wright function that appears suitable for a numerical evaluation in finite space-time domains. We also provide the general expressions for the Laplace transforms of the corresponding variance.

Then, in Section 4, we consider two case-studies for the fractional diffusion of distributed order: as a discrete distribution we take two distinct orders β_1, β_2 with $0 < \beta_1 < \beta_2 \leq 1$; as continuous distribution we take the uniform density with $0 < \beta < 1$. For these cases we provide the graphical representation of the fundamental solutions (in space at fixed times) and of the evolution in time of the corresponding variance.

Finally, in Section 5, the main conclusions are drawn and directions for future work are outlined.

In order to have a self-contained treatment, we have edited three Appendices: the Appendix A is devoted to the basic notions of fractional calculus, whereas Appendices B and C deal special functions of Mittag-Leffler and Exponential Integral type, respectively, in view of their relevance for our treatment.

2 Time-fractional diffusion of single order

2.1 The standard diffusion

The *standard diffusion equation* in re-scaled non-dimensional variables is known to be

$$\frac{\partial}{\partial t} u(x,t) = \frac{\partial^2}{\partial x^2} u(x,t) \,, \quad x \in \mathbf{R} \,, \quad t \in \mathbf{R}_0^+ \,, \tag{2.1}$$

with $u(x,t)$ as the field variable. We assume that $u(x,t)$ is subjected to the initial condition

$$u(x,0^+) = u_0(x) \,, \tag{2.2}$$

where $u_0(x)$ denotes a given ordinary or generalized function defined on $\mathbf{R}$, that we assume to be Fourier transformable in ordinary or generalized sense, respectively. We assume to work in a suitable space of generalized functions where it is possible to deal freely with delta functions, integral transforms of Fourier, Laplace and Mellin type, and fractional integrals and derivatives.

It is well known that the fundamental solution (or *Green function*) of Eq. (2.1) i.e. the solution subjected to the initial condition $u(x,0^+) = u_0(x) = \delta(x)$, and to the decay to zero conditions for $|x| \to \infty$, is the Gaussian *probability density function* (*pdf*)

$$u(x,t) = \frac{1}{2\sqrt{\pi}} t^{-1/2} \, \mathrm{e}^{-x^2/(4t)} \,, \tag{2.3}$$

that evolves in time with second moment growing linearly with time,

$$\mu_2(t) := \int_{-\infty}^{+\infty} x^2 \, u(x,t) \, dx = 2t \,, \tag{2.4}$$

consistently with a law of *normal diffusion*[4]. We note the *scaling property* of the Green function, expressed by the equation

[4] The centred second moment provides the variance usually denoted by $\sigma^2(t)$. It is a measure for the spatial spread of $u(x,t)$ with time of a random walking

$$u(x,t) = t^{-1/2}\, U(x/t^{1/2})\,, \quad \text{with} \quad U(x) := u(x,1)\,. \tag{2.5}$$

The function $U(x)$ depending on the single variable x turns out to be an even function $U(x) = U(|x|)$ and is called the *reduced Green function.* The variable $X := x/t^{1/2}$ acts as the similarity variable. It is known that the Cauchy problem $\{(2.1) - (2.2)\}$ is equivalent to the integro-differential equation

$$u(x,t) = u_0(x) + \int_0^t \left[\frac{\partial^2}{\partial x^2}\, u(x,\tau) \right] d\tau\,, \tag{2.6}$$

where the initial condition is incorporated.

2.2 The two forms of time-fractional diffusion

Now, by using the tools of the fractional calculus we can generalize the above Cauchy problem in order to obtain the so-called *time-fractional diffusion equation* in the two distinct (but mathematically equivalent) forms available in the literature, where the initial condition is understood as (2.2). For the essentials of fractional calculus we refer the interested reader to the Appendix A.

If β denotes a real number such that $0 < \beta < 1$ the two forms are as follows:

$$\frac{\partial}{\partial t} u(x,t) = {}_tD^{1-\beta}\, \frac{\partial^2}{\partial x^2}\, u(x,t)\,, \quad x \in \mathbf{R}\,,\; t \in \mathbf{R}_0^+\,; \quad u(x,0^+) = u_0(x)\,, \tag{2.7}$$

where ${}_tD^{1-\beta}$ denotes the *Riemann-Liouville* (R-L) time-derivative of order $1-\beta$ and

$${}_tD_*^{\beta}\, u(x,t) = \frac{\partial^2}{\partial x^2}\, u(x,t)\,, \quad x \in \mathbf{R}\,,\; t \in \mathbf{R}_0^+\,; \quad u(x,0^+) = u_0(x)\,, \tag{2.8}$$

where ${}_tD_*^{\beta}$ denotes the time derivative of order β intended in the *Caputo* sense. In analogy with the standard diffusion equation we can provide an integro-differential form that incorporates the initial condition (2.2): for this purpose we replace in (2.6) the ordinary integral with the Riemann-Liouville time-fractional integral ${}_tJ^{\beta}$ of order β namely,

$$u(x,t) = u_0(x) + {}_tJ^{\beta} \left[\frac{\partial^2}{\partial x^2}\, u(x,t) \right]. \tag{2.9}$$

In view of the definitions of ${}_tJ^{\beta}$, ${}_tD^{1-\beta} := {}_tD^1\, {}_tJ^{\beta}$ and ${}_tD_*^{\beta} := {}_tJ^{1-\beta}\, {}_tD^1$, see Appendix A and take there $m = 1$, the above equations read explicitly:

particle starting at the origin $x = 0$, pertinent to the solution of the diffusion equation (2.1) with initial condition $u(x,0) = \delta(x)$. The asymptotic behaviour of the variance as $t \to \infty$ is relevant to distinguish *normal diffusion* ($\sigma^2(t)/t \to c$, $c > 0$) from anomalous processes of *sub-diffusion* ($\sigma^2(t)/t \to 0$) and of *super-diffusion* ($\sigma^2(t)/t \to +\infty$).

$$\frac{\partial}{\partial t}u(x,t) = \frac{1}{\Gamma(\beta)}\frac{\partial}{\partial t}\left\{\int_0^t \left[\frac{\partial^2}{\partial x^2}u(x,\tau)\right]\frac{d\tau}{(t-\tau)^{1-\beta}}\right\},\; u(x,0^+) = u_0(x)\,, \tag{2.7'}$$

$$\frac{1}{\Gamma(1-\beta)}\int_0^t \left[\frac{\partial}{\partial \tau}\,u(x,\tau)\right]\frac{d\tau}{(t-\tau)^{\beta}} = \frac{\partial^2}{\partial x^2}\,u(x,t)\,,\; u(x,0^+) = u_0(x)\,, \tag{2.8'}$$

$$u(x,t) = u_0(x) + \frac{1}{\Gamma(\beta)}\int_0^t \left[\frac{\partial^2}{\partial x^2}\,u(x,\tau)\right]\frac{d\tau}{(t-\tau)^{1-\beta}}\,. \tag{2.9'}$$

The two Cauchy problems (2.7), (2.8) and the integro-differential equation (2.9) are equivalent[5]: for example, we derive (2.7) from (2.9) simply differentiating both sides of (2.9), whereas we derive (2.9) from (2.8) by fractional integration of order β. In fact, in view of the semigroup property (A.2) of the fractional integral, we note that

$${}_tJ^{\beta}\,{}_tD_*^{\beta}u(x,t) = {}_tJ^{\beta}\,{}_tJ^{1-\beta}\,{}_tD^{1}u(x,t) = {}_tJ^{1}\,{}_tD^{1}u(x,t) = u(x,t) - u_0(x)\,. \tag{2.10}$$

In the limit $\beta = 1$ we recover the well-known diffusion equation (2.1).

Eq. (2.7) can be put in a *conservative form* as a *continuity equation*,

$$\frac{\partial}{\partial t}\,u(x,t) + \frac{\partial}{\partial x}\,F\left[u(x,t)\right] = 0\,, \tag{2.11}$$

where F is the *flux* given by

$$F\left[u(x,t)\right] = -\frac{\partial}{\partial x}\left[{}_tD^{1-\beta}\,u(x,t)\right] = -\frac{\partial}{\partial x}\left\{\frac{1}{\Gamma(\beta)}\frac{\partial}{\partial t}\left[\int_0^t \frac{u(x,\tau)\,d\tau}{(t-\tau)^{1-\beta}}\right]\right\}. \tag{2.12}$$

For $\beta = 1$ in (2.12) we recover in the limit the standard *Fick law*

$$F\left[u(x,t)\right] = -\frac{\partial}{\partial x}\,u(x,t)\,, \tag{2.13}$$

which leads to the standard diffusion equation (2.1) by using the continuity law (2.11).

---

[5] The integro-differential equation (2.9) was investigated via Mellin transforms by Schneider & Wyss [SW89] in their pioneering 1989 paper. The time-fractional diffusion equation in the form (2.8) with the Caputo derivative has been preferred and investigated by several authors. From the earlier contributors let us quote Caputo himself [Cap69], Mainardi, see *e.g.* [Main94, Main96, Main97] and Gorenflo & Rutman [GR95]. In particular, Mainardi has expressed the fundamental solution in terms of a special function (of Wright type) of which he has studied the analytical properties and provided plots also for $1 < \beta < 2$, see also [GLM99, GLM00, MP03] and references therein. For the form (2.7) with the R-L derivative earlier contributors include Nigmatullin [Nig86], Giona & Roman [GR92], the group of Prof. Nonnenmacher, see *e.g.* [MGN94], and Saichev & Zaslavsky [SZ97]. The equivalence between the two forms (2.7) and (2.8) was also pointed out recently by Sokolov and Klafter, see *e.g.* [SK05].

We also note that Eq. (2.12) can be interpreted as a generalized Fick law[6] where (long) memory effects are taken into account through a time-fractional derivative of order $1-\beta$ in the *Riemann-Liouville* sense.

We observe that the form (2.7) of the time-fractional diffusion equation with the R-L fractional derivative has the advantage of being derived in a direct way from a conservation principle by introducing a generalized Fick's law: in addition it can be interpreted as a master evolution equation of a dynamical system where in the LHS the time derivative of the first order usually appears.

The form (2.8) with the Caputo derivative, however, has the advantage to be treated in a simpler way with the Laplace transform requiring as the initial value $u(x, 0^+)$ as in the standard case, see Eq. (A.13). We note how in its definition (A.6) (for $m = 1$) the first derivative is weighted by a memory function of power law type, that formally degenerates to a delta function ($\delta(t) = t_+^{-1}/\Gamma(0)$, see [GS64]) as soon as the order tends to 1 from below. We observe that the Caputo form can be obtained from the master integral equation of the Continuous Time Random Walk (CTRW) by a well scaled transition to the diffusion limit as shown by Gorenflo and Mainardi, [GM03, GM05], see also [MVG05, SGM04].

2.3 The fundamental solution

Let us consider from now on the Eq. (2.8) with $u_0(x) = \delta(x)$: the fundamental solution can be obtained by applying in sequence the Fourier and Laplace transforms to it. We write, for generic functions $v(x)$ and $w(t)$, these transforms as follows:

$$\begin{aligned} \mathcal{F}\{v(x);\kappa\} &= \widehat{v}(\kappa) := \int_{-\infty}^{+\infty} \mathrm{e}^{i\kappa x}\, v(x)\, dx\,, \quad \kappa \in \mathbf{R}\,, \\ \mathcal{L}\{w(t);s\} &= \widetilde{w}(s) := \int_0^{+\infty} \mathrm{e}^{-st}\, w(t)\, dt\,, \quad s \in \mathbf{C}\,. \end{aligned} \tag{2.14}$$

Then, in the Fourier-Laplace domain our Cauchy problem [(2.8) with $u(x, 0^+) = \delta(x)$], after applying formula (A.13) for the Laplace transform of the fractional derivative and observing $\widehat{\delta}(\kappa) \equiv 1$, appears in the form

$$s^\beta\, \widehat{\widetilde{u}}(\kappa, s) - s^{\beta-1} = -\kappa^2\, \widehat{\widetilde{u}}(\kappa, s)\,,$$

implying

$$\widehat{\widetilde{u}}(\kappa, s) = \frac{s^{\beta-1}}{s^\beta + \kappa^2}\,, \quad 0 < \beta \le 1\,, \qquad \Re(s) > 0\,, \quad \kappa \in \mathbf{R}\,. \tag{2.15}$$

---

[6] We recall that the Fick law is essentially a phenomenological law, which represents the simplest relationship between the flux F and the gradient of the concentration u. If u is a temperature, F is the heat-flux, so we speak of the Fourier law. In both cases the law can be replaced by a more suitable phenomenological relationship which may account for possible non-local, non-linear and memory effects, without violating the conservation law expressed by the continuity equation.

To determine the Green function $u(x,t)$ in the space-time domain we can follow two alternative strategies related to the order in carrying out the inversions in (2.15).
(S1) : invert the Fourier transform getting $\widetilde{u}(x,s)$ and then invert the remaining Laplace transform;
(S2) : invert the Laplace transform getting $\widehat{u}(\kappa,t)$ and then invert the remaining Fourier transform.

Strategy (S1): Recalling the Fourier transform pair, see *e.g.* [AS65],

$$\frac{a}{b+\kappa^2} \overset{\mathcal{F}}{\leftrightarrow} \frac{a}{2b^{1/2}}\,\mathrm{e}^{-|x|b^{1/2}}\,, \quad b>0\,, \tag{2.16}$$

and setting $a = s^{\beta-1}$, $b = s^{\beta}$ we get

$$\widetilde{u}(x,s) = \frac{s^{\beta/2-1}}{2}\,\mathrm{e}^{-|x|s^{\beta/2}}\,, \quad 0<\beta\le 1\,. \tag{2.17}$$

The strategy (S1) has been applied by Mainardi [Main94, Main96, Main97] to obtain the Green function in the form

$$u(x,t) = t^{-\beta/2}\,U\left(|x|/t^{\beta/2}\right)\,, \quad -\infty < x < +\infty\,, \quad t\ge 0\,, \tag{2.18}$$

where the variable $X := x/t^{\beta/2}$ acts as *similarity variable* and the function $U(x) := u(x,1)$ denotes the *reduced Green function.* Restricting from now on our attention to $x \ge 0$, the solution turns out as

$$\begin{aligned} U(x) = U(-x) &= \frac{1}{2}\,M_{\frac{\beta}{2}}(x) = \frac{1}{2}\sum_{k=0}^{\infty}\frac{(-x)^k}{k!\,\Gamma[-\beta k/2+(1-\beta/2)]} \\ &= \frac{1}{2\pi}\sum_{k=0}^{\infty}\frac{(-x)^k}{k!}\,\Gamma[(\beta(k+1)/2]\,\sin[(\pi\beta(k+1)/2]\,, \end{aligned} \tag{2.19}$$

where $M_{\frac{\beta}{2}}(x)$ is an an entire transcendental function (of order $1/(1-\beta/2)$) of the Wright type, see also [GLM99, GLM00] and [Pod99].

Strategy (S2): Recalling the Laplace transform pair, see e.g. [EMOT55, GM97, Pod99],

$$\frac{s^{\beta-1}}{s^{\beta}+c} \overset{\mathcal{L}}{\leftrightarrow} E_{\beta}(-ct^{\beta})\,, \quad c>0\,, \tag{2.20}$$

and setting $c = \kappa^2$ we get

$$\widehat{u}(\kappa,t) = E_{\beta}(-\kappa^2 t^{\beta})\,, \quad 0<\beta\le 1\,, \tag{2.21}$$

where E_{β} denotes the Mittag-Leffler function, see Appendix B.

The strategy (S2) has been followed by Gorenflo, Iskenderov & Luchko [GIL00] and by Mainardi, Luchko & Pagnini [MLP01] to obtain the Green functions of the more general space-time-fractional diffusion equations, and requires to invert the Fourier transform by using the machinery of the Mellin convolution and the Mellin-Barnes integrals. Restricting ourselves here to recall the final results, the reduced Green function for the time-fractional diffusion equation now appears, for $x \geq 0$, in the form:

$$U(x) = U(-x) = \frac{1}{\pi} \int_0^\infty \cos(\kappa x)\, E_\beta\left(-\kappa^2\right) d\kappa = \frac{1}{2x} \frac{1}{2\pi i} \int_{\gamma - i\infty}^{\gamma + i\infty} \frac{\Gamma(1-s)}{\Gamma(1-\beta s/2)}\, x^{\,s}\, ds, \tag{2.22}$$

with $0 < \gamma < 1$. By solving the Mellin-Barnes integrals using the residue theorem, we arrive at the same power series (2.19).

Both strategies allow us to prove that the Green function is non-negative and normalized, so it can be interpreted as a spatial probability density evolving in time with the similarity law (2.18). Although the two strategies are equivalent for yielding the required result, the second one appears more general and so more suitable to treat the more complicated case of fractional diffusion of distributed order, see the next Section.

It is relevant to point out, see *e.g.* [Main96, MLP01], that for $0 < \beta < 1$ as $|x| \to \infty$ the solution decays faster than exponential and slower than Gaussian. We have, for $x > 0$,

$$U(x) \sim A\, x^a\, \mathrm{e}^{-bx^c}, \quad x \to \infty, \tag{2.23}$$

with

$$A = \left\{2\pi(2-\beta)\, 2^{\beta/(2-\beta)}\, \beta^{(2-2\beta)/(2-\beta)}\right\}^{-1/2}, \tag{2.24}$$

$$a = \frac{2\beta - 2}{2(2-\beta)}, \quad b = (2-\beta)\, 2^{-2/(2-\beta)}\, \beta^{\beta/(2-\beta)}, \quad c = \frac{2}{2-\beta}. \tag{2.25}$$

We note in fact that c increases from 1 to 2 as β varies from 0 to 1; for $\beta = 1$ we recover the exact solution $U(x) = \exp(-x^2/4)/(2\sqrt{\pi})$, consistent with (2.3). Furthermore, the moments (of even order) of $u(x,t)$ are

$$\mu_{2n}(t) := \int_{-\infty}^{+\infty} x^{2n}\, u(x,t)\, dx = \frac{\Gamma(2n+1)}{\Gamma(\beta n + 1)}\, t^{\beta n}, \quad n = 0, 1, 2, \ldots, \; t \geq 0. \tag{2.26}$$

Of particular interest is the evolution of the variance $\sigma^2(t) = \mu_2(t)$ (the second centred moment); we get from (2.26):

$$\sigma^2(t) = 2\, \frac{t^\beta}{\Gamma(\beta+1)}, \quad 0 < \beta \leq 1, \tag{2.27}$$

so that for $\beta < 1$ we note a sub-linear growth in time, consistent with an anomalous process of *slow diffusion* in contrast with the law (2.3) of normal diffusion. Such result can also be obtained in a simpler way from the Fourier transform (2.21) noting that

$$\sigma^2(t) = -\frac{\partial^2}{\partial \kappa^2}\, \widehat{u}(\kappa = 0, t). \tag{2.28}$$

2.4 Graphical representation of the fundamental solutions

Let us consider the time-fractional diffusion equation of a single order $\beta = \beta_0$, whose fundamental solution has the peculiar property to be self-similar according to the similarity variable $x/t^{\beta_0/2}$. For this reason it is sufficient to consider the fundamental solution for $t = 1$, namely the reduced Green function $U(x)$, given by Eq. (2.19) in terms of the special function $M_{\beta_0/2}(x)$ of the Wright type. In Fig.1 we show the graphical representations of $U(x)$

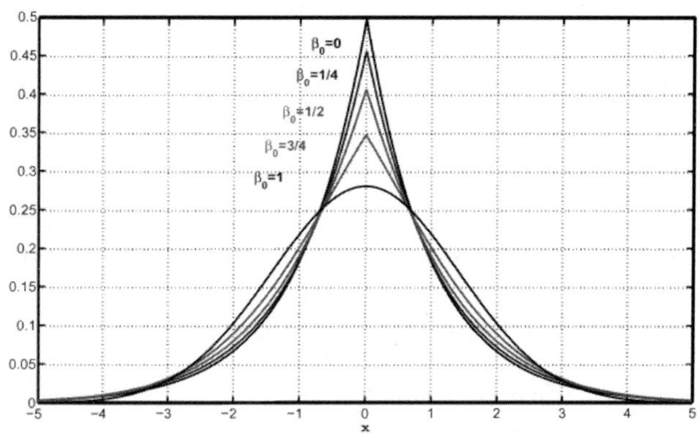

Fig. 1. Plots (in linear scales) of the reduced Green function $U(x) = \frac{1}{2}M_{\beta_0/2}(x))$ versus x (in the interval $|x| \leq 5$), for $\beta_0 = 0, 1/4, 1/2, 3/4, 1$.

for different orders β_0 ranging from $\beta_0 = 0$, for which we recover the Laplace *pdf*

$$U(x) = \frac{1}{2}\,\mathrm{e}^{-|x|}\,, \tag{2.29}$$

to $\beta_0 = 1$, for which we recover the Gaussian *pdf* (of variance $\sigma^2 = 2$)

$$U(x) = \frac{1}{2\sqrt{\pi}}\,\mathrm{e}^{-x^2/4}\,. \tag{2.30}$$

To visualize the decay of the queues of the above (symmetric) *pdf*'s as stated in Eqs. (2.23)-(2.25) we refer to Fig. 2, where we have adopted semi-logarithmic scales. In this case the decay-plot of the queues is ranging from a straight line ($\beta_0 = 0$) to a parabolic line ($\beta = 1$).

For more information about plots and properties of the M-Wright function we refer the reader to previous articles of our research group, see *e.g.* [Main96, Main97, MLP01, MP03].

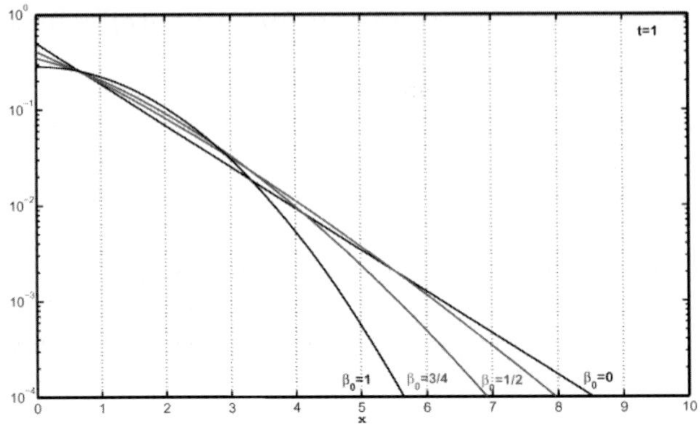

Fig. 2. Plots (in linear-logarithmic scales) of the reduced Green function $U(x) = \frac{1}{2}M_{\beta_0/2}(x))$ versus x (in the interval $0 \le x \le 10$), for $\beta_0 = 0, 1/2, 3/4, 1$.

3 Time-fractional diffusion equation of distributed order

3.1 The two forms for time-fractional diffusion

The time-fractional diffusion equations (2.7) and (2.8) can be generalized by using the notion of time-fractional derivative of distributed order[7]. For this purpose we need to consider a function $p(\beta)$ that acts as weight for the order of differentiation $\beta \in (0,1]$ such that

$$p(\beta) \ge 0\,, \quad \text{and} \quad \int_0^1 p(\beta)\, d\beta = c > 0\,. \tag{3.1}$$

The positive constant c can be taken as 1 if we like to assume the normalization condition for the integral. Clearly, some special conditions of regularity and behaviour near the boundaries will be required for the weight function $p(\beta)$[8]. Such function, that can be referred to as the *order density* if $c = 1$, is allowed to have δ-components if we are interested in a discrete distribution of orders.

Then, if we weight the time-fractional derivative in Eq. (2.7) (where it is intended in the R-L sense), and in Eq. (2.8) (where it is intended the C sense) by using the weight function $p(\beta)$ in (3.1), we finally obtain the *time-fractional diffusion equation of distributed order* in the two forms:

---

[7] We find an earlier idea of fractional derivative of distributed order in time in the 1969 book by Caputo [Cap69], that was later developed by Caputo himself, see [Cap95, Cap01], and by Bagley & Torvik, see [BT00].

[8] For the weight function $p(\beta)$ we conveniently require that its primitive $P(\beta) = \int_0^\beta p(\beta')\, d\beta'$ vanishes at $\beta = 0$ and is there continuous from the right, attains the value c at $\beta = 1$ and has at most finitely many (upwards) jump points in the half-open interval $0 < \beta \le 1$, these jump points allowing delta contributions to $p(\beta)$ (particularly relevant for discrete distributions of orders).

$$\frac{\partial}{\partial t}\, u(x,t) = \int_0^1 p(\beta)\, {}_tD^{1-\beta} \left[\frac{\partial^2}{\partial x^2}\, u(x,t) \right] d\beta\,, \quad x \in \mathbf{R},\ t \geq 0\,, \tag{3.2}$$

and

$$\int_0^1 p(\beta) \, \left[{}_tD_*^{\beta}\, u_*(x,t) \right] \, d\beta = \frac{\partial^2}{\partial x^2}\, u_*(x,t)\,, \quad x \in \mathbf{R},\ t \geq 0\,. \tag{3.3}$$

From now on we shall restrict our attention on the fundamental solutions of Eqs. (3.2)-(3.3) so we understand that these equations are subjected to the initial condition $u(x,0^+) = u_*(x,0^+) = \delta(x)$. Since for distributed order the solution depends on the selected approach (as we shall show hereafter), we now distinguish the fractional equations (3.2) and (3.3) and their fundamental solutions by decorating in the Caputo case the variable $u(x,t)$ with subscript $*$.

Diffusion equations of distributed order of C-type (3.3) have recently been discussed in [CGS02, CGSG03, CKS03, SCK04] and in [Nab04]. Diffusion equations of distributed order of R-L type (2.2) have been considered by Sokolov et al. [SCK04, SK05]. These authors have referred to Eqs. (3.3), (3.2) as to *normal* and *modified* forms of the time-fractional diffusion equation of distributed order, respectively. In their analysis they have pointed out the different evolutions of the variance corresponding to the case of a combination of two derivatives of order β_1, β_2 with $0 < \beta_1 < \beta_2 < 1$, although both forms exhibit slow diffusion. For the modified form with two fractional orders, recently Langlands [Lang06] has provided the fundamental solution as an infinite series of H-Fox functions.

As usual, we have considered the initial condition $u(x,0^+) = u_*(x,0^+) = \delta(x)$ in order to keep the probability meaning. Indeed, already in the paper [CGS02], it was shown that the Green function is non-negative and normalized, so allowing interpretation as a density of the probability at time t of a diffusing particle to be in the point x. The main interest of the authors in [CGS02, CGSG03, CKS03, SCK04] was devoted to the second moment of the Green function (the displacement variance) in order to show the sub-diffusive character of the related stochastic process by analyzing some interesting cases of the order-density function $p(\beta)$. In this paper, extending the approach by Naber [Nab04], we are interested to provide a general representation of the fundamental solution corresponding to a generic order-density $p(\beta)$.

For a thorough general study of fractional pseudo-differential equations of distributed order let us cite the paper by Umarov and Gorenflo [UG05]. For a relationship between the C fractional diffusion equation of distributed order (3.3) and the Continuous Time Random Walk (CTRW) models we may refer to the paper by Gorenflo and Mainardi [GM05]. Let us remark that the flux formula (2.12) for the fractional diffusion equations of a single order (2.7)-(2.8) can be generalized to hold for the R-L fractional diffusion equation of distributed order (3.2) as follows:

$$F\left[u(x,t)\right] = -\frac{\partial}{\partial x}\left\{\int_0^1 p(\beta)\,\left[{}_tD^{1-\beta}\,u(x,t)\right]\,d\beta\right\}. \tag{3.4}$$

3.2 The Fourier-Laplace transforms of the fundamental solutions

Let us now apply the Laplace transform to Eqs. (3.2)-(3.3) by using the rules (A.15) and (A.13) appropriate to the R-L and C derivatives, respectively, with $m = 1$. Introducing the relevant functions

$$A(s) = s\int_0^1 p(\beta)\,s^{-\beta}\,d\beta\,, \tag{3.5}$$

and

$$B(s) = \int_0^1 p(\beta)\,s^{\beta}\,d\beta\,, \tag{3.6}$$

we then get for the R-L and C cases, after simple manipulation, the Laplace transforms of the corresponding fundamental solutions:

$$\widehat{\widetilde{u}}(\kappa,s) = \frac{1}{s+\kappa^2 A(s)}\,, \tag{3.7}$$

and

$$\widehat{\widetilde{u}}_*(\kappa,s) = \frac{B(s)/s}{\kappa^2 + B(s)}\,. \tag{3.8}$$

We easily note that in the particular case $p(\beta) = \delta(\beta - \beta_0)$ we have in (3.5): $A(s) = s^{1-\beta_0}$, and in (3.6): $B(s) = s^{\beta_0}$. Then, Eqs. (3.7) and (3.8) provide the same result (2.15) and its consequences for the time-fractional diffusion of a single order $\beta = \beta_0$.

3.3 The inversion of the Laplace transforms

By inverting the Laplace transforms in (3.7) and (3.8) we obtain the remaining Fourier transforms of the fundamental solutions for the R-L and C time-fractional diffusion of distributed order.

Let us start with the R-L case. We get (in virtue of the Titchmarsh theorem on Laplace inversion) the representation

$$\widehat{u}(\kappa,t) = -\frac{1}{\pi}\int_0^\infty \mathrm{e}^{-rt}\,\mathrm{Im}\left\{\widehat{\widetilde{u}}\left(\kappa, r\mathrm{e}^{i\pi}\right)\right\}\,dr\,, \tag{3.9}$$

that requires the expression of $-\mathrm{Im}\left\{1/[s+\kappa^2 A(s)]\right\}$ along the ray $s = r\,\mathrm{e}^{i\pi}$ with $r > 0$ (the branch cut of the function $s^{-\beta}$). We write

$$A\left(r\,\mathrm{e}^{i\pi}\right) = \rho\,\cos(\pi\gamma) + i\rho\sin(\pi\gamma)\,, \tag{3.10}$$

where

$$\begin{cases} \rho = \rho(r) = \left|A\left(r\,\mathrm{e}^{i\pi}\right)\right|, \\ \gamma = \gamma(r) = \dfrac{1}{\pi}\arg\left[A\left(r\,\mathrm{e}^{i\pi}\right)\right]. \end{cases} \tag{3.11}$$

Then, after simple calculations, we get

$$\widehat{u}(\kappa, t) = \int_0^\infty \frac{\mathrm{e}^{-rt}}{r}\, H(r;\kappa)\, dr\,, \tag{3.12}$$

with

$$H(\kappa; r) = \frac{1}{\pi}\, \frac{\kappa^2\, r\, \rho\, \sin(\pi\gamma)}{r^2 - 2\kappa^2\, r\, \rho\, \cos(\pi\gamma) + \kappa^4 \rho^2} \geq 0\,. \tag{3.13}$$

Similarly for the C case we obtain

$$\widehat{u}_*(\kappa, t) = -\frac{1}{\pi}\int_0^\infty \mathrm{e}^{-rt}\, \mathrm{Im}\left\{\widetilde{\widehat{u}}_*\left(\kappa, r\mathrm{e}^{i\pi}\right)\right\} dr\,, \tag{3.14}$$

that requires the expression of $-\mathrm{Im}\left\{B(s)/[s(\kappa^2 + B(s))]\right\}$ along the ray $s = r\,\mathrm{e}^{i\pi}$ with $r > 0$ (the branch cut of the function s^β). We write

$$B\left(r\,\mathrm{e}^{i\pi}\right) = \rho_*\, \cos(\pi\gamma_*) + i\rho_*\, \sin(\pi\gamma_*)\,, \tag{3.15}$$

where

$$\begin{cases} \rho_* = \rho_*(r) = \left|B\left(r\,\mathrm{e}^{i\pi}\right)\right|, \\ \gamma_* = \gamma_*(r) = \dfrac{1}{\pi}\arg\left[B\left(r\,\mathrm{e}^{i\pi}\right)\right]. \end{cases} \tag{3.16}$$

After simple calculations we get

$$\widehat{u}_*(\kappa, t) = \int_0^\infty \frac{\mathrm{e}^{-rt}}{r}\, K(r;\kappa)\, dr\,, \tag{3.17}$$

with

$$K(\kappa; r) = \frac{1}{\pi}\, \frac{\kappa^2\, \rho_*\, \sin(\pi\gamma_*)}{\kappa^4 + 2\kappa^2\, \rho_*\, \cos(\pi\gamma_*) + \rho_*^2} \geq 0\,. \tag{3.18}$$

We note that the expressions of H and K are related through the transformation

$$\rho_* \Longleftrightarrow r/\rho\,, \quad \gamma_* \Longleftrightarrow 1 - \gamma\,. \tag{3.19}$$

3.4 The inversion of the Fourier transforms

Since $u(x,t)$ and $u_*(x,t)$ are symmetric in x, the inversion formula for the Fourier transforms in (3.12) and (3.17) yields

$$u(x,t) = \frac{1}{\pi}\int_0^{+\infty} \cos(\kappa x) \left\{\int_0^\infty \frac{\mathrm{e}^{-rt}}{r}\, H(\kappa, r)\, dr\right\} d\kappa\,, \tag{3.20}$$

and

$$u_*(x,t) = \frac{1}{\pi} \int_0^{+\infty} \cos(\kappa x) \left\{ \int_0^{\infty} \frac{\mathrm{e}^{-rt}}{r} K(\kappa, r)\, dr \right\} d\kappa \,. \qquad (3.21)$$

We note that the evaluation of the Fourier integral in Eq. (3.21) concerning the C case has been recently carried out by Mainardi and Pagnini [MP06] by the method of the Mellin transform. Referring the reader to [MP06] for details we can now state that for the C case the fundamental solution reads, taking as usual $x \geq 0$,

$$u_*(x,t) = \frac{1}{2\pi x} \int_0^{\infty} \frac{e^{-rt}}{r} F_*(\rho_*^{1/2} x)\, dr \,, \qquad (3.22)$$

where

$$F_*(\rho_*^{1/2} x) = \frac{1}{2\pi i} \int_{\sigma - i\infty}^{\sigma + i\infty} \Gamma(1-s) \sin(\pi \gamma_* s/2) (\rho_*^{1/2} x)^s ds \,. \qquad (3.23)$$

and $\rho_* = \rho_*(r)$, $\gamma_* = \gamma_*(r)$. In [MP06] the Authors have expressed the function F_* in terms of Fox-Wright functions by using the method of the Mellin-Barnes integrals; then the series expansion of F_* yields the required solution as

$$u_*(x,t) = \frac{1}{2\pi} \sum_{k=0}^{\infty} \frac{(-x)^k}{k!} \varphi_k(t) \,, \quad x \geq 0 \,, \qquad (3.24)$$

with

$$\varphi_k(t) = \int_0^{\infty} \frac{e^{-rt}}{r} \sin[\pi \gamma_* (k+1)/2]\, \rho_*^{(k+1)/2}\, dr \,. \qquad (3.25)$$

For numerical purposes we now prefer to find an alternative representation of the function F_* that we can get by taking inspiration from an exercise in the book by Paris and Kaminski, see [PK01]: p.89, Eq. (3.3.2). The new representation of F_* reads

$$\begin{aligned} F_*(\rho_*^{1/2} x) &= \mathrm{Im}\{\rho_*^{1/2} x\, \mathrm{e}^{i\pi\gamma_*/2} \mathrm{e}^{-e^{i\pi\gamma_*/2} \rho_*^{1/2} x}\} \\ &= \rho_*^{1/2} x\, \mathrm{e}^{-\rho_*^{1/2} x \cos(\pi\gamma_*/2)} \sin[\pi\gamma_*/2 - \rho^{1/2} x \sin(\pi\gamma_*/2)] \,. \end{aligned} \qquad (3.26)$$

As a matter of fact, from the numerical view-point the integral representation (3.22) with (3.26) is indeed more convenient than the series representation (3.24) with (3.25) that was provided in [MP06].

For the fundamental solution of the R-L case, we shall use the representation (3.22) with (3.26) of the C case by invoking the transformation (3.19).

3.5 The variance of the fundamental solutions

We now consider the evaluation of the variance of the fundamental solutions, that is, according to the two approaches:

$$\text{R-L} : \sigma^2(t) := \int_{-\infty}^{+\infty} x^2\, u(x,t)\, dx\,; \quad \text{C} : \sigma_*^2(t) := \int_{-\infty}^{+\infty} x^2\, u_*(x,t)\, dx\,. \quad (3.27)$$

Like for the single order case we can obtain these quantities (fundamental for classifying the type of diffusion) in a simpler way according

$$\text{R-L} : \sigma^2(t) = -\frac{\partial^2}{\partial \kappa^2}\, \widehat{u}(\kappa=0,t)\,; \quad \text{C} : \sigma_*^2(t) = -\frac{\partial^2}{\partial \kappa^2}\, \widehat{u}_*(\kappa=0,t)\,. \quad (3.28)$$

As a consequence of (3.28) we thus must invert only Laplace transforms as follows. We have, for κ near zero,
for <u>the R-L case</u> we get from Eq. (3.7),

$$\widetilde{\widehat{u}}(\kappa,s) = \frac{1}{s}\left(1 - \kappa^2\, \frac{A(s)}{s} + \ldots\right), \text{ so } \widetilde{\sigma^2}(s) = -\frac{\partial^2}{\partial \kappa^2}\widetilde{\widehat{u}}(\kappa=0,s) = \frac{2A(s)}{s^2}\,, \quad (3.29)$$

for <u>the C case</u> we get from Eq. (3.8)

$$\widetilde{\widehat{u}}_*(\kappa,s) = \frac{1}{s}\left(1 - \kappa^2\, \frac{1}{B(s)} + \ldots\right), \text{ so } \widetilde{\sigma^2}(s) = -\frac{\partial^2}{\partial \kappa^2}\widetilde{\widehat{u}}(\kappa=0,s) = \frac{2}{s\, B(s)}\,. \quad (3.30)$$

4 Examples of fractional diffusion of distributed order

We shall now concentrate our interest to choices of some typical weight functions $p(\beta)$ in (3.1) that characterizes the order distribution for the time-fractional diffusion equations of distributed order (3.2) and (3.3). This will allow us to compare the results for the R-L form and for the C form.

4.1 The fractional diffusion of double-order

First, we consider the choice

$$p(\beta) = p_1 \delta(\beta - \beta_1) + p_2 \delta(\beta - \beta_2)\,, \quad 0 < \beta_1 < \beta_2 \le 1\,, \quad (4.1)$$

where the constants p_1 and p_2 are both positive, conveniently restricted to the normalization condition $p_1 + p_2 = 1$.

Then for <u>the R-L case</u> we have

$$A(s) = p_1\, s^{1-\beta_1} + p_2\, s^{1-\beta_2}\,, \quad (4.2)$$

so that, inserting (4.2) in (3.7),

$$\widetilde{\widehat{u}}(\kappa,s) = \frac{1}{s[1 + \kappa^2(p_1\, s^{-\beta_1} + p_2\, s^{-\beta_2})]}\,, \quad (4.3)$$

Similarly, for the C case we have

$$B(s) = p_1\, s^{\beta_1} + p_2\, s^{\beta_2}\,, \tag{4.4}$$

so that, inserting (4.4) in (3.8),

$$\widehat{\widetilde{u}}_*(\kappa, s) = \frac{p_1\, s^{\beta_1} + p_2\, s^{\beta_2}}{s[\kappa^2 + p_1\, s^{\beta_1} + p_2\, s^{\beta_2}]}\,. \tag{4.5}$$

We leave as an exercise the derivation of the spectral functions $H(\kappa; r)$ and $K(\kappa; r)$ of the corresponding fundamental solutions, that are used for the numerical computation.

Let us now evaluate the second moments starting from the corresponding Laplace transforms (3.29) and (3.30) inserting the expressions of $A(s)$ and $B(s)$) provided by Eqs. (4.2) and (4.4), respectively.
For the R-L form we have

$$\widetilde{\sigma^2}(s) = 2\, p_1\, s^{-(1+\beta_1)} + 2\, p_2\, s^{-(1+\beta_2)}\,; \tag{4.6}$$

for the C form we have

$$\widetilde{\sigma^2}_*(s) = \frac{2}{p_1\, s^{(1+\beta_1)} + p_2\, s^{(1+\beta_2)}}\,, \tag{4.7}$$

Now the Laplace inversion yields:
for the R-L case, see and compare Sokolov et al [SCK04] and Langlands [Lang06],

$$\sigma^2(t) = 2\, p_1\, \frac{t^{\beta_1}}{\Gamma(\beta_1+1)} + 2\, p_2\, \frac{t^{\beta_2}}{\Gamma(\beta_2+1)} \sim \begin{cases} 2p_1\, \dfrac{t^{\beta_1}}{\Gamma(1+\beta_1)}\,; & t \to 0^+\,, \\ 2p_2\, \dfrac{t^{\beta_2}}{\Gamma(1+\beta_2)}\,, & t \to +\infty\,; \end{cases} \tag{4.8}$$

for the C case, see and compare Chechkin et al. [CGS02]

$$\sigma^2_*(t) = \frac{2}{p_2}\, t^{\beta_2}\, E_{\beta_2-\beta_1,\beta_2+1}\left(-\frac{p_1}{p_2}\, t^{\beta_2-\beta_1}\right) \sim \begin{cases} \dfrac{2}{p_2}\, \dfrac{t^{\beta_2}}{\Gamma(1+\beta_2)}\,, & t \to 0^+\,, \\ \dfrac{2}{p_1}\, \dfrac{t^{\beta_1}}{\Gamma(1+\beta_1)}\,, & t \to +\infty\,. \end{cases} \tag{4.9}$$

Then we see that for the R-L case we have an explicit combination of two power laws: the smallest exponent(β_1) dominates at small times whereas the largest exponent (β_2) dominates at large times. For the C case we have a Mittag-Leffler function in two parameters so we have a combination of two power laws only asymptotically for small and large times; precisely we get a behaviour opposite to the previous one, so the largest exponent(β_2) dominates at small times whereas the smallest exponent (β_1) dominates at large times.

We can derive the above asymptotic behaviours directly from the Laplace transforms (4.6)-(4.7) by applying the Tauberian theory for Laplace transforms[9]. In fact for the R-L case we note that for $A(s)$ in (4.2) $s^{1-\beta_1}$ is negligibly small in comparison with $s^{1-\beta_2}$for $s \to 0^+$ and, viceversa, $s^{1-\beta_2}$ is negligibly small in comparison to $s^{1-\beta_1}$for $s \to +\infty$. Similarly for the C case we note that for $B(s)$ in (4.4) s^{β_2} is negligibly small in comparison to $s^{\beta_1}$for $s \to 0^+$ and, viceversa, s^{β_1} is negligibly small in comparison $s^{\beta_2}$for $s \to +\infty$.

4.2 The fractional diffusion of uniformly distributed order

Second, we consider the choice

$$p(\beta) = 1\,, \quad 0 < \beta < 1\,. \tag{4.10}$$

For the R-L case we have

$$A(s) = s \int_0^1 s^{-\beta}\, d\beta = \frac{s-1}{\log s}\,, \tag{4.11}$$

hence, inserting (4.11) in (3.7)

$$\widehat{\widetilde{u}}(\kappa, s) = \frac{\log s}{s\log s + \kappa^2\,(s-1)}\,. \tag{4.12}$$

For the C case we have

$$B(s) = \int_0^1 s^{\beta}\, d\beta = \frac{s-1}{\log s}\,, \tag{4.13}$$

hence, inserting (4.13) in (3.8),

$$\widehat{\widetilde{u}}_*(\kappa, s) = \frac{1}{s}\frac{s-1}{\kappa^2\log s + s - 1} = \frac{1}{s} - \frac{1}{s}\frac{\kappa^2 \log s}{\kappa^2 \log s + s - 1}\,. \tag{4.14}$$

We leave as an exercise the derivation of the spectral functions $H(\kappa; r)$ and $K(\kappa; r)$ of the corresponding fundamental solutions, that are used for the numerical computation.

Let us now evaluate the second moments starting from the corresponding Laplace transforms (3.29) and (3.30) inserting the expressions of $A(s)$ and $B(s)$) provided by Eqs. (4.2) and (4.4), respectively. We note that for this special order distribution we have $A(s) = B(s)$.

For the R-L case we have

---

[9] According to this theory the asymptotic behaviour of a function $f(t)$ near $t = \infty$ and $t = 0$ is (formally) obtained from the asymptotic behaviour of its Laplace transform $\widetilde{f}(s)$ for $s \to 0^+$ and for $s \to +\infty$, respectively.

$$\widetilde{\sigma^2}(s) = 2\left[\frac{1}{s\log s} - \frac{1}{s^2\log s}\right]. \tag{4.15}$$

Then, by inversion, see Appendix C: Eq. (C.16), we get

$$\sigma^2(t) = 2\left[\nu(t,0) - \nu(t,1)\right] \sim \begin{cases} 2/\log(1/t), & t \to 0, \\ 2t/\log t, & t \to \infty, \end{cases} \tag{4.16}$$

where

$$\nu(t,a) = \int_0^\infty \frac{t^{a+\tau}}{\Gamma(a+\tau+1)}\,d\tau, \quad a > -1,$$

denotes a special function introduced in Appendix C along with its Laplace transform.

For the C case we have

$$\widetilde{\sigma_*^2}(s) = \frac{2}{s}\frac{\log s}{s-1}. \tag{4.17}$$

Then, by inversion, see Appendix C: Eqs. (C.11), (C.14), and compare with Chechkin et al. [CGS02], Eqs. (23)-(27),

$$\sigma_*^2(t) = 2\left[\log t + \gamma + e^t\,\mathcal{E}_1(t)\right] \sim \begin{cases} 2t\log(1/t), & t \to 0, \\ 2\log(t), & t \to \infty, \end{cases} \tag{4.18}$$

where $\mathcal{E}(t)$ denotes the exponential integral function and $\gamma = 0.57721...$ is the so-called Euler-Mascheroni constant.

4.3 Graphical representation of the fundamental solutions

For the general time-fractional diffusion equations of distributed order, namely (3.2) for the R-L form and (3.3) for the C form, we limit ourselves to a few cases selected from the examples treated above. Specifically we consider the case of two distinct, equally weighted, orders, see Eq. (4.1) with $p_1 = p_2 = 1/2$ and the case of a uniform distribution of orders, see Eq. (4.10).

We recall that, in contrast with the single order, for the distributed order the self-similarity of the fundamental solution is lost so we need to provide graphical representations for different times.

For the case of two orders, we chose $\{\beta_1 = 1/4, \beta_2 = 1\}$ in order to better contrast the different evolution of the fundamental solution for the R-L and the C forms. In Fig. 3 we exhibit the plots of the corresponding solution versus x (in the interval $|x| \leq 5$), at different times, selected as $t = 0.1$, $t = 1$ and $t = 10$. In this limited spatial range we can note how the time evolution of the *pdf* depends on the different time-asymptotic behaviour of the variance, for the two forms, as stated in Eqs. (4.8)-(4.9), respectively.

For the uniform distribution, we find it instructive to compare in Fig. 4 the solutions corresponding to R-L and C forms with the solutions of the

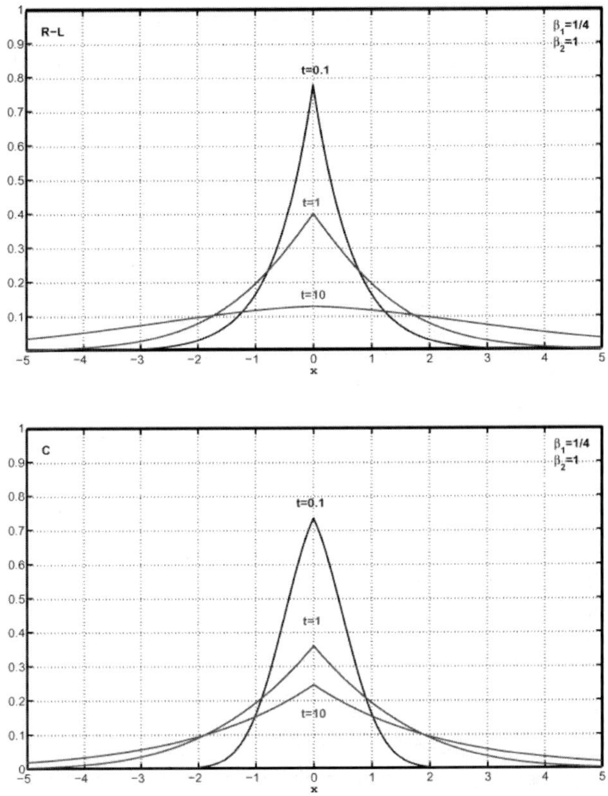

Fig. 3. Plots of the fundamental solution versus x (in the interval $|x| \leq 5$), for the case $\{\beta_1 = 1/4, \beta_2 = 1\}$ at times $t = 0.1, 1, 10$; top: R-L form, bottom: C form.)

fractional diffusion of a single order $\beta_0 = 1/4, 3/4, 1$ at fixed times, selected as $t = 1, 10$. We have skipped the order $\beta_0 = 1/2$ and the time $t = 0.1$ for a better view of the plots. Then in Figs 5,6 we compare the variance of for moderate times ($0 \leq t \leq 10$, using linear scales) and large times ($10^1 \leq t \leq 10^7$, using logarithmic scales), respectively. Here we have inserted the plot for $\beta_0 = 1/2$.

To interpret these asymptotic behaviours we observe that $\beta = 0$ is the smallest, $\beta = 1$ the largest relevant index for the constant order-density. Due to the logarithmic constituents in the R-L case the smallest β, namely 0, now plays the dominant role for $s \to \infty$ and $t \to 0$, see (4.11),(4.15) and (4.16), whereas the largest β, namely 1, is dominant for $s \to 0$ and $t \to \infty$. This situation is reversed for the C case, see (4.13),(4.17) and (4.18). We observe that in the R-L case the variance (the second moment) grows slightly slower than linearly for $t \to \infty$, but extremely slowly near $t = 0$. In the C case the variance exhibits a slightly super-linear growth near $t = 0$, but an extremely slow growth for $t \to \infty$.

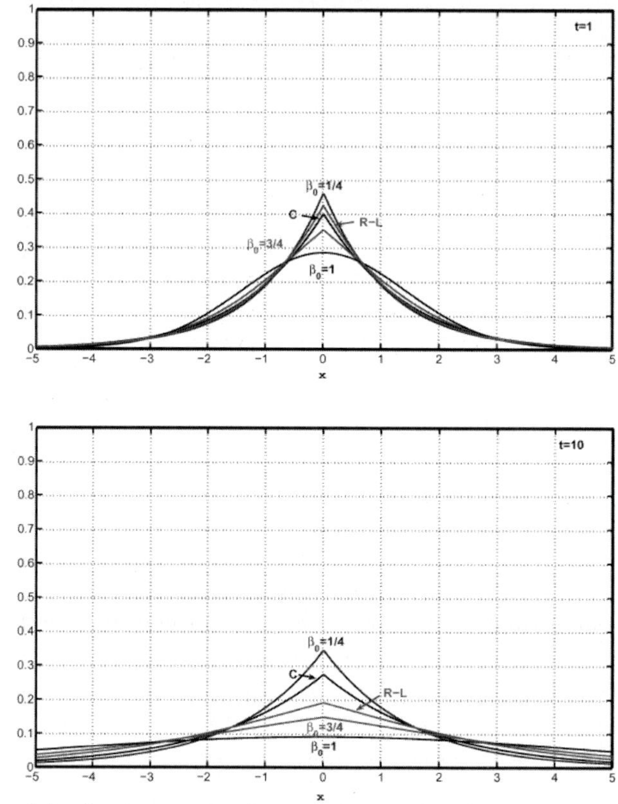

Fig. 4. Plots of the fundamental solution versus x (in the interval $|x| \leq 5$), for the uniform order distribution in R-L and C forms compared with the solutions of some cases of single order at $t = 1$ (top) and $t = 10$ (bottom).

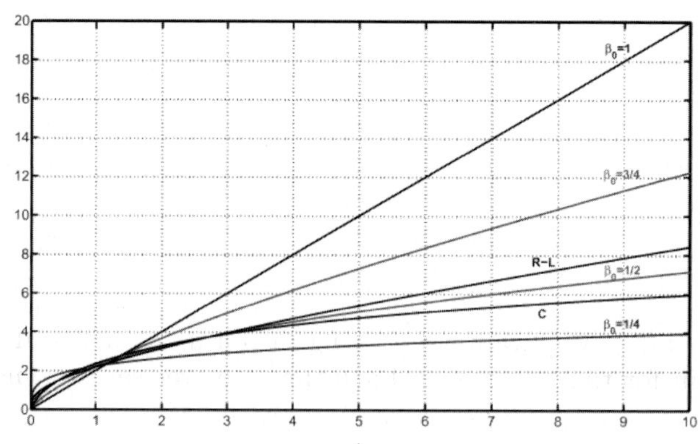

Fig. 5. Plots of the variance versus t in the interval $0 \leq t \leq 10$ (linear scales), for the uniform order distribution.

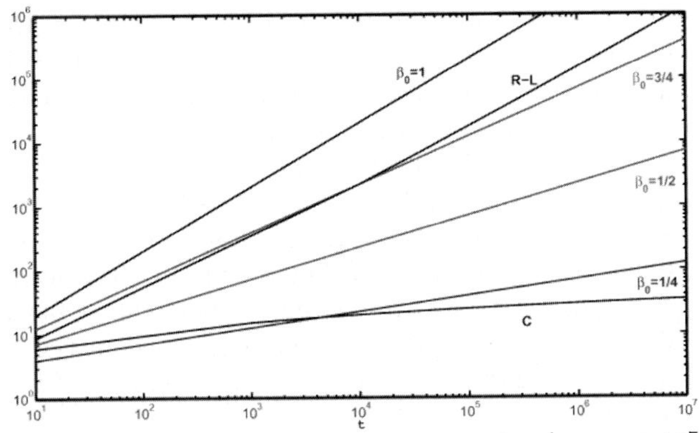

Fig. 6. Plots of the variance versus t in the interval $10^1 \leq t \leq 10^7$ (logarithmic scales) for the uniform order distribution

5 Conclusions and outlook

After outlining the basic theory of the Cauchy problem for the spatially one-dimensional and symmetric time-fractional diffusion equation (with its main equivalent formulations), we have paid special attention to transform methods for finding its fundamental solution or (exploiting self-similarity) the corresponding reduced Green function. We have stressed the importance of the transforms of Fourier, Laplace and Mellin and of the functions of Mittag-Leffler and Wright type, avoiding however the cumbersome H-Fox function notations.

A natural first step for construction of the fundamental solution consists in applying in either succession the transforms of Fourier in space and Laplace in time to the Cauchy problem. This yields in the Fourier-Laplace domain the solution in explicit form, but for the space-time domain we must invert both transforms in sequence for which there are two choices, both leading to the same power series in the spatial variable with time-dependent coefficients. The strategy, called by us strategy (S2), of first doing Laplace inversion and then the Fourier inversion yields the reduced Green function as a Mellin-Barnes integral form which, by the calculus of residues, the power series is obtained. This strategy can be adapted to the treatment of the more general case of the time-fractional diffusion equation of distributed order. Now the fundamental solution can be expressed as an integral over a Mellin-Barnes integral containing two parameters having the form of functionals of the order-density. Again for the fundamental solution a power series comes out whose coefficients, however, are time-dependent functionals of the order-density. But, if there is more than one time derivative-order present, self-similarity is lost. Finally, we have worked out how to express the fundamental solution in terms of an integral of Laplace type, more suitable for a numerical evaluation.

We have studied in detail and illustrated by graphics the time-fractional diffusion of a single order (where self-similarity holds true) and two simple but noteworthy case-studies of distributed order, namely the case of a superposition of two different orders β_1 and β_2 and the case of a uniform order distribution. In the first case one of the orders dominates the time-asymptotics near zero, the other near infinity, but β_1 and β_2 change their roles when switching from the (R-L) form to the (C) form of the time-fractional diffusion. The asymptotics for uniform order density is remarkably different, the extreme orders now being (roughly speaking) 0 and 1. We now meet super-slow and slightly super-fast time behaviours of the variance near zero and near infinity, again with the interchange of behaviours between the R-L and C form. We clearly see the above effects described in the figures at the end of Section 4, in particular the extremely slow growth of the variance as $t \to \infty$ for the C form.

More general studies are desirable for fractional diffusion equations of distributed order in time as well as in space. For the case of one single order in time and in space (the space-time-fractional diffusion equation) we refer the reader to the exhaustive paper by Mainardi, Luchko & Pagnini [MLP01].

As our emphasis in this paper is on pure analysis we have not touched the wide field of simulations of trajectories of a particle subjected to the random process modelled by the equations at hand.

Acknowledgements

This work has been carried out in the framework of a joint research project for *Fractional Calculus Modelling* (`www.fracalmo.org`). We thank M. Stojanović for discussions on some analytical details.

Appendix A: Essentials of fractional calculus

For a sufficiently well-behaved function $f(t)$ $(t \in \mathbf{R}^+)$ we may define the fractional derivative of order μ $(m-1 < \mu \le m\,,\ m \in \mathbf{N})$, see *e.g.* [GM97, Pod99], in two different senses, that we refer here as to *Riemann-Liouville* (R-L) derivative and *Caputo* (C) derivative, respectively. Both derivatives are related to the so-called Riemann-Liouville fractional integral of order $\alpha > 0$ defined as

$$ {}_tJ^{\alpha}\, f(t) := \frac{1}{\Gamma(\alpha)} \int_0^t (t-\tau)^{\alpha-1}\, f(\tau)\, d\tau\,, \quad \alpha > 0\,. \qquad (A.1)$$

We recall the convention ${}_tJ^0 = I$ (Identity operator) and the semigroup property

$$ {}_tJ^{\alpha}\, {}_tJ^{\beta} = {}_tJ^{\beta}\, {}_tJ^{\alpha} = {}_tJ^{\alpha+\beta}\,, \quad \alpha, \beta \ge 0\,. \qquad (A.2)$$

Furthermore

$$ {}_tJ^{\alpha}\, t^{\gamma} = \frac{\Gamma(\gamma+1)}{\Gamma(\gamma+1+\alpha)}\, t^{\gamma+\alpha}\,, \quad \alpha \geq 0\,, \quad \gamma > -1\,, \quad t > 0\,. \tag{A.3} $$

The fractional derivative of order $\mu > 0$ in the *Riemann-Liouville* sense is defined as the operator ${}_tD^{\mu}$ which is the left inverse of the Riemann-Liouville integral of order μ (in analogy with the ordinary derivative), that is

$$ {}_tD^{\mu}\, {}_tJ^{\mu} = I\,, \quad \mu > 0\,. \tag{A.4} $$

If m denotes the positive integer such that $m-1 < \mu \leq m$, we recognize from Eqs. (A.2) and (A.4) ${}_tD^{\mu}\, f(t) := {}_tD^{m}\, {}_tJ^{m-\mu}\, f(t)$, hence

$$ {}_tD^{\mu}\, f(t) = \begin{cases} \dfrac{d^m}{dt^m}\left[\dfrac{1}{\Gamma(m-\mu)}\displaystyle\int_0^t \dfrac{f(\tau)\, d\tau}{(t-\tau)^{\mu+1-m}}\right], & m-1 < \mu < m, \\ \dfrac{d^m}{dt^m} f(t)\,, & \mu = m. \end{cases} \tag{A.5} $$

For completion we define ${}_tD^0 = I$.

On the other hand, the fractional derivative of order $\mu > 0$ in the *Caputo* sense is defined as the operator ${}_tD_*^{\mu}$ such that ${}_tD_*^{\mu}\, f(t) := {}_tJ^{m-\mu}\, {}_tD^{m}\, f(t)$, hence

$$ {}_tD_*^{\mu}\, f(t) = \begin{cases} \dfrac{1}{\Gamma(m-\mu)}\displaystyle\int_0^t \dfrac{f^{(m)}(\tau)\, d\tau}{(t-\tau)^{\mu+1-m}}\,, & m-1 < \mu < m, \\ \dfrac{d^m}{dt^m} f(t)\,, & \mu = m. \end{cases} \tag{A.6} $$

Thus, when the order is not integer the two fractional derivatives differ in that the derivative of order m does not generally commute with the fractional integral.

We point out that the *Caputo* fractional derivative satisfies the relevant property of being zero when applied to a constant, and, in general, to any power function of non-negative integer degree less than m, if its order μ is such that $m-1 < \mu \leq m$. Furthermore we note that

$$ {}_tD^{\mu}\, t^{\gamma} = \frac{\Gamma(\gamma+1)}{\Gamma(\gamma+1-\mu)}\, t^{\gamma-\mu}\,, \quad \mu \geq 0\,, \quad \gamma > -1\,, \quad t > 0\,. \tag{A.7} $$

Gorenflo and Mainardi [GM97] have shown the essential relationships between the two fractional derivatives (when both of them exist),

$$ {}_tD_*^{\mu}\, f(t) = \begin{cases} {}_tD^{\mu}\left[f(t) - \displaystyle\sum_{k=0}^{m-1} f^{(k)}(0^+)\, \dfrac{t^k}{k!}\right], \\ {}_tD^{\mu}\, f(t) - \displaystyle\sum_{k=0}^{m-1} \dfrac{f^{(k)}(0^+)\, t^{k-\mu}}{\Gamma(k-\mu+1)}\,, \end{cases} \quad m-1 < \mu < m\,. \tag{A.8} $$

In particular, if $m = 1$ we have

$$
{}_tD_*^\mu\, f(t) = \begin{cases} {}_tD^\mu\left[f(t)-f(0^+)\right], \\ {}_tD^\mu\, f(t) - \dfrac{f(0^+)\,t^{-\mu}}{\Gamma(1-\mu)}, \end{cases} \quad 0<\mu<1\,. \tag{A.9}
$$

The *Caputo* fractional derivative, practically ignored in the mathematical treatises, represents a sort of regularization in the time origin for the *Riemann-Liouville* fractional derivative. We note that for its existence all the limiting values $f^{(k)}(0^+) := \lim_{t\to 0^+} f(t)$ are required to be finite for $k = 0, 1, 2.\dots m-1$.

We observe the different behaviour of the two fractional derivatives at the end points of the interval $(m-1, m)$ namely when the order is any positive integer: whereas ${}_tD^\mu$ is, with respect to its order μ, an operator continuous at any positive integer, ${}_tD_*^\mu$ is an operator left-continuous since

$$
\begin{cases} \lim\limits_{\mu\to(m-1)^+} {}_tD_*^\mu\, f(t) \;=\; f^{(m-1)}(t) - f^{(m-1)}(0^+)\,, \\ \lim\limits_{\mu\to m^-} {}_tD_*^\mu\, f(t) \;=\; f^{(m)}(t)\,. \end{cases} \tag{A.10}
$$

We also note for $m-1<\mu\le m$,

$$
{}_tD^\mu\, f(t) \;=\; {}_tD^\mu\, g(t) \iff f(t) = g(t) + \sum_{j=1}^{m} c_j\, t^{\mu-j}\,, \tag{A.11}
$$

$$
{}_tD_*^\mu\, f(t) \;=\; {}_tD_*^\mu\, g(t) \iff f(t) = g(t) + \sum_{j=1}^{m} c_j\, t^{m-j}\,. \tag{A.12}
$$

In these formulae the coefficients c_j are arbitrary constants. Last but not least, we point out the major utility of the Caputo fractional derivative in treating initial-value problems for physical and engineering applications where initial conditions are usually expressed in terms of integer-order derivatives. This can be easily seen using the Laplace transformation, according to which

$$
\mathcal{L}\left\{{}_tD_*^\mu\, f(t); s\right\} = s^\mu\, \widetilde{f}(s) - \sum_{k=0}^{m-1} s^{\mu-1-k}\, f^{(k)}(0^+)\,, \quad m-1<\mu\le m\,, \tag{A.13}
$$

where $\widetilde{f}(s) \;=\; \mathcal{L}\{f(t); s\} \;=\; \int_0^\infty \mathrm{e}^{-st}\, f(t)\, dt\,,\;\; s \in \mathbf{C}$, and $f^{(k)}(0^+) := \lim_{t\to 0^+} f(t)$. The corresponding rule for the Riemann-Liouville derivative is more cumbersome: for $m-1<\mu\le m$ it reads

$$
\mathcal{L}\left\{{}_tD^\mu\, f(t); s\right\} = s^\mu\, \widetilde{f}(s) - \sum_{k=0}^{m-1} \left[{}_tD^k\, {}_tJ^{(m-\mu)}\right] f(0^+)\, s^{m-1-k}\,, \tag{A.14}
$$

where, in analogy with (A.13), the limit for $t\to 0^+$ is understood to be taken after the operations of fractional integration and derivation. As soon as all the limiting values $f^{(k)}(0^+)$ are finite and $m-1 < \mu < m$, the formula (A.14) simplifies into

$$\mathcal{L}\left\{ {}_tD^{\mu}\, f(t); s\right\} = s^{\mu}\, \widetilde{f}(s)\,. \tag{A.15}$$

In the special case $f^{(k)}(0^+) = 0$ for $k = 0, 1, m-1$, we recover the identity between the two fractional derivatives, consistently with Eq. (A.8).

We remind that the Laplace transform rule (A.13) was practically the starting point of Caputo himself in defining his generalized derivative in the late sixties, [Cap67, Cap69]. Later, Caputo and Mainardi in 1971 [CM71a, CM71b] and Mainardi in the nineties, see e.g. [Main94, Main96], have followed the notation involving a convolution with the so-called Gel'fand-Shilov (generalized) function $\Phi_\lambda(t) := t_+^{\lambda-1}/\Gamma(\lambda)$ discussed in [GS64]. The notation here adopted was introduced in a systematic way by Gorenflo and Mainardi in their 1996 CISM lectures [GM97], partly based on the book on Abel Integral Equations by Gorenflo & Vessella [GV91] and on the article by Gorenflo & Rutman [GR95].

For further reading on the theory and applications of fractional calculus we recommend to consult in addition to the well-known books by Samko, Kilbas & Marichev [SKM93], by Miller & Ross [MR93], by Podlubny [Pod99], those appeared in the last few years, by Kilbas, Srivastava & Trujillo [KST06], by West, Bologna & Grigolini [WBG03], and by Zaslavsky [Zas05].

Appendix B: The Mittag-Leffler functions

B.1 The classical Mittag-Leffler function

Let us recall that the Mittag-Leffler function $E_\mu(z)$ ($\mu > 0$) is an entire transcendental function of order $1/\mu$, defined in the complex plane by the power series

$$E_\mu(z) := \sum_{k=0}^{\infty} \frac{z^k}{\Gamma(\mu\, k+1)}\,, \quad \mu > 0\,, \quad z \in \mathbf{C}\,. \tag{B.1}$$

It was introduced and studied by the Swedish mathematician Mittag-Leffler at the beginning of the XX century to provide a noteworthy example of entire function that generalizes the exponential (to which it reduces for $\mu = 1$). For details on this function we refer e.g. to [EMOT55, KST06, GM97, Pod99, SKM93]. In particular we note that the function $E_\mu(-x)$ ($x \geq 0$) turns a completely monotonic function of x if $0 < \mu \leq 1$. This property is still valid if we consider the variable $x = \lambda\, t^\mu$ where λ is a positive constant. Thus the function $E_\mu(-\lambda t^\mu)$ preserves the *complete monotonicity* of the exponential $\exp(-\lambda t)$: indeed it is represented in terms of a real Laplace transform (of a real parameter r) of a non-negative function (that we refer to as the spectral function)

$$E_\mu(-\lambda t^\mu) = \frac{1}{\pi}\int_0^\infty \frac{\mathrm{e}^{-rt}}{r}\,\frac{\lambda r^\mu \sin(\mu\pi)}{\lambda^2 + 2\lambda\, r^\mu \cos(\mu\pi) + r^{2\mu}}\,dr\,,\ t \ge 0\,,\ 0 < \mu < 1\,. \tag{B.2}$$

We note that as $\mu \to 1$ the spectral function tends to the generalized Dirac function $\delta(r-\lambda)$. We point out that the Mittag-Leffler function (B.2) starts at $t = 0$ as a stretched exponential and decreases for $t \to \infty$ like a power with exponent $-\mu$:

$$E_\mu(-\lambda t^\mu) \sim \begin{cases} 1 - \lambda\,\dfrac{t^\mu}{\Gamma(1+\mu)} \sim \exp\left\{-\dfrac{\lambda t^\mu}{\Gamma(1+\mu)}\right\}, & t \to 0^+\,, \\ \dfrac{t^{-\mu}}{\lambda\,\Gamma(1-\mu)}\,, & t \to \infty\,. \end{cases} \tag{B.3}$$

The noteworthy results (B.2) and (B.3) can also be derived from the Laplace transform pair

$$\mathcal{L}\{E_\mu(-\lambda t^\mu); s\} = \frac{s^{\mu-1}}{s^\mu + \lambda}\,. \tag{B.4}$$

In fact it it sufficient to apply the Titchmarsh theorem ($s = r\mathrm{e}^{i\pi}$) for deriving (B.2) and the Tauberian theory ($s \to \infty$ and $s \to 0$) for deriving (B.3).

If $\mu = 1/2$ we have for $t \ge 0$:

$$E_{1/2}(-\lambda\sqrt{t}) = \mathrm{e}^{\lambda^2 t}\,\mathrm{erfc}(\lambda\sqrt{t}) \sim 1/(\lambda\sqrt{\pi\, t})\,, \quad \text{as} \quad t \to \infty\,, \tag{B.5}$$

where erfc denotes the *complementary error* function, see e.g. [AS65].

B.2 The generalized Mittag-Leffler function

The Mittag-Leffler function in two parameters $E_{\mu,\nu}(z)$ ($\Re\{\mu\} > 0,\ \nu \in \mathbf{C}$) is defined by the power series

$$E_{\mu,\nu}(z) := \sum_{k=0}^{\infty} \frac{z^k}{\Gamma(\mu\, k + \nu)}\,, \qquad z \in \mathbf{C}\,. \tag{B.6}$$

It generalizes the classical Mittag-Leffler function to which it reduces for $\nu = 1$. It is an entire transcendental function of order $1/\Re\{\mu\}$ on which the reader can inform himself by again consulting *e.g.* [EMOT55, KST06, GM97, Pod99, SKM93].

With $\mu, \nu \in \mathbf{R}$ the function $E_{\mu,\nu}(-x)$ ($x \ge 0$) turns a completely monotonic function of x if $0 < \mu \le 1$ and $\nu \ge \mu > 0$, see *e.g.* [Sch96, MS97, MS01]. Again this property is still valid if we consider the variable $x = \lambda\, t^\mu$ where λ is a positive constant.

We point out the Laplace transform pair, see [Pod99],

$$\mathcal{L}\{t^{\nu-1}\, E_{\mu,\nu}(-\lambda t^\mu); s\} = \frac{s^{\mu-\nu}}{s^\mu + \lambda}\,, \quad \mu > 0\,,\ \nu > 0\,. \tag{B.7}$$

For $0 < \mu \le \nu < 1$ this Laplace transform pair can be used to derive for the function $E_{\mu,\nu}(-\lambda t^\mu)$ its asymptotic representations as $t \to 0$ and $t \to \infty$, by applying the Tauberian theory ($s \to \infty$, $s \to 0$). Indeed we have

$$E_{\mu,\nu}(-\lambda t^\mu) \sim \begin{cases} \dfrac{1}{\Gamma(\nu)}\left[1 - \lambda\dfrac{\Gamma(\nu)}{\Gamma(\nu+\mu)}\,t^\mu\right] \sim \dfrac{1}{\Gamma(\nu)}\exp\left\{-\dfrac{\lambda\Gamma(\nu)t^\mu}{\Gamma(1+\mu)}\right\}, & t \to 0^+, \\ \dfrac{1}{\lambda}\,\dfrac{t^{-\mu+\nu-1}}{\Gamma(\nu-\mu)}\,, & t \to \infty. \end{cases} \tag{B.8}$$

In particular, for $0 < \mu = \nu < 1$ we point out the noteworthy identity

$$t^{-(1-\mu)}\,E_{\mu,\mu}\left(-\lambda\,t^\mu\right) = -\frac{1}{\lambda}\,\frac{d}{dt}E_\mu\left(-\lambda\,t^\mu\right)\,. \tag{B.9}$$

Appendix C: The Exponential integral functions

C.1 Basic definitions and properties

The exponential integral function, that we denote by $\mathcal{E}_1(z)$, is defined as

$$\mathcal{E}_1(z) = \int_z^\infty \frac{\mathrm{e}^{-t}}{t}\,dt = \int_1^\infty \frac{\mathrm{e}^{-zt}}{t}\,dt\,. \tag{C.1}$$

We have used the letter $\mathcal{E}$ instead of E (commonly adopted in the literature) in order to avoid confusion with the Mittag-Leffler functions that play a more relevant role in fractional calculus. This function exhibits a branch cut along the negative real semi-axis and admits the representation

$$\mathcal{E}_1(z) = -\gamma - \log z - \sum_{n=1}^\infty \frac{z^n}{n\,n!}\,, \quad |\arg z| < \pi\,, \tag{C.2}$$

where $\gamma = 0.57721...$ is the so-called Euler-Mascheroni constant. The power series in the R.H.S. is absolutely convergent in all of $\mathbf{C}$ and represents the entire function called the *modified exponential integral*[10]

$$\mathrm{Ein}\,(z) := \int_0^z \frac{1-\mathrm{e}^{-\zeta}}{\zeta}\,d\zeta = -\sum_{n=1}^\infty \frac{z^n}{n\,n!}\,, \tag{C.3}$$

Thus, in view of (C.2) and (C.3), we write

[10] The Italian mathematicians Tricomi and Gatteschi have pointed out the major utility of the modified exponential integral, being an entire function, with respect to the exponential integral. This entire function was introduced by S.A. Schelkunoff in 1944. ["Proposed symbols for the modified cosine and integral exponential integral", *Quart. Appl. Math.* **2** (1944), p. 90]

$$\mathcal{E}_1(z) = -\gamma - \log z + \mathrm{Ein}\,(z)\,, \quad |\arg z| < \pi\,. \tag{C.4}$$

This relation is important for understanding the analytic properties of the classical exponential integral function in that it isolates the multi-valued part represented by the logarithmic function from the regular part represented by the entire function Ein (z). Furthermore, Ein (x) is an *increasing* function on $\mathbf{R}$ because

$$\frac{d}{dx}\,\mathrm{Ein}\,(x) = \frac{1-\mathrm{e}^{-x}}{x} > 0\,, \quad \forall x \in \mathbf{R}\,.$$

In $\mathbf{R}^+$ the function Ein (x) turns out to be a *Bernstein function*, which means that is positive, increasing, with the first derivative *completely monotonic.*

C.2 Asymptotic expansion of the exponential integral

The asymptotic behaviour as $z \to \infty$ of the exponential integrals can be obtained from the integral representation (C.1) noticing that

$$\mathcal{E}_1(z) := \int_z^\infty \frac{\mathrm{e}^{-t}}{t}\,dt = \mathrm{e}^{-z}\int_0^\infty \frac{\mathrm{e}^{-u}}{u+z}\,du\,. \tag{C.5}$$

In fact, by repeated partial integrations in the R.H.S., we get

$$\mathcal{E}_1(z) \sim \frac{\mathrm{e}^{-z}}{z}\sum_{n=0}^{\infty}(-1)^n\,\frac{n!}{z^n}\,, \quad z\to\infty\,, \quad |\arg z| \le \pi-\delta\,. \tag{C.6}$$

C.3 Laplace transform pairs related to exponential integrals

We now report a number of relevant Laplace transform pairs related to logarithmic and exponential integral functions.

Taking $t > 0$, the basic Laplace transforms pairs are

$$\mathcal{L}\{\log t; s\} = -\frac{\gamma + \log s}{s}\,, \quad \Re s > 0\,, \tag{C.7}$$

$$\mathcal{L}\{\mathcal{E}_1(t); s\} = \frac{\log\,(s+1)}{s}\,, \quad \Re s > -1\,, \tag{C.8}$$

The proof of (C.7) and (C.8) is found, for example, in the treatise by Ghizzetti & Ossicini, see [GO71], Eqs. [4.6.15-16]), pp. 104-105. We then easily derive

$$\mathcal{L}\{\gamma + \log t; s\} = -\frac{\log s}{s}\,, \quad \Re s > 0\,, \tag{C.9}$$

$$\mathcal{L}\{\gamma+\log t+\mathcal{E}_1(t); s\} = \frac{\log\,(s+1)}{s} - \frac{\log s}{s} = \frac{\log\,(1/s+1)}{s} \quad \Re s > 0\,. \tag{C.10}$$

$$\mathcal{L}\{\gamma + \log t + \mathrm{e}^t\,\mathcal{E}_1(t); s\} = \frac{\log s}{s-1} - \frac{\log s}{s} = \frac{\log s}{s(s-1)}\,, \quad \Re s > 0\,, \tag{C.11}$$

We outline the different asymptotic behaviour of the three functions $f_1(t) = \mathcal{E}_1(t)$, $f_2(t) = \mathrm{Ein}\,(t) = \gamma + \log t + \mathcal{E}_1(t)$ and $f_3(t) = \gamma + \log t + \mathrm{e}^t\,\mathcal{E}_1(t)$ for small argument ($t \to 0^+$) and large argument ($t \to +\infty$). By using Eqs. (C.2), (C.4) and (C.6), we have

$$f_1(t) = \mathcal{E}_1(t)) \sim \begin{cases} \log(1/t)\,, & t \to 0^+\,, \\ \mathrm{e}^{-t}/t\,, & t \to +\infty\,. \end{cases} \tag{C.12}$$

$$f_2(t) = \mathrm{Ein}\,(t) = \gamma + \log t + \mathcal{E}_1(t) \sim \begin{cases} t\,, & t \to 0^+\,, \\ \log t\,, & t \to +\infty\,. \end{cases} \tag{C.13}$$

$$f_3(t) = \gamma + \log t + \mathrm{e}^t\,\mathcal{E}_1(t) \sim \begin{cases} t\log(1/t)\,, & t \to 0^+\,, \\ \log t\,, & t \to +\infty\,. \end{cases} \tag{C.14}$$

We note that all the above asymptotic representations can be obtained from the Laplace transforms of the corresponding functions by invoking the Tauberian theory for *regularly varying functions* (power functions multiplied by *slowly varying functions*[11]), a topic adequately treated in the treatise on Probability by Feller, see [Fel71], Chapter XIII.5.

C.4 The $\nu(t)$ function and the related Laplace transform pair

In the third volume of the Handbook of the Bateman Project, in the Chapter XVIII devoted the Miscellaneous functions, see [EMOT55], §18.3 pp. 217-224, we find, in addition to the functions of the Mittag-Leffler and Wright type, the function

$$\nu(t,a) = \int_0^\infty \frac{t^{a+\tau}}{\Gamma(a+\tau+1)}\,d\tau\,, \quad a > -1\,. \tag{C.15}$$

Such special function is relevant for our purposes because of the Laplace transform pair, see [EMOT55], Eq. (18), p.222,

$$\mathcal{L}\{\nu(t,a;s\} = \frac{1}{s^{a+1}\log s}\,, \quad \Re s > 0\,, \tag{C.16}$$

[11] **Definition:** We call a (measurable) positive function $a(y)$, defined in a right neighbourhood of zero, *slowly varying at zero* if $a(cy)/a(y) \to 1$ with $y \to 0$ for every $c > 0$. We call a (measurable) positive function $b(y)$, defined in a neighbourhood of infinity, *slowly varying at infinity* if $b(cy)/a(y) \to 1$ with $y \to \infty$ for every $c > 0$. Examples: $(\log y)^\gamma$ with $\gamma \in \mathbf{R}$ and $\exp\,(\log y/\log\log y)$.

References

[MK00] Metzler, R., Klafter, J.: The random walk's guide to anomalous diffusion: a fractional dynamics approach. Phys. Reports, **339**, 1-77 (2000)

[MK04] Metzler, R., Klafter, J.: The restaurant at the end of the random walk: Recent developments in the description of anomalous transport by fractional dynamics. J. Phys. A. Math. Gen., **37**, R161-R208 (2004)

[PSW05] Piryatinska, A., Saichev, A.I., Woyczynski, W.A.: Models of anomalous diffusion: the subdiffusive case. Physica A, **349**, 375-420 (2005)

[Zas02] Zaslavsky, G.M.: Chaos, fractional kinetics and anomalous transport. Phys. Reports, **371**, 461-580 (2002)

[KS05] Klafter, J., Sokolov, I.M.: Anomalous diffusion spreads its wings. Physics World, **18**, 29-32 (2005)

[SK05] Sokolov, I.M., Klafter, J.: From diffusion to anomalous diffusion: a century after Einstein's Brownian motion. Chaos, **15**, 026103-026109 (2005)

[SW89] Schneider, W.R., Wyss, W.: Fractional diffusion and wave equations, J. Math. Phys., **30**, 134-144 (1989)

[Main94] Mainardi, F.: On the initial value problem for the fractional diffusion-wave equation. In: Rionero, S., Ruggeri, T. (ed) Waves and Stability in Continuous Media. World Scientific, Singapore (1994)

[Main96] Mainardi, F.: Fractional relaxation-oscillation and fractional diffusion-wave phenomena. Chaos, Solitons and Fractals, **7**, 1461–1477 (1996)

[Main97] Mainardi, F.: Fractional calculus: some basic problems in continuum and statistical mechanics. In: Carpinteri, A., Mainardi, F. (ed) Fractals and Fractional Calculus in Continuum Mechanics. Springer Verlag, Wien New-York (1997) [Reprinted in `http://www.fracalmo.org`]

[GR95] Gorenflo, R., Rutman, R.: On ultraslow and intermediate processes. In: Rusev, P., Dimovski, I., Kiryakova, V. (ed) Proc. Workshop on Transform Methods and Special Functions (Sofia 1994). Science Culture Technology, Singapore (1995)

[GLM99] Gorenflo, R., Luchko, Yu., Mainardi, F.: Analytical properties and applications of the Wright function. Fractional Calculus and Applied Analysis, **2**, 383-414 (1999)

[GLM00] Gorenflo, R., Luchko, Yu., Mainardi, F.: Wright functions as scale-invariant solutions of the diffusion-wave equation. J. Computational and Applied Mathematics, **118**, 175-191 (2000)

[MP03] Mainardi, F., Pagnini, G.: The Wright functions as solutions of the time-fractional diffusion equations: Appl. Math. and Comp., **141**, 51-62 (2003)

[Nig86] Nigmatullin, R.R.: The realization of the generalized transfer equation in a medium with fractal geometry. Phys. Stat. Sol. B, **133**, 425-430 (1986)

[GR92] Giona, M., Roman, H.E.: Fractional diffusion equation for transport phenomena in random media. Physica A, **185**, 87-97 (1992)

[MGN94] Metzler, R., Glöckle, W.G., Nonnenmacher, T.F.: Fractional model equation for anomalous diffusion. Physica A, **211**, 13–24 (1994)

[SZ97] Saichev, A., Zaslavsky, G.: Fractional kinetic equations: solutions and applications. Chaos, **7**, 753-764 (1997)

[GS64] Gel`fand, I.M., Shilov, G.E.: Generalized Functions. Vol. I, Academic Press, New York London (1964)

[GM03] Gorenflo, R., Mainardi, F.: Fractional diffusion processes: probability distributions and continuous time random walk. In: Rangarajan, G., Ding, M. (ed) Processes with Long Range Correlations. Springer Verlag, Berlin (2003) [Lecture Notes in Physics, No. 621]

[GM05] Gorenflo, R., Mainardi, F.: Simply and multiply scaled diffusion limits for continuous time random walks. In: Benkadda, S., Leoncini, X., Zaslavsky, G. (ed) Proceedings of the International Workshop on Chaotic Transport and Complexity in Fluids and Plasmas, Carry Le Rouet (France) 20-25 June 2004. IOP (Institute of Physics) Journal of Physics: Conference Series **7** (2005)

[MVG05] Mainardi, F., Vivoli, A., Gorenflo, R.: Continuous time random walk and time fractional diffusion: a numerical comparison between the fundamental solutions. Fluctuation and Noise Letters, **5**, L291-L297 (2005)

[SGM04] Scalas, E., Gorenflo, R., Mainardi, F.: Uncoupled continuous-time random walks: solution and limiting behaviour of the master equation. Physical Review E, **69**, 011107-1/8 (2004)

[AS65] Abramowitz, M., Stegun, I.A.: Handbook of Mathematical Functions. Dover, New York (1965)

[Pod99] Podlubny, I.: Fractional Differential Equations. Academic Press, San Diego (1999)

[EMOT55] Erdélyi, A., Magnus, W., Oberhettinger, F., Tricomi, F.G.: Miscellaneous Functions. In: Higher Transcendental Functions. Bateman Project, Vols. 1-3, McGraw-Hill, New York (1955)

[GM97] Gorenflo, R., Mainardi, F.: Fractional calculus: integral and differential equations of fractional order. In: Carpinteri, A., Mainardi, F. (ed) Fractals and Fractional Calculus in Continuum Mechanics. Springer Verlag, Wien New York (1997) [Reprinted in `http://www.fracalmo.org`]

[GIL00] Gorenflo, R., Iskenderov, A., Luchko, Yu.: Mapping between solutions of fractional diffusion-wave equations. Fractional Calculus and Applied Analysis, **3**, 75–86 (2000)

[MLP01] Mainardi, F., Luchko, Yu., Pagnini, G.: The fundamental solution of the space-time fractional diffusion equation. Fractional Calculus and Applied Analysis, **4**, 153-192 (2001) [Reprinted in `http://www.fracalmo.org`]

[Cap69] Caputo, M.: Elasticità e Dissipazione. Zanichelli, Bologna (1969) [in Italian]

[Cap95] Caputo, M.: Mean fractional-order derivatives differential equations and filters. Ann. Univ. Ferrara, Sez VII, Sc. Mat., **41**, 73-84 (1995)

[Cap01] Caputo, M.: Distributed order differential equations modelling dielectric induction and diffusion. Fractional Calculus and Applied Analysis, **4**, 421-442 (2001)

[BT00] Bagley, R.L., Torvik, P.J.: On the existence of the order domain and the solution of distributed order equations. Int. J. Appl. Math., **2**, 865-882, 965-987 (2000)

[CGS02] Chechkin, A.V., Gorenflo,R., Sokolov, I.M.: Retarding subdiffusion and accelerating superdiffusion governed by distributed-order fractional diffusion equations. Phys. Rev. E, **66**, 046129/1-6 (2002)

[CGSG03] Chechkin, A.V., Gorenflo, R., Sokolov, I.M., Gonchar, V.Yu.: Distributed order time fractional diffusion equation. Fractional Calculus and Applied Analysis, **6**, 259-279 (2003)

[CKS03] Chechkin, A.V., Klafter, J., Sokolov, I.M.: Fractional Fokker-Planck equation for ultraslow kinetics. Europhysics Lett., **63**, 326-332 (2003)

[SCK04] Sokolov, I.M., Chechkin, A.V., Klafter, J.: Distributed-order fractional kinetics. Acta Physica Polonica, **35**, 1323-1341 (2004)

[Nab04] Naber, M.: Distributed order fractional subdiffusion. Fractals, **12**, 23-32 (2004)

[Lang06] Langlands, T.A.M.: Solution of a modified fractional diffusion equation. Physica A, **367**, 136-144 (2006)

[UG05] Umarov, S., Gorenflo, R.: Cauchy and nonlocal multi-point problems for distributed order pseudo-differential equations: Part one. Journal for Analysis and its Applications (ZAA), **24**, 449-466 (2005)

[MP06] Mainardi, F., Pagnini, G.: The role of the Fox-Wright functions in fractional subdiffusion of distributed order. J. Computational and Appl. Mathematics. (2006), in press.

[PK01] Paris, R.B., Kaminski, D.: Asymptotic and Mellin-Barnes Integrals. Cambridge Univ. Press, Cambridge (2001)

[Cap67] Caputo, M.: Linear models of dissipation whose Q is almost frequency independent, Part II. Geophys. J. R. Astr. Soc., **13**, 529–539 (1967)

[CM71a] Caputo, M., Mainardi, F.: A new dissipation model based on memory mechanism. Pure and Applied Geophysics (Pageoph), **91**, 134–147 (1971)

[CM71b] Caputo, M., Mainardi, F.: Linear models of dissipation in anelastic solids. Rivista del Nuovo Cimento (Ser. II), **1**, 161–198 (1971)

[GV91] Gorenflo, R., Vessella, S.: Abel Integral Equations: Analysis and Applications. Springer Verlag, Berlin (1991)

[SKM93] Samko, S.G., Kilbas, A.A., Marichev, O.I.: Fractional Integrals and Derivatives: Theory and Applications. Gordon and Breach, New York (1993)

[MR93] Miller, K.S., Ross, B.: An Introduction to the Fractional Calculus and Fractional Differential Equations. Wiley, New York (1993)

[KST06] Kilbas, A.A., Srivastava, H.M., Trujillo, J.J.: Theory and Applications of Fractional Differential Equations. Elsevier, Amsterdam (2006)

[WBG03] West, B.J., Bologna, M., Grigolini, P.: Physics of Fractal Operators. Springer Verlag, New York (2003)

[Zas05] Zaslavsky, G.M.: Hamiltonian Chaos and Fractional Dynamics. Oxford University Press, Oxford (2005)

[Sch96] Schneider, W.R.: Completely monotone generalized Mittag-Leffler functions. Expositiones Mathematicae, **14**, 3-16 (1996)

[MS97] Miller, K.S., Samko, S.G.: A note on the complete monotonicity of the generalized Mittag-Leffler function. Real Anal. Exchange, **23**, 753-755 (1997)

[MS01] Miller, K.S., Samko, S.G.: Completely monotonic functions. Integral Transforms and Special Functions, **12**, 389-402 (2001)

[GO71] Ghizzetti, A., Ossicini, A.: Trasformate di Laplace e Calcolo Simbolico. UTET, Torino (1971)

[Fel71] Feller, W.: An Introduction to Probability Theory and its Applications. Vol. 2, Wiley, New York (1971)

[Djr66] Djrbashian, M.M.: Integral Transforms and Representations of Functions in the Complex Plane. Nauka, Moscow (1966) [in Russian] [There is also the transliteration as Dzherbashian]

[MG00] Mainardi, F., Gorenflo, R.: On Mittag-Leffler type functions in fractional evolution processes. J. Comput. and Appl. Mathematics, **118**, 283-299 (2000)

[Mari83] Marichev, O.I.: Handbook of Integral Transforms of Higher Transcendental Functions, Theory and Algorithmic Tables. Chichester, Ellis Horwood (1983)

[Tem96] Temme, N.M.: Special Functions: An Introduction to the Classical Functions of Mathematical Physics. Wiley, New York (1996)

Neutrices and Generalized Functions

The composition and neutrix composition of distributions

Brian Fisher

Department of Mathematics, University of Leicester, Leicester, U.K. fbr@le.ac.uk

Let F be a distribution in $\mathcal{D}'$ and let f be a locally summable function. The neutrix composition $F(f(x))$ is said to exist and be equal to the distribution h if the neutrix limit of the sequence $\{F_n(f(x))\}$ is h, where $F_n(x) = F(x)*\delta_n(x)$ for $n = 1, 2, \ldots$ and $\{\delta_n(x)\}$ is a certain regular sequence converging to the Dirac delta funcion. In particular, the composition $F(f(x))$ is said to exist and be equal to the distribution h if the sequence $\{F_n(f(x))\}$ converges to h in the normal sense. Some results are proved.

1 Introduction

In the following, we let $\mathcal{D}$ be the space of infinitely differentiable functions with compact support, let $\mathcal{D}[a, b]$ be the space of infinitely differentiable functions with support contained in the interval $[a, b]$ and let $\mathcal{D}'$ be the space of distributions defined on $\mathcal{D}$.

Now let $\rho(x)$ be a function in $\mathcal{D}$ having the following properties:
(i) $\rho(x) = 0$ for $|x| \geq 1$,
(ii) $\rho(x) \geq 0$,
(iii) $\rho(x) = \rho(-x)$,
(iv) $\displaystyle\int_{-1}^{1} \rho(x)\, dx = 1$.

Putting $\delta_n(x) = n\rho(nx)$ for $n = 1, 2, \ldots,$ we have

$$\begin{aligned}\lim_{n\to\infty} \langle \delta_n(x), \varphi(x)\rangle &= \lim_{n\to\infty} \int_{-1/n}^{1/n} \delta_n(x)\varphi(x)\, dx = \lim_{n\to\infty} \int_{-1}^{1} \rho(t)\varphi(t/n)\, dt \\ &= \varphi(0) = \langle \delta(t), \varphi(t)\rangle,\end{aligned}$$

for arbitrary φ in $\mathcal{D}$. It follows that $\{\delta_n(x)\}$ is a regular sequence of infinitely differentiable functions converging to the Dirac delta-function $\delta(x)$.

Kenan Taş et al. (eds), Mathematical Methods in Engineering, 59–69.

If now $f(x)$ is an infinitely differentiable function having a single simple root at the point $x = x_0$ with $f'(x) > 0$, then putting $t = f(x)$ and $\psi(x) = f'(x)\varphi(f(x))$, we have

$$\int_{-\infty}^{\infty} \delta_n(t)\varphi(t)\,dt = \int_{-\infty}^{\infty} \delta_n(f(x))f'(x)\varphi(f(x))\,dx$$
$$= \int_{-\infty}^{\infty} \delta_n(f(x))\psi(x)\,dx = \langle \delta_n(f(x)), \psi(x)\rangle.$$

We therefore define the distribution $\delta(f(x))$ by

$$\langle \delta(f(x)), \psi(x)\rangle = \lim_{n\to\infty} \int_{-\infty}^{\infty} \delta_n(f(x))\psi(x)\,dx$$
$$= \varphi(0) = \frac{\psi(x_0)}{f'(x_0)} = \frac{1}{f'(x_0)}\langle \delta(x - x_0), \psi(x)\rangle.$$

If $f(x_0) < 0$, we would have had

$$\langle \delta(f(x)), \psi(x)\rangle = \frac{1}{|f'(x_0)|}\langle \delta(x - x_0), \psi(x)\rangle$$

and so

$$\delta(f(x)) = \frac{1}{|f'(x_0)|}\delta(x - x_0) \tag{1}$$

in either case. This is of course in agreement with Gel'fand and Shilov [GS64].

Differentiating (1), we get

$$\delta'(f(x)) = \frac{1}{|f'(x_0)|}\frac{1}{f'(x)}\frac{d}{dx}\delta(x - x_0)$$

and more generally, we have

$$\delta^{(r)}(f(x)) = \frac{1}{|f'(x_0)|}\left[\frac{1}{f'(x)}\frac{d}{dx}\right]^r \delta(x - x_0) \tag{2}$$

for $r = 0, 1, 2, \ldots$.

In order to generalize (2), the following definition was given in [Fis83].

Definition 1. *Let f be an infinitely differentiable function. We say that the distribution $\delta^{(r)}(f(x))$ exists and is equal to h on the open interval (a, b) if*

$$\operatorname*{N-lim}_{n\to\infty} \int_{-\infty}^{\infty} \delta_n^{(r)}(f(x))\varphi(x)\,dx = \langle h(x), \varphi(x)\rangle$$

for all φ in $\mathcal{D}[a, b]$, with $-\infty < a < b < \infty$, where N is the neutrix, see [vdCo59], having domain N' the positive integers and range N'' the real numbers, with negligible functions which are finite linear sums of the functions

$$n^\lambda \ln^{r-1} n,\ \ln^r n : \qquad \lambda > 0,\ r = 1, 2, \ldots$$

and all functions which converge to zero in the usual sense as n tends to infinity.

Note that taking the neutrix limit of a function $f(n)$, is equivalent to taking the usual limit of Hadamard's finite part of $f(n)$.

The following theorem was then proved.

Theorem 1. *The distribution $\delta^{(r)}(x^s)$ exists and*

$$\delta^{(r)}(x^{2s}) = 0,$$
$$\delta^{(r)}(x^{2s-1}) = \frac{r!}{s(rs+s-1)!}\delta^{(rs+s-1)}(x)$$

for $r = 0, 1, 2, \ldots$ and $s = 1, 2, \ldots$.

Definition 1 was later generalized with the following definition in [Fis85] and was originally called the composition of distributions.

Definition 2. *Let F be a distribution in $\mathcal{D}'$ and let f be a locally summable function. We say that the neutrix composition $F(f(x))$ exists and is equal to h on the open interval (a, b), with $-\infty < a < b < \infty$, if*

$$\underset{n\to\infty}{\text{N}-\lim} \int_{-\infty}^{\infty} F_n(f(x))\varphi(x)dx = \langle h(x), \varphi(x)\rangle$$

for all φ in $\mathcal{D}[a, b]$, where $F_n(x) = F(x) * \delta_n(x)$ for $n = 1, 2, \ldots$.

In particular, we say that the composition $F(f(x))$ exists and is equal to h on the open interval (a, b) if

$$\lim_{n\to\infty} \int_{-\infty}^{\infty} F_n(f(x))\varphi(x)dx = \langle h(x), \varphi(x)\rangle$$

for all φ in $\mathcal{D}[a, b]$.

The following theorem was then proved in [Fis85].

Theorem 2. *The neutrix composition $\delta^{(s)}(\operatorname{sgn} x|x|^\lambda)$ exists and*

$$\delta^{(s)}(\operatorname{sgn} x|x|^\lambda) = 0$$

for $s = 0, 1, 2, \ldots$ and $(s+1)\lambda = 1, 3, \ldots$ and

$$\delta^{(s)}(\operatorname{sgn} x|x|^\lambda) = \frac{(-1)^{(s+1)(\lambda+1)}s!}{\lambda[(s+1)\lambda - 1]!}\delta^{((s+1)\lambda-1)}(x)$$

for $s = 0, 1, 2, \ldots$ and $(s+1)\lambda = 2, 4, \ldots$.

2 Main Results

We need the following lemma which can be easily proved by induction:

Lemma 1.

$$\int_{-1}^{1} v^i \rho^{(r)}(v)\,dv = \begin{cases} 0, & 0 \le i < r, \\ (-1)^r r!, & i = r \end{cases}$$

for $r = 0, 1, 2, \ldots$.

We now prove the following theorem.

Theorem 3. *The compositions* $\delta^{(2s-1)}(\operatorname{sgn} x|x|^{1/s})$ *and* $\delta^{(s-1)}(|x|^{1/s})$ *exist and*

$$\delta^{(2s-1)}(\operatorname{sgn} x|x|^{1/s}) = \tfrac{1}{2}(2s)!\delta'(x), \tag{3}$$

$$\delta^{(s-1)}(|x|^{1/s}) = (-1)^{s-1}\delta(x) \tag{4}$$

for $s = 1, 2, \ldots$.

Proof. We will first of all prove (3). To do this, we first of all evaluate

$$\lim_{n\to\infty} \langle \delta_n^{(2s-1)}(\operatorname{sgn} x|x|^{1/s}), \varphi(x)\rangle, \tag{5}$$

for an arbitrary function $\varphi(x)$ in $\mathcal{D}$, where we may suppose that the support of $\varphi(x)$ is in the interval $[-a, a]$ with $a > 1$. By Taylor's Theorem we have

$$\varphi(x) = \varphi(0) + x\varphi'(0) + \frac{x^2}{2}\varphi''(\xi x),$$

where $0 < \xi < 1$ and so

$$\begin{aligned} \lim_{n\to\infty} \langle \delta_n^{(2s-1)}(\operatorname{sgn} x|x|^{1/s}), \varphi(x)\rangle = {} & \lim_{n\to\infty} \varphi(0) \int_{-a}^{a} \delta_n^{(2s-1)}(\operatorname{sgn} x|x|^{1/s})\,dx \\ & + \lim_{n\to\infty} \varphi'(0) \int_{-a}^{a} x\delta_n^{(2s-1)}(\operatorname{sgn} x|x|^{1/s})\,dx \\ & + \lim_{n\to\infty} \int_{-a}^{a} \frac{x^2}{2}\delta_n^{(2s-1)}(\operatorname{sgn} x|x|^{1/s})\varphi''(\xi x)\,dx \\ = {} & \lim_{n\to\infty} \varphi'(0) \int_{-n^{-s}}^{n^{-s}} x\delta_n^{(2s-1)}(\operatorname{sgn} x|x|^{1/s})\,dx \\ & + \lim_{n\to\infty} \int_{-n^{-s}}^{n^{-s}} \frac{x^2}{2}\delta_n^{(2s-1)}(\operatorname{sgn} x|x|^{1/s})\varphi''(\xi x)\,dx. \end{aligned} \tag{6}$$

Making the substitution $n|x|^{1/s} = u$, we have

$$\int_{-n^{-s}}^{n^{-s}} x\delta_n^{(2s-1)}(\operatorname{sgn} x|x|^{1/s})\,dx = s\int_{-1}^{1} u^{2s-1}\rho^{(2s-1)}(u)\,du = -\tfrac{1}{2}(2s)!, \qquad (7)$$

on using the lemma.

Next, we have

$$\int_{-n^{-s}}^{n^{-s}} |x^2\delta_n^{(2s-1)}(\operatorname{sgn} x|x|^{1/s})|\,dx = sn^{-s}\int_{-1}^{1} |u^{3s-1}\rho^{(2s-1)}(u)|\,du = O(n^{-s})$$

and so

$$\lim_{n\to\infty}\int_{-n^{-s}}^{n^{-s}} x^2\delta_n^{(2s-1)}(\operatorname{sgn} x|x|^{1/s})\varphi''(\xi x)\,dx = 0. \qquad (8)$$

It now follows from (5) to (8) that

$$\lim_{n\to\infty}\langle\delta_n^{(2s-1)}(\operatorname{sgn} x|x|^{1/s}),\varphi(x)\rangle = -\tfrac{1}{2}(2s)!\varphi'(0) = \tfrac{1}{2}(2s)!\langle\delta'(x),\varphi(x)\rangle,$$

proving (3).

To prove (4), we will evaluate

$$\lim_{n\to\infty}\langle\delta_n^{(s-1)}(|x|^{1/s}),\varphi(x)\rangle, \qquad (9)$$

where $\varphi(x)$ is an arbitrary function in $\mathcal{D}[-a,a]$, with

$$\varphi(x) = \varphi(0) + x\varphi'(\xi x),$$

where $0 < \xi < 1$. We have

$$\begin{aligned}\langle\delta_n^{(s-1)}(|x|^{1/s}),\varphi(x)\rangle &= \varphi(0)\int_{-a}^{a}\delta_n^{(s-1)}(|x|^{1/s})\,dx + \int_{-a}^{a} x\delta_n^{(s-1)}(|x|^{1/s})\varphi'(\xi x)\,dx \\ &= \varphi(0)\int_{-n^{-s}}^{n^{-s}}\delta_n^{(s-1)}(|x|^{1/s})\,dx \\ &\quad + \int_{-n^{-s}}^{n^{-s}} x\delta_n^{(s-1)}(|x|^{1/s})\varphi'(\xi x)\,dx, \qquad (10)\end{aligned}$$

where

$$\begin{aligned}\int_{-n^{-s}}^{n^{-s}} x\delta_n^{(s-1)}(|x|^{1/s})\,dx &= s\int_{-1}^{1} u^{s-1}\rho^{(s-1)}(u)\,du \\ &= (-1)^{s-1}s!, \qquad (11)\end{aligned}$$

on using the lemma.

Next, we have

$$\int_{-n^{-s}}^{n^{-s}} |x\delta_n^{(s-1)}(|x|^{1/s})|\,dx = sn^{-s}\int_{-1}^{1}|u^{2s-1}\rho^{(s-1)}(u)|\,du = O(n^{-s})$$

and so

$$\lim_{n\to\infty}\int_{-n^{-s}}^{n^{-s}} x\delta_n^{(s-1)}(|x|^{1/s})\varphi'(\xi x)\,dx = 0. \tag{12}$$

It now follows from (8) to (12) that

$$\lim_{n\to\infty}\langle\delta_n^{(s-1)}(|x|^{1/s}),\varphi(x)\rangle = (-1)^{s-1}s!\varphi(0) = (-1)^{s-1}s!\langle\delta(x),\varphi(x)\rangle,$$

proving (4). This completes the proof of the theorem.

Theorem 4. *The neutrix compositions* $\delta^{(s)}[\ln(1\pm x_+)]$ *exist and*

$$\delta^{(s)}[\ln(1+x_+)] = \sum_{k=0}^{s}\sum_{i=0}^{k}\binom{k}{i}\frac{(-1)^{s+i}(i+1)^s}{2k!}\delta^{(k)}(x), \tag{13}$$

$$\delta^{(s)}[\ln(1-x_+)] = -\sum_{k=0}^{s}\sum_{i=0}^{k}\binom{k}{i}\frac{(-1)^{s+k+i}(i+1)^s}{2k!}\delta^{(k)}(x) \tag{14}$$

for $s = 0,1,2,\ldots$.

Proof. To prove (13), we will evaluate

$$\operatorname*{N-lim}_{n\to\infty}\langle\delta_n^{(s)}[\ln(1+x_+)],\varphi(x)\rangle, \tag{15}$$

where $\varphi(x)$ is again an arbitrary function in $\mathcal{D}[-a,a]$. By Taylor's Theorem we have

$$\varphi(x) = \sum_{k=0}^{s}\frac{x^k}{k!}\varphi^{(k)}(0) + \frac{x^{s+1}}{(s+1)!}\varphi^{(s+1)}(\xi x),$$

where $0<\xi<1$ and so

$$\begin{aligned}\operatorname*{N-lim}_{n\to\infty}\langle\delta_n^{(s)}[\ln(1+x_+)],\varphi(x)\rangle &= \operatorname*{N-lim}_{n\to\infty}\sum_{k=0}^{s}\frac{\varphi^{(k)}(0)}{k!}\int_{-a}^{a}x^k\delta_n^{(s)}[\ln(1+x_+)]\,dx\\ &\quad+\operatorname*{N-lim}_{n\to\infty}\int_{-a}^{a}\frac{x^{s+1}}{(s+1)!}\delta_n^{(s)}[\ln(1+x_+)]\varphi^{(s+1)}(\xi x)\,dx\\ &= \operatorname*{N-lim}_{n\to\infty}\sum_{k=0}^{s}\frac{\varphi^{(k)}(0)}{k!}\int_{0}^{e^{1/n}-1}x^k\delta_n^{(s)}[\ln(1+x)]\,dx\\ &\quad+\operatorname*{N-lim}_{n\to\infty}\int_{0}^{e^{1/n}-1}\frac{x^{s+1}}{(s+1)!}\delta_n^{(s)}[\ln(1+x)]\varphi^{(s+1)}(\xi x)\,dx\\ &\quad+\operatorname*{N-lim}_{n\to\infty}\sum_{k=0}^{s}\frac{\varphi^{(k)}(0)}{k!}\int_{-a}^{0}x^k\delta_n^{(s)}(0)\,dx\\ &\quad+\operatorname*{N-lim}_{n\to\infty}\int_{-a}^{0}\frac{x^{s+1}}{(s+1)!}\delta_n^{(s)}(0)\varphi^{(s+1)}(\xi x)\,dx.\end{aligned} \tag{16}$$

Making the substitution $n\ln(1+x) = u$, we have

$$\begin{aligned}\int_0^{e^{1/n}-1} x^k \delta_n^{(s)}[\ln(1+x)]\,dx &= n^s \int_0^1 (e^{u/n}-1)^k e^{u/n} \rho^{(s)}(u)\,du \\ &= n^s \sum_{i=0}^{k} \int_0^1 \binom{k}{i} (-1)^{k+i} e^{(i+1)u/n} \rho^{(s)}(u)\,du \end{aligned} \tag{17}$$

and so

$$\begin{aligned}\operatorname*{N-lim}_{n\to\infty} \int_0^{e^{1/n}-1} x^k \delta_n^{(s)}[\ln(1+x)]\,dx &= \sum_{i=0}^{k} \int_0^1 \binom{k}{i} \frac{(-1)^{k+i}}{s!} (i+1)^s u^s \rho^{(s)}(u)\,du \\ &= \sum_{i=0}^{k} \binom{k}{i} \frac{(-1)^{s+k+i}(i+1)^s}{2} \end{aligned} \tag{18}$$

on using the lemma, for $k = 0, 1, 2, \ldots, s$.

When $k = s+1$, we have

$$\begin{aligned}\left| \int_0^{e^{1/n}-1} x^{s+1} \delta_n^{(s)}[\ln(1+x)]\,dx \right| &\leq n^s \int_0^1 (e^{u/n}-1)^{s+1} e^{u/n} |\rho^{(s)}(u)|\,du \\ &= n^s \int_0^1 [u/n + O(n^{-2})]^{s+1} e^{u/n} |\rho^{(s)}(u)|\,du \\ &= O(n^{-1})\end{aligned}$$

and so

$$\lim_{n\to\infty} \int_0^{e^{1/n}-1} x^{s+1} \delta_n^{(s)}[\ln(1+x)] \varphi^{(s+1)}(\xi x)\,dx = 0. \tag{19}$$

Next we have

$$\int_{-a}^{0} x^k \delta_n^{(s)}(0)\,dx = n^{s+1} \rho^{(s)}(0) \int_{-a}^{0} x^k\,dx$$

and it follows that

$$\operatorname*{N-lim}_{n\to\infty} \int_{-a}^{0} x^k \delta_n^{(s)}(0)\,dx = 0 \tag{20}$$

for $k = 0, 1, 2, \ldots, s$.

When $k = s+1$, we have

$$\int_{-a}^{0} x^{s+1} \delta_n^{(s)}(0) \varphi^{(s+1)}(\xi x)\,dx = n^{s+1} \rho^{(s)}(0) \int_{-a}^{0} x^{s+1} \varphi^{(s+1)}(\xi x)\,dx$$

and so

$$\operatorname*{N-lim}_{n\to\infty} \int_{-a}^{0} x^k \delta_n^{(s)}(0) \varphi^{(s+1)}(\xi x)\,dx = 0. \tag{21}$$

It now follows from (15) to (21) that

$$\begin{aligned}\operatorname*{N-lim}_{n\to\infty}\langle\delta_n^{(s)}[\ln(1+x_+)],\varphi(x)\rangle &= \sum_{i=0}^{k}\binom{k}{i}\frac{(-1)^{s+k+i}(i+1)^s}{2k!}\varphi^{(k)}(0)\\ &= \sum_{i=0}^{k}\binom{k}{i}\frac{(-1)^{s+i}(i+1)^s}{2k!}\langle\delta^{(k)}(x),\varphi(x)\rangle\end{aligned}$$

and equation (13) follows.

We now prove (14). Using the above working, it follows that

$$\begin{aligned}\operatorname*{N-lim}_{n\to\infty}\langle\delta_n^{(s)}[\ln(1-x_+)],\varphi(x)\rangle &= \operatorname*{N-lim}_{n\to\infty}\sum_{k=0}^{s}\frac{\varphi^{(k)}(0)}{k!}\int_0^{1-e^{-1/n}}x^k\delta_n^{(s)}[\ln(1-x)]\,dx\\ &\quad+\operatorname*{N-lim}_{n\to\infty}\int_0^{1-e^{-1/n}}\frac{x^{s+1}}{(s+1)!}\delta_n^{(s)}[\ln(1-x)]\varphi^{(s+1)}(\xi x)\,dx\\ &\quad+\operatorname*{N-lim}_{n\to\infty}\sum_{k=0}^{s}\frac{\varphi^{(k)}(0)}{k!}\int_{-a}^{0}x^k\delta_n^{(s)}(0)\,dx\\ &\quad+\operatorname*{N-lim}_{n\to\infty}\int_{-a}^{0}\frac{x^{s+1}}{(s+1)!}\delta_n^{(s)}(0)\varphi^{(s+1)}(\xi x)\,dx.\end{aligned}\tag{22}$$

Making the substitution $n\ln(1-x)=u$, we have

$$\begin{aligned}\int_0^{1-e^{-1/n}}x^k\delta_n^{(s)}[\ln(1-x)]\,dx &= -n^s\int_0^1(1-e^{u/n})^k e^{u/n}\rho^{(s)}(u)\,du\\ &= -n^s\sum_{i=0}^{k}\int_0^1\binom{k}{i}(-1)^i e^{(i+1)u/n}\rho^{(s)}(u)\,du\end{aligned}$$

and it follows that

$$\operatorname*{N-lim}_{n\to\infty}\int_0^{1-e^{1/n}}x^k\delta_n^{(s)}[\ln(1-x)]\,dx = -\sum_{i=0}^{k}\binom{k}{i}\frac{(-1)^{s+i}(i+1)^s}{2},\tag{23}$$

for $k=0,1,2,\ldots,s$.

Further, as above, we have

$$\operatorname*{N-lim}_{n\to\infty}\int_{-a}^{0}x^k\delta_n^{(s)}(0)\varphi^{(s+1)}(\xi x)\,dx = 0\tag{24}$$

for $k=0,1,2,\ldots,s$ and

$$\lim_{n\to\infty}\int_0^{1-e^{-1/n}}x^{s+1}\delta_n^{(s)}[\ln(1-x)]\varphi^{(s+1)}(\xi x)\,dx = 0,\tag{25}$$

$$\operatorname*{N-lim}_{n\to\infty}\int_{-a}^{0}x^{s+1}\delta_n^{(s)}\ln[(\ln 1-x)]\varphi^{(s+1)}(\xi x)\,dx = 0.\tag{26}$$

Equation (14) now follows as above from (22) to (26), completing the proof of the theorem.

Corollary 1. *The neutrix compositions $\delta^{(s)}[\ln(1 \pm x_-)]$ exist and*

$$\delta^{(s)}[\ln(1+x_-)] = \sum_{k=0}^{s}\sum_{i=0}^{k}\binom{k}{i}\frac{(-1)^{s+k+i}(i+1)^s}{2k!}\delta^{(k)}(x), \tag{27}$$

$$\delta^{(s)}[\ln(1-x_-)] = -\sum_{k=0}^{s}\sum_{i=0}^{k}\binom{k}{i}\frac{(-1)^{s+i}(i+1)^s}{2k!}\delta^{(k)}(x) \tag{28}$$

for $s = 0, 1, 2, \ldots$.

Proof. Equations (27) and (28) follow on replacing x by $-x$ in (13) and (14).

Theorem 5. *The neutrix compositions $\delta^{(s)}[\ln(1 \pm |x|)]$ exist and*

$$\delta^{(s)}[\ln(1+|x|)] = \sum_{k=0}^{s}\sum_{i=0}^{k}\binom{k}{i}\frac{(-1)^{s+i}[1+(-1)^k](i+1)^s}{2k!}\delta^{(k)}(x), \tag{29}$$

$$\delta^{(s)}[\ln(1-|x|)] = -\sum_{k=0}^{s}\sum_{i=0}^{k}\binom{k}{i}\frac{(-1)^{s+k+i}[1+(-1)^k](i+1)^s}{2k!}\delta^{(k)}(x) \tag{30}$$

for $s = 0, 1, 2, \ldots$.

In particular, the compositions $\delta[\ln(1 \pm |x|)]$ exist and

$$\delta[\ln(1 \pm |x|)] = \pm\delta(x). \tag{31}$$

Proof. To prove (29), we will evaluate

$$\begin{aligned}
\mathrm{N-}\lim_{n\to\infty}\langle\delta_n^{(s)}[\ln(1+|x|)],\varphi(x)\rangle &= \mathrm{N-}\lim_{n\to\infty}\sum_{k=0}^{s}\frac{\varphi^{(k)}(0)}{k!}\int_{-a}^{a}x^k\delta_n^{(s)}[\ln(1+|x|)]\,dx \\
&\quad + \mathrm{N-}\lim_{n\to\infty}\int_{-a}^{a}\frac{x^{s+1}}{(s+1)!}\delta_n^{(s)}[\ln(1+|x|)]\varphi^{(s+1)}(\xi x)\,dx \\
&= \mathrm{N-}\lim_{n\to\infty}\sum_{k=0}^{s}\frac{[1+(-1)^k]\varphi^{(k)}(0)}{k!}\int_0^{e^{1/n}-1}x^k\delta_n^{(s)}[\ln(1+x)]\,dx \\
&\quad + \mathrm{N-}\lim_{n\to\infty}\int_0^{e^{1/n}-1}\frac{[1-(-1)^s]x^{s+1}}{(s+1)!}\delta_n^{(s)}[\ln(1+x)]\varphi^{(s+1)}(\xi x)\,dx
\end{aligned}$$

for arbitrary $\varphi(x)$ in $\mathcal{D}[-a,a]$.

Using (18) and (19), we see that

$$\begin{aligned}
\mathrm{N-}\lim_{n\to\infty}\langle\delta_n^{(s)}[\ln(1+|x|)],\varphi(x)\rangle &= \sum_{k=0}^{s}\sum_{i=0}^{k}\frac{(-1)^{s+k+i}[1+(-1)^k](i+1)^s}{2k!}\varphi^{(k)}(0) \\
&= \sum_{k=0}^{s}\sum_{i=0}^{k}\binom{k}{i}\frac{(-1)^{s+i}[1+(-1)^k](i+1)^s}{2k!}\langle\delta^{(k)}(x),\varphi(x)\rangle,
\end{aligned}$$

proving (29).

To prove (30), note that

$$\mathop{\rm N-lim}_{n\to\infty}\langle\delta_n^{(s)}[\ln(1-|x|)],\varphi(x)\rangle=$$

$$=\mathop{\rm N-lim}_{n\to\infty}\sum_{k=0}^{s}\frac{[1+(-1)^k]\varphi^{(k)}(0)}{2k!}\int_0^{1-e^{-1/n}}x^k\delta_n^{(s)}[\ln(1-x)]\,dx$$

$$+\mathop{\rm N-lim}_{n\to\infty}\int_0^{1-e^{-1/n}}\frac{[1-(-1)^s]x^{s+1}}{(s+1)!}\delta_n^{(s)}[\ln(1-x)]\varphi^{(s+1)}(\xi x)\,dx$$

and it follows from (23) and (25) that

$$\mathop{\rm N-lim}_{n\to\infty}\langle\delta_n^{(s)}[\ln(1-|x|)],\varphi(x)\rangle=-\sum_{k=0}^{s}\sum_{i=0}^{k}\binom{k}{i}\frac{(-1)^{s+i}[1+(-1)^k](i+1)^s}{2k!}\varphi^{(k)}(0)$$

$$=-\sum_{k=0}^{s}\sum_{i=0}^{k}\binom{k}{i}\frac{(-1)^{s+k-i}[1+(-1)^k](i+1)^s}{2k!}\langle\delta^{(k)}(x),\varphi(x)\rangle,$$

proving (30).

To prove (31), note that when $s=0$, the above neutrix limits in fact exist as normal limits. Thus $\delta[(1\pm|x|)]$ exist as compositions. This completes the proof of the theorem.

Theorem 6. *The neutrix compositions $\delta^{(2s-1)}[\operatorname{sgn}x\ln(1\pm|x|)]$ exist and*

$$\delta^{(2s-1)}[\operatorname{sgn}x\ln(1+|x|)]=-\sum_{k=1}^{s}\sum_{i=1}^{2k-1}\binom{2k-1}{i}\frac{(-1)^i(i+1)^s}{(2k-1)!}\delta^{(2k-1)}(x),\quad(32)$$

$$\delta^{(2s-1)}[\operatorname{sgn}x\ln(1-|x|)]=\sum_{k=1}^{s}\sum_{i=1}^{2k-1}\binom{2k-1}{i}\frac{(-1)^{k+i}(i+1)^{2s-1}}{(2k-1)!}\delta^{(2k-1)}(x)\quad(33)$$

for $s=1,2,\ldots$.

In particular, the compositions $\delta'[\operatorname{sgn}x\ln(1\pm|x|)]$ exist and

$$\delta'[\operatorname{sgn}x\ln(1\pm|x|)]=\pm2\delta'(x).\quad(34)$$

Proof. To prove (32), note that with

$$\varphi(x)=\sum_{k=0}^{2s-1}\frac{x^k}{k!}\varphi^{(k)}(0)+\frac{x^{2s}}{(2s)!}\varphi^{(2s)}(\xi x),$$

we have

$$\langle\delta_n^{(2s-1)}[\operatorname{sgn}x\ln(1+|x|)],\varphi(x)\rangle=\sum_{k=0}^{2s-1}\frac{\varphi^{(k)}(0)}{k!}\int_{-a}^{a}x^k\delta_n^{(s)}[\operatorname{sgn}x\ln(1+|x|)]\,dx$$

$$+ \int_{-a}^{a} \frac{x^{2s}}{(2s)!} \delta_n^{(2s-1)}[\operatorname{sgn} x \ln(1+|x|)] \varphi^{(2s)}(\xi x)\, dx$$

$$= 2 \sum_{k=1}^{s} \frac{\varphi^{(2k-1)}(0)}{(2k-1)!} \int_0^{e^{1/n}-1} x^{2k-1} \delta_n^{(2s-1)}[\ln(1+x)]\, dx$$

$$+ \int_{1-e^{1/n}}^{e^{1/n}-1} \frac{x^{2s}}{(2s)!} \delta_n^{(2s)}[\operatorname{sgn} x \ln(1+|x|)] \varphi^{(2s)}(\xi x)\, dx. \tag{35}$$

Equation (32) follows on using the above working and (33) follows similarly. Equations (34) follow on noting that usual limits exist when $s = 1$. This completes the proof of the theorem.

For further results on the neutrix composition of distributions, see [FJÖ02, FKSN05, FT05a, FT05b, FT06].

References

[vdCo59] van der Corput, J.G.: Introduction to the neutrix calculus. J. Analyse Math., **7**, 291–398 (1959)

[Fis83] Fisher, B.: On defining the distribution $\delta^{(r)}(f(x))$. Rostock. Math. Kolloq., **23**, 73-80 (1983)

[Fis85] Fisher, B.: On defining the change of variable in distributions. Rostock. Math. Kolloq., **28**, 33-40 (1985)

[FJÖ02] Fisher, B., Jolevska-Tuneska, B., Özçağ, E.: Further results on the composition of distributions. Integral Transforms Spec. Funct., **13(2)**, 109-116 (2002)

[FKSN05] Fisher, B., Kananthai, A., Sritanatana, G., Nonlaopon, K.: The composition of the distributions $x_-^{ms} \ln x_-$ and $x_+^{r-p/m}$. Integral Transforms Spec. Funct., **16(1)**, 13-20 (2005)

[FT05a] Fisher, B., Taş, K.: On the composition of the distributions x_+^{-r} and x_+^{μ}. Indian J. Pure Appl. Math., **36(1)**, 11-22 (2005)

[FT05b] Fisher, B., Taş, K.: On the composition of the distributions $x^{-1} \ln |x|$ and x_+^r. Integral Transforms Spec. Funct., **16(7)**, 533-543 (2005)

[FT06] Fisher, B., Taş, K.: On the composition of the distributions x_+^{λ} and x_+^{μ}. J. Math. Anal. Appl., **318(1)**, 102-111 (2006)

[GS64] Gel'fand, I.M., Shilov, G.E.: Generalized Functions. Vol.I. Academic Press, New York London (1964)

A review on the products of distributions

C.K. Li*

Department of Mathematics and Computer Science, Brandon University, Brandon, Manitoba, Canada R7A 6A9 lic@brandonu.ca

The problem of defining products of distributions has been open and an active research area since Schwartz introduced the theory of distribution around 1950. The inherent difficulties of obtaining products have never prevented their appearance in literature, as they are needed in quantum field and differential equations with distribution involved. The objective of this paper is to recollect various approaches, which include sequential and complex analysis methods, to tackling products of distributions in one or multiple variables, as well as particular generalized functions defined on certain manifolds.

1 Introduction

The singular function $\delta(x)$, which is widely used in physics, was introduced by Dirac in 1920 as follows:

(i) $\delta(x) = 0$ for $x \neq 0$,
(ii) $\delta(x) = \infty$ for $x = 0$,
(iii) $\int_{-\infty}^{\infty} \delta(x)f(x)dx = f(0)$,

where $f(x)$ is continuous.

It is clear to see that the above definition of $\delta(x)$ contradicts with the integral theory in terms of Lebesgue sense, and hence it can not be properly defined within the framework of classical function theory.

Schwartz [Sch59] established the theory of distributions by treating singular functions as linear continuous functionals on the testing function space whose elements have compact support. However, one of the weakest points in distribution is the lack of definitions for products, convolutions and compositions of distributions in general, although they are in great demand for both

* This research is supported by NSERC and BURC.

Kenan Taş et al. (eds), Mathematical Methods in Engineering, 71–96.

quantum field theory and in seeking a weak solution of a differential equation. In elementary particle physics (see e.g. [Gasi66], p.141), one found the need to evaluate δ^2 when calculating the transition rates of certain particle interactions. In 1972, Antosik, Mikusinski and Sikorski [AMS73] introduced a definition for a product of distributions using delta sequences. However, δ^2 as a product of δ with itself was shown not to exist. In 1965, Bremermann [Bre65] used the Cauchy representations of distributions with compact support to define $\sqrt{\delta_+}$ and $\log \delta_+$. Unfortunately, his definition did not carry over to $\sqrt{\delta}$ and $\log \delta$. Based on the work of Bremermann, B. Li [Li78] defined the products of distributions by utilizing a nonstandard analytic representation of distributions, and the product may or may not be a distribution. In 1992, Embacher, Grübl and Oberguggenberger [EGO92] studied particular products of distributions and found applications to quantum electrodynamics. Many attempts have been made to define multiplications of distributions including works by Rosinger (Generalized Solutions of Nonlienar Partial Differential Equations) and Columbeau (New Generalized Functions and Multiplications of Distributions).

The sequential approach has been one of the main tools in dealing with products of distributions. There are many ways to construct a sequence of regular functions which converge to the δ function in sense of distribution. All that is needed is that the corresponding ordinary functions $f_\nu(x)$ form what we shall call a delta-convergent sequence, which means that they must possess the following two properties [GS64].

(i) For any $M > 0$ and for $|a| \leq M$ and $|b| \leq M$, the quantities

$$\left| \int_a^b f_\nu(\eta) d\eta \right|$$

must be bounded by a constant independent of a, b, or ν (in other words, depending only on M).

(ii) For any fixed non-vanishing a and b, we must have

$$\lim_{\nu \to \infty} \int_a^b f_\nu(\eta) d\eta = \begin{cases} 1 \text{ for } a < 0 < b, \\ 0 \text{ otherwise.} \end{cases}$$

Fisher [Fis71] through [FÖG05], with his collaborators, has actively used Jones' δ-sequence $\delta_n(x) = n\rho(nx)$ for $n = 1, 2, \ldots$, where $\rho(x)$ is a fixed infinitely differentiable function on R with the following properties:

(i) $\rho(x) \geq 0$,
(ii) $\rho(x) = 0$ for $|x| \geq 1$,
(iii) $\rho(x) = \rho(-x)$,
(iv) $\int_{-1}^{1} \rho(x) dx = 1$,

and the concept of neutrix limit [vdC60] of van der Corput to deduce numerous products, powers, convolutions and compositions of distributions on R since 1969. The technique of neglecting appropriately defined infinite quantities and resulting finite values extracted from the divergent integral is usually referred to as the Hadamard finite part. In fact Fisher's method in the computation can be regarded as a particular application of the neutrix calculus. This is a general principle for the discarding of unwanted infinite quantities from asymptotic expansions and has been exploited in context of distribution by Fisher in connection with the problem of distributional multiplication, convolution and composition.

In order to extend such an approach from one-dimensional to m-dimensional, C. Li ([Li00] through [CL91] and [LK98] through [Li01b]) built several workable δ-sequences on R^m for the commutative and non-commutative neutrix products such as $r^{-k} \cdot \nabla\delta$ as well as $r^{-k} \cdot \triangle^l \delta$ with the help of the well-known Pizetti's formula. One of Li's δ-sequences was given in [LF:90] as follows.

Let $r = (x_1^2 + \cdots + x_m^2)^{\frac{1}{2}}$ and let $\rho(s)$ be a fixed infinitely differentiable function defined on $R^+ = [0, \infty)$ having the properties:

(a) $\rho(s) \geq 0$,
(b) $\rho(s) = 0$ for $s \geq 1$,
(c) $\displaystyle\int_{R^m} \delta_n(x)dx = 1$,

where $\delta_n(x) = c_m n^m \rho(n^2 r^2)$ and c_m is the constant satisfying (c).

It follows that $\{\delta_n(x)\}$ is a regular δ-sequence of infinitely differentiable functions converging to $\delta(x)$ in the Schwartz space of distributions.

On the other hand, Aguirre [Agui03a] used the Laurent series expansion of r^λ and derived a more general product $r^{-k} \cdot \nabla(\triangle^l \delta)$ by calculating the residue of r^λ. His approach is another interesting example of using complex analysis to obtain products of distributions on R^m after Bremermann and B. Li.

The problem of defining products of distributions on manifolds (unit sphere as a particular example) has been a serious challenge since Gel'fand introduced special types of generalized functions, such as P_+^λ and $\delta^{(k)}(P)$. Aguirre [Agui03b] employed the Taylor expansion of distribution $\delta^{(k-1)}(m^2 + P)$ and gave a meaning of the product $\delta^{(k-1)}(m^2+P) \cdot \delta^{(l-1)}(m^2+P)$. Recently, C. Li and Aguirre [CL05a] and [LA04] derived the product $f(r) \cdot \delta^{(k)}(r-1)$ on unit sphere for any C^∞ function f and obtained several new results related to $\delta(x)$ on even-dimension spaces by complex analysis method. Furthermore, they [AL05] studied a more general product $f(H) \cdot \delta^{(k)}(H)$, where H is a regular hypersurface, and computed the product $f(P) \cdot \delta^{(k)}(P)$ by the substitutions.

In Section 2, we begin to provide the five definitions for the commutative and noncommutative products of one variable using the δ-sequence, and further discuss relations among them with a couple of typical proofs. The complex analysis approaches are presented in Section 3 and we especially utilize the Laurent series and Gel'fand's identities to compute natural products of distributions, which are simpler in calculation and identical with ones obtained in Section 2. Sections 3, 4 and 5 mainly consider the products of distributions with multiple variables by several techniques including the substitutions, some of which are recent work of the author and Aguirre.

2 The products of one variable

Now let $\mathcal{D}$ be the testing function space of infinitely differentiable functions of a single variable with compact support, and let $\mathcal{D}'$ be the space of distributions defined on $\mathcal{D}$. The definition of the product of a distribution and an infinitely differentiable function is the following (see for example [GS64]).

Definition 1. *Let f be a distribution in $\mathcal{D}'$ and let g be an infinitely differentiable function. Then the product fg is defined by*

$$(fg,\ \phi) = (f,\ g\phi)$$

for all functions ϕ in $\mathcal{D}$.

It follows from the above that

$$f(x)\delta^{(r)}(x) = \sum_{k=0}^{r}(-1)^{r-k}\binom{r}{k} f^{(r-k)}(0)\delta^{(k)}(x)$$

if f is infinitely differentiable. In particular, we have

$$\sin x\ \delta^{(r)}(x) = \sum_{k=0}^{r}(-1)^{r-k}\binom{r}{k} \sin[\frac{\pi}{2}(r-k)]\delta^{(k)}(x).$$

An extension of the product of a distribution and an infinitely differentiable function was given by Fisher in [Fis71].

Definition 2. *Let f and g be distributions in $\mathcal{D}'$ for which on the interval (a,b), f is the k-th derivative of a locally summable function F in $L^p(a,b)$ and $g^{(k)}$ is a locally summable function in $L^q(a,b)$ with $1/p+1/q=1$. Then the product $fg=gf$ of f and g is defined on the interval (a,b) by*

$$fg = \sum_{i=0}^{k}\binom{k}{i}(-1)^i\left[Fg^{(i)}\right]^{(k-i)}.$$

To further extend Definition 2, Fisher [Fis74] used Jones' δ-sequence $\delta_n(x) = n\rho(nx)$ defined as in the introduction and the concept of neutrix limit of van der Corput (in order to abandon unwanted infinite quantities from asymptotic expressions) to give the following definition for the commutative product of distributions.

Let f be an arbitrary distribution in $\mathcal{D}'$, we define

$$f_n(x) = (f * \delta_n)(x) = (f(t),\ \delta_n(x-t))$$

for $n = 1, 2, \ldots$. It follows that $\{f_n(x)\}$ is a regular sequence of infinitely differentiable functions converging to the distribution $f(x)$ in $\mathcal{D}'$.

Definition 3. *Let f and g be distributions in $\mathcal{D}'$ and let $f_n = f * \delta_n$ and $g_n = g * \delta_n$. The commutative neutrix product $f \cdot g$ of f and g exists and is equal to h if*

$$N - \lim_{n\to\infty} (f_n g_n,\ \phi) = (h,\ \phi)$$

for all functions ϕ in $\mathcal{D}$, where N is the neutrix (see [vdC60]) having domain $N' = \{1, 2, \ldots\}$ and range the real numbers, with negligible functions that are finite linear sums of the functions

$$n^\lambda \ln^{r-1} n,\quad \ln^r\quad (\lambda > 0,\ r = 1, 2, \ldots)$$

and all functions of n that converge to zero in the normal sense as n tends to infinity.

In 2005, Fisher, Özçağ and Gülen [FÖG05] employed the above definition to imply the following theorem.

Theorem 1. *Let f be a function which is infinitely differentiable on an open interval containing the origin and let $f_+(x) = H(x)f(x)$ and $f_-(x) = H(-x)f(x)$. Then the commutative neutrix products $f_+(x)\cdot\delta^{(r)}(x)$ and $f_-(x)\cdot \delta^{(r)}(x)$ exist and*

$$f_+(x)\cdot\delta^{(r)}(x) = f_-(x)\cdot\delta^{(r)}(x) = \sum_{k=0}^{r} \frac{(-1)^{r-k}}{2}\binom{r}{k} f^{(r-k)}(0)\delta^{(k)}(x)$$

where $r = 0, 1, 2, \ldots$.

Proof. (Outline) By Taylor's Theorem, we have

$$[f_+(x)]_n = f_+(x) * \delta_n(x) = \sum_{i=0}^{r} \frac{f^{(i)}(0)}{i!}(x_+^i)_n + \frac{[f^{(r+1)}(\xi x)x_+^{r+1}]_n}{(r+1)!},\quad 0<\xi<1$$

where $(x_+^i)_n = x_+^i * \delta_n(x)$ and $[f^{(r+1)}(\xi x)x_+^{r+1}]_n = [f^{(r+1)}(\xi x)x_+^{r+1}] * \delta_n(x)$.

Let ϕ be an arbitrary function in $\mathcal{D}$. We obtain

$$\phi(x) = \sum_{k=0}^{r} \frac{\phi^{(k)}(0)}{k!}x^k + \frac{\phi^{(r+1)}(\eta x)}{(r+1)!}x^{r+1},$$

where $0 < \eta < 1$. Then

$$\begin{aligned}([f_+(x)]_n\delta_n^{(r)}(x), \phi(x)) &= \sum_{k=0}^{r}\sum_{i=0}^{r} \frac{f^{(i)}(0)\phi^{(k)}(0)}{k!i!} \int_{-1/n}^{1/n} x^k(x_+^i)_n\delta_n^{(r)}(x)\,dx \\ &+ \frac{1}{(r+1)!}\sum_{i=0}^{r}\int_{-1/n}^{1/n} x^{r+1}\phi^{(r+1)}(\eta x)(x_+^i)_n\delta_n^{(r)}(x)\,dx \\ &+ \frac{1}{(r+1)!}\int_{-1/n}^{1/n} \phi(x)[f^{(r+1)}(\xi x)x_+^{r+1}]_n\,dx\end{aligned}$$

and it follows from complicated computations [FÖG05] that

$$\begin{aligned}N-\lim_{n\to\infty}([f_+(x)]_n\delta_n^{(r)}(x), \phi(x)) &= \sum_{k=0}^{r} \frac{(-1)^r}{2}\binom{r}{k} f^{(r-k)}(0)\phi^{(k)}(0) \\ &= \sum_{k=0}^{r} \frac{(-1)^{r-k}}{2}\binom{r}{k} f^{(r-k)}(0)(\delta^{(k)}(x), \phi(x)).\end{aligned}$$

Since f is an infinitely differentiable function, we know that

$$f(x)\,\delta^{(r)}(x) = \sum_{k=0}^{r}(-1)^{r-k}\binom{r}{k} f^{(r-k)}(0)\delta^{(k)}(x)$$

and since

$$f(x) = f_+(x) + f_-(x),$$

the result follows immediately. This completes the proof of the theorem.

To overcome difficulties arising from computing the products by Definition 3, Cheng and C. Li [CL91] and [CL88] provided a variant for the commutative neutrix products as follows.

Definition 4. *Let f and g be distributions in $\mathcal{D}'$ and let $f_n = f * \delta_n$ and $g_n = g * \delta_n$. The commutative neutrix product $f \diamond g$ of f and g exists and is equal to h if*

$$N-\lim_{n\to\infty} \frac{1}{2}\{(f_n g,\ \phi) + (f g_n,\ \phi)\} = (h,\ \phi)$$

for all functions ϕ in $\mathcal{D}$.

With Definition 4, C. Li recently studied the commutative neutrix products $f_+(x) \diamond \delta^{(r)}(x)$ and $f_-(x) \diamond \delta^{(r)}(x)$ under the weaker condition that f is only r-th differentiable.

Theorem 2. *Let f be a function which is r-th differentiable on an open interval containing the origin. Then the commutative neutrix products $f_+(x) \diamond \delta^{(r)}(x)$ and $f_-(x) \diamond \delta^{(r)}(x)$ exist and*

$$f_+(x) \diamond \delta^{(r)}(x) = f_-(x) \diamond \delta^{(r)}(x) = \sum_{k=0}^{r} \frac{(-1)^{r-k}}{2} \binom{r}{k} f^{(r-k)}(0)\delta^{(k)}(x)$$

where $r = 0, 1, 2, \ldots$.

Proof. For any $\phi \in \mathcal{D}$, we use integration by parts repeatedly

$$(f_+(x)(\delta^{(r)}(x) * \delta_n(x)),\ \phi(x)) = \int_0^\infty \delta_n^{(r)}(x) f(x)\phi(x)dx$$
$$= -f(0)\phi(0)\delta_n^{(r-1)}(0) + (-1)^2(f(0)\phi(0))'\delta_n^{(r-2)}(0) + \ldots$$
$$+(-1)^r \int_0^\infty (f(x)\phi(x))^{(r)}\delta_n(x)dx.$$

It follows that

$$N - \lim_{n\to\infty} (f_+(x)(\delta^{(r)}(x) * \delta_n(x)),\ \phi(x)) = \frac{1}{2}(-1)^r(f(0)\phi(0))^{(r)}$$
$$= \frac{1}{2}(-1)^r \sum_{k=0}^{r} \binom{r}{k} f^{(r-k)}(0)\phi^{(k)}(0)$$

by noting that $\int_0^\infty \delta_n(x)dx = \frac{1}{2}$.

On the other hand, we have

$$((f_+(x) * \delta_n(x))\delta^{(r)}(x),\ \phi(x)) = (\delta^{(r)}(x), \int_{-\infty}^{\infty} H(t)f(t)\delta_n(x-t)dt\ \phi(x))$$
$$= (-1)^r \left(\int_0^\infty f(t)\delta_n(x-t)dt\ \phi(x)\right)^{(r)}\Bigg|_{x=0}$$
$$= (-1)^r \sum_{k=0}^{r} \binom{r}{k} \int_0^\infty f(t)\delta_n^{(r-k)}(-t)dt\ \phi^{(k)}(0).$$

Making the substitution $u = -t$, we arrive at

$$\lim_{n\to\infty} \int_0^\infty f(t)\delta_n^{(r-k)}(-t)dt = \lim_{n\to\infty} \int_{-\infty}^0 f(-u)\delta_n^{(r-k)}(u)du = \frac{f^{(r-k)}(0)}{2}$$

due to $\int_{-\infty}^0 \delta_n(x)dx = \frac{1}{2}$. It follows that

$$\lim_{n\to\infty} ((f_+(x) * \delta_n(x))\delta^{(r)}(x),\ \phi(x)) = \frac{1}{2}(-1)^r \sum_{k=0}^{r} \binom{r}{k} f^{(r-k)}(0)\phi^{(k)}(0).$$

Adding the two terms above clearly gives us

$$f_+(x) \diamond \delta^{(r)}(x) = \sum_{k=0}^{r} \frac{(-1)^{r-k}}{2} \binom{r}{k} f^{(r-k)}(0)\delta^{(k)}(x).$$

To obtain the product $f_-(x) \diamond \delta^{(r)}(x)$, we note that

$$(H(x)f(-x)) \diamond \delta^{(r)}(x) = \sum_{k=0}^{r} \frac{1}{2} \binom{r}{k} f^{(r-k)}(0)\delta^{(k)}(x)$$

and replace x by $-x$ to get

$$f_-(x) \diamond \delta^{(r)}(x) = \sum_{k=0}^{r} \frac{(-1)^{r-k}}{2} \binom{r}{k} f^{(r-k)}(0)\delta^{(k)}(x)$$

by using $\delta^{(k)}(x) = (-1)^k \delta^{(k)}(-x)$ for $k = 0, 1, \ldots, r$. This completes the proof.

As a remark, we are able to deduce Theorem 2 with a very simple approach if f is infinitely differentiable. Since every distribution concentrated on a point is a linear combination of the delta function and its derivatives according to [GS64], we claim

$$\begin{aligned} f_+(x) \diamond \delta^{(r)}(x) &= b_0\delta(x) + b_1\delta'(x) + \cdots + b_l\delta^{(l)}(x) + \cdots \quad \text{and} \\ f_-(x) \diamond \delta^{(r)}(x) &= a_0\delta(x) + a_1\delta'(x) + \cdots + a_s\delta^{(s)}(x) + \cdots . \end{aligned}$$

As $f(x)$ is infinitely differentiable, we have

$$\lim_{x \to 0^+} f_+^{(i)}(x) = \lim_{x \to 0^-} f_-^{(i)}(x) = f^{(i)}(0) \tag{1}$$

for all $i = 0, 1, \ldots$.

Clearly, all coefficients b_l and a_l are determined by the same formula which only involves the values of $f(x)$ and its derivatives at $x = 0$ due to Equation (1). Thus, $f_+(x) \diamond \delta^{(r)}(x) = f_-(x) \diamond \delta^{(r)}(x)$.

By Theorem 5 below, we know that

$$\begin{aligned} & f(x) \diamond \delta^{(r)}(x) = (f_+(x) + f_-(x)) \diamond \delta^{(r)}(x) = f(x)\delta^{(r)}(x) \\ &= \sum_{k=0}^{r} (-1)^{r-k} \binom{r}{k} f^{(r-k)}(0)\delta^{(k)}(x). \end{aligned}$$

This concludes our remark.

As another example, we are going to compute the product $\frac{\psi(x)}{x} \diamond \delta(x)$ where $\psi(x)$ is a differentiable function on R and the distribution $\frac{\psi(x)}{x}$ is given by

$$(\frac{\psi(x)}{x}, \phi(x)) = \int_0^\infty \frac{\phi(x)\psi(x) - \phi(-x)\psi(-x)}{x} dx.$$

Obviously the above integral exists as ϕ has a bounded support.

Theorem 3. *The product* $\frac{\psi(x)}{x} \diamond \delta(x)$ *exists and*

$$\frac{\psi(x)}{x} \diamond \delta(x) = -\frac{\psi(0)}{2}\delta'(x) + \psi'(0)\delta(x).$$

Proof. By the mean value theorem, we get

$$\begin{aligned}(\frac{\psi(x)}{x}\delta_n(x),\ \phi(x)) &= \int_0^\infty \frac{\delta_n(x)\phi(x)\psi(x) - \delta_n(-x)\phi(-x)\psi(-x)}{x}dx \\ = \int_0^\infty \delta_n(x)\frac{\phi(x)\psi(x) - \phi(-x)\psi(-x)}{x}dx &= \frac{\phi(\zeta)\psi(\zeta) - \phi(-\zeta)\psi(-\zeta)}{2\zeta}\end{aligned}$$

where $0 < \zeta < 1/n$.

It follows that

$$\begin{aligned}&\lim_{n\to\infty}(\frac{\psi(x)}{x}\delta_n(x),\ \phi(x)) = \lim_{\zeta\to 0}\frac{\phi(\zeta)\psi(\zeta) - \phi(-\zeta)\psi(-\zeta)}{2\zeta} \\ &= \phi'(0)\psi(0) + \phi(0)\psi'(0).\end{aligned}$$

On the other hand, we have

$$\frac{\psi(x)}{x} * \delta_n(x) = (\frac{\psi(t)}{t},\ \delta_n(x-t))$$

and

$$\begin{aligned}&((\frac{\psi(x)}{x} * \delta_n(x))\delta(x),\ \phi(x)) = (\delta(x),\ (\frac{\psi(x)}{x} * \delta_n(x))\phi(x)) \\ &= (\frac{\psi(t)}{t},\ \delta_n(-t))\phi(0) = \int_0^\infty \delta_n(t)\frac{\psi(t) - \psi(-t)}{t}dt\ \phi(0).\end{aligned}$$

It follows that

$$\lim_{n\to\infty}((\frac{\psi(x)}{x} * \delta_n(x))\delta(x),\ \phi(x)) = \psi'(0)\phi(0).$$

Adding the two terms, we complete the proof.

Except the commutative products of distributions which naturally extend the classical concepts in multiplications, Fisher [Fis82b] also defined the noncommutative neutrix product of distributions.

Definition 5. *Let* f *and* g *be distributions in* $\mathcal{D}'$ *and let* $g_n = g * \delta_n$. *The noncommutative neutrix product* $f \circ g$ *of* f *and* g *exists and is equal to* h *if*

$$N - \lim_{n\to\infty}(fg_n,\ \phi) = (h,\ \phi)$$

for all functions ϕ *in* $\mathcal{D}$.

In 2005, Fisher and Taş [FT05b] used the above definition to derive the following result.

Theorem 4. *The non-commutative neutrix product* $(x^r \ln^p |x|) \circ x^{-s}$ *exists and*

$$(x^r \ln^p |x|) \circ x^{-s} = x^{r-s} \ln^p |x|$$

for $r - s = -2, -3, \ldots$ *and* $p = 0, 1, 2, \ldots$.

The product of Definition 5 is not symmetric, and hence $f \circ g \neq g \circ f$ in general.

Theorem 5. *Definition 4 extends Definition 1 and 2.*

Proof. Let g be a C^∞ function. Obviously, $g_n\phi$ has an uniform support and converges to $g\phi$ in $\mathcal{D}$. For any f in $\mathcal{D}'$, we imply that

$$\begin{aligned}(f \diamond g,\, \phi) &= N - \lim_{n\to\infty} \frac{1}{2}\{(f_n g,\, \phi) + (f g_n,\, \phi)\}\\ &= N - \lim_{n\to\infty} \frac{1}{2}\{(f_n,\, g\phi) + (f,\, g_n\phi)\} = (f,\, g\phi).\end{aligned}$$

To see Definition 4 is an extension of Definition 2, we let f and g be functions given in Definition 2. Then we have

$$\lim_{n\to\infty} (f g_n,\, \phi) = \lim_{n\to\infty} (f_n g,\, \phi) = (\sum_{i=0}^{k} \binom{k}{i} (-1)^i \left[F g^{(i)}\right]^{(k-i)},\, \phi)$$

by Theorem 2 in [Fis80]. This completes the proof.

At this moment, it is not quite clear if Definition 4 is equivalent to Definition 3, although both define the commutative neutrix products. In fact,

$$\begin{aligned}N - \lim_{n\to\infty} \frac{1}{2}\{(f_n g,\, \phi) + (f g_n,\, \phi)\} - N - \lim_{n\to\infty} (f_n g_n,\, \phi)\\ = N - \lim_{n\to\infty} \frac{1}{2}\{(f_n(g - g_n) + g_n(f - f_n),\, \phi)\}.\end{aligned}$$

One can neither prove the above being zero nor non-zero by a counter example. The author welcomes and appreciates any discussion from interested readers.

However, we are able to claim that the two definitions are different if we remove the neutrix limits from both. Indeed, we have $\delta'(x) \diamond \delta'(x) = 0$ on subspace $\mathcal{D}_0(R) = \{\phi(x) \in \mathcal{D}(R) \mid \phi(0) = 0\}$ because

$$\begin{aligned}(\delta_n'(x)\delta'(x),\, \phi(x)) &= -\, (\delta_n'(x)\phi(x))'|_{x=0}\\ &= -(\delta_n''(0)\phi(0) + \delta_n'(0)\phi'(0)) = 0\end{aligned}$$

for all n. But it is impractical to compute the product $\delta'(x) \cdot \delta'(x)$ without the neutrix limit since $(\delta_n'(x)\delta_n'(x),\, \phi(x))$ diverges.

3 The complex analysis approaches

The goal of this section is to obtain natural products of distributions with the help of the Laurent series and some of Gel'fand's identities [GS64] and[?], which is a recent work of C. Li and Aguirre. This approach is not only simpler than the sequential method for some products, but also without recourse to Van der Corput's neutrix limit nor to a delta sequence (an idea that requires a bit more machinery).

We define the locally summable functions x_+^λ and x_-^λ for $\lambda > -1$ by

$$x_+^\lambda = \begin{cases} x^\lambda \text{ if } x > 0 \\ 0 \ \text{ if } x < 0 \end{cases} \quad \text{and} \quad x_-^\lambda = \begin{cases} |x|^\lambda \text{ if } x < 0 \\ 0 \quad \text{ if } x > 0. \end{cases}$$

The distributions x_+^λ and x_-^λ are then defined inductively for $\lambda < -1$ and $\lambda \neq -2, -3, \ldots$ by

$$(x_+^\lambda)' = \lambda x_+^{\lambda-1} \quad \text{and} \quad (x_-^\lambda)' = -\lambda x_-^{\lambda-1}.$$

It follows that if r is a positive integer and $-r-1 < \lambda < -r$, then

$$(x_+^\lambda, \phi(x)) = \int_0^\infty x^\lambda \left[\phi(x) - \sum_{i=0}^{r-1} \frac{\phi^{(i)}(0)}{i!} x^i\right] dx \quad \text{and}$$
$$(x_-^\lambda, \phi(x)) = \int_{-\infty}^0 |x|^\lambda \left[\phi(x) - \sum_{i=0}^{r-1} \frac{\phi^{(i)}(0)}{i!} x^i\right] dx.$$

Theorem 6. *The products $x_+^\lambda \cdot x_-^{-k-\lambda}$ and $x_-^\lambda \cdot x_+^{-k-\lambda}$ exist and*

$$x_+^\lambda \cdot x_-^{-k-\lambda} = -\frac{\pi \ cosec \lambda\pi}{2(k-1)!} \delta^{(k-1)}(x), \tag{2}$$
$$x_-^\lambda \cdot x_+^{-k-\lambda} = \frac{(-1)^k \ \pi \ cosec\lambda\pi}{2(k-1)!} \delta^{(k-1)}(x) \tag{3}$$

where $\lambda \neq 0, \pm 1, \pm 2, \ldots$ and $k = 1, 2, \ldots$.

Proof. The following two formulas can be found in [GS64]

$$(x - i0)^{-k} = x^{-k} + i\pi \frac{(-1)^{k-1}}{(k-1)!} \delta^{(k-1)}(x) \quad \text{and}$$
$$(x - i0)^\lambda = x_+^\lambda + e^{-\lambda\pi i} x_-^\lambda.$$

Furthermore, $(x - i0)^\lambda$ is an entire function of λ.

Using the following Gel'fand's identities [GS64]

$$x_+^\lambda = \frac{(-1)^{n-1}}{(n-1)!(\lambda+n)}\,\delta^{(n-1)}(x) + F_{-n}(x_+,\lambda),$$
$$x_-^\lambda = \frac{1}{(n-1)!(\lambda+n)}\,\delta^{(n-1)}(x) + F_{-n}(x_-,\lambda),$$
$$e^{\pm i\lambda\pi} = (-1)^n[1 \pm (\lambda+n)\pi + \cdots]$$

where $F_{-n}(x_+,\lambda)$ and $F_{-n}(x_-,\lambda)$ are the regular parts of the Laurent expansions of x_+^λ and x_-^λ respectively, we arrive at

$$\lim_{\lambda\to -k}(x-i0)^\lambda = \lim_{\lambda\to -k}(x_+^\lambda + e^{-\lambda\pi i}x_-^\lambda)$$
$$= x^{-k} + i\pi\frac{(-1)^{k-1}}{(k-1)!}\delta^{(k-1)}(x) = (x-i0)^{-k}.$$

It follows that

$$(x-i0)^\lambda \cdot (x-i0)^\mu = (x-i0)^{\lambda+\mu}.$$

In particular, we have

$$(x-i0)^\lambda \cdot (x-i0)^{-\lambda-k} = (x-i0)^{-k}$$

by letting $\mu \to -\lambda - k$.

Hence, we come to

$$x^{-k} + i\pi\frac{(-1)^{k-1}}{(k-1)!}\delta^{(k-1)}(x) = (x-i0)^{-k} = (x-i0)^\lambda(x-i0)^{-\lambda-k}$$
$$= (x_+^\lambda + e^{-\lambda\pi i}x_-^\lambda)(x_+^{-\lambda-k} + e^{(\lambda+k)\pi i}x_-^{-\lambda-k})$$
$$= [x_+^{-k} + (-1)^k x_-^{-k}] + [(-1)^k x_+^\lambda \cdot x_-^{-\lambda-k} + x_-^\lambda \cdot x_+^{-\lambda-k}]\cos\lambda\pi$$
$$+ i[(-1)^k x_+^\lambda \cdot x_-^{-\lambda-k} - x_-^\lambda \cdot x_+^{-\lambda-k}]\sin\lambda\pi$$

which clearly implies

$$x^{-k} = x_+^{-k} + (-1)^k x_-^{-k},$$
$$(-1)^k x_+^\lambda \cdot x_-^{-\lambda-k} + x_-^\lambda \cdot x_+^{-\lambda-k} = 0 \quad \text{and}$$
$$[(-1)^k x_+^\lambda \cdot x_-^{-\lambda-k} - x_-^\lambda \cdot x_+^{-\lambda-k}]\sin\lambda\pi = \pi\frac{(-1)^{k-1}}{(k-1)!}\delta^{(k-1)}(x).$$

Therefore, we obtain

$$2(-1)^k x_+^\lambda \cdot x_-^{-\lambda-k}\sin\lambda\pi = \pi\frac{(-1)^{k-1}}{(k-1)!}\delta^{(k-1)}(x).$$

This completes the proof of Equation (2), and Equation (3) follows from

$$x_-^\lambda \cdot x_+^{-\lambda-k} = (-1)^{k+1}x_+^\lambda \cdot x_-^{-\lambda-k} = \frac{(-1)^k\,\pi\,\mathrm{cosec}\lambda\pi}{2(k-1)!}\,\delta^{(k-1)}(x).$$

Remark: Equation (2) was first obtained by Fisher in [Fis82b] with a longer and more complex proof by Definition 5. His work was based on the δ-sequence and the neutrix limit.

Theorem 7. *The products* $x_+^r \cdot \delta^{(r+k-1)}(x)$ *and* $\delta^{(r+k-1)}(x) \cdot x_+^r$ *exist and*

$$x_+^r \cdot \delta^{(r+k-1)}(x) = \delta^{(r+k-1)}(x) \cdot x_+^r = \frac{(-1)^r (r+k-1)!}{2(k-1)!} \delta^{(k-1)}(x)$$

for $r = 0, 1, 2, \ldots$ *and* $k = 1, 2, \ldots$. *In particular, we have* $H(x) \cdot \delta^{(k-1)}(x) = \delta^{(k-1)}(x)/2$.

Proof. Let s be a positive integer. By the following two identities

$$\frac{\pi}{\sin \lambda\pi} = \frac{\Gamma(1-\lambda)\,\Gamma(1+\lambda)}{\lambda} \quad \text{if } \lambda \text{ is near } -s \text{ and } \lambda \neq -s,$$

$$\lim_{\lambda \to -s} \frac{x_-^\lambda}{\Gamma(\lambda+1)} = (-1)^{s-1} \delta^{(s-1)}(x)$$

as well as Equation (3), we have

$$(-1)^{s-1} \delta^{(s-1)}(x) \cdot x_+^{s-k} = \frac{(-1)^{k+1}\,(s-1)!}{2(k-1)!} \delta^{(k-1)}(x)$$

It follows from setting $r = s - k$ that

$$\delta^{(r+k-1)}(x) \cdot x_+^r = \frac{(-1)^r (r+k-1)!}{2(k-1)!} \delta^{(k-1)}(x).$$

With a very similar argument, we can show that

$$x_+^r \cdot \delta^{(r+k-1)}(x) = \frac{(-1)^r (r+k-1)!}{2(k-1)!} \delta^{(k-1)}(x).$$

This completes the proof of Theorem 7.

Theorem 8. *The product* $x_+^{-k} \cdot \delta^{(p)}(x)$ *exists and*

$$x_+^{-k} \cdot \delta^{(p)}(x) = \frac{(-1)^k p!}{2(p+k)!} \delta^{(k+p)}(x)$$

for $p = 0, 1, 2, \ldots$ *and* $k = 1, 2, \ldots$. *In particular, we have* $H(x) \cdot \delta^{(p)}(x) = \delta^{(p)}(x)/2$ *again by letting* $k \to 0$.

Proof. From the Laurent series of x_+^λ

$$x_+^\lambda = \frac{(-1)^{k-1}}{(k-1)!(\lambda+k)} \delta^{(k-1)}(x) + x_+^{-k} + (\lambda+k) x_+^{-k} \ln x_+ + \cdots,$$

we have

$$\delta^{(p-1)}(x) = (-1)^{p-1}(p-1)! \lim_{\mu \to -p} (\mu + p)\, x_+^{\mu}.$$

We define distribution x_+^{-k} as the regular part of the Laurent expansion of x_+^{λ} about $\lambda = -k$ by

$$x_+^{-k} = \lim_{\lambda \to -k} \frac{\partial}{\partial \lambda} \left[(\lambda + k)\, x_+^{\lambda}\right].$$

Hence, we get

$$x_+^{-k} \cdot \delta^{(p-1)}(x) = (-1)^{p-1}(p-1)! \lim_{\lambda \to -k} \lim_{\mu \to -p} \frac{\partial}{\partial \lambda}[(\lambda + k)\,(\mu + p)\, x_+^{\lambda+\mu}],$$

due to the facts x_+^{λ}, x_+^{μ} and $x_+^{\lambda+\mu}$ are analytic for $\lambda,\ \mu,\ \lambda + \mu \neq -1, -2, \ldots$ and

$$x_+^{\lambda+\mu} = x_+^{\lambda} \cdot x_+^{\mu}.$$

Obviously,

$$(\lambda + k)\,(\mu + p) = \frac{1}{2}[(\lambda + k + \mu + p)^2 - (\lambda + k)^2 - (\mu + p)^2].$$

It follows that

$$\begin{aligned} x_+^{-k} \cdot \delta^{(p-1)}(x) = {} & \frac{(-1)^{p-1}(p-1)!}{2} \lim_{\lambda \to -k} \lim_{\mu \to -p} \frac{\partial}{\partial \lambda}[(\lambda + k + \mu + p)^2 x_+^{\lambda+\mu}] \\ & - \frac{(-1)^{p-1}(p-1)!}{2} \lim_{\lambda \to -k} \lim_{\mu \to -p} \frac{\partial}{\partial \lambda}[(\lambda + k)^2 x_+^{\lambda+\mu}] \\ & - \frac{(-1)^{p-1}(p-1)!}{2} \lim_{\lambda \to -k} \lim_{\mu \to -p} \frac{\partial}{\partial \lambda}[(\mu + p)^2 x_+^{\lambda+\mu}] \\ = {} & I_1 + I_2 + I_3. \end{aligned}$$

Since

$$x_+^{\lambda+\mu} = \frac{(-1)^{k+p-1}}{(k+p-1)!\,(\lambda + \mu + k + p)}\, \delta^{(k+p-1)}(x) + x_+^{-k-p} + \cdots,$$

we have

$$I_1 = \frac{(-1)^k (p-1)!}{2(p+k-1)!}\, \delta^{(k+p-1)}(x).$$

As for I_2, we arrive at

$$I_2 = -\frac{(-1)^k (p-1)!}{2(p+k-1)!}\, \delta^{(k+p-1)}(x).$$

due to the following fact

$$\lim_{\lambda \to -k} \lim_{\mu \to -p} \frac{\partial}{\partial \lambda} \left\{ \frac{(\lambda+k)^2}{\lambda+\mu+k+p} \right\} = 1.$$

With a very similar argument, we can show that

$$I_3 = \frac{(-1)^k (p-1)!}{2(p+k-1)!} \, \delta^{(k+p-1)}(x).$$

Replacing $p-1$ by p, we complete the proof of Theorem 8.

4 The products of m variables

Let us consider the functional r^λ (see [GS64]) defined by

$$(r^\lambda, \phi) = \int_{R^m} r^\lambda \phi(x) dx \tag{4}$$

where $R_e(\lambda) > -m$ and $\phi(x) \in \mathcal{D}_m$ (the Schwartz testing function space). Because the derivative

$$\frac{\partial}{\partial \lambda}(r^\lambda, \phi) = \int r^\lambda \ln r \, \phi(x) dx$$

exists, the functional r^λ is an analytic function of λ for $R_e(\lambda) > -m$.

For $R_e(\lambda) \leq -m$, we should use the following identity (5) to define its analytic continuation. For $R_e(\lambda) > 0$, we could deduce

$$\triangle(r^{\lambda+2}) = (\lambda+2)(\lambda+m) r^\lambda$$

simply by calculating the left-hand side. By iteration we find for any integer k that

$$r^\lambda = \frac{\triangle^k r^{\lambda+2k}}{(\lambda+2)\cdots(\lambda+2k)(\lambda+m)\cdots(\lambda+m+2k-2)}. \tag{5}$$

On making following substitution of spherical coordinates in (4),

$$\begin{aligned}
x_1 &= r\cos\theta_1, \\
x_2 &= r\sin\theta_1\cos\theta_2, \\
x_3 &= r\sin\theta_1\sin\theta_2\cos\theta_3, \\
&\cdots\cdots \\
x_{m-1} &= r\sin\theta_1\sin\theta_2\cdots\sin\theta_{m-2}\cos\theta_{m-1}, \\
x_m &= r\sin\theta_1\sin\theta_2\cdots\sin\theta_{m-2}\sin\theta_{m-1},
\end{aligned}$$

we come to

$$(r^\lambda, \phi) = \int_0^\infty r^\lambda \left\{ \int_{r=1} \phi(r\omega) d\omega \right\} r^{m-1} dr \tag{6}$$

where $d\omega$ is the hypersurface element on the unit sphere. The integral appearing in the above integrand can be written in the form

$$\int_{r=1} \phi(r\omega)d\omega = \int_{\Omega} \phi(r\omega)d\omega = \Omega_m S_\phi(r) \tag{7}$$

where Ω_m is the hypersurface area of the unit sphere imbedded in Euclidean space of m dimensions, and $S_\phi(r)$ is the mean value of ϕ on the sphere of radius r.

It was proven in [GS64] that $S_\phi(r)$ is infinitely differentiable for $r \geq 0$, bounded support, and

$$\begin{aligned} S_\phi(r) &= \phi(0) + \frac{1}{2!}S''_\phi(0)r^2 + \cdots + \frac{1}{(2k)!}S^{(2k)}_\phi(0)r^{2k} + \cdots \\ &= \sum_{k=0}^{\infty} \frac{\triangle^k \phi(0) r^{2k}}{2^k k! m(m+2)\cdots(m+2k-2)}, \end{aligned}$$

which is the well-known Pizetti's formula and it plays an important role in computing some neutrix products of multiple variables [Li00] through [LF:90] and [CL91].

In 1991, Aguirre expressed distribution $\delta^{(k)}(r-c)$ in terms of an infinite sum of linear combinations of $\triangle^l \delta$. Please refer to reference [Agui91] for detail.

Theorem 9. *The non-commutative neutrix product $\triangle r^{-k} \cdot \delta$ exists. Furthermore*

$$\triangle r^{-2k} \cdot \delta = \frac{k(2k+2-m)}{2^k(k+1)!\, m(m+2)\cdots(m+2k)} \triangle^{k+1}\delta \tag{8}$$

$$\triangle r^{1-2k} \cdot \delta = 0 \tag{9}$$

where k is any positive integer.

Proof. We note that r^{-k} is a locally summable function on R^m for $k = 1, 2, \ldots, m-1$. With Definition 5, we naturally consider

$$\begin{aligned} I &= (\triangle r^{-k} \cdot \delta_n,\ \phi) = (r^{-k},\ \triangle(\delta_n \phi)) \\ &= (r^{-k},\ \triangle\delta_n \phi) + (r^{-k},\ \delta_n \triangle\phi) + 2\sum_{i=1}^{m} (r^{-k},\ D_i\delta_n D_i\phi) \\ &= I_1 + I_2 + I_3. \end{aligned}$$

Clearly,

$$(r^{-k} \cdot \triangle\delta,\ \phi) = N - \lim_{n\to\infty} I_1 = N - \lim_{n\to\infty} (r^{-k},\ \triangle\delta_n\phi)$$

and using several results in [LF:90] and [CL91], we obtain

$$(r^{-2k} \cdot \triangle\delta,\ \phi) = \frac{(\triangle^{k+1}\delta,\ \phi)}{2^k(k+1)!\,(m+2)(m+4)\cdots(m+2k)}, \tag{10}$$

$$(r^{1-2k} \cdot \triangle\delta,\ \phi) = 0, \tag{11}$$

which indeed hold for any positive integer k.

It follows that

$$N - \lim_{n\to\infty} I_2 = N - \lim_{n\to\infty} (r^{-k},\ \delta_n \triangle\phi) = \frac{S^{(k)}_{\triangle\phi}(0)}{k!}$$

and applying Pizetti's formula, we have

$$N - \lim_{n\to\infty} (r^{-2k},\ \delta_n \triangle\phi) = \left(\frac{\triangle^{k+1}\delta}{2^k k!\, m(m+2)\cdots(m+2k-2)},\ \phi\right), \tag{12}$$

$$N - \lim_{n\to\infty} (r^{1-2k},\ \delta_n \triangle\phi) = (0,\ \phi), \tag{13}$$

which are again true for any positive integer k.

Putting $\psi_i = x_i D_i \phi$, we deduce that

$$I_3 = 4c_m n^{m+2} \Omega_m \sum_{i=1}^{m} \int_0^{\frac{1}{n}} r^{m-k-1} \rho'(n^2r^2) S_{\psi_i}(r) dr$$

and by Taylor's formula, we obtain

$$S_{\psi_i}(r) = \sum_{j=0}^{k+1} \frac{S^{(j)}_{\psi_i}(0)}{j!} r^j + \frac{S^{(k+2)}_{\psi_i}(0)}{(k+2)!} r^{k+2} + \frac{S^{(k+3)}_{\psi_i}(\zeta r)}{(k+3)!} r^{k+3},$$

where $0 < \zeta < 1$. Hence

$$\begin{aligned} I_3 &= 4c_m \Omega_m n^{m+2} \sum_{i=1}^{m} \sum_{j=0}^{k+1} \frac{S^{(j)}_{\psi_i}(0)}{j!} \int_0^{\frac{1}{n}} r^{m-k-1} \rho'(n^2r^2)\, r^j\, dr \\ &+ 4c_m \Omega_m n^{m+2} \sum_{i=1}^{m} \int_0^{\frac{1}{n}} r^{m-k-1}\, \rho'(n^2r^2) \frac{S^{(k+2)}_{\psi_i}(0)}{(k+2)!}\, r^{k+2} dr \\ &+ 4c_m \Omega_m n^{m+2} \sum_{i=1}^{m} \int_0^{\frac{1}{n}} r^{m-k-1}\, \rho'(n^2r^2) \frac{S^{(k+3)}_{\psi_i}(\zeta r)}{(k+3)!}\, r^{k+3} dr \\ &= I_1' + I_2' + I_3' \end{aligned}$$

respectively. Employing the substitution $t = nr$, we get

$$I_1' = 4c_m \Omega_m \sum_{i=1}^{m} \sum_{j=0}^{k+1} n^{k+2-j} \frac{S_{\psi_i}^{(j)}(0)}{j!} \int_0^1 t^{m+j-k-1} \rho'(t^2)\, dt,$$

thus

$$N - \lim_{n\to\infty} I_1' = 0.$$

As for

$$I_2' = 4c_m \Omega_m \sum_{i=1}^{m} \frac{S_{\psi_i}^{(k+2)}(0)}{(k+2)!} \int_0^1 t^{m+1} \rho'(t^2)\, dt$$

integrating by parts, we have

$$\begin{aligned} 4c_m \Omega_m \int_0^1 t^{m+1} \rho'(t^2)\, dt &= 2c_m \Omega_m \int_0^1 t^m \, d\, \rho(t^2) \\ &= -2c_m \Omega_m \cdot m \int_0^1 t^{m-1} \rho(t^2)\, dt \\ &= -2m \int_{R^m} \delta_n(x) dx = -2m. \end{aligned}$$

Hence

$$I_2' = -2m \sum_{i=1}^{m} \frac{S_{\psi_i}^{(k+2)}(0)}{(k+2)!} = -\frac{2m}{(k+2)!} \sum_{i=1}^{m} S_{\psi_i}^{(k+2)}(0).$$

Putting

$$\mathrm{M} = \sup \left\{ \left| S_{\psi_i}^{(k+3)}(r) \right| : r \in R^+ \quad \text{and} \quad 1 \le i \le m \right\},$$

we obtain that

$$|I_3'| \le 4c_m \Omega_m \frac{mM}{n(k+3)!} \int_0^1 t^{m+2} \, |\rho'(t^2)|\, dt \to 0$$

as $n \to \infty$.

Hence it follows from above that

$$N - \lim_{n\to\infty} I_3 = I_2' = -\frac{2m}{(k+2)!} \sum_{i=1}^{m} S_{\psi_i}^{(k+2)}(0)$$

which can be extended to the case $k \ge m$ by utilizing the normalization procedure of $\mu(x) x_+^{\lambda}$ [CL91].

On using Pizetti's formula and Lemma 2 in [Li01b], we come to

$$-\frac{2m}{(2k+2)!}\sum_{i=1}^{m} S_{\psi_i}^{(2k+2)}(0) = \frac{-2m\sum_{i=1}^{m}\triangle^{k+1}\psi_i(0)}{2^{k+1}(k+1)!\,m(m+2)\cdots(m+2k)}$$
$$= \frac{2m\sum_{i=1}^{m}(D_i(x_i\triangle^{k+1}\delta),\ \phi)}{2^{k+1}(k+1)!\,m(m+2)\cdots(m+2k)}$$
$$= \frac{-4m(k+1)(\triangle^{k+1}\delta,\ \phi)}{2^{k+1}(k+1)!\,m(m+2)\cdots(m+2k)} \tag{14}$$

by substituting $\psi_i = x_i D_i \phi$ back and obviously

$$-\frac{2m}{(2k+1)!}\sum_{i=1}^{m} S_{\psi_i}^{(2k+1)}(0) = 0. \tag{15}$$

Therefore the result follows from Equation (10) to (15).

5 The products by the Laurent series

Using the Laurent expansion of r^λ at $\lambda = -m - 2j$

$$r^\lambda = \frac{\Omega_m}{(2j)!(\lambda+m+2j)}\delta^{(2j)}(r) + \Omega_m r^{-2j-m} + \Omega_m(\lambda+m+2j)r^{-2j-m}\ln r + \cdots,$$

Aguirre [Agui03a] derived the following identity

$$\delta^{(2j)}(r) = \frac{(2j)!}{\Omega_m}\lim_{\lambda\to -m-2j}(\lambda+m+2j)r^\lambda$$
$$= \frac{(2j)!}{\Omega_m}\operatorname{Res}_{\lambda=-m-2j} r^\lambda = \frac{(2j)!\Gamma(\frac{m}{2})}{2^{2j}j!\Gamma(\frac{m}{2}+j)}\triangle^j\delta(x) \tag{16}$$

from the following fact in [GS64]

$$\operatorname{Res}_{\lambda=-m-2j}(r^\lambda,\ \phi) = \frac{\Omega_m\Gamma(\frac{m}{2})}{2^{2j}j!\Gamma(\frac{m}{2}+j)}(\triangle^j\delta,\ \phi).$$

Theorem 10. *The power $\delta^2(x) = 0$ in space of even dimension.*

Proof. It follows from identity (16) that $\delta(r) = \delta(x)$ by setting $j = 0$. Since m is even, there exists a positive integer j such that $m = 2j$. Thus

$$\begin{aligned}
\delta^2(x) &= \delta(r)\cdot\delta(r) = \frac{1}{\Omega_m}\lim_{\lambda\to -m}(\lambda+m)r^{\lambda}\cdot\frac{1}{\Omega_m}\lim_{\lambda\to -m}(\lambda+m)r^{\lambda}\\
&= \frac{1}{\Omega_m^2}\lim_{\lambda\to -m}(\lambda+m)^2r^{2\lambda} = \frac{1}{\Omega_m^2}\lim_{s\to -m-m}(\frac{s}{2}+m)^2r^{s} \qquad \text{set } s=2\lambda\\
&= \frac{1}{4\Omega_m^2}\lim_{s\to -m-m}(s+m+m)^2r^{s} = \frac{1}{4\Omega_m^2}\lim_{s\to -m-2j}(s+m+2j)^2r^{s}\\
&= \frac{1}{4\Omega_m^2}\lim_{s\to -m-2j}(s+m+2j)\lim_{s\to -m-2j}(s+m+2j)r^{s}\\
&= \frac{1}{4\Omega_m}\lim_{s\to -m-2j}(s+m+2j)\frac{\delta^{(m)}(r)}{m!} = 0.
\end{aligned}$$

On the other hand, we can follow a different approach to show that $\delta^2(x)=0$ for $x\in R$ by applying the Hilbert transform [KL93]

$$\phi(z) = \frac{1}{\pi i}\int_{-\infty}^{+\infty}\frac{\phi(t)}{t-z}dt, \qquad \text{where } \phi\in\mathcal{D}(R)$$

where $I_m z > 0$.

Indeed from Cauchy's representation of distribution, we have

$$\begin{aligned}
(\delta^2(x),\ \phi(x)) &= \lim_{\epsilon\to 0^+} R_e(\delta^2(z-i\epsilon),\phi(z))\\
&\triangleq \lim_{\epsilon\to 0^+} R_e\frac{1}{(2\pi i)^2}\oint_{|z-i\epsilon|=\frac{\epsilon}{2}}\frac{\phi(z)}{(z-i\epsilon)^2}dz.
\end{aligned}$$

By Cauchy's integral formula, we come to

$$(\delta^2(x),\ \phi(x)) = \lim_{\epsilon\to 0^+} R_e\frac{1}{2\pi i}\frac{\phi'(i\epsilon)}{(2-1)!} = R_e\frac{1}{2\pi i}\phi'(0) = 0.$$

Therefore $\delta^2(x)=0$.

In 1992, Koh and C. Li [KL92] chose a fixed δ-sequence and used the concept of neutrix limit to define the distribution δ^k for some k.

Theorem 11. *For $k\in(0,1)$, $\delta^k(x)=0$. For $l=1,2,3,\ldots$, $\delta^{2l}(x)=0$. For $l=0,1,2,3,\ldots$, $\delta^{2l+1}(x)=C_l\delta^{(2l)}(x)$ where*

$$C_l = \frac{1}{2^{2l}l!(2l+1)^{\frac{2l+1}{2}}\pi^l}.$$

In 2001, Özçağ defined the kth powers of the δ distribution for negative integers. Please refer to reference [Öz01] for detail.

Theorem 12. *The product $r^{-2m-n}\cdot\delta^{(2s)}(r)$ exists and $r^{-2m-n}\cdot\delta^{(2s)}(r)=0$ for s, $m=0,1,2,\ldots$ and $n=1,2,\ldots$.*

Proof. From the Laurent series of r^λ

$$r^\lambda = \frac{a_{-1}}{\lambda+n+2m} + a_0 + a_1(\lambda+n+2m) + \cdots$$

where $a_{-1} = \dfrac{\Omega_n \delta^{(2m)}(r)}{(2m)!}$, $a_0 = \Omega_n r^{-2m-n}$ and $a_1 = \Omega_n r^{-2m-n} \ln r$ (Ω_n is the hypersurface area of the unit sphere).

The distribution r^{-2m-n} as the regular part of the Laurent expansion of r^λ about $\lambda = -n-2m$ is defined by

$$r^{-2m-n} = \frac{1}{\Omega_n} \lim_{\lambda\to-n-2m} \frac{\partial}{\partial\lambda}\,[(\lambda+n+2m)\,r^\lambda]. \tag{17}$$

Clearly, for $s = 0, 1, \ldots$, we have

$$\delta^{(2s)}(r) = \frac{(2m)!}{\Omega_n} \lim_{\mu\to-n-2s} [(\mu+n+2s)\,r^\mu] \tag{18}$$

from the the Laurent series of r^μ.
It follows that

$$\begin{aligned}
&r^{-2m-n}\cdot\delta^{(2s)}(r)\\
&= \frac{(2m)!}{\Omega_n^2} \lim_{\lambda\to-n-2m}\lim_{\mu\to-n-2s} \frac{\partial}{\partial\lambda}\,[(\lambda+n+2m)\,r^\lambda\cdot(\mu+n+2s)\,r^\mu]\\
&= \frac{(2m)!}{\Omega_n^2} \lim_{\lambda\to-n-2m}\lim_{\mu\to-n-2s} \frac{\partial}{\partial\lambda}\,[(\lambda+n+2m)(\mu+n+2s)\,r^{\lambda+\mu}].
\end{aligned}$$

Applying the following two identities

$$\begin{aligned}
&(\lambda+n+2m)(\mu+n+2s)\\
&= \frac{1}{2}\left\{(\lambda+\mu+n+2m+n+2s)^2 - (\lambda+n+2m)^2 - (\mu+n+2s)^2\right\},\\
&r^{\lambda+\mu} = \frac{b_{-1}}{\lambda+\mu+n+2m+2s} + b_0 + b_1(\lambda+\mu+n+2m+2s) + \cdots,
\end{aligned}$$

we come to

$$\begin{aligned}
&r^{-2m-n}\cdot\delta^{(2s)}(r)\\
&= \frac{(2m)!}{\Omega_n^2} \lim_{\lambda\to-n-2m}\lim_{\mu\to-n-2s} \frac{\partial}{\partial\lambda}\left\{\frac{(\lambda+\mu+n+2m+n+2s)^2}{2(\lambda+\mu+n+2m+2s)}\,b_{-1} + \cdots\right\}\\
&\quad - \frac{(2m)!}{\Omega_n^2} \lim_{\lambda\to-n-2m}\lim_{\mu\to-n-2s} \frac{\partial}{\partial\lambda}\left\{\frac{(\lambda+n+2m)^2}{2(\lambda+\mu+n+2m+2s)}\,b_{-1} + \cdots\right\}\\
&\quad - \frac{(2m)!}{\Omega_n^2} \lim_{\lambda\to-n-2m}\lim_{\mu\to-n-2s} \frac{\partial}{\partial\lambda}\left\{\frac{(\mu+n+2s)^2}{2(\lambda+\mu+n+2m+2s)}\,b_{-1} + \cdots\right\}\\
&= I_1 + I_2 + I_3.
\end{aligned}$$

By direct computation, we obtain

$$\lim_{\lambda\to -n-2m}\lim_{\mu\to -n-2s}\frac{\partial}{\partial\lambda}\left\{\frac{(\lambda+\mu+n+2m+n+2s)^2}{2(\lambda+\mu+n+2m+2s)}\,b_{-1}\right\}=0 \tag{19}$$

and the rest in I_1 is zero since there is only one n in the denominators after taking the partial derivative, which will never vanish after the two limits, while all numerators disappear.

Similarly, we get $I_2 = I_3 = 0$. This completes the proof of the theorem.

6 The products by the substitutions

Let $H(x_1, x_2, \ldots, x_m)$ be any sufficiently smooth function such that on $H = 0$ we have

$$\text{grad} H \neq 0$$

which means that there are no singular points on $H = 0$. Then the generalized function $\delta(H)$ can be defined in the following way.

$$(\delta(H),\, \phi) = \int_{P=0} \psi(0, u_2, \ldots, u_m) du_2 \cdots du_m$$

where $\phi_1(u_1, \ldots, u_m) = \phi(x_1, \ldots x_m)$ and $\psi = \phi_1(u)D\binom{x}{u}$.

Similarly, we shall define

$$(\delta^{(k)}(H),\, \phi) = (-1)^k \int_{P=0} \psi_{u_1}^{(k)}(0, u_2, \ldots, u_m) du_2 \cdots du_m.$$

As an example, we consider the generalized function $\delta(\alpha_1 x_1 + \cdots + \alpha_m x_m)$, where $\sum_{i=1}^m \alpha_i^2 = 1$. The equation

$$\alpha_1 x_1 + \cdots + \alpha_m x_m = 0$$

determines a hypersurface which passes through the origin and is orthogonal to the unit vector α. Making the substitution

$$u_1 = \alpha_1 x_1 + \cdots + \alpha_m x_m, \quad u_2 = x_2, \cdots, u_m = x_m,$$

we thus arrive at

$$(\delta(\alpha_1 x_1 + \cdots + \alpha_m x_m), \phi) = \int_{\sum \alpha_i x_i = 0} \phi\, du_2 \cdots du_m.$$

Theorem 13. *Let f be a C^∞ function and let H be defined as above. Then the product $f(H) \cdot \delta^{(k)}(H)$ exists and*

$$f(H) \cdot \delta^{(k)}(H) = \sum_{i=0}^{k} \binom{k}{i} (-1)^i f^{(i)}(0) \delta^{(k-i)}(H).$$

Proof. Using the substitutions $u_1 = H(x_1, \ldots, x_m)$, $u_2 = x_2, \ldots, u_m = x_m$, we arrive at

$$(f(H) \cdot \delta^{(k)}(H), \phi) = (-1)^k \int_{H=0} \frac{\partial^k}{\partial u_1^k} \left\{ f(u_1)\phi_1 D\binom{x}{u} \right\}\Big|_{u_1=0} du_2 \cdots du_m$$

and

$$\frac{\partial^k}{\partial u_1^k} \left\{ f(u_1)\phi_1 D\binom{x}{u} \right\}\Big|_{u_1=0} = \sum_{i=0}^{k} \binom{k}{i} f^{(i)}(0)\, D_{u_1}^{k-i} \phi_1 D\binom{x}{u}\Big|_{u_1=0}.$$

Hence

$$\begin{aligned}
&(f(H) \cdot \delta^{(k)}(H), \phi) \\
&= (-1)^k \sum_{i=0}^{k} \binom{k}{i} f^{(i)}(0) \int_{H=0} \frac{\partial^{k-i}}{\partial u_1^{k-i}} \phi_1 D\binom{x}{u}\Big|_{u_1=0} du_2 \cdots du_m \\
&= \sum_{i=0}^{k} \binom{k}{i} (-1)^i f^{(i)}(0) (\delta^{(k-i)}(H), \phi)
\end{aligned}$$

which completes the proof of the theorem.

In particular, we have

$$\begin{aligned}
H \cdot \delta'(H) &= -\delta(H), \\
H^2 \cdot \delta'(H) &= 0.
\end{aligned}$$

Assume that both $p > 1$ and $q > 1$. Let P be a quadratic form defined by $P(x) = P(x_1, x_2, \ldots, x_m) = x_1^2 + \cdots + x_p^2 - x_{p+1}^2 - \cdots - x_{p+q}^2$ with $p+q = m$, then the $P = 0$ hypersurface is a hypercone with a singular point (the vertex) at the origin.

We start by assuming that $\phi(x)$ vanishes in a neighborhood of the origin. The distribution $\delta^{(k)}(P)$ is defined by

$$(\delta^{(k)}(P), \phi) = (-1)^k \int \frac{\partial^k}{\partial P^k} \{ \frac{1}{2} \phi\, (r^2 - P)^{\frac{1}{2}(q-2)} \}\Big|_{P=0} r^{p-1} dr d\Omega^{(p)} d\Omega^{(q)},$$

which is convergent.

Furthermore, if we transform from P to $s = \sqrt{r^2 - P}$ we note that $\partial/\partial P = -(2s)^{-1}\partial/\partial s$, and we may write this in the form

$$(\delta^{(k)}(P), \phi) = \int \left[(\frac{\partial}{2s\, \partial s})^k \{ s^{q-2} \frac{\phi}{2} \} \right]_{s=r} r^{p-1} dr\, d\Omega^{(p)}\, d\Omega^{(p)}.$$

Let us now define

$$\psi(r,s) = \int \phi d\Omega^{(p)}\, d\Omega^{(p)},$$

Hence,

$$(\delta^{(k)}(P), \phi) = \int_0^\infty \left[\left(\frac{\partial}{2s\,\partial s}\right)^k \{ s^{q-2} \frac{\psi(r,s)}{2} \} \right]_{s=r} r^{p-1} dr.$$

Theorem 14. *The product P^n and $\delta^{(k)}(P)$ exists and*

$$P^n \cdot \delta^{(k)}(P) = \begin{cases} n! \binom{k}{n} \delta^{(k-n)}(P) & \text{if } k \geq n, \\ 0 & \text{if } k < n. \end{cases}$$

7 Conclusion

As outlined in the introduction, defining products of distribution on manifolds has been a serious challenge. However, there is one special case worth considering here:

Let $f(t)$ be a distribution of one variable and let $u \in C^\infty(R^m)$ and the manifold $u(x) = 0$ has no any critical point. According to Leray [Ler57], we have

$$(f(u(x)), \phi(x)) = (f(t), \psi(t)) \qquad \phi \in \mathcal{D}(R^m)$$

where $\psi(t) = \int_{u(x)=t} \phi(x) d\,s.$

A problem of interest is to show that $\psi(t) \in \mathcal{D}$ under certain conditions restricted on the manifold, which can lead to study products of the distributions on the manifold $u(x) = 0$.

References

[Sch59] Schwartz, L.: Théorie des distributions. Vols.I, II, Hermann, Paris (1957)

[Gasi66] Gasiorowicz, S.: Elementary particle physics. J. Wiley and Sons Inc., New York (1966)

[AMS73] Antosik, P., Mikusinski, J. and Sikorski, R.: Theory of distributions, the sequential approach. PWN-Polish Scientific Publishers, Warsawa (1973)

[Bre65] Bremermann, J.H.: Distributions, complex variables, and Fourier transforms. Addison-Wesley, Reading, Massachusetts (1965)

[Li78] Li, B.H.: Non-standard analysis and multiplication of distributions. Sci. Sinica, **21(5)**, 561–585 (1978)

[EGO92] Embacher, H.G., Grübl, G., Oberguggenberger, M.: Z. Anal. Anw., **11**, 437–454 (1992)

[GS64] Gel'fand, I.M., Shilov, G.E.: Generalized functions. Vol. I. Academic Press, New York London (1964)

[Fis71] Fisher, B.: The product of distributions. Quart. J. Math. Oxford, **22**, 291–298 (1971)

[Fis74] Fisher, B.: The neutrix distribution product $x_+^{-r}\delta^{(r-1)}$. Studia Sci. Math. Hungar., **9**, 439–441 (1974)

[Fis82a] Fisher, B.: On defining the convoltion of distributions. Math. Nachr., **106**, 261–269 (1982)

[Fis82b] Fisher, B.: A non-commutative neutrix product of distributions. Math. Nachr., **108**, 117–127 (1982)

[Fis80] Fisher, B.: On defining the product of distributions. Math. Nachr., **99**, 239–249 (1980)

[KF03] Kilicman, A., Fisher, B.: On the Fresnel integrals and the convolution. Int. J. Math. Math. Sci., **41**, 2635-2643 (2003)

[FT05a] Fisher, B., Taş, K.: The convolution of functions and distributions. J. Math. Anal. Appl., **306(1)**, 364–274 (2005)

[FT06a] Fisher, B., Taş, K.: On the composition of the distributions x_+^λ and x_+^μ. J. Math. Anal. Appl., **318(1)**, 102–111 (2006)

[FT05b] Fisher, B., Taş, K.: On the non-commutative neutrix product of the distributions $x^r \ln^p |x|$ and x^{-s}. Integral Transform Spec. Funct., **16(2)**, 131–138 (2005)

[FT06b] Fisher, B., Taş, K.: On the commutative product of distributions. J. Korean Math. Soc., **43(2)**, 271–281 (2006)

[FN98] Fisher, B., Nicholas, J.D.: Some results on the commutative neutrix product of distributions. J. Anal., **6**, 33–44 (1998)

[Fis72] Fisher, B.: The product of the distributions $x_+^{-r-\frac{1}{2}}$ and $x_-^{-r-\frac{1}{2}}$. Proc. Cambridge Philos. Soc., **71**, 123–130 (1972)

[FL01] Fisher, B., Li, C.K.: On the cosine and sine integrals. Int. J. Appl. Math., **7(4)**, 419–437 (2001)

[FL93] Fisher, B., Li, C.K.: A commutative neutrix convolution product of distributions. Zb. Rad. Prirod.-Mat. Fak. Ser. Mat., **23(1)**, 13–27 (1993)

[FKL00] Fisher, B., Kilicman, A., Li, C.K.: An extension of a result on the non-commutative neutrix convolution product of distributions. Int. J. Appl. Math., **3(1)**, 71–80 (2000)

[Fis70] Fisher, B.: The generalized function $(x + i0)^\lambda$. Proc. Camb. Phil. Soc., **68**, 707–708 (1970)

[FÖG05] Fisher, B., Özçağ, E., Gülen, Ü.: A theorem on the commutative neutrix product of distributions. Sarajevo J. Math., **1**, 235–242 (2005)

[vdC60] van der Corput, J.G.: Introduction to the neutrix calculus. J. Analyse Math., **7**, 291–398 (1959)

[Li00] Li, C.K.: The product of r^{-k} and $\nabla\delta$. Int. J. Math. Math. Sci., **24**, 361–369 (2000)

[Li01a] Li, C.K.: A note on the product $r^{-k} \cdot \nabla(\triangle r^{2-m})$. Integral Transform. Spec. Func., **12**, 341–348 (2001)

[LF:90] Li, C.K., Fisher, B.: Examples of the neutrix product of distributions on R^m. Rad. Mat., **6**, 129–137 (1990)

[CL05a] Li, C.K.: The products on the unit sphere and even-dimension spaces. J. Math. Anal. Appl., **305(1)**, 97–106 (2005)

[Li05b] Li, C.K.: An approach for distributional products on R^m. Integral Transforms Spec. Funct., **16(2)**, 139–151 (2005)

[CL91] Cheng, L.Z. and Li, C.K.: A commutative neutrix product of distributions on R^m. Math. Nachr., **151**, 345–355 (1991)

[CL88] Cheng, L.Z. and Li, C.K.: The product of generalized functions. J. Math. Res. Exposition, **8(4)**, 543–546 (1988)

[LK98] Li, C.K., Koh, E.L.: The neutrix convolution product in $Z'(m)$ and the exchange formula. Int. J. Math. Math. Sci., **21(4)**, 695–700 (1998)

[LZ04] Li, C.K., Zou, V.: On defining the product $r^{-k} \cdot \nabla^l \delta$. Int. J. Math. Math. Sci., **16(13-16)**, 833–845 (2004)

[Li01b] Li, C.K.: The sequential approach to the product of distribution. Int. J. Math. Math. Sci., **28(12)**, 743–751 (2001)

[KL93] Koh, E.L., Li, C.K.: On defining the generalized functions $\delta^\alpha(z)$ and $\delta^n(x)$. Int. J. Math. Math. Sci., **16(4)**, 749–754 (1993)

[Li03] Li, C.K.: The neutrix square of δ. Int. J. Appl. Math., **12(2)**, 115–124 (2003)

[LA04] Li, C.K., Aguirre, M.A.: The distributional products by the Laurent series. submitted.

[KL92] Koh, E.L., Li, C.K.: On the distributions δ^k and $(\delta')^k$. Math. Nachr., **157**, 243–258 (1992)

[AL05] Aguirre, M.A., Li, C.K.: The distributional products of particular distributions. to appear in Applied Mathematics and Computation.

[Agui03a] Aguirre, M.A.: A convolution product of $(2j)$-th derivative of Diracs delta in r and multiplicative distributional product between r^{-k} and $\nabla(\triangle^j \delta)$. Int. J. Math. Math. Sci., **13**, 789–799 (2003)

[Agui03b] Aguirre, M.A.: The expansion in series (of Taylor Types) of $(k-1)$ derivatrive of Diracs delta in $m^2 + P$. Integral Transform. Spec. Func., **14**, 117–127 (2003)

[Agui91] Aguirre, M.A.: The series expansion of $\delta^{(k)}(r-c)$. Mathematicae Notae, **35**, 53-61 (1991)

[Öz01] Özçağ, E.: Defining the kth powers of the Dirac-delta distribution for negative integers. Appl. Math. Lett., **14**, 419–423 (2001)

[Ler57] Leray, J.: Hyperbolic differential eqautions. The Institute for Advanced Study, Princeton New Jersey (1957)

Some remarks on the incomplete gamma function ⋆

Emin Özçağ[1], İnci Ege[1], Haşmet Gürçay[1] and Biljana Jolevska-Tuneska[2]

[1] Department of Mathematics, Hacettepe University, 06532-Beytepe, Ankara, Turkey ozcag1@hacettepe.edu.tr

[2] Faculty of Electrical Engineering, Karpos II bb, Skopje, Republic of Macedonia biljana@etf.ukim.edu.mk

The incomplete Gamma function $\gamma(\alpha, x)$ is defined for $\alpha > 0$ and $x \geq 0$ by

$$\gamma(\alpha, x) = \int_0^x u^{\alpha-1} e^{-u}\, du$$

and by using the recurrence formula

$$\gamma(\alpha + 1, x) = \alpha\gamma(\alpha, x) - x^{\alpha} e^{-x}$$

the definition of $\gamma(\alpha, x)$ can be extended to negative, non integer value of α. Recently Fisher et al. [FJK03] defined $\gamma(-m, x)$ for $m = 0, 1, 2, \ldots$. In this paper we consider the derivatives of the incomplete Gamma function $\gamma(\alpha, x)$ and the derivatives of locally summable function $\gamma(\alpha, x_+) = H(x)\gamma(\alpha, x)$ for negative integers, where $H(x)$ denotes the Heaviside function.

1 Introduction

The incomplete gamma function and its complement are defined by integrals

$$\gamma(a, z) = \int_0^z e^{-t} t^{a-1}\, dt \tag{1}$$

and

$$\Gamma(a, z) = \int_z^{\infty} e^{-t} t^{a-1}\, dt \tag{2}$$

respectively for $z \in C \setminus R^-$ and $Ra > 0$.

⋆ This research was supported by TUBITAK, project number TBAG-U/133 (105T057) (Turkey) and the Ministry of Education of Macedonia

Kenan Taş et al. (eds), Mathematical Methods in Engineering, 97–108.

For $a = 0$, the exponential integral in (2) occurs in applications most notably in quantum-mechanical electronic structure calculations. The interesting expansion formula for the integral in (2) in a series of incomplete gamma functions were presented by W. Gautschi et al., see [GHT03] and by S. Lin et al., see [LCS05].
The simple asymptotic expansions of $\gamma(a, z)$ for large a and z were recently considered by C. Ferreira et al., see [FLS05].

In this paper we consider the derivatives of the incomplete gamma function of the real variable for negative integers. Also we consider the derivatives of the locally summable function $\gamma(\alpha, x_+) = H(x)\gamma(\alpha, x)$ for negative integers.

The technique of neglecting appropriately defined infinite quantities was devised by Hadamard and the resulting finite value extracted from the divergent integral is usually referred to as the Hadamard finite part.
Using the concepts of the neutrix and the neutrix limit due to van der Corput [vdC59], Fisher gave the general principle for the discarding of unwanted infinite quantities from asymptotic expansions and has been exploited in context of distributions, see [Fis87].

In the following we let N be the neutrix (see [vdC59]) having domain $N' = \{\epsilon : 0 < \epsilon < \infty\}$ and range N'' the real numbers, with negligible functions finite linear sums of the functions

$$\epsilon^{\lambda} \ln^{r-1} \epsilon, \quad \ln^{r} \epsilon \qquad (\lambda < 0, \quad r = 1, 2, \ldots)$$

and all functions $f(\epsilon)$ which converge to zero in the normal sense as ϵ tends to zero.

If $f(\epsilon)$ is a real (or complex) valued function defined on N' and if it is possible to find a constant c such that $f(\epsilon) - c$ is in N, then c is called the neutrix limit of $f(\epsilon)$ as $\epsilon \to 0$ and we write $\mathrm{N}\!-\!\lim_{\epsilon\to 0} f(\epsilon) = c$.

Note that if a function $f(\epsilon)$ tends to c in the normal sense as ϵ tends to zero, it converges to c in the neutrix sense.
The reader may find the general definition of the neutrix limit with some examples in [vdC59, Fis87].
In the following we apply Fisher's principle to the incomplete Gamma function to define the derivative of the incomplete Gamma function for negative integers.

The gamma function $\Gamma(x)$ is usually defined for $x > 0$ by

$$\Gamma(x) = \int_0^{\infty} t^{x-1} e^{-t}\, dt \tag{3}$$

The integral only converging for $x > 0$, see [EMOT53, GS64].
It follows from equation (3) that

$$\Gamma(x+1) = x\Gamma(x)$$

for $x > 0$ and this equation is used to define $\Gamma(x)$ for negative, non-integer values of x.

Using the regularization, Gelfand and Shilov [GS64] define the gamma function

$$\Gamma(x) = \int_0^1 t^{x-1}\Big[e^{-t} - \sum_{i=0}^{n-1}(-1)^i \frac{t^i}{i!}\Big]\,dt + \int_1^\infty t^{x-1}e^{-t}\,dt + \sum_{i=0}^{n-1}\frac{(-1)^i}{i!(x+i)}$$

for $x > -n, \quad x \neq 0, -1, -2, \ldots, -n+1$ and

$$\Gamma(x) = \int_0^\infty t^{x-1}\Big[e^{-t} - \sum_{i=0}^{n-1}(-1)^i \frac{t^i}{i!}\Big]\,dt$$

for $-n < x < -n+1$.
It was proved in [FK03] that

$$\Gamma(x) = \mathrm{N}\!-\!\lim_{\epsilon\to 0}\int_\epsilon^\infty t^{x-1}\,e^{-t}\,dt$$

$x \neq 0, -1, -2, \ldots$ and

$$\Gamma(-m) = \mathrm{N}\!-\!\lim_{\epsilon\to 0}\int_\epsilon^\infty t^{-m-1}\,e^{-t}\,dt$$

$$= \int_1^\infty t^{-m-1}e^{-t}\,dt + \int_0^1 t^{-m-1}\Big[e^{-t} - \sum_{i=0}^{m}\frac{(-t)^i}{i!}\Big]\,dt - \sum_{i=0}^{m-1}\frac{(-1)^i}{i!(m-i)} \tag{4}$$

for $m = 1, 2, \ldots$.
Fisher and Kuribayashi [FK03] proved the existence of $\Gamma^{(r)}(0)$ and they then defined $\Gamma^{(r)}(0)$ by the equation

$$\begin{aligned}\Gamma^{(r)}(0) &= \mathrm{N}\!-\!\lim_{\epsilon\to 0}\int_\epsilon^\infty t^{-1}\ln^r te^{-t}\,dt \\ &= \int_1^\infty t^{-1}\ln^r te^{-t}\,dt + \int_0^1 t^{-1}\ln^r t[e^{-t}-1]\,dt \end{aligned} \tag{5}$$

for $r = 0, 1, 2, \ldots$. This suggested that $\Gamma^{(r)}(-m)$ be defined by

$$\Gamma^{(r)}(-m) = \mathrm{N}\!-\!\lim_{\epsilon\to 0}\int_\epsilon^\infty t^{-m-1}\ln^r te^{-t}\,dt$$

$$\begin{aligned}= \int_1^\infty t^{-m-1}\ln^r te^{-t}\,dt + \int_0^1 t^{-m-1}\ln^r t\Big[e^{-t} - \sum_{i=0}^{m}\frac{(-t)^i}{i!}\Big]\,dt + \\ - \sum_{i=0}^{m-1}\frac{(-1)^i}{i!}r!(m-i)^{-r-1}\end{aligned} \tag{6}$$

for $r = 0, 1, 2, \ldots$ and $m = 1, 2, \ldots$.

2 The incomplete gamma function $\gamma(\alpha, x)$ and its derivatives

The incomplete Gamma function $\gamma(\alpha, x)$ is defined for $\alpha > 0$ and $x \geq 0$ by

$$\gamma(\alpha, x) = \int_0^x u^{\alpha-1} e^{-u}\, du \tag{7}$$

see [EMOT53, FJK03, GS64], the integral diverging for $\alpha \leq 0$.
The incomplete gamma function can be defined for $\alpha < 0$ and $\alpha \neq -1, -2, -3, \ldots$ by using the recurrence formula

$$\gamma(\alpha + 1, x) = \alpha\gamma(\alpha, x) - x^{\alpha} e^{-x}.$$

On integrating by parts, we see that

$$\gamma(\alpha, x) = \int_0^x u^{\alpha-1}(e^{-u} - 1)\, du + \alpha^{-1} e^{-x}$$

and more generally if $-m < \alpha < -m + 1$ and $x > 0$, we have by induction

$$\gamma(\alpha, x) = \int_0^x u^{\alpha-1}\Big[e^{-u} - \sum_{i=0}^{m-1} \frac{(-u)^i}{i!}\Big]\, du + \sum_{i=0}^{m-1} \frac{(-1)^i x^{\alpha+i}}{(\alpha+i)i!}. \tag{8}$$

It follows that

$$\lim_{x\to\infty} \gamma(\alpha, x) = \Gamma(\alpha) \tag{9}$$

for $\alpha \neq 0, -1, -2, \ldots$.
On using equation (8), the incomplete gamma function $\gamma(\alpha, x)$ is defined by

$$\gamma(\alpha, x) = \mathrm{N}\!-\!\lim_{\epsilon\to 0} \int_\epsilon^x u^{\alpha-1} e^{-u}\, du$$

for $\alpha \neq 0, -1, -2, \ldots$, $x > 0$ and the function $\gamma(-m, x)$ is defined by

$$\begin{aligned}\gamma(-m, x) &= \mathrm{N}\!-\!\lim_{\epsilon\to 0} \int_\epsilon^x u^{-m-1} e^{-u}\, du \\ &= \int_0^x u^{-m-1}\Big[e^{-u} - \sum_{i=0}^{m} \frac{(-u)^i}{i!}\Big]\, du - \sum_{i=0}^{m-1} \frac{(-1)^i x^{i-m}}{i!(m-i)} + \frac{(-1)^m \ln x}{m!} \\ &= \int_1^x u^{-m-1} e^{-u}\, du + \int_0^1 u^{-m-1}\Big[e^{-u} - \sum_{i=0}^{m} \frac{(-u)^i}{i!}\Big]\, du - \sum_{i=0}^{m-1} \frac{(-1)^i}{i!(m-i)} \end{aligned} \tag{10}$$

for $m = 1, 2, \ldots$, and $x > 0$ see [FJK03].

In particular when $m = 0$, the equation (10) becomes

$$\gamma(0,x) = \mathrm{N-}\lim_{\epsilon\to 0}\int_\epsilon^x u^{-1}e^{-u}\,du = \int_0^x u^{-1}(e^{-u}-1)\,du + \ln x.$$

It follows from equation (4) that

$$\lim_{x\to\infty}\gamma(-m,x) = \Gamma(-m) \tag{11}$$

for $m = 1, 2, \ldots$.

On integrating by parts we have

$$\begin{aligned}\gamma(-m,x) &= -\frac{1}{m}\int_1^x e^{-u}u^{-m}\,du - \frac{1}{m}e^{-x}x^{-m} + \frac{1}{m}\sum_{i=0}^{m}\frac{(-1)^i}{i!} - \\ &\quad -\frac{1}{m}\int_0^1 u^{-m}\Big[e^{-u} - \sum_{i=0}^{m-1}\frac{(-u)^i}{i!}\Big]\,du - \sum_{i=0}^{m-1}\frac{(-1)^i}{i!(m-i)} \\ &= -\frac{1}{m}\int_1^x u^{-m}e^{-u}\,du - \frac{1}{m}\int_0^1 u^{-m}\Big[e^{-u} - \sum_{i=0}^{m-1}\frac{(-u)^i}{i!}\Big]\,du + \\ &\quad +\frac{1}{m}\sum_{i=0}^{m-2}\frac{(-1)^i}{i!(m-1-i)} + \frac{(-1)^m}{mm!} - \frac{1}{m}e^{-x}x^{-m} \\ &= -\frac{1}{m}\gamma(-m+1,x) + \frac{(-1)^m}{mm!} - \frac{1}{m}e^{-x}x^{-m}\end{aligned}$$

and so the interesting formula

$$\gamma(-m,x) + \frac{1}{m}\gamma(-m+1,x) = \frac{(-1)^m}{mm!} - \frac{1}{m}e^{-x}x^{-m} \tag{12}$$

is obtained for $m = 1, 2, \ldots$.

It follows from equation (8) that

$$\begin{aligned}\gamma'(\alpha,x) &= \int_0^x u^{\alpha-1}\ln u\Big[e^{-u} - \sum_{i=0}^{m-1}\frac{(-u)^i}{i!}\Big]\,du + \\ &\quad + \sum_{i=0}^{m-1}(-1)^i\frac{(\alpha+i)x^{\alpha+i}\ln x - x^{\alpha+i}}{(\alpha+i)^2 i!}\end{aligned}$$

for $-m < \alpha < -m+1$.

On the other hand, we have

$$\int_\epsilon^x u^{\alpha-1}\ln ue^{-u}\,du =$$

$$= \int_\epsilon^x u^{\alpha-1}\ln u\Big[e^{-u} - \sum_{i=0}^{m-1}\frac{(-u)^i}{i!}\Big]\,du + \sum_{i=0}^{m-1}\frac{(-1)^i}{i!}\int_\epsilon^x u^{\alpha+i-1}\ln u\,du$$

$$= \int_\epsilon^x u^{\alpha-1}\ln u\Big[e^{-u} - \sum_{i=0}^{m-1}\frac{(-u)^i}{i!}\Big]\,du + \sum_{i=0}^{m-1}\frac{(-1)^i}{i!(\alpha+i)}[x^{\alpha+i}\ln x - \epsilon^{\alpha+i}\ln\epsilon] -$$

$$-\sum_{i=0}^{m-1}\frac{(-1)^i}{i!(\alpha+i)^2}[x^{\alpha+i} - \epsilon^{\alpha+i}].$$

Thus

$$\gamma'(\alpha,x) = \mathrm{N}\!-\!\lim_{\epsilon\to 0}\int_\epsilon^x u^{\alpha-1}\ln ue^{-u}\,du$$

for $\alpha \neq 0,-1,-2,\ldots$.
More generally it can be shown that

$$\gamma^{(r)}(\alpha,x) = \mathrm{N}\!-\!\lim_{\epsilon\to 0}\int_\epsilon^x u^{\alpha-1}\ln^r ue^{-u}\,du \tag{13}$$

for $\alpha \neq 0,-1,-2,\ldots, r = 0,1,2,\ldots$ and $x > 0$.
This suggests the following definition.

Definition 1. *The rth derivative of incomplete gamma function, $\gamma^{(r)}(-m,x)$ is defined by*

$$\gamma^{(r)}(-m,x) = \mathrm{N}\!-\!\lim_{\epsilon\to 0}\int_\epsilon^x u^{-m-1}\ln^r ue^{-u}du \tag{14}$$

for $r,m = 0,1,2,\ldots$ and $x > 0$ provided that the neutrix limit exists.

Equation (13) will then define $\gamma^{(r)}(\alpha,x)$ for all α and $r = 0,1,2,\ldots$.
Before proving the neutrix limit above exists we need the following lemma.

Lemma 1.

$$\int \ln^r u\,du = \sum_{i=0}^{r-1}\frac{(-1)^i r!}{(r-i)!}u\ln^{r-i}u + (-1)^r r!u \tag{15}$$

and

$$\int u^{-s-1}\ln^r u\,du = -\sum_{i=0}^{r-1}\frac{r!}{(r-1)!}s^{-i-1}u^{-s}\ln^{r-i}u - r!s^{-r-1}u^{-s} \tag{16}$$

for $r,s = 1,2,\ldots$.

Proof. Equation (15) follows by induction and equation(16) follows on using equation (15) and making the substitution $w = u^{-s}$. □

Theorem 1. *The functions $\gamma^{(r)}(0,x)$ and $\gamma^{(r)}(-m,x)$ exist and*

$$\gamma^{(r)}(0,x) = \int_1^x u^{-1}\ln^r ue^{-u}\,du + \int_0^1 u^{-1}\ln^r u[e^{-u}-1]\,du \tag{17}$$

for $r = 0,1,2,\ldots$ and

$$\gamma^{(r)}(-m,x) = \int_1^x u^{-m-1}\ln^r ue^{-u}\,du +$$

$$+\int_0^1 u^{-m-1}\ln^r u\Big[e^{-u} - \sum_{i=0}^{m}\frac{(-u)^i}{i!}\Big]\,du - \sum_{i=0}^{m-1}\frac{(-1)^i}{i!}r!(m-i)^{-r-1} \tag{18}$$

for $r,m = 1,2,\ldots$.

Proof. We have

$$\int_\epsilon^x u^{-1}\ln^r ue^{-u}\,du =$$

$$= \int_1^x u^{-1}\ln^r ue^{-u}\,du + \int_\epsilon^1 u^{-1}\ln^r u[e^{-u}-1]\,du + \int_\epsilon^1 u^{-1}\ln^r u\,du$$

$$= \int_1^x u^{-1}\ln^r ue^{-u}\,du + \int_\epsilon^1 u^{-1}\ln^r u[e^{-u}-1]\,du - \frac{\ln^{r+1}\epsilon}{r+1}.$$

for $r = 0,1,2,\ldots$. It follows that

$$\operatorname*{N-lim}_{\epsilon\to 0}\int_\epsilon^x u^{-1}\ln^r ue^{-u}\,du =$$

$$= \int_1^x u^{-1}\ln^r ue^{-u}\,du + \int_0^1 u^{-1}\ln^r u[e^{-u}-1]\,du.$$

and so equation (17) follows.

Now let us consider

$$\int_\epsilon^x u^{-m-1}\ln^r u\,e^{-u}\,du = \int_1^x u^{-m-1}\ln^r ue^{-u}\,du +$$

$$+\int_\epsilon^1 u^{-m-1}\ln^r u\Big[e^{-u} - \sum_{i=0}^{m}\frac{(-u)^i}{i!}\Big]\,du + \sum_{i=0}^{m}\frac{(-1)^i}{i!}\int_\epsilon^1 u^{-m+i-1}\ln^r u\,du$$

$$= \int_1^x u^{-m-1}\ln^r ue^{-u}\,du + \int_\epsilon^1 u^{-m-1}\ln^r u\Big[e^{-u} - \sum_{i=0}^{m}\frac{(-u)^i}{i!}\Big]\,du +$$

$$+\sum_{i=0}^{m-1}\sum_{j=0}^{r-1}\frac{(-1)^i r!}{i!(r-j)}(m-i)^{-j-1}\epsilon^{i-m}\ln^{r-j}\epsilon -$$

$$-\sum_{i=0}^{m-1}\frac{(-1)^i r!}{i!}(m-i)^{-r-1}(1-\epsilon^{i-m}) - \frac{(-1)^m}{m!(r+1)}\ln^{r+1}\epsilon.$$

Thus

$$\gamma^{(r)}(-m,x) = \mathrm{N}\!-\!\lim_{\epsilon\to 0}\int_\epsilon^x u^{-m-1}\ln^r u e^{-u}\,du = \int_1^x u^{-m-1}\ln^r u e^{-u}\,du +$$

$$+\int_0^1 u^{-m-1}\ln^r u\Big[e^{-u} - \sum_{i=0}^{m}\frac{(-u)^i}{i!}\Big]\,du - \sum_{i=0}^{m-1}\frac{(-1)^i}{i!}r!(m-i)^{-r-1}.$$

for $r, m = 1, 2, \ldots$. Equation (18) follows. □

Moreover it follows from equations (6) and (18) that

$$\lim_{x\to\infty}\gamma^{(r)}(-m,x) = \Gamma^{(r)}(-m).$$

3 The locally summable function $\gamma(\alpha, x_+)$

The incomplete gamma function $\gamma(\alpha, x_+)$ is defined as locally summable function on the real line for $\alpha > 0$ by

$$\gamma(\alpha, x_+) = H(x)\gamma(\alpha, x) = \int_0^{x_+} u^{\alpha-1}e^{-u}\,du \tag{19}$$

see [FJK03] and can be defined as a distribution for $\alpha < 0$ and $\alpha \neq -1, -2, \ldots$ by the formula

$$\gamma(\alpha+1, x_+) = \alpha\gamma(\alpha, x_+) - x_+^\alpha e^{-x}. \tag{20}$$

It was proved in [FJK03] that

$$\gamma(\alpha, x_+) = \int_0^{x_+} u^{\alpha-1}\Big[e^{-u} - \sum_{i=0}^{r-1}\frac{(-u)^i}{i!}\Big]\,du + \sum_{i=0}^{r-1}\frac{(-1)^i}{(\alpha+i)i!}x_+^{\alpha+i} \tag{21}$$

if $-m < \alpha < -m+1$ for $m = 1, 2, \ldots$.

The distribution $\gamma(\alpha, x_+)$ is defined by

$$\langle\gamma(\alpha, x_+), \varphi(x)\rangle = \mathrm{N}\!-\!\lim_{\epsilon\to 0}\int_\epsilon^\infty \varphi(x)\int_\epsilon^x u^{\alpha-1}e^{-u}\,du\,dx \tag{22}$$

for arbitrary$\varphi \in \mathcal{D}$ and if $-m < \alpha < -m+1$ for $m = 1, 2, \ldots$, see [FJK03]. This suggested the following definition of $\gamma(-m, x_+)$.

Definition 2. *The distribution $\gamma(-m, x_+)$ is defined by*

$$\langle\gamma(-m, x_+), \varphi(x)\rangle = \mathrm{N}\!-\!\lim_{\epsilon\to 0}\int_\epsilon^\infty \varphi(x)\int_\epsilon^x u^{-m-1}e^{-u}\,du\,dx \tag{23}$$

for arbitrary $\varphi \in \mathcal{D}$ and $m = 0, 1, 2, \ldots$,see [FJK03].

We now need the following definitions.
The distribution x_+^{-m} is defined by

$$x_+^{-m} = \frac{(-1)^{m-1}}{(m-1)!}(\ln x_+)^{(m)}$$

see [Fis87]. The definition of x_+^{-m} here is not the same as Gelfand and Shilov's definition of x_+^{-m} see [GS64], which we will denote by $F(x_+, -m)$.

It is shown that

$$x_+^{-m} = F(x_+, -m) + \frac{(-1)^m \phi(m-1)}{(m-1)!}\delta^{(m-1)}(x) \tag{24}$$

for $m = 1, 2, \ldots$, where

$$\phi(m) = \begin{cases} 0, & m = 0, \\ \sum_{i=1}^{m} i^{-1}, & m > 0. \end{cases}$$

It was proved that

$$\mathrm{N-}\lim_{\epsilon\to 0}\int_\epsilon^\infty x^\alpha \varphi(x)\, dx = \langle x_+^\alpha, \varphi(x)\rangle$$

if $-r < \alpha < -r+1$ for $r = 1, 2, \ldots$ and

$$\mathrm{N-}\lim_{\epsilon\to 0}\int_\epsilon^\infty x^{-m} \varphi(x)\, dx = \langle F(x_+, -m), \varphi(x)\rangle$$

for arbitrary $\varphi \in \mathcal{D}$ and $m = 1, 2, \ldots$, see [Fis87].

Theorem 2.

$$\begin{aligned}\langle \gamma^{(r)}(\alpha, x_+), \varphi(x)\rangle &= (-1)^r\, \mathrm{N-}\lim_{\epsilon\to 0}\int_\epsilon^\infty \varphi^{(r)}(x)\int_\epsilon^x u^{\alpha-1}e^{-u}\, du\, dx \\ &= (-1)^r \langle \gamma(\alpha, x_+), \varphi^{(r)}(x)\rangle \end{aligned} \tag{25}$$

for arbitrary $\varphi \in \mathcal{D}$ and if $-m < \alpha < -m+1$ for $m = 1, 2, \ldots$.

Proof.

$$\begin{aligned}&\mathrm{N-}\lim_{\epsilon\to 0}(-1)^r\int_\epsilon^\infty \varphi^{(r)}(x)\int_\epsilon^x u^{\alpha-1}e^{-u}\, du\, dx = \\ &\quad = (-1)^r \lim_{\epsilon\to 0}\int_\epsilon^\infty \varphi^{(r)}(x)\int_\epsilon^x u^{\alpha-1}\Big[e^{-u} - \sum_{i=0}^{r-1}\frac{(-u)^i}{i!}\Big]\, du\, dx + \\ &\qquad + (-1)^r\, \mathrm{N-}\lim_{\epsilon\to 0}\sum_{i=0}^{r-1}\int_\epsilon^\infty \frac{(-1)^i[x_+^{\alpha+i} - \epsilon^{\alpha+i}]}{i!(\alpha+i)}\varphi^{(r)}(x)\, du\, dx\end{aligned}$$

On using Taylor's theorem, we have

$$\begin{aligned}\mathrm{N}\!-\!\lim_{\epsilon\to 0}\int_\epsilon^\infty & \epsilon^{\alpha+i}\varphi^{(r)}(x)\,dx \\ &= \mathrm{N}\!-\!\lim_{\epsilon\to 0}\epsilon^{\alpha+1}[\psi(\infty)-\psi(\epsilon)] \\ &= -\,\mathrm{N}\!-\!\lim_{\epsilon\to 0}\epsilon^{\alpha+i}\sum_{j=0}^{r-2}\frac{\epsilon^j\psi^{(j)}(0)}{j!} - \lim_{\epsilon\to 0}\frac{\epsilon^{r+\alpha+i-1}\psi^{(r-1)}(\xi x)}{(r-1)!} \\ &= 0\end{aligned}$$

where $\psi(x)$ is the primitive of $\varphi^{(r)}(x)$.
Thus

$$\begin{aligned}\mathrm{N}\!-\!\lim_{\epsilon\to 0}(-1)^r\int_\epsilon^\infty \varphi^{(r)}(x)\int_\epsilon^x u^{\alpha-1}e^{-u}\,du\,dx &= \\ = (-1)^r\int_0^\infty \varphi^{(r)}(x)\int_0^x u^{\alpha-1}\Big[e^{-u}-\sum_{i=0}^{r-1}\frac{(-u)^i}{i!}\Big]\,du\,dx &+ \\ +(-1)^r\sum_{i=0}^{r-1}\frac{(-1)^i}{i!(\alpha+i)}\langle x_+^{\alpha+i},\varphi^{(r)}(x)\rangle & \\ = (-1)^r\langle\gamma(\alpha,x_+),\varphi^{(r)}(x)\rangle. \ \square &\end{aligned}$$

Theorem 2 suggests the following definition.

Definition 3. *The distribution $\gamma^{(r)}(-m,x_+)$ is defined by*

$$\begin{aligned}\langle\gamma^{(r)}(-m,x_+),\varphi(x)\rangle &= (-1)^r\,\mathrm{N}\!-\!\lim_{\epsilon\to 0}\int_\epsilon^\infty \varphi^{(r)}(x)\int_\epsilon^x u^{-m-1}e^{-u}\,du\,dx \\ &= (-1)^r\langle\gamma(-m,x_+),\varphi^{(r)}(x)\rangle \qquad (26)\end{aligned}$$

for arbitrary $\varphi\in\mathcal{D}$ and $r,m=0,1,2,\ldots$.

Theorem 3. *The following equations*

$$\begin{aligned}\langle\gamma^{(r)}(0,x_+),\varphi(x)\rangle &= (-1)^r\int_0^\infty \varphi^{(r)}(x)\int_0^x u^{-1}(e^{-u}-1)\,du\,dx + \\ &+(-1)^r\int_0^\infty \ln x\varphi^{(r)}(x)\,dx \\ &= (-1)^r\int_0^\infty \gamma(0,x_+)\varphi^{(r)}(x)\,dx \qquad (27)\end{aligned}$$

and

$$\begin{aligned}
&\langle \gamma^{(r)}(-m, x_+), \varphi(x) \rangle = \\
&= (-1)^r \int_0^\infty \varphi^{(r)}(x) \int_0^x u^{-m-1} \Big[e^{-u} - \sum_{i=0}^{m} \frac{(-u)^i}{i!} \Big]\, du\, dx + \\
&\qquad -(-1)^r \sum_{i=0}^{m-1} \frac{(-1)^i}{i!(m-i)} \langle F(x_+, -m+i), \varphi^{(r)}(x) \rangle + \\
&\qquad + \sum_{i=0}^{m-1} \frac{(-1)^m}{i!(m-i)(m-i)!} \langle \delta^{(m-i-1)}(x), \varphi^{(r)}(x) \rangle + \\
&\qquad + \frac{(-1)^{m+r}}{m!} \langle \ln x_+, \varphi^{(r)}(x) \rangle
\end{aligned} \tag{28}$$

hold for arbitrary $\varphi \in \mathcal{D}$ *and* $m = 1, 2, \ldots$ *and* $r = 0, 1, 2, \ldots$.

Proof. When $m = 0$ we have on using equation (26)

$$\begin{aligned}
&\langle \gamma^{(r)}(0, x_+), \varphi(x) \rangle = \\
&= (-1)^r \operatorname*{N-lim}_{\epsilon \to 0} \int_\epsilon^\infty \varphi^{(r)}(x) \int_\epsilon^x u^{-1} e^{-u}\, du\, dx \\
&= (-1)^r \lim_{\epsilon \to 0} \int_\epsilon^\infty \varphi^{(r)}(x) \int_\epsilon^x u^{-1}(e^{-u} - 1)\, du\, dx + \\
&\qquad +(-1)^r \operatorname*{N-lim}_{\epsilon \to 0} \int_\epsilon^\infty \varphi^{(r)}(x) \int_\epsilon^x u^{-1}\, du\, dx \\
&= (-1)^r \int_0^\infty \varphi^{(r)}(x) \int_0^x u^{-1}(e^{-u} - 1)\, du\, dx + (-1)^r \int_0^\infty \ln x \varphi^{(r)}(x)\, dx \\
&= (-1)^r \langle \gamma(0, x_+), \varphi^{(r)}(x) \rangle.
\end{aligned}$$

More generally we have

$$\begin{aligned}
&\langle \gamma^{(r)}(-m, x_+), \varphi(x) \rangle = \\
&(-1)^r \lim_{\epsilon \to 0} \int_\epsilon^\infty \varphi^{(r)}(x) \int_\epsilon^x u^{-m-1} \Big[e^{-u} - \sum_{i=0}^{m} \frac{(-u)^i}{i!} \Big]\, du\, dx + \\
&+(-1)^r \operatorname*{N-lim}_{\epsilon \to 0} \sum_{i=0}^{m} \int_\epsilon^\infty \varphi^{(r)}(x) \int_\epsilon^x \frac{(-1)^i u^{-m+i-1}}{i!}\, du\, dx \\
&= (-1)^r \lim_{\epsilon \to 0} \int_\epsilon^\infty \varphi^{(r)}(x) \int_\epsilon^x u^{-m-1} \Big[e^{-u} - \sum_{i=0}^{m} \frac{(-u)^i}{i!} \Big]\, du\, dx + \\
&+(-1)^r \operatorname*{N-lim}_{\epsilon \to 0} \sum_{i=0}^{m} (-1)^i \int_\epsilon^\infty \frac{x^{-m+i} - \epsilon^{-m+i}}{i!(m-i)} \varphi^{(r)}(x)\, dx
\end{aligned} \tag{29}$$

Let ψ be again the primitive of $\varphi^{(r)}(x)$. Now by Taylor's theorem we have

$$\begin{aligned}
&\mathrm{N}-\lim_{\epsilon\to 0}\int_{\epsilon}^{\infty}\epsilon^{-m+i}\varphi^{(r)}(x)\,dx = \mathrm{N}-\lim_{\epsilon\to 0}\epsilon^{-m+i}[\psi(\infty)-\psi(\epsilon)] \\
&= \mathrm{N}-\lim_{\epsilon\to 0}\epsilon^{-m+i}\sum_{j=0}^{m-i}\frac{\epsilon^{j}\psi^{(j)}(0)}{j!} + \lim_{\epsilon\to 0}\frac{\epsilon\psi^{(m-i+1)}(\xi x)}{(m-i+1)!} \\
&= \frac{\psi^{(m-i)}(0)}{(m-i)!} = \frac{\varphi^{(r+m-i-1)}(0)}{(m-i)!} \\
&= \frac{(-1)^{r}}{(m-i)!}\langle\delta^{(m-i-1)}(x),\varphi^{(r)}(x)\rangle. \qquad (30)
\end{aligned}$$

Equation (28) now follows from equations (29) and (30) . □

References

[vdC59] van der Corput, J. G.: Introduction to the Neutrix Calculus. J. Analyse Math., **7**, 291–398 (1959)

[EMOT53] Erdelyi, A., Magnus, W., Oberhettinger F., Tricomi, F.G.: Higher Transcendental Functions. Vol.I. McGraw-Hill, New York London Toronto (1953)

[FLS05] Ferreira, C., Lopez, J.L., Sinusia, E.P.: Incomplete gamma functions for large values of their variables. Adv. Appl. Math., **34**, 467–485 (2005)

[Fis87] Fisher, B.: Neutrices and distributions. Bulgarian Academy of Sciences, 169–175 (1987)

[FK03] Fisher, B., Kuribayashi, Y.: Neutrices and the Gamma function. J. Fac. Ed. Tottori Univ. Mat. Sci., **36(1-2)**, 1–7 (1987)

[FJK03] Fisher, B., Jolevska-Tuneska B., Kiliçman, A.: On defining the incomplete Gamma function. Integral Trans. Spec. Funct., **14(4)**, 293–299 (2003)

[Fis04] Fisher, B.: On defining the incomplete gamma function $\gamma(-m, x_-)$. Integral Trans. Spec. Funct., **15(6)**, 467–476 (2004)

[GHT03] Gautschi, W., Harris F.E., Temme, N.M.: Expansions of the exponential integral in Incomplete gamma functions. Appl. Math.Lett., **16** 1095–1099 (2003)

[GS64] Gel'fand I.M., Shilov, G.E.: Generalized Functions. Vol.I. Academic Press, New York London (1964)

[LCS05] Lin, S.D., Chao Y.S., Srivastava, H.M.: Some expansions of the exponential integral in series of the incomplete gamma function. Appl. Math. Lett., **18**, 513–520 (2005)

Boundary Value Problems

One-dimensional wave propagation in functionally graded cylindrical layered media

Ibrahim Abu-Alshaikh

Department of Mathematics, Fatih University, 34500 Istanbul, Turkey
aibrahim@fatih.edu.tr

In this study, the numerical solution of one-dimensional wave equation in multilayered cylindrical media is investigated. The multilayered medium consists of N different layers of Functionally Graded Material, i.e., it is assumed that the stiffness and the density of each layer are varying continuously in the radial direction but isotropic and homogeneous in the circumferential and axial directions. The inner surface of the layered medium is assumed to be subjected to a uniform dynamic in-plane time-dependent normal stress; whereas, the outer surface of the layered medium is assumed free of surface traction or fixed. The method of characteristics is employed to obtain the numerical solutions of this initial-boundary value problem. The obtained numerical results reveal clearly the scattering effects caused by the reflections and refractions of waves at the boundaries and at the interfaces of the layers and the effects of non-homogeneity in the wave profiles. Furthermore, based on the results obtained from this paper, one may conclude that when the inner surface is stiffer than the outer surface, the stress-wave levels throughout the functionally graded cylindrical layers become less than the load applied at the inner surface.

1 Introduction

Functionally graded materials (FGMs) are a new generation of engineering materials which are continuously changing their thermal and mechanical properties at the macroscopic or continuum scale [YKHS90]. FGMs are increasingly expected to be used in structural applications where high strength-to-weight and stiffness-to-weight ratios are required. Example applications include pressure vessels and pipes in nuclear reactors can be found in the review papers [Tan95, Nod99]. In such applications the metallic-rich region of a functionally graded material is exposed to low temperature with a gradual micro structural transition in the direction of the temperature gradient, while a ceramic-rich region is exposed to high temperature.

Kenan Taş et al. (eds), Mathematical Methods in Engineering, 111–121.

Due to the fact that the material properties of FGMs are functions of one or more space variable, wave propagation problems related to FGMs are generally difficult to analyze without employing some numerical approaches. Numerical solutions of one-dimensional stress wave propagation in an FGM plate subjected to shear or normal tractions are discussed in [HL02, LHL99, HLLO00, CE99]. In these studies, the material properties are assumed to be vary in the thickness direction and the FGM plate is divided into; linearly inhomogeneous elements [HL02] or quadratic inhomogeneous layer elements [LHL99, HLLO00], whereas in [CE99], the material properties of the FGM plate are assumed to be functions with arbitrary power throughout its thickness direction.

In this paper, the method of characteristics is employed to obtain the solutions. This method has been employed effectively in investigating one and two-dimensional transient wave propagation problems in multilayered plane, cylindrical and spherical homogeneous layered media [TCZ91, MT90, Weg93]. In these references, the multilayered medium consists of N layers of isotropic, homogeneous and linearly elastic or viscoelastic material with one or two relaxation times. It is well known that the characteristic manifold for one-dimensional wave propagation through homogeneous plane, spherical or cylindrical media consists of straight lines in the solution region and the canonical equations holding on them are ODEs which can be integrated accurately using a numerical method, such as, implicit trapezoidal rule formula [TCZ91, MT90, Weg93]. However, for functionally graded materials, the characteristic manifold consists of nonlinear curves in the rt-plane and the canonical equations can be integrated approximately along the characteristic curves by employing a small time-step discretization along the t-axis (here t and r denote time and space variable, respectively). This step-by-step numerical technique is capable of describing the sharp variation of disturbance in the neighborhood of the wave front without showing any sign of instability. Hence, and as will be shown in this study, the method of characteristics can be used conveniently for one-dimensional transient wave propagation through functionally graded materials, where every FGM layer is subdivided into a finite number of linearly homogeneous thin elements [AK06].

2 Formulation of the problem

In the present paper, the dynamic response of layered composites consisting of N isotropic, elastic and functionally graded cylindrical layers (non-homogeneous) will be investigated. For this cylindrical composite, it is referred to cylindrical coordinate system where the distance normal to the layering is measured by r. The body is assumed to be subjected to uniform time-dependent dynamic input at its inner boundary($r = R_i$), that is, the dynamic input is normal traction in the in-plane direction of this plane-stain problem. The outer surface ($r = R_o$)of the body is assumed to be free of surface traction, fixed or subjected to a load similar to that applied at the inner boundary $(r = R_i)$. Moreover, the body is assumed to be initially at rest and the layers of the composite body are assumed to be perfectly bonded to each other at the interfaces. Under these boundary, initial and interface conditions, the responses of the bodies are axisymmetrical, that is all the field variables are functions of r and t. Moreover, the only non vanishing displacement component is u_r, that is the displacement component in the direction normal to the layering $(r - direction)$. Thus, the displacement vector for a typical cylindrical layer can be expressed as:

$$u_r = u_r(r,t), \quad u_\theta = u_z = 0. \tag{1}$$

where u_θ and u_z are the displacements in the circumferential and axial directions, respectively. In view of Equation (1), the stress equation of motion, the strain-displacement relations and the stress-strain relations in cylindrical coordinates can be adapted directly from [Eri67] as

$$\begin{aligned} &\tfrac{\partial \tau_{rr}}{\partial r} + \tfrac{\tau_{rr}-\tau_{\theta\theta}}{r} = \rho \tfrac{\partial^2 u_r}{\partial t^2}, \ \varepsilon_{rr} = \tfrac{\partial u_r}{\partial r}, \ \varepsilon_{\theta\theta} = \tfrac{u_r}{r}, \\ &\tau_{rr} = (2\mu + \lambda)\,\varepsilon_{rr} + \lambda\varepsilon_{\theta\theta}, \ \tau_{\theta\theta} = (2\mu + \lambda)\,\varepsilon_{\theta\theta} + \lambda\varepsilon_{rr}, \\ &\tau_{zz} = \lambda\varepsilon_{\theta\theta} + \lambda\varepsilon_{rr}, \end{aligned} \tag{2}$$

where all other stress and strain components are zero and v_r is the particle velocity in r-direction, i.e.,

$$\frac{\partial u_r}{\partial t} - v_r = 0 \tag{3}$$

In Equations (2) the stiffness $c = 2\mu + \lambda$ and the mass density ρ of the medium are assumed to be vary continuously in $r-$direction, but homogeneous and isotropic in θ and $z-$directions, that is

$$\begin{aligned} &c = c_0\,(a + br)^m = (2\mu_0 + \lambda_0)(a + br)^m, \\ &\rho = \rho_0 (a + b\,r)^n, \ \mu = \mu_0\,(a + br)^m, \ \lambda = \lambda_0\,(a + br)^m, \end{aligned} \tag{4}$$

where a, b, m and n are dimensionless constants representing the gradients of the typical FGM layer. $c_0 = 2\mu_0 + \lambda_0$ and ρ_0 are, respectively, the stiffness and mass density at a specified surface of the typical FG layer. Similar forms of Equation (4) with $a = 1$ and $m = n = 1$ were used by Liu *et al.* [LHL99], with $a = 1$ and $m = n = 2$ by Han *et al.* [HL02, HLLO00] and with $a = 1$ by Chiu and Erdogan [CE99], in investigating one-dimensional transient wave propagation in an FGM plate subjected to a uniform pressure wavelet at one of its outer boundaries. This general form of Equation (4) is selected because it is suitable for a multilayered medium that consists of more than one repeated FGM layer [AK06].

In view of Equation(4), the constitutive equations, Equations (2), can be combined in one equivalent equation (one-dimensional wave equation), in terms of one independent variable, i.e., in terms of the displacement component u_r, as

$$c\frac{\partial^2 u_r}{\partial r^2}+\left(\frac{dc}{dr}+\frac{c}{r}\right)\frac{\partial u_r}{\partial r}+\left(\frac{d\lambda}{dr}-\frac{c}{r}\right)\frac{u_r}{r}=\rho\frac{\partial^2 u_r}{\partial t^2}, \tag{5}$$

where $0 \leq t,\ \ R_i \leq r \leq R_o$. In this paper, Equation (5) is required to be solved, satisfying boundary, initial and interface conditions. The boundary condition at the inner surface ($r = R_i$) of the multilayered medium is a time-dependent pressure pulse defined as

$$\tau_{rr}(R_i,t)=c\frac{\partial u_r(R_i,t)}{\partial r}+\lambda\frac{u_r(R_i,t)}{r}=-p_0 f(t), \tag{6}$$

where p_0 is the intensity of the applied load and $f(t)$is a prescribed function of t. The outer surface ($r = R_o$) is assumed to be either free of surface traction, fixed or it can be assumed to be subjected to the same load applied at the inner surface, Equation (6). Hence, the free or fixed boundary conditions can be written, respectively, as

$$\tau_{rr}(R_o,t)=0 \ \ or \ \ u_r(R_o,t)=0. \tag{7}$$

In view of Equations (3-4), the governing field equations, Equations (2), are to be applied to each layer and the solutions will be required to satisfy the interface continuity conditions of the displacement (u_r) and normal stress (τ_{rr}), the boundary conditions at inner and outer surfaces, Equations (6-7), and quiescent initial condition.

3 Solution of the problem

The solution is obtained by employing the method of characteristics. This numerical technique involves first rewriting the constitutive hyperbolic PDE, Equation (5), in view of Equations (2-4) as a system of first order PDEs in matrix form as:

$$\underline{A}\,\underline{U}_{,t}+\underline{B}\,\underline{U}_{,r}+\underline{F}=\underline{0} \tag{8}$$

where$\underline{A}$ being a (5x5) identity matrix and $\underline{B}$ is (5x5) square matrix with the following nonzero elements:

$$B_{15}=-c,\quad B_{25}=-\lambda,\quad B_{35}=-1,\quad B_{53}=\frac{-c}{\rho},\quad B_{54}=\frac{-\lambda}{\rho\, r}, \tag{9}$$

where $c = 2\mu + \lambda$. Furthermore, $\underline{F}$ is a five-dimensional column vector with the elements

$$\begin{array}{l} F_1=-(\lambda/r)v_r,\ F_2=-(c/r)v_r,\ \ F_3=0,\ \ F_4=-v_r, \\ F_5=-\frac{1}{\rho}\left(\frac{dc}{dr}\right)\varepsilon_{rr}-\frac{1}{\rho r}\frac{d\lambda}{dr}u_r+\frac{\lambda}{r^2\rho}u_r-\frac{1}{\rho r}\tau_{rr}+\frac{1}{\rho r}\tau_{\theta\theta}, \end{array} \tag{10}$$

and $\underline{U}$ is a five-dimensional column vector containing the unknown field variables:

$$\underline{U} = \left[\tau_{rr} \quad \tau_{\theta\theta} \quad \varepsilon_{rr} \quad u_r \quad v_r\right]^T \quad , \tag{11}$$

where the letter T over a vector (or a matrix) quantity denotes its transpose. In Equation (8), comma denotes partial differentiation with respect to the corresponding field variables presented in the column vector $\underline{U}$, i.e.,

$$\underline{U}_{,t} = \frac{\partial \underline{U}}{\partial t}, \quad \underline{U}_{,r} = \frac{\partial \underline{U}}{\partial r}. \tag{12}$$

Before establishing the canonical form of the governing equations, we will establish the characteristic curves along which these equations are valid. These curves are governed by the characteristic equation, which can be written as [CH66]

$$\det(\underline{B} - V\underline{A}) = 0, \tag{13}$$

where $V = (dr/dt)$defines the characteristic curves on the $(r-t)$plane. Now, by substituting the identity matrix $\underline{A}$ and the matrix $\underline{B}$ given in Equation (9) into (13), the characteristic equation can be expressed as

$$V^3\left(-V^2 + \frac{c}{\rho}\right) = 0, \tag{14}$$

and its corresponding eigenvalues can be found as:

$$V_1 = c_p,\ V_2 = -c_p,\ V_3 = 0,\ V_4 = 0,\ V_5 = 0, \tag{15}$$

where, in terms of Equation (4),

$$c_p = \sqrt{\frac{c}{\rho}} = \sqrt{\frac{\lambda + 2\mu}{\rho}} = \sqrt{\frac{(2\mu_0 + \lambda_0)(a + br)^m}{\rho_0(a + br)^n}}, \tag{16}$$

where c_p is the dilatational (pressure or longitudinal) wave velocity. The waves generated by c_p propagate in the direction perpendicular to the layering. We should observe further, that the dilatational wave is the only wave generated in the cylindrical domain, because the inner surface of this domain is subjected only to a uniform radial pressure.

The characteristic curves are defined as [CH66] by:

$$\frac{dr}{dt} = V_1 = c_p \ \ along \ \ C^{(1)}, \frac{dr}{dt} = V_2 = -c_p \ \ along \ \ C^{(2)}, \tag{17}$$
$$\frac{dr}{dt} = V_j = 0; \quad j = 2, 4, 5 \ \ along \ \ C^{(3)}, C^{(4)}, C^{(5)}.$$

Integration of Equation (17), gives the families of the characteristic curves $C^{(i)}(i = 1,\ 2,\ 3,\ 4,\ 5)$ as

$$C^{(1)} : t = \int \frac{dr}{\sqrt{\frac{(2\mu_0+\lambda_0)(a+br)^m}{\rho_0(a+br)^n}}}, \quad C^{(2)} : t = \int \frac{dr}{-\sqrt{\frac{(2\mu_0+\lambda_0)(a+br)^m}{\rho_0(a+br)^n}}} \tag{18}$$
$$C^{(3)} : r = constant, C^{(4)} : r = constant, C^{(5)} : r = constant.$$

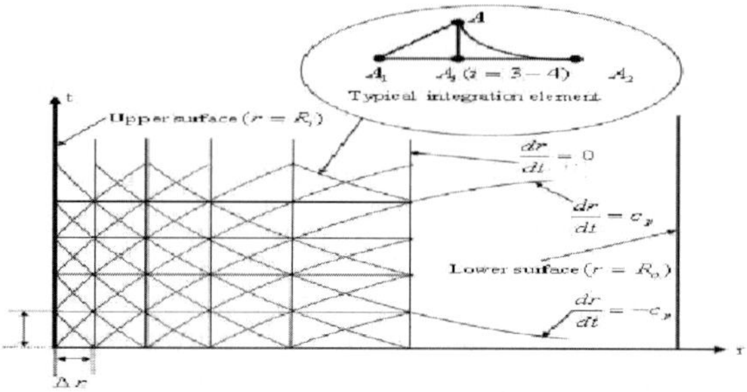

Fig. 1. Network of characteristic curves on the $(r-t)$ plane.

Here, we note that $C^{(i)}\quad (i=1,2)$ describe two families of curves with slopes c_p and $-c_p$ respectively, whereas $C^{(3)}$, $C^{(4)}$ and $C^{(5)}$ describe one family of straight lines which is parallel to the t-axis in the $(r-t)$ plane, Figure 1.

The next step in establishing the canonical form of the governing PDEs is finding the left-hand eigenvectors $\underline{l}_i\ \ (i=1,\ 2,\ 3,\ 4,5)$ defined as

$$(\underline{B}^T - V_i\underline{A}^T)\underline{l}_i^T = \underline{0}. \tag{19}$$

Now, in view of the identity matrix $\underline{A}$, the matrix $\underline{B}$ given in Equation (9), and Equations (15-16), Equation (19) can be used to find the linearly independent set of the eigenvectors $\underline{l}_i\ \ (i=1,\ 2,\ 3,\ 4,5)$. To obtain the canonical equations, we multiply Equation (8) by l_i^T and use (19); noting that, $V_i = \frac{dr}{dt}$ and the total derivative of $\underline{U}$ with respect to t is $\frac{d\underset{\sim}{U}}{dt} = \underset{\sim}{U}_{,t} + \frac{dr}{dt}\underset{\sim}{U}_{,r}$, thus the canonical equations can be written, respectively, along the curves (18) as:

$$\underset{\sim}{l_i^T}\,\underset{\sim}{A}\,\frac{d\underset{\sim}{U}}{dt} + \underset{\sim}{l_i^T}\,\underset{\sim}{F} = 0 \tag{20}$$

Now, in view of the identity matrix $\underline{A}$ and the column vector $\underline{F}$ given in Equation (10), and due to the set of the eigenvectors $\underline{l}_i\ \ (i=1,\ 2,\ 3,\ 4,5)$ that can be found from Equation (19), Equation (20) leads to five canonical equations which can be written explicitly as:

$$\begin{aligned}
&-c_p\frac{d\varepsilon_{rr}}{dt} - \frac{\lambda}{\rho r c_p}\frac{du_r}{dt} + \frac{dv_r}{dt} + \frac{\lambda}{\rho\, r c_p}v_r - \frac{1}{\rho}\frac{dc}{dr}\varepsilon_{rr} + u_r\left(\frac{\lambda}{r^2\rho} - \frac{1}{\rho r}\frac{d\lambda}{dr}\right) - \\
&\frac{1}{\rho r}\tau_{rr} + \frac{1}{\rho r}\tau_{\theta\theta} = 0,\ \ c_p\frac{d\varepsilon_{rr}}{dt} + \frac{\lambda}{\rho r c_p}\frac{du_r}{dr} + \frac{dv_r}{dr} + \frac{-\lambda}{\rho r c_p}v_r - \frac{1}{\rho}\frac{dc}{dr}\varepsilon_{rr} + \\
&u_r\left(\frac{\lambda}{r^2\rho} - \frac{1}{\rho r}\frac{d\lambda}{dr}\right) - \frac{1}{\rho r}\tau_{rr} + \frac{1}{\rho r}\tau_{\theta\theta} = 0,\ \ \frac{d\tau_{rr}}{dt} - c\frac{d\varepsilon_{rr}}{dt} - \frac{\lambda}{r}v_r = 0, \\
&\frac{d\tau_{rr}}{dt} - c\frac{d\varepsilon_{rr}}{dt} - \frac{\lambda}{r}v_r = 0,\ \ \frac{d\tau_{rr}}{dt} - \frac{c}{\lambda}\frac{d\tau_{\theta\theta}}{dt} - \frac{\lambda}{r}v_r + \frac{c^2}{r\lambda}v_r = 0, \\
&\frac{du_r}{dr} - v_r = 0.
\end{aligned} \tag{21}$$

These equations are valid along the characteristic curves defined by Equation (17) as $V_i = \frac{dr}{dt}\ \ (i=1,2,3,4,5)$, respectively. Thus, the system of governing PDEs, (8), is transformed into a set of ODEs, (21), which are valid along the characteristic curves given in Equation (17). A through description of the method of characteristics is given by Courant and Hilbert [CH66].

Our aim now, is to solve the canonical equations, (21), by integrating them numerically along the characteristic curves. For this purpose, the trapezoidal technique will be used employing the typical integration element shown inside Figure 1. In this typical element, A and $A_i$are consecutive points along the characteristic curves defined, respectively, at current and previous time steps as shown in Figure 1. In other words, using this typical integration element the field variables at a specific point along any line parallel to the r- axis in the solution region can be found in terms of the known field variables defined on the previous time step. For this purpose, we refer to the network of the characteristic curves shown is Figure 1. To compute the components of the unknown vector $\{U_j(j = 1-5)\}$, given in Equation (11), at every intersection point between the characteristic curves on the $r-t$ plane: we start our solution on the network from the r- axis, where the values of all field variables are zero due to zero initial conditions, and advance into the solution region by computing U_j at the intersection points of the network between the inner and the outer boundary along the lines$t = \Delta t$, $t = 2\Delta t$, $t = 3\Delta t$, ...,$t = J_{\max}\Delta t$ etc, respectively. In this computational process, the inner layer is considered to be layer 1, while the outer layer is considered to be the last layer (N th layer). To explain this numerical procedure in more details, we refer to four different locations of the typical integration element:

When the typical integration element is located at the inner boundary then the first integrated canonical equation, which is valid along the curve $A - A_1$ is replaced by the boundary condition applied at that boundary. Second, if the integration element is an interior element, then the procedure involves the determination of the values of the unknown vector at a point A in terms of their values at A_1, A_2 and A_i, $(i = 3-4)$using the five integrated canonical equations. Third, if a point A of an integration element is located at an interface between two different FGM layers and the material properties change sharply at that interface, then the first two equations of the integrated form of the canonical equations will be replaced by the interface continuity conditions. In this case the number of field variables becomes double at that point. Furthermore, we should note that for N layers $(L = 1, 2, ..., N)$ we have $(N - 1)$interfaces. Finally, the second equation of the integrated canonical equations which is valid along the curve $A - A_2$ is replaced by the boundary condition applied at the outer boundary, if the typical integration element lies at that boundary.

The numerical procedure discussed above is repeated as we proceed along the $t-$axis, for example along the line $t = 2\Delta t$ instead of using the initial conditions along the line $t = 0$, we use the field variable which is already evaluated in the previous step along the line $t = \Delta t$. This process is repeated until getting results for a sufficient value of t, for example $t = J_{\max}\Delta t$ where $J_{\max}$ is the maximum number of intervals considered in the t-direction. The code of the numerical procedure is written in Fortran 90, and the details of the process of integration of the canonical equations can be given in a more extended version of this paper.

4 Numerical results and discussion

In this section, our aim is to present only two numerical examples related to the problem formulated in the previous Sections. In these examples, the numerical computations have been carried out and the results are displayed in terms of non-dimensional quantities. These dimensionless quantities are taken in terms of the thickness of the first layer ($h_{(1)}$), density of the top surface of first FG layer ($\rho_0^{(1)}$) and dilatational wave velocity ($c_p^{(1)}$) at the top of the first FG layer, i.e., these values are considered to be unity where the other material properties are computed in terms of these of the three quantities. Furthermore, we should also note that the non-dimensional quantities are prescribed by putting bars over them. We further refer to the quantities pertaining to layers 1 and 2 by putting subscripts or superscripts 1 and 2 in parenthesis, respectively. For example, $\bar{\rho}_0^{(1)}$ and $\bar{\rho}_0^{(2)}$ denote the dimensionless mass densities, at the top surface of layers 1 and 2, respectively, and $h_{(1)}$ and $h_{(2)}$ represent the thicknesses of layers 1 and 2, respectively.

In the first example, we consider the solutions of Equation (5) for which λ and c are assumed to be constants through the thickness of the cylindrical layer, that is, in Equation (4) $a = 1$, $m = 1$, $n = 1$ and b is assumed to be zero. In this example, the following values for the non-dimensional material and geometrical properties for layers 1 and 2 are taken as [TCZ91]:

$$\begin{aligned} &\bar{\mu}_0^{(1)} = 0.254, \quad \bar{\mu}_0^{(2)} = 0.964, \quad \bar{\lambda}_0^{(1)} = 0.492, \quad \bar{\lambda}_0^{(2)} = 0.972 \\ &\bar{\rho}_0^{(1)} = 1, \quad \bar{\rho}_0^{(2)} = 2.9, \quad \bar{h}_1 = 1, \quad \bar{R}_i = 1, \quad \bar{h}_2 = 1, \quad \bar{R}_o = 6 \end{aligned} \tag{22}$$

In the second example we consider similar properties of (22), but λ and μ are assumed to be linear functions in $r-$direction and uniform in the other two directions. That is, in Equation (4) the dimensionless quantities$a$,b,m and n are taken as follows:

$$a = \frac{5}{6}, \quad b = \frac{1}{6}, \quad m = 1, \quad n = 1, \tag{23}$$

In this case we noted from the above non-dimensional quantities that the cylindrical layered media is assumed to be made of functionally graded material whereas the properties are assumed to be varying linearly through thickness direction.

In examples 1 and 2, the inner surface ($\bar{r} = 1$) is assumed to be subjected to uniform pressure with an initial ramp, that is

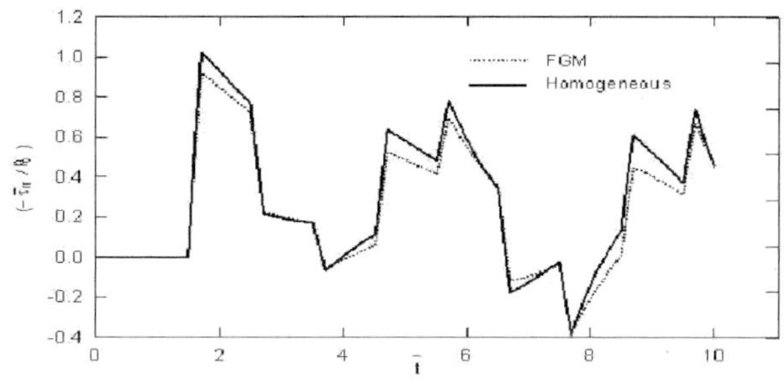

Fig. 2. Time variation of the normal stress $\frac{\tau_{rr}}{p_0}$ at $\bar{r} = 2.5$ for three pairs of alternating layered cylindrical medium.

$$f(t) = \left\{ \begin{array}{ll} 5t & \text{if } t \leq 0.2 \\ 1 & \text{if } t > 0.2 \end{array} \right\} \tag{24}$$

On the other hand, the outer surface ($\bar{r} = 7$) is assumed to be fixed, that is $u_r = 0$. The numerical results presented for these examples have been obtained for three pairs of alternating layers. The innermost layer is taken as layer 1, whereas the outermost layer is taken as layer 2, with the layer sequence, starting from the innermost layer, as 1/2/1/2/1/2. In Figures (2-3) the variations of the dimensionless normal stresses τ_{rr}/p_0 and $\tau_{\theta\theta}/p_0$ with non-dimensional time at $\bar{r} = 2.5$ are shown for the layered cylindrical composite. In these figures, solid curves are given for the cases where the non homogeneity effects are neglected, that is, for $a = 1$ and $b = 0$ .Our solutions presented in Figures (2-3) at $\bar{r} = 2.5$ for the homogeneous case fit exactly those solutions presented in [TCZ91]. These results give us more confidence of the method applied in this paper. On the other hand, solutions presented by dashed curves are devoted for FGM composites with properties given in Equations (23). From Figures (2-3) one can see clearly that the stress level for the homogeneous material are greater than those correspond to FGM composite; this is due to the fact that the outer boundary of the FGM composite is stiffer than the inner boundary. The curves of Figures (2-3), further show the effects of reflections and refractions from the inner and outer boundaries and from the interfaces. These effects can be noticed from the sudden changes of stress levels. We note that large changes in stress levels are due to the reflections and refractions from the outer and inner boundaries, whereas small changes in stress levels are due to reflections and refractions from the interfaces between layers.

Based on the results obtained from this study, one may conclude that; at a specified location the amplitudes of the resultant stress-waves become less than that applied at the inner boundary when the inner surface is stiffer than the outer surface and they become greater when the outer surface of an FGM layer is stiffer than the inner surface. Finally we can conclude that the method

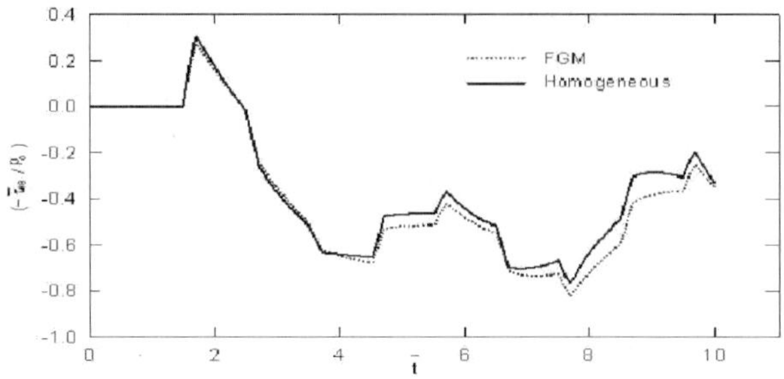

Fig. 3. Time variation of the normal stress $\frac{\tau_{\theta\theta}}{p_0}$ at $\bar{r} = 2.5$ for three pairs of alternating layered cylindrical medium.

of characteristics can be combined with Fourier transform to investigate two-dimensional transient response in multilayered FG media.

References

[YKHS90] Noda, N., Tsuji, T.: Steady thermal stresses in a plate of a functionally gradient material Yamanouchi. In: M., Koizumi, M., Hirai, M., Shiota, I. (ed.) Proc. of the 1st Int. Symp. on Functionally Graded Materials(Sendai, Japan) (1990)

[Tan95] Tanigawa, Y.: Some basic thermoelastic problems for non-homogeneous structural materials. Appl. Mech. Rev., **48**, 287–300 (1995)

[Nod99] Noda, N.: Thermal stresses in functionally graded material. J. Ther. Stre., **22**, 477–512 (1999)

[HL02] Han, X., Liu, G.R.: Effects of SH waves in a functionally graded plate. Mech. Res. Comm., **29**, 327–338 (2002)

[LHL99] Liu, G.R., Han, X., Lam, K.Y.: Stress Waves in functionally gradient materials and its use for material characterization. Compos. Part B: Eng., **30** 383–394 (1999)

[HLLO00] Han, X., Liu, G.R., Lam, K.Y., Ohyoshi, T.: A quadratic layer element for analyzing stress waves in FGMs and its application in material characterization, J. Sound and Vib., **239**, 307–321 (2000)

[CE99] Chiu, T.C., Erdogan, F.: One-dimensional wave propagation in a functionally graded elastic medium. J. Sound Vib., **222**, 453–487 (1999)

[TCZ91] Turhan, D., Celep, Z., Zain-eddin, I.K.: Transient wave propagation in layered media conducting heat. J. Sound Vib., **144**, 247–261 (1991)

[MT90] Mengi, Y., Tanrikulu, A.K.: A numerical technique for two- dimensional transient wave propagation analyses. Commun. Appl. Num. Meth., **6**, 623–632 (1990)

[Weg93] Wegner, J.L.: Propagation of waves from a spherical cavity in an unbounded linear viscolastic solid. Int. J. Eng. Sci., **31**, 493–508 (1993)

[AK06] Abu-Alshaikh, I., Köklüce, B.: One dimensional transient dynamic in functionally graded media. J. Eng. Math., **54**, 17–30 (2006)

[Eri67] Eringen, A.C.: Mechanics of Continua. John Wiley & Sons, New York (1967)

[CH66] Courant, R., Hilbert, D.: Methods of Mathematical Physics. Vol. II, Inter Science Publishers, New York (1966)

Piecewise constant control of boundary value problem for linear impulsive differential systems

J. O. Alzabut

Department of Mathematics and Computer Science, Faculty of Arts and Sciences, Çankaya University, 06530 Ankara, Turkey jehad@cankaya.edu.tr

A piecewise constant control that solves the boundary value problem for linear impulsive differential systems is considered. We establish a necessary and sufficient conditions for the existence of such control. Moreover, a result that explicitly characterizes the solving control is presented.

1 Introduction

Many dynamical systems are characterized by the fact that at certain moments in their evolution they undergo rapid changes. Most notably this takes place in certain biological systems, population systems and even in control systems such as pulse frequency modulated control system. In modelling such systems, it is more tractable and convenient to neglect the duration of these rapid changes and assume the state changes by jumps. The mathematical models of such processes are described by the so called impulsive differential equations, i.e., an ordinary differential equations together with relations defining the jump conditions [BS89, SP95]. More specifically, the model is given by the system

$$\begin{cases} x'(t) = f(t,x), & h(t,x) \neq 0 \\ \Delta x(\theta_i) = I(t,x), & h(t,x) = 0 \end{cases} \tag{1}$$

where $t \in \mathbb{R}$ is the time variable, $x \in \mathbb{R}^n$ is the state vector, $f : \mathbb{R} \times \mathbb{R}^n \to \mathbb{R}^n$ and $I : \mathbb{R} \times \mathbb{R}^n \to \mathbb{R}^n$ defines the jump conditions. A point (t, x) is the extended phase space follows the solution trajectory of the differential system and as soon as it hits the surface σ of equation $h(t,x) = 0$ the system performs an instantaneous jump of size $I(t,x)$.

In this paper, we deal solely with deterministic, linear impulsive differential systems whose instants of impulse effect are fixed. Namely, system of the form

$$\begin{cases} x'(t) = A(t)x(t), & t \neq \theta_i, \\ \Delta x(\theta_i) = C_i x(\theta_i), \end{cases} \tag{2}$$

Kenan Taş et al. (eds), Mathematical Methods in Engineering, 123–129.

where A, C_i are $n \times n$ matrices and $\{\theta_i\}$ is an increasing sequence such that $\lim_{i\to\infty} \theta_i = \infty$.

The wide application prospects of impulsive differential systems in real life problems have attracted the attention of many authors over the last two decades. A large number of papers and monographs investigating such equations have been documented by addressing various methods and using different techniques, see [SP95, HLL89, Nie02, BHHL89, BS95, Akh03] and the list of references therein.

To the best of our knowledge, the problem of the control of impulsive differential systems was first considered at the beginning of nineties, we name here the papers [LMS93, BS94, APT95]. The last few years has witnessed the appearance of several results concerning the controllability of impulsive differential systems mostly in their linear, quasilinear and rarely in their nonlinear forms, see for instance [AS99, GQY02, AZS02, AZ04]. In this paper and unlike most existing results, a piecewise constant control that solves the boundary value problem for linear impulsive differential systems is considered. We establish a necessary and sufficient conditions for the existence of such control. Moreover, a result that explicitly characterizes the solving control is presented.

2 Preliminaries

Let α, β be fixed real numbers such that $\alpha < \beta$. We denote the set of piecewise constant functions $\varphi : [\alpha, \beta] \to \mathbb{R}^n$ by $PWC^n[\alpha, \beta]$ and the set of sequences $\xi_i \in \mathbb{R}^n, i = 1, 2, \ldots, p$ by $S_i[1, p]$ where p is fixed positive integer number. Define the space $\coprod_i^n[\alpha, \beta] = PWC^n[\alpha, \beta] \times S_i[1, p]$ whose elements are represented by the pair $\{\varphi, \xi\}$. Let

$$< \{\varphi, \xi\}, \{w, v\} >= \sum_{s=1}^{r-1} \varphi_s^T \int_{t_s}^{t_s+1} w dt + \sum_{i=1}^{p} \xi_i^T v_i \tag{3}$$

be an inner product defined in $\coprod_i^n[\alpha, \beta]$.

Consider the control system

$$\begin{cases} x'(t) = A(t)x(t) + B(t)u(t) + f(t), & t \neq \theta_i \\ \Delta x(\theta_i) = C_i x(\theta_i) + D_i v_i + g_i, \end{cases} \tag{4}$$

with the boundary conditions

$$x(\alpha) = a, \quad x(\beta) = b, \quad a, b \in \mathbb{R}^n, \tag{5}$$

where the following conditions are assumed:

(i) $\det(I + C_i) \neq 0,\ i = 1, \ldots, p$;
(ii) A and B are respectively $n \times n$ and $n \times m$ continuous matrices;
(iii) C_i and D_i are respectively $n \times n$ and $n \times m$ constant matrices, $i = 1, \ldots, p$;

(iv)$\{u, v_i\} \in \coprod_i^m[\alpha,\beta]$ and $\{f, g_i\} \in \coprod_i^n[\alpha,\beta]$.

By $\Delta x(t)$ we mean the difference $x(t^+) - x(t)$. The solutions of equation (4) are functions $x : [\alpha,\beta] :\longrightarrow \mathbb{R}^n$ which are piecewise absolutely continuous, continuous on the left with jumps at the points θ_i, $i = 1,\dots,p$.

We say that the control problem ©$_1$ is solvable if for each $\{f, g_i\} \in \coprod_i^n[\alpha,\beta]$ and every $a, b \in \mathbb{R}^n$ there exists a control $\{u, v\} \in \coprod_i^m[\alpha,\beta]$ such that the problem (4), (5) has a solution. In case that $x(\alpha) = 0, x(\beta) = 0$, the problem is called control problem ©$_2$.

In the sequel, we shall denote by $\{\tilde{\varphi}, \tilde{\varphi}\}$ an element $\{\tilde{\varphi}(t), \tilde{\varphi}(\theta_i)\}$ for an arbitrary function $\tilde{\varphi}(t) : [\alpha,\beta] \to \mathbb{R}^n$.

Lemma 1. *©$_1$ is solvable if and only if ©$_2$ is solvable.*

Proof. Let ©$_1$ is solvable. Since ©$_2$ is a particular case of ©$_1$, ©$_2$ is solvable. Suppose that ©$_2$ is solvable. Let $\psi(t)$ be the Lagrange polynomial such that $\psi(\alpha) = a, \psi(\beta) = b$. Replacing $x(t)$ by $z(t) + \psi(t)$ in (4), (5), we see that $z(t)$ satisfies

$$\begin{cases} z'(t) = A(t)z(t) + B(t)u(t) + [f(t) - \psi'(t) + A(t)\psi(t)], & t \neq \theta_i \\ \Delta z(\theta_i) = C_i z(\theta_i) + D_i v_i + [g_i + C_i\psi(\theta_i)], \\ z(\alpha) = z(\beta) = 0. \end{cases} \tag{6}$$

This problem is solvable due to the assumption. □

It is known [SP95] that the adjoint system of (2) is

$$\begin{cases} y'(t) = -A^T(t)y(t), \quad t \neq \theta_i \\ \Delta y(\theta_i) = -(I + C_i^T)^{-1} C_i^T y(\theta_i), \end{cases} \tag{7}$$

The following lemma has been proved in [SP95] for $\{F, G\} \in \prod_i^n[\alpha,\beta] = L_2^n[\alpha,\beta] \times S_i[1,p]$ where $L_2^n[\alpha,\beta]$ denotes the set of square integrable functions $\phi : [\alpha,\beta] \to \mathbb{R}^n$. It is clear that, it is also valid for $\{F, G\} \in \coprod_i^n[\alpha,\beta]$. □

Lemma 2. *Let $\{F, G\} \in \coprod_i^n[\alpha,\beta]$. Then the boundary value problem*

$$\begin{cases} x'(t) = A(t)x(t) + F(t), \quad t \neq \theta_i \\ \Delta x(\theta_i) = C_i x(\theta_i) + G_i, \\ x(\alpha) = x(\beta) = 0, \end{cases} \tag{8}$$

is solvable if and only if for any solution y of (7), the following relation holds:

$$< \{F, G\}, \{y, y\} >= 0, \tag{9}$$

where $< , >$ is defined by (3).

3 The main results

Theorem 1. *The control problem* $©_1$ *is solvable if and only if the trivial solution of equation (7) is the only solution satisfying the condition*

$$< \{Bu, Dv\}, \{y, y\} >= 0, \tag{10}$$

for all $\{u, v\} \in \coprod_i^m [\alpha, \beta]$.

Proof. NECESSITY. Assume the contrary. The control problem $©_1$ is solvable and there exists a nontrivial solution of equation (7) that satisfies (10). It is easy to show that there exists $\{f, g\} \in \coprod_i^n [\alpha, \beta]$ for which the relation $< \{f, g\}, \{y, y\} >\neq 0$ is true. Let us fix this element. Then by adding the last equality to relation (10), we get $< \{Bu + f, Dv + g\}, \{y, y\} >\neq 0$ which contradicts the existence of a solution of the boundary value problem.

SUFFICIENCY. Let $Y(t) = (y_1, y_2, \ldots, y_n)$ be a fundamental matrix of solutions of equation (7) and $c \in \mathbb{R}^n$. According to the condition of the theorem, the infinite system of equations

$$< \{Bu, Dv\}, \{Yc, Yc\} >= 0, \tag{11}$$

admits only the trivial solution $c = 0$ for all $\{u, v\} \in \coprod_i^m [\alpha, \beta]$. Let us show that there exists n elements $\{u^k, v^k\} \in \coprod_i^m [\alpha, \beta]$ for which the matrix

$$N = \Big(\sum_{s=0}^{r-1} (u_s^k)^T \int_{t_s}^{t_{s+1}} (B, y_j) \mathrm{d}t + \sum_{i=1}^{p} (v_i^k)^T (D_i, y_j)\Big)_{jk}, \; j, k = 1, \ldots, n$$

is nonsingular. Assume the contrary. Without loss of generality, we can assume that the last row of the matrix N linearly depends on the other rows. Denote by c^* a nontrivial solution of the system

$$< \{Bu^k, Dv^k\}, \{Yc, Yc\} >= 0, \quad k = 1, \ldots, n-1. \tag{12}$$

Since, for any $\{u, v\} \in \coprod_i^m [\alpha, \beta]$, there exist constants μ_k, $k = 1, \ldots, n-1$ for which the equality

$$\Big(\sum_{s=0}^{r-1} (u_s)^T \int_{t_s}^{t_{s+1}} (B, y_j) \mathrm{d}t + \sum_{i=1}^{p} (v_i)^T (D_i, y_j)\Big) =$$
$$\sum_{k=1}^{n-1} \mu_k \Big(\sum_{s=0}^{m-1} (u_s^k)^T \int_{t_s}^{t_{s+1}} (B, y_j) \mathrm{d}t + \sum_{i=1}^{p} (v_i^k)^T (D_i, y_j)\Big),$$

$j = 1, \ldots, n$ is true, it follows from (12) that equality (11) holds for the nonzero vector c^*. Hence, the matrix N must be nonsingular. Let us consider the boundary value problem

$$\begin{cases} x'(t) = A(t)x(t) + f(t) - B(t)\sum_{k=1}^{n} c_k u^k(t), \quad t \neq \theta_i \\ \Delta x(\theta_i) = C_i x(\theta_i) + g_i - D_i \sum_{k=1}^{n} c_k v_i^k, \\ x(\alpha) = x(\beta) = 0, \end{cases} \tag{13}$$

where $\{u, v\} \in \coprod_i^m [\alpha, \beta]$. By virtue of Lemma 2, for the solvability of this problem, it suffices that there exist solutions of the system

$$\sum_{k=1}^{n} c_k \Big(\sum_{s=0}^{r-1} (u_s^k)^T \int_{t_s}^{t_{s+1}} (B, y_j) \mathrm{d}t + \sum_{i=1}^{p} (v_i^k)^T (D_i, y_j) \Big) =< \{f, g\}, \{y_j, y_j\} >,$$

$j = 1, \ldots, n$, which is true due to the non singularity of the matrix N. □

Corollary 1. *The control problem ©$_1$ is solvable if and only if for any $t \in [\alpha, \beta]$ and $i = 1, \ldots, p$, the following relations hold:*

$$\det(B^T(t)Y(t)) \neq 0, \quad \det(D_i^T(t)Y(\theta_i)) \neq 0.$$

Let Γ be the Gram matrix defined as

$$\Gamma = \int_{\alpha}^{\beta} Y_i^T B B^T Y_i \mathrm{d}t + \sum_{i=1}^{p} Y_i^T D_i D_i^T Y_i.$$

Theorem 2. *The control problem ©$_1$ is solvable if and only if the Gram matrix Γ is nonsingular.*

Proof. In view of Theorem 1, system (10) has only the trivial solution $c = 0$. Setting $\{u, v\} = \{B^T(t)Yc, D_i^T Yc\}$ in this equation we find that the system $c^T\Gamma = 0$ has also only a trivial solution. If this equation $c^T\Gamma = 0$ has only a trivial solution then system (10) has only the solution $c = 0$. The proof is complete. □

Let

$$K = \Gamma^{-1}\Big[Y^T(\beta)b - Y^T(\alpha)a - \int_{\alpha}^{\beta} Y^T(t)f(t)\mathrm{d}t - \sum_{i=1}^{p} Y^T(\theta_i)g_i\Big].$$

Theorem 3. *If the control problem ©$_1$ is solvable then the control which is given by the formulas*

$$U(t) = B^T(t)Y(t)K, \quad V_i = D_i^T(t)Y^T(\theta_i)K, \tag{14}$$

solves ©$_1$.

Proof. Changing the variables $x = z + \psi(t)$ in the boundary value problem where $\psi(t)$ is the Lagrange polynomial satisfying the boundary conditions and $\psi(t_i) = 0, i = 1, \ldots, p$, we obtain

$$\begin{cases} z'(t) = A(t)z(t) + B(t)u(t) + f(t) + [\psi'(t) - A(t)\psi(t)], & t \neq \theta_i \\ \Delta z(\theta_i) = C_i z(\theta_i) + D_i v_i + g_i, \\ z(\alpha) = z(\beta) = 0. \end{cases} \tag{15}$$

By virtue of Lemma 2, for the solvability of this problem, it is necessary and sufficient that for all $\{u, v\} \in \coprod_i^m [\alpha, \beta]$ the conditions

$$\int_\alpha^\beta Y^T(t)[B(t)u(t) + f(t)]\mathrm{d}t + \sum_{i=1}^p Y^T(\theta_i)[D_i v_i + g_i] = \int_\alpha^\beta Y^T(t)[\psi'(t) - A(t)\psi(t)]\mathrm{d}t, \tag{16}$$

be satisfied. Integrating by parts, we conclude that the relation

$$\int_\alpha^\beta Y^T(t)[B(t)u(t)+f(t)]\mathrm{d}t+\sum_{i=1}^p Y^T(\theta_i)[D_i v_i+g_i] = Y^T(\beta)b-Y^T(\alpha)a, \tag{17}$$

is the necessary and sufficient condition for the solvability of the control problem ©$_1$. By substituting the expressions

$$U = B^T(t)Y(t)c, \quad V_i = D_i^T Y(\theta_i)c, \tag{18}$$

into (17), we obtain a system of linear equations with respect to the vector c

$$\Big[\int_\alpha^\beta Y^T(t)B(t)B^T(t)Y(t)\mathrm{d}t + \sum_{i=1}^p Y^T(\theta_i)D_i D_i^T Y(\theta_i)\Big]c = Y^T(\beta)b - Y^T(\alpha)a - \int_\alpha^\beta Y^T(t)f(t)\mathrm{d}t - \sum_{i=1}^p Y^T(\theta_i)\beta_i. \tag{19}$$

Substituting the solutions of system (19) into (18), we get the expressions (14).

Acknowledgments

The author expresses his sincere thank and gratefulness to Prof. Dr. M. U. Akhmet for his precious guidance, motivation and encouragement.

References

[BS89] Bainov D.D., Simeonov P.S.: Systems with Impulse Effect: Stability, Theory and Applications. Wiley, New York (1989)

[SP95] Samoilenko A.M., Perestyuk N.A.: Impulsive differential equations. World Scientific Publishing Co., Singapore New Jersey London Hong Kong (1995)

[HLL89] Hu S., Lakshmikantham V., Leela S.: Impulsive Differential Systems and the Pulse Phenomena. J. Math. Anal. App., **137**, 605–612 (1989)

[Nie02] Nieto J.J.: Periodic Boundary Value Problems for First Order Impulsive Ordinary Differential Equations. Nonlinear Anal., **51**, 1223–1232 (2002)

[BHHL89] Bainov D., Hristova S., Hu S., Lakshmikantham V.: Periodic Boundary Value Problems for Systems of First order Impulsive Differential Equations. Differential Integral Equations, **2**, 37–43 (1989)

[BS95] Bainov D.D., Simeonov P.S.: Impulsive Differential Equations, Asymptotic Properties of the solution. World Scientific Publishers, Singapore (1995)

[Akh03] Akhmet M.U.: On the General Problrm of Stability for Impulsive Differential Equations. J. Math. Anal. Appl. **288(1)**, 182–196 (2003)

[LMS93] Leela, S., Marae F.A., Sivasundaram S.: Controllability of Impulsive Differential Equations. J. Math. Anal. Appl., **177(1)**, 24–30 (1993)

[BS94] Benzaid Z., Sznaier M.: Constrained Controllability of Linear Impulse Differential System. IEEE Trans. Automat. Contr., **39**, 1064–1066 (1994)

[APT95] Akhmetov M.U., Perestyuk N.A., Tleubergenova M.A.: Control over linear pulse systems, Ukrain. Math. Zh., **47(3)**, 307–314 (1995)

[AS99] Akhmetov M.U., Sejilova R.: The Control of the Boundary Value problem for linear impulsive integro-differential systems. J. Math. Anal. Appl., **236**, 312–326 (1999)

[GQY02] Guan, Z.H., Qian, T.H., Yu, X.: Controllability and Observability of Linear Time Varying Impulsive Systems. IEEE Trans. Circuits Syst. I., **49(8)**, 1198–1207 (2002)

[AZS02] Akhmetov, M.U., Zafer, A., Sejilova, R.D.: The Control of Boundary Value Problems for quasiinear impulsive integro-differential systems. Nonlinear Analysis, **48**, 271–286 (2002)

[AZ04] Akhmet M.U., Zafer, A.: Controllability of Two-Point Nonlinear Boundary Value Problems by the Numerical-Analytical Method. Appl. Math. Comput., **151**, 729–744 (2004)

On nonlocal boundary value problems for hyperbolic-parabolic equations

Allaberen Ashyralyev[1] and Yildirim Ozdemir[1,2]

[1] Department of Mathematics, Fatih University, Istanbul, Turkey
aashyr@fatih.edu.tr, yozdemir@fatih.edu.tr
[2] Department of Mathematics, Gebze Institute of Technology, Kocaeli, Turkey

A numerical method is proposed for solving the hyperbolic-parabolic partial differential equations with nonlocal boundary condition. The first and second order of accuracy difference schemes are presented. The method is illustrated by numerical examples.

1 Introduction

It is known that some problems in fluid mechanics and mathematical biology lead to partial differential equations of the hyperbolic-parabolic type. Methods of solutions of the nonlocal boundary value problems for hyperbolic-parabolic differential equations have been studied extensively by many researchers (see, e.g., [KL01, Nak95, Ram06, Sha04, LCS06] and the references given therein).

It is known (see, for example, [Kre66, AY01, AM98, AO99]) that various nonlocal boundary value problems for the hyperbolic-parabolic equations can be reduced to the nonlocal boundary value problem

$$\begin{cases} \frac{d^2u(t)}{dt^2} + Au(t) = f(t), (0 \leq t \leq 1), u(-1) = \alpha u(\mu) + \beta u'(\lambda) + \varphi, \\ \frac{du(t)}{dt} + Au(t) = g(t), (-1 \leq t \leq 0), |\alpha|, |\beta| \leq 1, 0 < \mu, \lambda \leq 1 \end{cases} \quad (1)$$

for differential equations of mixed type in a Hilbert space H with self-adjoint positive definite operator A.

We are interested in studying the stability of solutions of the problem (1) for $\beta \neq 0$. In the paper [AO05] the first and second order of accuracy difference schemes approximately solving the boundary value problem (1) are presented. The stability estimates for the solution of these difference schemes are established.

In the present paper we consider the applications of these results to the numerical solutions of the difference schemes of the nonlocal boundary-value problems for the multidimensional hyperbolic-parabolic equations. The stability estimates for the solutions of the difference schemes of the nonlocal

Kenan Taş et al. (eds), Mathematical Methods in Engineering, 131–140.

boundary-value problems for the multidimensional hyperbolic-parabolic equations are obtained. A procedure of modified Gauss elimination method is used for solving these difference schemes in the case of a one-dimensional hyperbolic-parabolic partial differential equations.

2 The stability estimates

Let Ω be the unit open cube in the n-dimensional Euclidean space R^n $(x = (x_1, \ldots, x_n) : 0 < x_k < 1, 1 \le k \le n)$ with boundary $S, \overline{\Omega} = \Omega \cup S$. In $[0,1] \times \Omega$ we consider the mixed problem for hyperbolic-parabolic equation

$$\begin{cases} u_{tt} - \sum\limits_{r=1}^{n} (a_r(x) u_{x_r})_{x_r} = f(t,x), 0 \le t \le 1, x \in \Omega, \\ u_t - \sum\limits_{r=1}^{n} (a_r(x) u_{x_r})_{x_r} = g(t,x), -1 \le t \le 0, x \in \Omega, \\ u(-1,x) = u(1,x) + u_t(1,x) + \varphi(x), x \in \overline{\Omega}, \\ u(t,x) = 0, x \in S, -1 \le t \le 1, \end{cases} \tag{2}$$

where $a_r(x), (x \in \Omega), \varphi(x)$ $(x \in \overline{\Omega})$ and $f(t,x)$ $(t \in (0,1),\ x \in \Omega), g(t,x)$ $(t \in (-1,0),\ x \in \Omega)$ are given smooth functions and $a_r(x) \ge a > 0$.

The discretization of problem (2) is carried out in two steps. In the first step let us define the grid sets

$$\widetilde{\Omega}_h = \{x = x_m = (h_1 m_1, \cdots, h_n m_n), m = (m_1, \cdots, m_n)\ ,$$

$$0 \le m_r \le N_r, h_r N_r = L, r = 1, \cdots, n\}\,,$$

$$\Omega_h = \widetilde{\Omega}_h \cap \Omega, S_h = \widetilde{\Omega}_h \cap S.$$

We introduce the Hilbert space $L_{2h} = L_2(\widetilde{\Omega}_h)$ of the grid functions $\varphi^h(x) = \{\varphi(h_1 m_1, \cdots, h_n m_n)\}$ defined on $\widetilde{\Omega}_h$, equipped with the norm

$$\|\varphi^h\|_{L_2(\widetilde{\Omega}_h)} = \left(\sum_{x \in \overline{\Omega_h}} \left|\varphi^h(x)\right|^2 h_1 \cdots h_n \right)^{1/2}.$$

To the differential operator A generated by the problem (2) we assign the difference operator A_h^x by the formula

$$A_h^x u_x^h = -\sum_{r=1}^{n} \left(a_r(x) u^h_{\underline{x}_r} \right)_{x_r, j_r} \tag{3}$$

acting in the space of grid functions $u^h(x)$, satisfying the conditions $u^h(x) = 0$ for all $x \in S_h$. It is known that A_h^x is a self-adjoint positive definite operator in $L_2(\widetilde{\Omega}_h)$. With the help of A_h^x we arrive at the nonlocal boundary-value problem

$$\begin{cases} \frac{d^2u^h(t,x)}{dt^2} + A_h^x u^h(t,x) = f^h(t,x), 0 \le t \le 1, x \in \widetilde{\Omega}_h, \\ \frac{du^h(t,x)}{dt} + A_h^x u^h(t,x) = f^h(t,x), -1 \le t \le 0, x \in \widetilde{\Omega}_h, \\ u^h(-1,x) = u^h(1,x) + \frac{du^h(1,x)}{dt} + \varphi^h(x), x \in \widetilde{\Omega}_h, \\ u^h(0+,x) = u^h(0-,x), \frac{du^h(0+,x)}{dt} = \frac{du^h(0-,x)}{dt}, x \in \widetilde{\Omega}_h \end{cases} \tag{4}$$

for an infinite system of ordinary differential equations.

In the second step we replace problem (4) by the difference schemes of paper [AO05], we obtain the first order of accuracy difference scheme

$$\begin{cases} \frac{u_{k+1}^h(x) - 2u_k^h(x) + u_{k-1}^h(x)}{\tau^2} + A_h^x u_{k+1}^h(x) = f_k^h(x), \\ f_k^h(x) = \{f(t_{k+1}, x_n)\}_1^{M-1}, t_{k+1} = (k+1)\tau, \\ 1 \le k \le N-1, N\tau = 1, x \in \widetilde{\Omega}_h, \\ \frac{u_k^h(x) - u_{k-1}^h(x)}{\tau} + A_h^x u_k^h(x) = g_k^h(x), \\ g_k^h(x) = \{g(t_k, x_n)\}_1^{M-1}, t_k = k\tau, -N+1 \le k \le -1, x \in \widetilde{\Omega}_h, \\ u_{-N}^h(x) = u_N^h(x) + \frac{u_N^h(x) - u_{N-1}^h(x)}{\tau} + \varphi^h(x), x \in \widetilde{\Omega}_h, \\ \frac{u_1^h(x) - u_0^h(x)}{\tau} = -A_h^x u_0^h(x) + g_0^h(x), g_0^h(x) = g^h(0,x), x \in \widetilde{\Omega}_h, \end{cases} \tag{5}$$

and two second order of accuracy difference schemes

$$\begin{cases} \frac{u_{k+1}^h(x) - 2u_k^h(x) + u_{k-1}^h(x)}{\tau^2} + A_h^x u_k^h(x) + \frac{\tau^2}{4}\left(A_h^x\right)^2 u_{k+1}^h(x) = f_k^h(x), \\ f_k^h(x) = f(t_k, x), \quad t_k = k\tau, \ 1 \le k \le N-1, x \in \widetilde{\Omega}_h, \\ \left(\frac{I + \tau^2 A_h^x}{\tau}\right)\left(u_1^h(x) - u_0^h(x)\right) = Z_1, \\ Z_1 = \frac{\tau}{2}\left(f^h(0,x) - A_h^x u_0^h(x)\right) + \left(g^h(0,x) - A_h^x u_0^h(x)\right), x \in \widetilde{\Omega}_h, \\ \frac{u_k^h(x) - u_{k-1}^h(x)}{\tau} + A_h^x\left(I + \frac{\tau}{2}A_h^x\right)u_k^h(x) = \left(I + \frac{\tau}{2}A_h^x\right)g_k^h(x), \\ g_k^h(x) = g^h\left(t_k - \frac{\tau}{2}, x\right), \quad t_k = k\tau, \ -(N-1) \le k \le 0, x \in \widetilde{\Omega}_h, \\ u_{-N}^h(x) = \alpha\left(u_0^h(x) + \mu\left(-A_h^x u_0^h(x) + g_0^h(x)\right)\right) \\ +\beta\left(-A_h^x u_0^h(x) + g_0^h(x)\right. \\ \left.+\lambda\left(-A_h^x u_0^h(x) + f_0^h(x)\right)\right) + \varphi^h(x), \mu \le 2\tau, \lambda \le 2\tau, x \in \widetilde{\Omega}_h, \\ u_{-N}^h(x) = \alpha\left(u_{[\mu/\tau]}^h(x) + (\mu - [\frac{\mu}{\tau}]\tau)\frac{u_{[\mu/\tau]}^h(x) - u_{[\mu/\tau]-1}^h(x)}{\tau}\right) \\ +\beta\left(-A_h^x u_0^h(x) + g_0^h(x) + \lambda\left(-A_h^x u_0^h(x) + f_0^h(x)\right)\right) \\ +\varphi^h(x), 2\tau < \mu, \lambda \le 2\tau, x \in \widetilde{\Omega}_h, \\ u_{-N}^h(x) = \alpha\left(u_0^h(x) + \mu\left(-A_h^x u_0^h(x) + g_0^h(x)\right)\right) \\ +\beta\left(\frac{u_{[\lambda/\tau]}^h(x) - u_{[\lambda/\tau]-1}^h(x)}{\tau} + (\lambda - [\frac{\lambda}{\tau}]\tau + \frac{\tau}{2})(f_{[\frac{\lambda}{\tau}]}^h(x) - A_h^x u_{[\frac{\lambda}{\tau}]}^h(x))\right) \\ +\varphi^h(x), \mu \le 2\tau, 2\tau < \lambda, x \in \widetilde{\Omega}_h, \\ u_{-N}^h(x) = \alpha\left(u_{[\mu/\tau]}^h(x) + (\mu - [\frac{\mu}{\tau}]\tau)\frac{u_{[\mu/\tau]}^h(x) - u_{[\mu/\tau]-1}^h(x)}{\tau}\right) \\ +\beta\left(\frac{u_{[\lambda/\tau]}^h(x) - u_{[\lambda/\tau]-1}^h(x)}{\tau} + (\lambda - [\frac{\lambda}{\tau}]\tau + \frac{\tau}{2})(f_{[\frac{\lambda}{\tau}]}^h(x) - A_h^x u_{[\frac{\lambda}{\tau}]}^h(x))\right) \\ +\varphi^h(x), 2\tau < \mu, 2\tau < \lambda, x \in \widetilde{\Omega}_h. \end{cases} \tag{6}$$

$$\begin{cases} \frac{u_{k+1}^h(x)-2u_k^h(x)+u_{k-1}^h(x)}{\tau^2} + \frac{1}{2}A_h^x u_k^h(x) \\ +\frac{1}{4}A_h^x\left(u_{k+1}^h(x)+u_{k-1}^h(x)\right) = f_k^h(x), \\ f_k^h(x) = f(t_k,x), \quad t_k = k\tau,\ 1\le k\le N-1, x\in\widetilde{\Omega}_h, \\ \left(\frac{I+\tau^2A_h^x}{\tau}\right)\left(u_1^h(x)-u_0^h(x)\right) = Z_1, \\ Z_1 = \frac{\tau}{2}\left(f^h(0,x)-A_h^x u_0^h(x)\right) + \left(g^h(0,x)-A_h^x u_0^h(x)\right), x\in\widetilde{\Omega}_h, \\ \frac{u_k^h(x)-u_{k-1}^h(x)}{\tau} + A_h^x\left(I+\frac{\tau}{2}A_h^x\right)u_k^h(x) = \left(I+\frac{\tau}{2}A_h^x\right)g_k^h(x), \\ g_k^h(x) = g^h\left(t_k-\frac{\tau}{2},x\right), \ t_k=k\tau, \ -(N-1)\le k\le 0, x\in\widetilde{\Omega}_h, \\ u_{-N}^h(x) = \alpha\left(u_0^h(x)+\mu\left(-A_h^x u_0^h(x)+g_0^h(x)\right)\right) \\ +\beta\left(-A_h^x u_0^h(x)+g_0^h(x)\right. \\ \left.+\lambda\left(-A_h^x u_0^h(x)+f_0^h(x)\right)\right)+\varphi^h(x), \mu\le 2\tau, \lambda\le 2\tau, x\in\widetilde{\Omega}_h, \\ u_{-N}^h(x) = \alpha\left(u_{[\mu/\tau]}^h(x)+(\mu-[\frac{\mu}{\tau}]\tau)\frac{u_{[\mu/\tau]}^h(x)-u_{[\mu/\tau]-1}^h(x)}{\tau}\right) \\ +\beta\left(-A_h^x u_0^h(x)+g_0^h(x)+\lambda\left(-A_h^x u_0^h(x)+f_0^h(x)\right)\right) \\ +\varphi^h(x), 2\tau<\mu, \lambda\le 2\tau, x\in\widetilde{\Omega}_h, \\ u_{-N}^h(x) = \alpha\left(u_0^h(x)+\mu\left(-A_h^x u_0^h(x)+g_0^h(x)\right)\right) \\ +\beta\left(\frac{u_{[\lambda/\tau]}^h(x)-u_{[\lambda/\tau]-1}^h(x)}{\tau}+\left(\lambda-[\frac{\lambda}{\tau}]\tau+\frac{\tau}{2}\right)\left(f_{[\frac{\lambda}{\tau}]}^h(x)-A_h^x u_{[\frac{\lambda}{\tau}]}^h(x)\right)\right) \\ +\varphi^h(x), \mu\le 2\tau, 2\tau<\lambda, x\in\widetilde{\Omega}_h, \\ u_{-N}^h(x) = \alpha\left(u_{[\mu/\tau]}^h(x)+(\mu-[\frac{\mu}{\tau}]\tau)\frac{u_{[\mu/\tau]}^h(x)-u_{[\mu/\tau]-1}^h(x)}{\tau}\right) \\ +\beta\left(\frac{u_{[\lambda/\tau]}^h(x)-u_{[\lambda/\tau]-1}^h(x)}{\tau}+\left(\lambda-[\frac{\lambda}{\tau}]\tau+\frac{\tau}{2}\right)\left(f_{[\frac{\lambda}{\tau}]}^h(x)-A_h^x u_{[\frac{\lambda}{\tau}]}^h(x)\right)\right) \\ +\varphi^h(x), 2\tau<\mu, 2\tau<\lambda, x\in\widetilde{\Omega}_h. \end{cases} \tag{7}$$

Theorem 1. *Let τ and $|h|$ be a sufficiently small numbers. Then the solutions of difference schemes (5), (6) and (7) satisfy the following stability estimates:*

$$\max_{-N\le k\le N}\left\|u_k^h\right\|_{L_{2h}} \le M_1\left[\left\|f_1^h\right\|_{L_{2h}} + \max_{2\le k\le N-1}\left\|\left(f_k^h-f_{k-1}^h\right)\tau^{-1}\right\|_{L_{2h}}\right.$$

$$\left. + \left\|g_0^h\right\|_{L_{2h}} + \max_{-N+1\le k\le 0}\left\|\left(g_k^h-g_{k-1}^h\right)\tau^{-1}\right\|_{L_{2h}} + \left\|\varphi^h\right\|_{L_{2h}}\right],$$

$$\max_{-N+1\le k\le N}\left\|\tau^{-1}\left(u_k^h-u_{k-1}^h\right)\right\|_{L_{2h}} + \max_{-N\le k\le N}\sum_{r=1}^n\left\|\left(u_k^h\right)_{x_r,j_r}\right\|_{L_{2h}}$$

$$\le M_1\left[\left\|f_1^h\right\|_{L_{2h}} + \max_{2\le k\le N-1}\left\|\left(f_k^h-f_{k-1}^h\right)\tau^{-1}\right\|_{L_{2h}}\right.$$

$$\left. + \left\|g_0^h\right\|_{L_{2h}} + \max_{-N+1\le k\le 0}\left\|\left(g_k^h-g_{k-1}^h\right)\tau^{-1}\right\|_{L_{2h}} + \sum_{r=1}^n\left\|\left(\varphi^h\right)_{-x_r,j_r}\right\|_{L_{2h}}\right],$$

$$\max_{1\le k\le N-1}\left\|\tau^{-2}\left(u_{k+1}^h-2u_k^h+u_{k-1}^h\right)\right\|_{L_{2h}}$$

$$+\max_{-N\leq k\leq N}\sum_{r=1}^{n}\left\|\left(u_k^h\right)_{-x_rx_r,j_r}\right\|_{L_{2h}}+\max_{-N+1\leq k\leq 0}\left\|\tau^{-1}\left(u_k^h-u_{k-1}^h\right)\right\|_{L_{2h}}$$

$$\leq M_1\left[\sum_{r=1}^{n}\left\|\left(f_1^h\right)_{-x_r,j_r}\right\|_{L_{2h}}+\left\|\tau^{-1}\left(f_2^h-f_1^h\right)\right\|_{L_{2h}}\right.$$

$$+\max_{2\leq k\leq N-1}\left\|\tau^{-2}\left(f_{k+1}^h-2f_k^h+f_{k-1}^h\right)\right\|_{L_{2h}}$$

$$+\sum_{r=1}^{n}\left\|\left(g_0^h\right)_{-x_r,j_r}\right\|_{L_{2h}}+\left\|\tau^{-1}\left(g_0^h-g_{-1}^h\right)\right\|_{L_{2h}}$$

$$\left.+\max_{-N+1\leq k\leq -1}\left\|\tau^{-2}\left(g_{k+1}^h-2g_k^h+g_{k-1}^h\right)\right\|_{L_{2h}}+\sum_{r=1}^{n}\left\|\left(\varphi^h\right)_{-x_rx_r,j_r}\right\|_{L_{2h}}\right].$$

Here M_1 does not depend on τ, h, $\varphi^h(x)$ and $f_k^h(x), 1\leq k\leq N-1, g_k^h, -N+1\leq k\leq 0$.

The proof of Theorem 1 is based on the abstract Theorem of paper [AO05], and the symmetry properties of the difference operator A_h^x defined by the formula (3).

3 Numerical analysis

We have not been able to a sharp estimate for the constants figuring in the stability inequality. Therefore, we will give the following results of numerical experiments of the nonlocal boundary value problem

$$\begin{cases}\frac{\partial^2 u(t,x)}{\partial t^2}-\frac{\partial^2 u(t,x)}{\partial x^2}=e^{\left(1-t^2\right)}\left(-2+4t^2+\pi^2\right)\sin\pi x,\\ 0<t<1, 0<x<1,\\ \frac{\partial u(t,x)}{\partial t}-\frac{\partial^2 u(t,x)}{\partial x^2}=e^{\left(1-t^2\right)}\left(-2t+\pi^2\right)\sin\pi x,\\ -1<t<0, 0<x<1,\\ u(0+,x)=u(0-,x), u_t(0+,x)=u_t(0-,x), 0\leq x\leq 1,\\ u(-1,x)=u\left(1,x\right)+u_t(1,x)+2\sin\pi x, 0\leq x\leq 1,\\ u(t,0)=u(t,1)=0, -1\leq t\leq 1,\end{cases}\tag{8}$$

for hyperbolic-parabolic equation.

First, applying the first order of accuracy difference scheme (5), we present the first order of accuracy difference scheme for the approximate solution of the problem (8). Then we have $(2N+1)\times(2N+1)$ system of linear equations and we will write them in the matrix form

$$\begin{cases}A\,U_{n+1}+B\,U_n+C\,U_{n-1}=D\varphi_n, \quad 0\leq n\leq M,\\ U_0=\widetilde{0},\ \ U_M=\widetilde{0},\end{cases}$$

where

$$\begin{cases} \beta_{n+1} = -(C + D\alpha_n + E\beta_{n-1} + E\alpha_{n-1}\alpha_n)^{-1}(A), \\ \alpha_{n+1} = -(C + D\alpha_n + E\beta_{n-1} + E\alpha_{n-1}\alpha_n)^{-1} \\ \times(B + D\beta_n + E\alpha_{n-1}\beta_n), \\ \gamma_{n+1} = (C + D\alpha_n + E\beta_{n-1} + E\alpha_{n-1}\alpha_n)^{-1} \\ \times(R\varphi_n - D\gamma_n - E\alpha_{n-1}\gamma_n - E\gamma_{n-1}), n = 1, \cdots, M-1. \end{cases} \tag{13}$$

Here

$$\alpha_1 = \begin{bmatrix} 0 & \dots & 0 \\ \dots & \dots & \dots \\ 0 & \dots & 0 \end{bmatrix}, \beta_1 = \begin{bmatrix} 0 & \dots & 0 \\ \dots & \dots & \dots \\ 0 & \dots & 0 \end{bmatrix}, \gamma_1 = \begin{bmatrix} 0 \\ \dots \\ 0 \end{bmatrix},$$

$$\gamma_2 = \begin{bmatrix} 0 \\ \dots \\ 0 \end{bmatrix}, \alpha_2 = \begin{bmatrix} \frac{4}{5} & \dots & 0 \\ \dots & \dots & \dots \\ 0 & \dots & \frac{4}{5} \end{bmatrix}, \beta_2 = \begin{bmatrix} -\frac{1}{5} & \dots & 0 \\ \dots & \dots & \dots \\ 0 & \dots & -\frac{1}{5} \end{bmatrix},$$

$$\begin{cases} U_M = \tilde{0}, \\ U_{M-1} = [(\beta_{M-2} + 5I) - (4I - \alpha_{M-2})\alpha_{M-1}]^{-1} \\ \times[(4I - \alpha_{M-2})\gamma_{M-1} - \gamma_{M-2}], \\ U_{M-2} = [(4I - \alpha_{M-2})]^{-1}[(\beta_{M-2} + 5I)U_{M-1} + \gamma_{M-2}]. \end{cases}$$

Third, applying the second order of accuracy difference scheme (7), we present the following second order of accuracy difference schemes for the approximate solutions of the problem (8). So, we have again the $(2N+1) \times (2N+1)$ system of linear equations and it can be written in the same matrix form (12), where

$$A = \begin{bmatrix} 0\,0\,0\,0 \,.\, 0\,0\,0\,0 \\ 0\,x\,0\,0 \,.\, 0\,0\,0\,0 \\ .\ .\ .\ .\ .\ .\ .\ .\ . \\ 0\,0\,0\,x \,.\, 0\,0\,0\,0 \\ .\ .\ .\ .\ .\ .\ .\ .\ . \\ 0\,0\,0\,0 \,.\, 0\,y\,0\,0 \\ .\ .\ .\ .\ .\ .\ .\ .\ . \\ 0\,0\,0\,0 \,.\, 0\,0\,0\,y \\ 0\,0\,0\,0 \,.\, 0\,0\,0\,0 \end{bmatrix}, B = \begin{bmatrix} 0\,0\,0\,0 \,.\, 0\,0\,0\,0 \\ 0\,z\,0\,0 \,.\, 0\,0\,0\,0 \\ .\ .\ .\ .\ .\ .\ .\ .\ . \\ 0\,0\,0\,z \,.\, 0\,0\,0\,0 \\ 0\,0\,0\,0 \,.\, w\,v\,0\,0 \\ .\ .\ .\ .\ .\ .\ .\ .\ . \\ 0\,0\,0\,0 \,.\, 0\,0\,w\,v \\ 0\,0\,0 \,.\, m\,n \,.\, 0\,0 \end{bmatrix}, C = \begin{bmatrix} 1\,0\,0\,0 \,.\, 0\,r\,s\,t \\ 0\,a\,0\,0 \,.\, 0\,0\,0\,0 \\ .\ .\ .\ .\ .\ .\ .\ .\ . \\ 0\,0\,0\,a \,.\, 0\,0\,0\,0 \\ .\ .\ .\ .\ .\ .\ .\ .\ . \\ 0\,0\,0\,0 \,.\, b\,c\,0\,0 \\ .\ .\ .\ .\ .\ .\ .\ .\ . \\ 0\,0\,0\,0 \,.\, 0\,0\,b\,c \\ 0\,0 \,.\, p\,q\,p \,.\, 0\,0 \end{bmatrix},$$

and

$$D = B, E = A,$$

$$R = \begin{bmatrix} 1 & \dots & 0 \\ \dots & \dots & \dots \\ 0 & \dots & 1 \end{bmatrix}, U_s = \begin{bmatrix} U_s^{-N} \\ \dots \\ U_s^{N} \end{bmatrix}, wheres = n \pm 2, n \pm 1, n, \varphi_n = \begin{bmatrix} \varphi_n^{-N} \\ \dots \\ \varphi_n^{N} \end{bmatrix}.$$

Here

$$x = \frac{\tau}{2h^4}, y = \frac{\tau^2}{4h^4}, z = -\frac{1}{h^2} - \frac{2\tau}{h^4}, w = -\frac{1}{h^2}, v = -\frac{\tau^2}{h^4}, m = \frac{\tau}{h^2} - \frac{\tau}{2h^2} - \frac{1}{h^2}$$

$$n=\frac{1}{\tau}+\frac{2\tau}{h^2},a=\frac{3\tau}{h^4}+\frac{1}{\tau}+\frac{2}{h^2},b=-\frac{2}{\tau^2}+\frac{2}{h^2},$$

$$c=\frac{1}{\tau^2}+\frac{3\tau^2}{2h^4},p=\frac{1}{\tau}+\frac{2\tau}{h^2},q=-\frac{1}{\tau}-\frac{2}{h^2}+\frac{\tau}{h^2}+\frac{2}{h^2},r=-\frac{1}{2\tau},s=\frac{2}{\tau},t=-\frac{3}{2\tau}-1,$$

$$\varphi_n^k=\begin{cases}2\sin(\pi x_n),g(t_k-\frac{\tau}{2},x_n)-\frac{\tau}{2h^2}[g(t_k-\frac{\tau}{2},x_{n+1})\\+g(t_k-\frac{\tau}{2},x_n)+g(t_k-\frac{\tau}{2},x_{n-1})],\\-N+1\le k\le 0,\\f(t_k,x_n),1\le k\le N-1,\\0,k=N.\end{cases}$$

To solve this difference equation we have applied the same modified Gauss elimination method (13), where

$$\begin{cases}U_M=\tilde{0},\\U_{M-1}=[(\beta_{M-2}+5I)-(4I-\alpha_{M-2})\alpha_{M-1}]^{-1}\\\times[(4I-\alpha_{M-2})\gamma_{M-1}-\gamma_{M-2}],\\U_{M-2}=(4I-\alpha_{M-2})^{-1}(\beta_{M-2}+5I)U_{M-1}+\gamma_{M-2}].\end{cases}$$

Now, we will give the results of the numerical analysis. For their comparison, the errors computed by

$$E_M^N=\max_{1\le k\le N-1}\left(\sum_{n=1}^{M-1}\left|u\left(t_k,x_n\right)-u_n^k\right|^2h\right)^{\frac{1}{2}}$$

of the numerical solutions, where $u\left(t_k,x_n\right)$ represents the exact solution and u_n^k represents the numerical solution at $\left(t_k,x_n\right)$. The result are shown in the following table.

Table 1. Comparison of the errors of different difference schemes for $N=M=30$.

Difference schemes	E_M^N
The first order of accuracy difference scheme (5)	0.0370
The second order of accuracy difference scheme (6)	0.0069
The second order of accuracy difference scheme (7)	0.0059

Thus, the second order of accuracy difference schemes are more accurate comparing with the first order of accuracy difference scheme.

References

[KL01] Korzyuk, V.I., Lemeshevsky, S.V.: Problems on conjugation of polytypic equations. Mathematical Modelling and Analysis, **6(1)**, 106-116 (2001)

[Nak95] Nakhushev, A.M.: Equations of Mathematical Biology. Textbook for Universities, Vysshaya Shkola, Moscow (1995) (Russian)

[Ram06] Ramos, J.I.: Linearly-implicit, approximate factorization, exponential methods for multi-dimensional reaction–diffusion equations. Applied Mathematics and Computation, **174(2)**, 1609-1633 (2006)

[Sha04] Shakhmurov, V.B.: Coercive boundary value problems for regular degenerate differential-operator equations. Journal of Mathematical Analysis and Applications, **292(2)**, 605-620 (2004)

[LCS06] Liu, X.Z., Cui, X., Sun, J.G.: FDM for multi-dimensional nonlinear coupled system of parabolic and hyperbolic equations. Journal of Computational and Applied Nathematics, **186(2)**, 432-449 (2006)

[Kre66] Krein, S.G.: Linear Differential Equations in Banach space. Nauka, Moscow (1966) (Russian)

[AY01] Ashyralyev, A., Yurtsever, A.: On a nonlocal boundary value problem for semilinear hyperbolic-parabolic equations. Nonlinear Analysis, Theory, Methods and Applications, **47**, 3585-3592 (2001)

[AM98] Ashyralyev, A., Muradov, I.: On difference schemes second order of accuracy for hyperbolic-parabolic equations. In: Muradov, A.N.(ed) Modeling the Processes in Exploration of Gas Deposits and Applied Problems of Theoretical Gas Hydrodynamics. Ilym, Ashgabat (1998) (Russian)

[AO99] Ashyralyev, A., Orazov, M.B.: The theory of operators and the stability of difference schemes for partial differential equations mixed types. Firat University, Fen ve Muh. Bilimleri Dergisi, **11(3)**, 249-252 (1999)

[AO05] Ashyralyev, A., Ozdemir, Y.: Stability of difference schemes for hyperbolic-parabolic equations. Computers and Mathematics with Applications, **50(2)**, 1443-1476 (2005)

On asymptotical behavior of solution of Riccati equation arising in linear filtering with shifted noises

Agamirza E. Bashirov[1,2] and Zeka Mazhar[3]

[1] Department of Mathematics, Eastern Mediterranean University, Gazimagusa, Mersin 10, Turkey agamirza.bashirov@emu.edu.tr
[2] Institute of Cybernetics, National Academy of Sciences, F. Agayev St. 9, Az1141, Baku, Azerbaijan
[3] Department of Mathematics, Eastern Mediterranean University, Gazimagusa, Mersin 10, Turkey zeka.mazhar@emu.edu.tr

In this paper we consider a linear signal system together with the two linear observation systems. The observation systems differ from each other by the noise processes. The noise of one of them is a constant shift in time of the signal noise. In the other one the shift is neglected. Respectively, we consider two best estimates of the signal corresponding to two different observation systems. The following problem is investigated: whether the effect of the shift on the best estimate becomes negligible as time increases. This leads to a comparison of the asymptotical behaviors of the solutions of respective Riccati equations. It is proved that under a certain relation between the parameters, the effect of the shift is not negligible.

1 Introduction

Kalman filtering for both independent and correlated white noises (see, for example, [Ben92, CP78, LS98, Dav77, FR75]) and its modification to colored noises (see [BJ68]) are very powerful method of estimation in engineering, especially, in space engineering (see [BJ68, CJ04]). However, a detailed study of the nature of noises arising in guidance and control of spacecrafts shows that, more adequately, the noises disturbing the signal and the observations are shifted in time for some small value, while this shift is neglected in space engineering.

Indeed, let ε be the time needed for electromagnetic signals to run the path ground radar–satellite–ground radar. Assume that the control action u changes the parameter x of the satellite in accordance with the linear equation $x' = ax + bu$ if noise effects and the distance to the satellite are neglected.

Kenan Taş et al. (eds), Mathematical Methods in Engineering, 141–149.

Then at the time t the ground radar detects the signal $z(t) = x(t-\varepsilon/2)+w(t)$ that is the useful information $x(t-\varepsilon/2)$ about the parameter of the satellite at $t-\varepsilon/2$ corrupted by white noise $w(t)$ due to atmospheric propagation. Furthermore, the parameter of the satellite at $t-\varepsilon/2$ is changed by the control action $u(t-\varepsilon)$ that is sent by the ground radar at the time moment $t-\varepsilon$. This control passing through the atmosphere is corrupted by the noise $w(t-\varepsilon)$. Hence, the equation for the parameter of the satellite must be written as $x'(t-\varepsilon/2) = ax(t-\varepsilon/2)+b[u(t-\varepsilon)+w(t-\varepsilon)]$. Substituting $\tilde{x}(t) = x(t-\varepsilon/2)$ and $\tilde{u}(t) = u(t-\varepsilon)$, we obtain the state-observation system

$$\begin{cases} \tilde{x}'(t) = a\tilde{x}(t) + b\tilde{u}(t) + bw(t-\varepsilon), \\ z(t) = \tilde{x}(t) + w(t), \end{cases}$$

disturbed by shifted white noises with the state noise delaying the observation noise. Since the Earth orbiting satellites have a nearly constant distance from the Earth, the value ε of the shift for them is time independent.

New applications of the GPS such as measuring vertical and horizontal ground deformations aimed to study volcanos and earthquakes want getting a centimeter (or millimeter) accuracy of satellites' positions. Among different ways toward this aim, one may be the use of Kalman type optimal filter for shifted white noises, obtained in [Bash03] and [Bash05] (abbreviate this filter as KF*). Note that in this case the observations are more informative than in the case of correlated noises since they depend on the future of the signal noise as well. Proper filtering with respect to such observations should produce an improvement in comparison with the Kalman filter for correlated white noises (abbreviate this filter as KF).

Thus, we have two filters KF and KF*. The first one is easy in its realization, successfully tested in many applications and produces the best estimate if the shift in the model is neglected. But for the model with shifted noises it produces an estimate which is not the best one, being perhaps close to it. On the other hand, the second one produces the best estimate for the model with shifted noises, but it needs relatively more calculations for its realization and not yet used in applications. Whether the replacement of KF by KF* is reasonable? For this, let $\hat{x}_y(t)$ and $\hat{x}_z(t)$ be the estimates of the signal process $x(t)$ in accordance with KF* and KF, respectively. Denote $i(t) = \mathbf{E}[\hat{x}_y(t)-\hat{x}_z(t)]^2$, where $\mathbf{E}$ is a symbol for expectation, and call it an *improvement process*. From engineering point of view, regarding the guidance and control of satellites, the asymptotical behavior of $i(t)$ should be important since once a satellite is established on its approximate position in the orbit, $\lim_{t\to\infty} i(t)$ will say whether the improvement is valid at further time moments. If $\lim_{t\to\infty} i(t) = 0$, then the improvement provided by KF* in comparison with KF becomes negligible for large time moments and, therefore, this case does not support the replacement. Unlike, if $\lim_{t\to\infty} i(t) > 0$ or $\lim_{t\to\infty} i(t)$ does not exist, then the best estimate $\hat{x}_y(t)$ non-negligibly deviates from the estimate $\hat{x}_z(t)$ for large t and, hence, the replacement is recommended.

In this paper we study $\lim_{t\to\infty} i(t)$, and use the respective Riccati equations of KF and KF*. While KF is well discussed in the existing literature, KF* was found recently (see [Bash03, Bash05] together with Remark 1 in this paper). We proved that under certain relation between the parameters of the system, $\hat{x}_y(t)$ non-negligibly deviates from the estimate $\hat{x}_z(t)$ for large t. Moreover, numerical study of the respective Riccati equations shows that the error of estimation of KF* is greater than the same of KF. This also supports the replacement of KF by KF* because the greater is the error by KF*, less reliable is the estimate by KF. Finally, note that the results of this paper are obtained for one dimensional systems. As far as the deviation of $\hat{x}_y(t)$ from $\hat{x}_z(t)$ is detected in an easy case, it should expectedly more valid for complicated cases as well.

2 Description of the problem

We will set the problem in one-dimensional case while the results can be extended to multidimensional case as well. Consider the one-dimensional linear signal system

$$x'(t) = ax(t) + bw(t),\ x(0) = x_0,\ t > 0, \tag{1}$$

and the two one-dimensional linear observation systems

$$z'(t) = cx(t) + w(t),\ z(0) = 0,\ t > 0, \tag{2}$$

$$y'(t) = cx(t) + w(t+\varepsilon),\ y(0) = 0,\ t > 0, \tag{3}$$

where $x(t)$ is a signal process, $y(t)$ and $z(t)$ are observation processes, $w(t)$ is a Gaussian white noise process with the mean $\mathbf{E}w(t) = 0$ and with the covariance $\operatorname{cov}(w(t), w(s)) = \delta(t-s)$, δ is the Dirac's delta function, $\varepsilon > 0$, a, b, c are real numbers, x_0 is a Gaussian random variable with the mean $\mathbf{E}(x_0) = 0$ and with the variance p_0, x_0 and $w(t)$, $t \geq 0$, are independent.

Let $\hat{x}_z(t)$ and $\hat{x}_y(t)$ be the best estimates of the signal $x(t)$ based on the observations $z(s)$, $0 \leq s \leq t$, and $y(s)$, $0 \leq s \leq t$, respectively. Here, $\hat{x}_z(t)$ is the output of the well-known KF for the correlated white noises with the error of estimation

$$e_z(t) = \mathbf{E}[\hat{x}_z(t) - x(t)]^2 = f(t),$$

where $f(t)$ is a solution of the Riccati equation

$$f'(t) = 2(a - bc)f(t) - c^2 f(t)^2,\ f(0) = p_0,\ t > 0. \tag{4}$$

Adapting the results from [Bash05] for the estimate $\hat{x}_y(t)$, we can deduce that $\hat{x}_y(t)$ is the output of the KF* with the error of estimation

$$e_y(t) = \mathbf{E}[\hat{x}_y(t) - x(t)]^2 = p(t),$$

where $p(t)$ is a solution of the equation

$$\begin{cases} p'(t) = 2ap(t) + 2q(t,0) + b^2\chi_{(0,\varepsilon]}(t) - c^2p(t)^2, \\ p(0) = p_0, \ t > 0, \end{cases} \tag{5}$$

with $\chi_{(0,\varepsilon]}(t)$ being the indicator function of the interval $(0,\varepsilon]$. Here $q(t,\theta)$ is a solution of

$$\begin{cases} \left(\frac{\partial}{\partial t} + \frac{\partial}{\partial \theta}\right)q(t,\theta) = aq(t,\theta) + r(t,0,\theta) - c^2p(t)q(t,\theta), \\ q(0,\theta) = 0, \ -\varepsilon \le \theta \le 0, \\ q(t,-\varepsilon) = -bcp(t), \ t > 0, \end{cases} \tag{6}$$

with $r(t,\tau,\theta)$ being a solution of

$$\begin{cases} \left(\frac{\partial}{\partial t} + \frac{\partial}{\partial \tau} + \frac{\partial}{\partial \theta}\right)r(t,\tau,\theta) = -c^2q(t,\tau)q(t,\theta), \\ r(0,\tau,\theta) = 0, \ -\varepsilon \le \theta \le 0, \ -\varepsilon \le \tau \le 0, \\ r(t,-\varepsilon,\theta) = -bc(q(t,-\varepsilon) + q(t,\theta)), \ -\varepsilon \le \theta \le 0, \ t > 0, \\ r(t,\tau,-\varepsilon) = -bc(q(t,-\varepsilon) + q(t,\tau)), \ -\varepsilon \le \tau \le 0, \ t > 0. \end{cases} \tag{7}$$

Remark 1. There is a misprint in the formula (18) from Bashirov [Bash05]. The boundary condition

$$R\big(t,\tau,t-\lambda^{-1}(t)\big) = -Q^{\mathrm{T}}(t,\tau)C^{\mathrm{T}}F^{\mathrm{T}}$$

in this formula must be read as

$$R\big(t,\tau,t-\lambda^{-1}(t)\big) = -FCQ\big(t,t-\lambda^{-1}(t)\big) - Q^{\mathrm{T}}(t,\tau)C^{\mathrm{T}}F^{\mathrm{T}}.$$

Respectively, the boundary conditions

$$R(t,\tau,-\varepsilon) = -Q^{\mathrm{T}}(t,\tau)C^{\mathrm{T}}F^{\mathrm{T}},$$

and

$$R\big(t,\tau,t-c^{-1}t\big) = -Q^{\mathrm{T}}(t,\tau)C^{\mathrm{T}}F^{\mathrm{T}},$$

in the formulae (27) and (32) from Bashirov [Bash05] must also be read as

$$R(t,\tau,-\varepsilon) = -FCQ(t,-\varepsilon) - Q^{\mathrm{T}}(t,\tau)C^{\mathrm{T}}F^{\mathrm{T}},$$

and

$$R\big(t,\tau,t-c^{-1}t\big) = -FCQ\big(t,t-c^{-1}t\big) - Q^{\mathrm{T}}(t,\tau)C^{\mathrm{T}}F^{\mathrm{T}}.$$

3 The stability of the improvement

It is natural to call the mean square difference

$$i(t) = \mathbf{E}[\hat{x}_y(t) - \hat{x}_z(t)]^2$$

as an *improvement* provided by KF* in comparison with KF. We say that the improvement $i(t)$ is *unstable* if $\lim_{t\to\infty} i(t) = 0$. Otherwise, we say that it is *stable*. Note that the stability of the improvement should not be confused with the stability of the signal system or filters.

Lemma 1. *Let $f(t)$ and $p(t)$ be solutions of the equations (4) and (5), respectively. The following statements hold:*

(a) $\left(\sqrt{f(t)} - \sqrt{p(t)}\right)^2 \le i(t) \le \left(\sqrt{f(t)} + \sqrt{p(t)}\right)^2$.

(b) *If both $\lim_{t\to\infty} f(t)$ and $\lim_{t\to\infty} p(t)$ exist and equal to 0, then the improvement $i(t)$ is unstable.*

(c) *If both $\lim_{t\to\infty} f(t)$ and $\lim_{t\to\infty} p(t)$ exist and equal to different values, then the improvement $i(t)$ is stable.*

(d) *If $\lim_{t\to\infty} f(t)$ exists, while $\lim_{t\to\infty} p(t)$ does not exist, then the improvement $i(t)$ is stable.*

Proof. By Cauchy–Schwarz inequality,

$$\begin{aligned} i(t) &= \mathbf{E}[\hat{x}_z(t) - \hat{x}_y(t)]^2 = \mathbf{E}[(\hat{x}_z(t) - x(t)) - (\hat{x}_y(t) - x(t))]^2 \\ &= f(t) + p(t) - 2\mathbf{E}[(\hat{x}_z(t) - x(t))(\hat{x}_y(t) - x(t))] \\ &\ge f(t) + p(t) - 2\sqrt{\mathbf{E}[\hat{x}_z(t) - x(t)]^2\mathbf{E}[\hat{x}_y(t) - x(t)]^2} \\ &= f(t) + p(t) - 2\sqrt{f(t)p(t)} = \left(\sqrt{f(t)} - \sqrt{p(t)}\right)^2, \end{aligned}$$

and

$$\begin{aligned} i(t) &= f(t) + p(t) - 2\mathbf{E}[(\hat{x}_z(t) - x(t))(\hat{x}_y(t) - x(t))] \\ &\le f(t) + p(t) + 2|\mathbf{E}[(\hat{x}_z(t) - x(t))(\hat{x}_y(t) - x(t))]| \\ &\le f(t) + p(t) + 2\sqrt{\mathbf{E}[\hat{x}_z(t) - x(t)]^2\mathbf{E}[\hat{x}_y(t) - x(t)]^2} \\ &= f(t) + p(t) + 2\sqrt{f(t)p(t)} = \left(\sqrt{f(t)} + \sqrt{p(t)}\right)^2, \end{aligned}$$

proving part (a). Part (b) follows from part (a) by the squeeze principle. Also, parts (c) and (d) follow from the first inequality in part (a).

Parts (c) and (d) of Lemma 1 give sufficient conditions for stability of the improvement, while Lemma 1(b) presents a sufficient condition for being unstable. Also, Lemma 1(a) tells us that the expression $(\sqrt{f(t)} - \sqrt{p(t)})^2$ is a lower bound of $i(t)$ and in case of stability it can be used to approximate the improvement from below at different instants.

Concerning the Riccati equation (4), it has a trivial solution $f(t) = 0$ if $p_0 = 0$. In case $p_0 > 0$, its solution can explicitly be expressed as

$$f(t) = \frac{p_0}{1 + p_0 c^2 t} \tag{8}$$

if $a = bc$, and

$$f(t) = \left[\frac{c^2}{2(a-bc)} + \left(\frac{1}{p_0} - \frac{c^2}{2(a-bc)}\right)e^{-2(a-bc)t}\right]^{-1} \tag{9}$$

if $a \ne bc$. One particular subcase of (9) is $f(t) \equiv p_0$ if $2(a - bc) = p_0 c^2$. Hence, the following asymptotical behavior of $f(t)$ can easily be deduced:

$$\lim_{t\to\infty} f(t) = \begin{cases} 0, & \text{if } a \le bc \text{ or } p_0 = 0, \\ 2(a-bc)/c^2, & \text{if } a > bc \text{ and } p_0 > 0. \end{cases} \tag{10}$$

Numerical investigation of the equations (5)–(7) allows to conjecture that

$$\lim_{t\to\infty} p(t) \begin{cases} = 0, \text{ if } a \le bc \text{ or } p_0 = 0, \\ > 0, \text{ if } a > bc \text{ and } p_0 > 0. \end{cases}$$

Therefore, in the next section in order to prove that there are the values of the parameters a, b and c, which make the improvement $i(t)$ stable, we will concentrate on the case when $p_0 > 0$ and $a > bc$, assuming that

$$\lim_{t\to\infty} p(t) = \lim_{t\to\infty} f(t) = 2(a-bc)/c^2. \tag{11}$$

Then we will deduce a necessary condition for this assumption. The negation of the necessary condition will produce a sufficient condition for the improvement $i(t)$ to be stable.

4 The system Riccati equation (5)–(7)

Let $p_0 > 0$ and $a > bc$ and assume that (11) holds. To investigate $\lim_{t\to\infty} p(t)$ we need an explicit representation for the solution $(p(t), q(t,\theta), r(t,\tau,\theta))$ of the system (5)–(7). While it will not be used in the sequel, it is interesting to note that $r(t,\tau,\theta) = 0$ if $0 \le t \le \max(\varepsilon+\theta, \varepsilon+\tau)$, and

$$r(t,\tau,\theta) = -bc \begin{Bmatrix} q(t-\theta-\varepsilon, \tau-\theta-\varepsilon) + q(t-\theta-\varepsilon, -\varepsilon), & -\varepsilon \le \theta \le \tau \le 0 \\ q(t-\tau-\varepsilon, \theta-\tau-\varepsilon) + q(t-\tau-\varepsilon, -\varepsilon), & -\varepsilon \le \tau \le \theta \le 0 \end{Bmatrix}$$
$$-c^2 \int_{\max(t-\theta-\varepsilon, t-\tau-\varepsilon)}^{t} q(s, s-t+\tau) q(s, s-t+\theta)\, ds$$

if $t > \max(\varepsilon+\theta, \varepsilon+\tau)$. Moreover, $r(t,\tau,\theta)$ is a continuous kernel of the nonnegative integral operator (see [Bash05]) and, therefore, it satisfies $r(t,\tau,\theta) = r(t,\theta,\tau)$ and

$$r(t,\tau,\theta) \ge 0,\ t \ge 0,\ -\varepsilon \le \theta \le 0,\ -\varepsilon \le \tau \le 0, \tag{12}$$

that will be used later.

Furthermore, the solution of the equation (6) can be represented as $q(t,\theta) = 0$ if $0 \le t \le \varepsilon + \theta$, and

$$q(t,\theta) = -bce^{a(\varepsilon+\theta) - c^2 \int_{t-\varepsilon-\theta}^{t} p(\alpha)\, d\alpha} p(t-\theta-\varepsilon)$$
$$+ \int_{t-\theta-\varepsilon}^{t} e^{a(t-s) - c^2 \int_s^t p(\alpha)\, d\alpha} r(s, 0, s-t+\theta)\, ds \tag{13}$$

if $t > \varepsilon + \theta$.

Regarding $p(t)$, it satisfies the initial condition $p(0) = p_0$ together with the differential equation $p'(t) = 2ap(t) + b^2 - c^2p(t)^2$, if $0 < t \le \varepsilon$, and

$$p'(t) = 2(a - bc)p(t) - c^2p(t)^2 + 2[q(t,0) - q(t,-\varepsilon)]$$

if $t > \varepsilon$. Therefore, for a given $\delta > 0$, we can represent $p(t)$ in the form

$$\begin{aligned} p(t) &= e^{2(a-bc)\delta - c^2 \int_{t-\delta}^{t} p(\alpha)\, d\alpha} p(t-\delta) \\ &\quad +2 \int_{t-\delta}^{t} e^{2(a-bc)(t-s) - c^2 \int_{s}^{t} p(\alpha)\, d\alpha} [q(s,0) - q(s,-\varepsilon)]\, ds \end{aligned} \tag{14}$$

for sufficiently large t.

Applying the assumption (11) to (14), we obtain

$$\lim_{t\to\infty} \int_{t-\delta}^{t} [q(s,0) - q(s,-\varepsilon)]\, ds = 0.$$

Since $\delta > 0$ is arbitrary and $q(t,0) - q(t,-\varepsilon)$ is continuous in t,

$$\lim_{t\to\infty} [q(t,0) - q(t,-\varepsilon)] = 0. \tag{15}$$

By (6) and (13),

$$\begin{aligned} q(t,0) - q(t,-\varepsilon) &= bc\Big[p(t) - e^{a\varepsilon - c^2 \int_{t-\varepsilon}^{t} p(\alpha)\, d\alpha} p(t-\varepsilon)\Big] \\ &\quad + \int_{t-\varepsilon}^{t} e^{a(t-s) - c^2 \int_{s}^{t} p(\alpha)\, d\alpha} r(s,0,s-t)\, ds. \end{aligned}$$

Hence, from (15) and by the assumption (11),

$$\lim_{t\to\infty} \int_{t-\varepsilon}^{t} e^{(2bc-a)(t-s)} r(s,0,s-t)\, ds = \frac{2b(a-bc)}{c}\Big[e^{(2bc-a)\varepsilon} - 1\Big].$$

Thus, by (12),

$$\frac{2b(a-bc)}{c}\Big[e^{(2bc-a)\varepsilon} - 1\Big] \ge 0. \tag{16}$$

Since are in the case $a > bc$, we find out that the inequality (16) does not hold if $0 < 2bc < a$. Hence, the following is proved.

Theorem 1. *If $0 < 2bc < a$ and $p_0 > 0$, then the improvement $i(t)$ is stable.*

5 Concluding remarks

In this paper the asymptotical behaviors of solutions of two related Riccati equations are compared. These solutions represent the errors of estimation

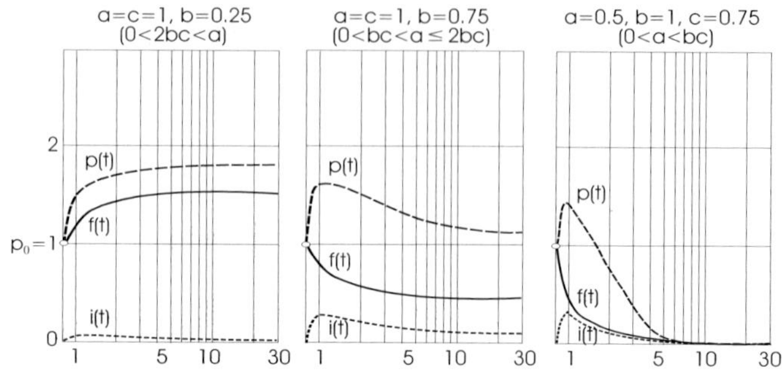

Fig. 1. Graphs of $f(t)$, $p(t)$ and $i(t)$ in different cases.

by KF* and KF, producing the best estimates when the noises in the linear filtering problem are shifted in time and not shifted, respectively. The aim was to detect the cases of the parameters a, b and c for which the best estimates by KF* and KF asymptotically deviate from each other, and wherethrough to support the significance of KF* for models with shifted noises. In the paper we prove that if $0 < 2bc < a$ and $p_0 > 0$ the improvement $i(t)$ provided by KF* in comparison with KF is asymptotically non-negligible while it may be small. The graphs of functions $f(t)$, $p(t)$ and $i(t)$ from Fig. 1 (left), obtained by numerical methods, are typical for the case $0 < 2bc < a$. Numerical study of other cases allows to conjecture that $i(t)$ is still asymptotically non-negligible in the case $0 < bc < a \leq 2bc$ (see Fig. 1 (center)), while it is negligible when $0 < a < bc$ (Fig. 1 (right)). One more observation from Fig. 1 is that in any case $p(t) > f(t)$, i.e., the error of estimation by KF* is greater than the same of KF. Since in applications we never get the exact values of unknown parameters and make decision about them on the base of estimations and respective errors, this tells us that the preciseness of estimation by KF is indeed less reliable. Although the results of the paper are obtained for a one dimensional system, the complication of the system should expectedly make the deviation of $\hat{x}_y(t)$ from $\hat{x}_z(t)$ more valid. All this supports KF* in applications required delicate estimations, especially for positioning satellites with extreme precision.

References

[Ben92] Bensoussan, A.: Stochastic Control of Partially Observable Systems. Cambridge Univ Press, Cambridge (1992)

[CP78] Curtain, R.F., Pritchard, A.J.: Infinite Dimensional Linear Systems Theory, Lect. Notes Contr. and Inf. Sci. Vol.8. Springer, Berlin (1978)

[LS98] Liptser, R.S., Shiryaev, A.N.: Statistics of Random Processes 2: Applications. Springer, New York (1998)

[Dav77] Davis, M.H.A.: Linear Estimation and Stochastic Control. Chapman and Hall, London (1977)

[FR75] Fleming, W.H., Rishel R.W.: Deterministic and Stochastic Optimal Control. Springer, New York (1975)

[BJ68] Bucy, R.S., Joseph, P.D.: Filtering for Stochastic Processes with Application to Guidance. Interscience, New York (1968)

[CJ04] Crassidis, J.L., Junkins, J.L.: Optimal Estimation of Dynamic Systems. Chapman and Hall, Boca Raton (2004)

[Bash03] Bashirov, A.E.: Partially Observable Linear Systems under Dependent Noises. Birkhäuser Verlag, Basel (2003)

[Bash05] Bashirov, A.E.: Filtering for linear systems with shifted noises. Inter. J. Contr., **78(7)**, 521–529 (2005)

Isomorphism classes of ordinary elliptic curves over fields of characteristic 3

Murat Cenk[1] and Ferruh Özbudak[2]

[1] Department of Mathematics and Computer Science, Faculty of Arts and Sciences, Çankaya University, Ankara, Turkey mcenk@cankaya.edu.tr

[2] Department of Mathematics and Institute of Applied Mathematics, Middle East Technical University, Ankara, Turkey ozbudak@metu.edu.tr

Ordinary elliptic curves over fields of characteristic 3 can be represented by $y^2 = x^3 + ax^2 + b$ where $a, b \neq 0 \in F_{q=3^n}$. In this paper we count the number of different isomorphism classes of ordinary elliptic curves over finite fields of characteristic three. We show there are $(2q-2)$ different isomorphism classes.

1 Introduction

Elliptic curves have been studied in some branch of mathematics such as algebraic geometry and number theory. In 1984 Lenstra [Len87] showed how elliptic curves are used for factorization of integers and then after proposing elliptic curve cryptography by N. Koblitz [Kob87] and V. Miller [Mill86] independently in 1985, elliptic curves have been playing important role in cryptography. Elliptic curve cryptography has some advantage with respect to other cryptosystems such as cryptography based on multiplicative group of a finite field and based on systems integer factorization. Lack of sub-exponential algorithms for solving discrete logarithm problem based on elliptic group is main advantage of this system. In order to chose appropriate elliptic curves in cryptography one should know isomorphism classes of elliptic curves. In cryptography elliptic curves over finite fields of characteristic 2 and over prime fields are often used. The prime fields should contain bigger than 2^{160} elements to construct secure cryptosystem. Isomorphism classes of elliptic curves over characteristic 2 and prime fields having characteristic larger than 3 can be seen in [Men93, MV90]. In this paper we classify isomorphism classes of ordinary elliptic curves over fields of characteristic 3.

The rest of this paper is organized as follows. In section 2 we give some basic definitions, theorems and notation about elliptic curves over finite fields. In section 3 the previous works done by mainly A. Menezes and S. Vanstone [Men93, MV90] are investigated and results of those works are explained. In

Kenan Taş et al. (eds), Mathematical Methods in Engineering, 151–158.

section 4 we classify isomorphism classes of ordinary elliptic curves over fields of characteristic 3 and finally we give the conclusion.

2 Elliptic Curves over Finite Fields

Let $\mathbb{F}_q$ denote the field containing q elements, where q is a prime power and let $\mathbb{K}$ is field and $\overline{\mathbb{K}}$ be its algebraic closure. The projective plane $P^2(\mathbb{K})$ over $\mathbb{K}$ is the set of equivalence classes of the relation $\sim$ acting on $K^3 \setminus \{0,0,0\}$. The equivalence classes can be given by the following relation:

$$(x_1, y_1, z_1) \sim (x_2, y_2, z_2) \Leftrightarrow \exists u \in \mathbb{K}^* \ni x_1 = ux_2, \quad y_1 = uy_2, \quad z_1 = uz_2.$$

It is denoted the equivalence class containing (x, y, z) by $(x : y : z)$. A Weierstrass equation is a homogeneous equation of the degree 3 and it is

$$Y^2Z + a_1XYZ + a_3YZ^2 = X^3 + a_2X^2Z + a_4XZ^2 + a_6Z^3,$$

where $a_1, a_2, a_3, a_4, a_6 \in \overline{\mathbb{K}}$. The Weierstrass equation is defined as smooth or non-singular if for all projective points $P = (X : Y : Z) \in P^2(\overline{\mathbb{K}})$ satisfying

$$F(X, Y, Z) = Y^2Z + a_1XYZ + a_3YZ^2 - X^3 - a_2X^2Z - a_4XZ^2 - a_6Z^3 = 0,$$

at least one of the three partial derivatives $\frac{\partial F}{\partial X}, \frac{\partial F}{\partial Y}, \frac{\partial F}{\partial Z}$ is non-zero at P. If all derivatives are zero at some point P, then P is called a singular point, and the Weierstrass equation is singular.

An elliptic curve E is the set of all solutions in $P^2(\overline{\mathbb{K}})$ of a smooth Weierstrass equation. There is exactly one point in E with Z-coordinate equal to 0, namely $(0 : 1 : 0)$. This point is the point at the infinite and denoted by O.

Weierstrass equation for an elliptic curve can be written by using non-homogeneous (affine) coordinates $x = X/Z$, $y = Y/Z$,

$$y^2 + a_1xy + a_3y = x^3 + a_2x^2 + a_4x + a_6. \tag{1}$$

Then an elliptic curve E is the set of solutions to equation (1) in the affine plane $A^2(\overline{\mathbb{K}} = \overline{\mathbb{K}} \times \overline{\mathbb{K}},)$ together with the extra point at the infinity O. If $a_1, a_2, a_3, a_4, a_6 \in \mathbb{K}$, then E is said to be defined over $\mathbb{K}$, and we write this by $E/\mathbb{K}$.

In this paper we investigate the isomorphism classes of elliptic curves and the following result relates the notion of isomorphism of elliptic curves to the coefficients of the Weierstrass equations that define the curves .

Theorem 1. *[Men93] Two elliptic curves $E_1/\mathbb{K}$ and $E_2/\mathbb{K}$ given by the equations*

$$y^2 + a_1xy + a_3y = x^3 + a_2x^2 + a_4x + a_6$$

$$y^2 + \overline{a_1}xy + \overline{a_3}y = x^3 + \overline{a_2}x^2 + \overline{a_4}x + \overline{a_6}$$

are isomorphic over $\mathbb{K}$, *denoted* $E_1/\mathbb{K} \cong E_2/\mathbb{K}$, *if and only if there exists* $u, r, s, t \in \mathbb{K}$, $u \neq 0$, *such that the change of variables*

$$(x, y) \longrightarrow (u^2x + r, u^3y + u^2sx + t) \tag{2}$$

transforms equation E_1 *to equation* E_2. *The relationship of isomorphism is an equivalence relation.*

The change of variables (2) is said to be admissible change of variables. Notice that if $E_1 \cong E_2$ over $\mathbb{K}$, and if the change of variables (2) transforms equation E_1 to equation E_2, then the change of variables

$$(x, y) \longrightarrow (u^{-2}(x - r), u^{-3}(y - sx - t + rs)) \tag{3}$$

transforms equation E_2 to equation E_1, and (3) is also an admissible change of variables. Moreover,

$$\phi : (x, y) \longrightarrow (u^{-2}(x - r), u^{-3}(y - sx - t + rs)) \tag{4}$$

maps the points of E_1 onto E_2, while

$$\psi : (x, y) \longrightarrow (u^2x + r, u^3y + u^2sx + t) \tag{5}$$

maps E_2 onto E_1.

Now, if $E_1 \cong E_2$ over $\mathbb{K}$, then the change of variables (2) transforms equation E_1 to equation E_2. The next theorem is now clearly equivalent to Theorem 1.

Theorem 2. *[Men93] Two elliptic curves* $E_1/\mathbb{K}$ *and* $E_2/\mathbb{K}$ *are isomorphic over* $\mathbb{K}$ *if and only if there exists* $u, r, s, t \in \mathbb{K}$, $u \neq 0$, *that satisfy*

$$\begin{aligned}
u\overline{a_1} &= a_1 + 2s \\
u^2\overline{a_2} &= a_2 - sa_1 + 3r - s^2 \\
u^3\overline{a_3} &= a_3 + ra_1 + 2t \\
u^4\overline{a_4} &= a_4 - sa_3 + 2ra_2 - (t + rs)a_1 + 3r^2 - 2st \\
u^6\overline{a_6} &= a_6 + ra_4 + r^2a_2 + r^3 - ta_3 - t^2 - rta_1.
\end{aligned}$$

The discriminant and $j-$invariant are important terms for elliptic curves. Let E be a curve given by a non-homogeneous Weierstrass equation (1). We have some formulas:

$$\begin{aligned}
d_2 &= a_1^2 + 4a_2 \\
d_4 &= 2a_4 + a_1a_3 \\
d_6 &= a_3^2 + 4a_6 \\
d_8 &= a_1^2a_6 + 4a_2a_6 - a_1a_3a_4 + a_2a_3^2 - a_4^2 \\
c_4 &= d_2^2 - 24d_4 \\
\Delta &= -d_2^2d_8 - 8d_4^3 - 27d_6^2 + 9d_2d_4d_6 \\
j(E) &= c_4/\Delta.
\end{aligned}$$

The quantity Δ is called the discriminant of the Weierstrass equation and $j(E)$ is $j-$invariant of E if $\Delta \neq 0$. The following theorem is important for analyzing isomorphism classes.

Theorem 3. *[Men93] If two elliptic curves $E_1/\mathbb{K}$ and $E_2/\mathbb{K}$ are isomorphic over $\mathbb{K}$, then $j(E_1) = j(E_2)$. The converse is also true if $\mathbb{K}$ is an algebraically closed field.*

3 Isomorphism classes of elliptic curves over $\mathbb{F}_q$, char$(\mathbb{F}_q) \neq 3$

In this section we will give the isomorphism classes of elliptic curves over finite fields of characteristic not 3. This section contains previous works and we will not give details of theorems. One can find details in [Men93, MV90].

Isomorphism classes of curves over $\mathbb{F}_q$, where char$(\mathbb{F}_q) \neq 2, 3$ is given by following theorem.

Theorem 4. *[Men93]The number of isomorphism classes of elliptic curves over the finite field $\mathbb{F}_q$, char$(\mathbb{F}_q) > 3$, is 2q+6, 2q+2, 2q+4, 2q, for $q \equiv 1, 5, 7, 11$ (mod12) respectively.*

Isomorphism classes of elliptic curves over finite fields of characteristic 2 were studied by A. Menezes and S. Vanstone and they classified the isomorphism of curves. Now we will give a brief summary of their work, for details see [MV90]. If the theorem (2) is applied to non-supersingular elliptic curves defined over $\mathbb{F}_{2^m}$, one can find sufficient and necessary conditions for isomorphism classes. Let E_1, E_2 be non-supersingular elliptic curves defined over $\mathbb{F}_{2^m}$ and by the equations

$$E_1 : y^2 + xy = x^3 + a_2x^2 + a_6 \quad (a_6 \neq 0)$$

$$E_2 : y^2 + xy = x^3 + \overline{a_2}x^2 + \overline{a_6} \quad (\overline{a_6} \neq 0)$$

Applying theorem (2) it is found that $E_1 \cong E_2$ if and only if $a_6 = \overline{a_6}$ and if there exits $s \in \mathbb{F}_{2^m}$ such that $\overline{a_2} = a_2 + s + s^2$. This leads the following theorem:

Theorem 5. *[Men93] Let γ be an element of $\mathbb{F}_{2^m}$ such that $Tr(\gamma) = 1$ (if m is odd, we can take $\gamma = 1$). A set of representatives of the isomorphism classes is*

$$\{y^2 + xy = x^3 + a_2x^2 + a_6 | a_6 \in \mathbb{F}_{2^m}^*, a_2 \in \{0, \gamma\}\}.$$

Isomorphism classes of supersingular curves over $\mathbb{F}_{2^m}$, depends on m whether it is odd or even. Now assume that m is odd. In this case we can write ant supersingular elliptic curve over $\mathbb{F}_{2^m}$ with m odd has the form

$$E : y^2 + y = x^3 + a_4x + a_6. \tag{6}$$

If $\overline{E}$ is the curve given by

$$\overline{E} : y^2 + y = x^3 + \overline{a_4}x + \overline{a_6},$$

then by theorem (2) $E \cong \overline{E}$ over $\mathbb{F}_{2^m}$ if and only if if there exist $s, t \in \mathbb{F}_{2^m}$ such that

$$s^4 + s + a_4 + \overline{a_4} = 0 \quad (7)$$

$$t^2 + t + s^6 + a_4 s^2 + a_6 + \overline{a_6} = 0. \quad (8)$$

Now consider the curve

$$E : y^2 + y = x^3.$$

Assume that $E \cong E_1$ over $\mathbb{F}_{2^m}$. Then we have the following equations.

$$s^4 + s + a_4 = 0 \quad (9)$$

$$t^2 + t + s^6 + a_6 = 0. \quad (10)$$

Equation (9) has exactly two solutions in $\mathbb{F}_{2^m}$, s_1 and $(s_1 + 1)$. If (s_1, t_1) is a solution to (10) then $Tr(s_1^6 + a_6 = 0)$. So $Tr((s_1 + 1)^6 + a_6) = 1$, and there are two solutions (s, t) to (9) and (10). We see that there are $q^2/2$ curves isomorphic to E_1 since there are q^2 admissible change of variables.

If we investigate the curves isomorphic to curves E_2 and E_3 where

$$E_2 : y^2 + y = x^3 + x$$

$$E_3 : y^2 + y = x^3 + x + 1,$$

we conclude following theorem. For details see [Men93].

Theorem 6. *[Men93] There are 3 isomorphism classes of supersingular elliptic curves over $\mathbb{F}_{2^m}$, where m is odd. A representative form each class is*

1. $y^2 + y = x^3$
2. $y^2 + y = x^3 + x$
3. $y^2 + y = x^3 + x + 1$.

There are 7 isomorphism classes of supersingular elliptic curves over $\mathbb{F}_{2^m}$, where m is even. Let $E/\mathbb{F}_{2^m}$ be the curve

$$E : y^2 + a_3 y = x^3 + a_4 x + a_6, \quad (a_3 \neq 0).$$

Consider three types of curves:
Type I: a_3 is not a cube.
Type II:a_3 is a cube, and $Te(a_4) \neq 0$.
Type III: a_3 is a cube, and $Te(a_4) \neq 0$,
where Te denote the function $Te : \mathbb{F}_{2^m} \longrightarrow F_4$ and this function is defined by

$$Te : \alpha \longrightarrow \alpha + \alpha^{2^2} + \alpha^{2^4} + ... + \alpha^{2^{m-2}}.$$

Now we can give the following theorem:

Theorem 7. *[Men93] There are seven isomorphism classes of supersingular elliptic curves over $\mathbb{F}_{2^m}$, where m is even. Let γ be a non-cube in $\mathbb{F}_{2^m}$. Let $\alpha, \beta, \delta, \omega \in \mathbb{F}_{2^m}$ be such that $Tr(\gamma^{-2}\alpha) = 1$, $Te(\delta) \neq 0$ and $Tr(\omega) = 1$. Then a representative from each class is:*

1. $y^2 + \gamma y = x^3$ *(Type I)*
2. $y^2 + \gamma y = x^3 + \alpha$ *(Type I)*
3. $y^2 + \gamma^2 y = x^3$ *(Type I)*
4. $y^2 + \gamma^2 y = x^3 + \beta$ *(Type I)*
5. $y^2 + y = x^3 + \delta x$ *(Type II)*
6. $y^2 + y = x^3$ *(Type III)*
7. $y^2 + y = x^3 + \omega$ *(Type III)*

4 Isomorphism classes of elliptic curves over $\mathbb{F}_q$, char$(\mathbb{F}_q) = 3$

Elliptic curves are written in Weierstrass form using affine coordinates as

$$y^2 + a_1 xy + a_3 y = x^3 + a_2 x^2 + a_4 x + a_6. \tag{11}$$

Let E be an elliptic curve over finite field $\mathbb{F}_q$, where $q = 3^n$. In this section we will assume field that we are studying has characteristic 3. We know that if characteristic of $\mathbb{F}_q$ is not 2, then admissible change of variables

$$(x, y) \longrightarrow (x, y - \frac{a_1}{2}x - \frac{a_3}{2})$$

transforms E to the curve

$$E'/\mathbb{F}_q : y^2 = x^3 + b_2 x^2 + b_4 x + b_6.$$

Note that $E \cong E'$ over $\mathbb{F}_q$. So elliptic curves over finite fields of characteristic 3 can be written as

$$E/\mathbb{F}_q : y^2 = x^3 + ax^2 + bx + c.$$

The j-invariant of E is $(a^6)/(2b^3 + a^2b^2 + 2a^3c)$. It is well-known that if char$(\mathbb{F}_q) = 2$ or char$(\mathbb{F}_q) = 3$, then elliptic curve is supersingular if and only if $j(E) = 0$. If $a = 0$ then $j(E) = 0$ and so E is supersingular since $char(\mathbb{F}_q) = 3$. If $a \neq 0$ then E is non-supersingular or ordinary. Assume that we have an ordinary elliptic curve

$$E/\mathbb{F}_q : y^2 = x^3 + ax^2 + bx + c.$$

If we apply admissible change of variables

$$(x, y) \longrightarrow (x + r, y), \quad r = \frac{b}{a} \in \mathbb{F}_q$$

we get

$$y^2 = x^3 + ax^2 + (2\frac{b^2}{a} + \frac{b^3}{a^3} + c).$$

Therefore we can say ordinary curves are of the form

$$y^2 = x^3 + ax^2 + b, \quad a, b \in \mathbb{F}_q, \quad a \neq 0, \quad b \neq 0. \tag{12}$$

Now we can give the following lemma which classifies the isomorphisms of some ordinary elliptic curves over $\mathbb{F}_q$.

Lemma 1. *Let $E(\alpha)$ be an ordinary elliptic curve of the form*

$$y^2 = x^3 + x^2 + \alpha, \tag{13}$$

where $\alpha \neq 0 \in \mathbb{F}_q$. For each $\alpha \in \mathbb{F}_{||} \setminus \{0\}$, there are exactly $(q-1)/2$ ordinary elliptic curves isomorphic to the ordinary elliptic curve $E(\alpha)$. For $\alpha_1, \alpha_2 \in \mathbb{F}_{||} \setminus \{0\}$, the ordinary elliptic curves $E(\alpha_1)$ and $E(\alpha_2)$ are different isomorphism classes if $\alpha_1 \neq \alpha_2$.

Proof. Assume that we have two elliptic curves

$$E(\alpha) : y^2 = x^3 + x^2 + \alpha$$
$$E(\beta) : y^2 = x^3 + x^2 + \beta$$

over $\mathbb{F}_{3^n}$ where $\alpha, \beta \in \mathbb{F}_{3^n}$. If $E_1 \cong E_2$ then from Theorem 2 $\alpha = \beta$. So if $\alpha \neq \beta$ we have $E_1 \ncong E_2$ and they are in different isomorphism classes. On the other hand if we have ordinary elliptic curves

$$E_1 : y^2 = x^3 + x^2 + \alpha$$
$$E_2 : y^2 = x^3 + a_2x^2 + a_6$$

over $\mathbb{F}_{3^n}$ where $a_2, a_6 \in \mathbb{F}_{3^n}$ with $E_1 \cong E_2$ then from Theorem 2 we obtain $u^2 = a_2^{-1}$, and $a_2^3 = a_6\alpha$. Therefore u must be quadratic residue in $\mathbb{F}_{3^n}$. We have $(q-1)/2$ elements which are quadratic residue and for each possible a_2 there exists unique a_6. So there are $(q-1)/2$ elliptic curves isomorphic to E_1.

□

Therefore we have isomorphism classes of the form (13) and each class has $(q-1)/2$ curves. We count $(q-1)(q-1)/2 = (q-1)^2/2$ curves. We have totally $(q-1)^2$ curves of the form (12) and we count half of them.

Now assume that $E_1/\mathbb{F}_{3^n} : y^2 = x^3 + ax^2 + b$ where a is not quadratic residue in $\mathbb{F}_{3^n}$. E_1 does not belong to any isomorphism classes of form (13) since $\sqrt{a} \notin \mathbb{F}_{3^n}$. In addition, if we have another curve $E_2/\mathbb{F}_{3^n} : y^2 = x^3 + ax^2 + c$ such that $E_1 \cong E_2$ then we get $b = c$. So if $b \neq c$, $E_1 \ncong E_2$ and therefore for each $b \in \mathbb{F}_{3^n}$ elliptic curve $E/\mathbb{F}_{3^n} : y^2 = x^3 + ax^2 + b$ where a is not quadratic residue in $\mathbb{F}_{3^n}$ is in different isomorphism class. To find number of curves in each isomorphism classes consider

$$E_1/\mathbb{F}_{3^n} : y^2 = x^3 + ax^2 + b$$

$$E_2/\mathbb{F}_{3^n} : y^2 = x^3 + cx^2 + d$$

where a, c are not quadratic residue in $\mathbb{F}_{3^n}$. Then applying theorem (2) we obtain a/c is quadratic residue so we have $(q-1)/2$ such a $c \in \mathbb{F}_{3^n}$. (The reason can be explained as follows: Let g be generator of $\mathbb{F}_{3^n}^*$ if $a = g^x, c = g^y \Rightarrow ac = g^{x+y}$ where x is odd number since $\sqrt{a} \notin \mathbb{F}_{3^n}$ and therefore y must be odd number so that ac^{-1} will be quadratic residue $\mathbb{F}_{3^n}$.) Then for each value for c we have $d = bc^3/a^3$ which is unique. So isomorphism classes of each curve $E/\mathbb{F}_{3^n} : y^2 = x^3 + ax^2 + b$ where a is not quadratic residue in $\mathbb{F}_{3^n}$ has $(q-1)/2$ curves. There are $(q-1)$ such curve therefore we count more $(q-1)^2/2$ curves. Therefore we count all $(q-1)^2$ curves and we find isomorphism classes of ordinary curves over fields of characteristic 3. So we can give the following theorem.

Theorem 8. *There are* $(2q-2)$ *different isomorphism classes of (12) and each class has* $(q-1)/2$ *curves. A complete list of representatives of distinct isomorphism classes of ordinary elliptic curves of characteristic 3 are*

$$\{y^2 = x^3 + ax^2 + b| \quad a, b \in \mathbb{F}_{3^n}, \sqrt{a} \notin \mathbb{F}_{3^n}\} \cup \{y^2 = x^3 + x^2 + \alpha| \quad \alpha \in \mathbb{F}_{3^n}\}.$$

5 Conclusion

In this paper we have investigated the isomorphism classes of ordinary elliptic curves over finite fields of characteristics three. We have shown that there are $2q-2$ different isomorphism classes and we have pointed representative forms of isomorphism class.

References

[Len87] Jr. Lenstra, H.W.: Factoring integers with elliptic curves. Ann. of Math., **126(2)**, 649-673 (1987)

[Kob87] Koblitz, N.: Elliptic curve cryptosystems. Mathematics of Computation, **48**, 203-209 (1987)

[Mill86] Miller, V.: Uses of elliptic curves in cryptography, Advances in Cryptology-CRYPTO 85. Lecture Notes in Computer Science, **218**, 417-426 (1986)

[Men93] Menezes, A.: Elliptic Curve Public Key Cryptosystems. Kluwer Academic Publishers, Boston Dordrecht London 1993.

[MV90] Menezes, A., Vanstone, S.: Isomorphism classes of Elliptic Curves over Finite Fields of Characteristic 2. Utilitas Mathematica, **38**, 135-154 (1990)

Hidden symmetries of two dimensional superintegrable systems

Özlem Defterli[1] and Dumitru Baleanu[1,2]

[1] Department of Mathematics and Computer Science, Faculty of Arts and Science, Çankaya University Ankara, Turkey defterli@cankaya.edu.tr
[2] Institute of Space Sciences, P.O. BOX, MG-23, R 76900 Magurele-Bucharest, Romania baleanu@venus.nipne.ro, dumitru@cankaya.edu.tr

Classification of the invariants of two - dimensional superintegrable systems is presented. The hidden symmetries associated to the existence of Killing - Yano tensors are investigated.

1 Introduction

The concept of superintegrability in two-dimensions is one of the today's hot topics in applied mathematics. It is analyzed in details that superintegrability for two-dimensional spaces with constant curvature and non-constant curvature when all the independent integrals are the polynomials that are quadratic or linear in canonical momenta. The classification of superintegrable systems for these two-dimensional Riemannian spaces have done by several authors [RS99, KMP00, KKPM01, STW01, TTW01, CZ02, BD04a, Eva90, GPS95].

A D-dimensional Hamiltonian system H is said to be Liouville integrable [Arn89, Golds80] if there exists a system of D functionally independent functions in involution. The system is superintegrable [Eva90, Wojciech83] if a further m integrals $\{Y_1, \dots, Y_m, 1 \leq m \leq n-1\}$ exist such that the set of constants $\{I_1 = H, I_2, \dots, I_n, Y_1, \dots, Y_m\}$ are functionally independent. The additional integrals have vanishing Poisson bracket with H, but not necessarily with each other or with the I_i's. A classical Hamiltonian system is maximally superintegrable if $m = n-1$. Then, there are $2n-1$ functionally independent integrals of motion. [FMSUW65, Hieta84, Hieta87, Hol82, LV95]. Darboux and Koenigs classified the two dimensional Riemannian spaces with the infinitesimal distance $ds^2 = g_{ij}(u)du^i du^j$, $u = (u^1, u^2)$ which admits more than one quadratic constant of motion [Koen72].By assuming the existence of a second order Killing tensor, Koenigs found all free Hamiltonians $H = g^{ij}p_i p_j$ which admit at least two extra functionally independent constants of motion having the form $\lambda = a^{ij}(u)p_i p_j$, $a^{ij} = a^{ji}$. For these type of two-dimensional

Kenan Taş et al. (eds), Mathematical Methods in Engineering, 159–166.

Riemannian spaces the superintegrability means the existence of $2n-1$ functionally independent second order quantum mechanical observables for the Schrödinger equation in n dimensions and additive separation of variables of the Hamilton - Jacobi equation corresponding to the classical Hamiltonian that describes the system [KKW02].

The free classical and quantum systems in Darboux spaces are all quadratically superintegrable by construction, which means that the corresponding Hamiltonian allows separation of variables in at least two coordinate systems: they have three functionally independent integrals of motion [KKMW03]. These kind of two-dimensional Riemannian spaces are classified by the following four types of infinitesimal distances [KKW02]

$$ds^2 = (x+y)dxdy, \qquad ds^2 = [\frac{a}{(x-y)^2} + b]dxdy, \tag{1}$$

$$ds^2 = (ae^{-\frac{x+y}{2}} + be^{-x-y})dxdy, \qquad ds^2 = \frac{a(e^{\frac{x-y}{2}} + e^{\frac{y-x}{2}}) + b}{(e^{\frac{x-y}{2}} - e^{\frac{y-x}{2}})^2}dxdy.$$

These spaces are called Darboux spaces and they are denoted by D_1, D_2, D_3 and D_4 respectively.

Killing-Yano (KY) tensors were introduced by Yano in [KSHM80] from mathematical point of view. A KY tensor of order two, $f_{\mu\nu}$, is an antisymmetric tensor defined as

$$f_{\mu\nu;\lambda} + f_{\lambda\nu;\mu} = 0, \tag{2}$$

where ; represents the covariant derivative. The physical meaning of KY tensors was clarified by Gibbons [GRH93], namely KY tensors are the generators of non-generic symmetries on manifolds admitting these tensors. Killing tensors are related to the constants of motion and so it is very important in solving the geodesic equation. A second order Killing tensor, $K_{\mu\nu}$, which is a symmetric tensor can be obtained from a KY tensor of order two by using the relation

$$K_{\mu\nu} = f_{\mu\lambda} f^{\lambda}_{\nu}. \tag{3}$$

where

$$K_{\mu\nu} = g^{\alpha\beta} f_{\mu\alpha} f_{\beta\nu}, \tag{4}$$

without solving the Killing equation

$$K_{\nu\lambda;\mu} + K_{\lambda\mu;\nu} + K_{\mu\nu;\lambda} = 0. \tag{5}$$

Obtaining Killing tensors by solving the partial differential equations in (5) is the standard method. Then from the Killing tensors we can calculate the corresponding constants of motion by the following formula

$$I = K_{\mu\nu} p^{\mu} p^{\nu}. \tag{6}$$

For the Darboux spaces D_1, D_2, D_3 and D_4 being the examples of two - dimensional superintegrable systems, some of the generic and non-generic symmetries were investigated [KKW02, DB04, Def04, BD04b].

In this work, Killing and KY tensors for the above defined Darboux spaces are investigated. From these Killing tensors the constants of motions corresponding to each Darboux's space are calculated and the results are compared.

2 The invariants of Darboux spaces

The free classical systems in Darboux spaces are all quadratically superintegrable by construction. They have three functionally independent integrals of motion.

The infinitesimal distance for the Darboux spaces of type one, given in (1), is

$$ds^2 = (x + y)dxdy. \tag{7}$$

If we do the change of coordinates as $(x, y) \rightarrow (u, v)$, where $x = u + iv$, $y = u - iv$ and $i^2 = -1$, then (7) can be rewritten as [KKW02]

$$ds^2 = 2u(d_u^2 + d_v^2). \tag{8}$$

The free Hamiltonian of this system in the coordinates (u, v) is

$$H = \frac{(p_u^2 + p_v^2)}{4u}. \tag{9}$$

The three constants of motions corresponding to the free Hamiltonian (9) are

$$\begin{aligned} K &= p_v, \\ X_1 &= p_u p_v - \frac{v}{2u}(p_u^2 + p_v^2), \\ X_2 &= p_v(vp_u - up_v) - \frac{v^2}{4u}(p_u^2 + p_v^2), \end{aligned} \tag{10}$$

and they satisfy the relation

$$4HX_2 + X_1^2 + K^4 = 0.$$

For the Darboux spaces of type two in (1), the infinitesimal distance

$$ds^2 = [\frac{a}{(x - y)^2} + b]dxdy, \tag{11}$$

becomes

$$ds^2 = \frac{(d_u^2 + d_v^2)(u^2+1)}{u^2}, \tag{12}$$

in the coordinates $x = 1/2(v+iu)$, $y = 1/2(v-iu)$. The Hamiltonian for this space is

$$H = \frac{(u^2)(p_u^2 + p_v^2)}{u^2+1}, \tag{13}$$

and the three constants of motions associated with the Hamiltonian (13) are as follows [KKMW03]

$$\begin{aligned} K &= p_v, \\ X_1 &= \frac{2v(p_v^2 - u^2 p_u^2)}{u^2+1} + 2up_u p_v, \\ X_2 &= \frac{(v^2-u^4)p_v^2 + u^2(1-v^2)p_u^2}{u^2+1} + 2uvp_u p_v, \end{aligned} \tag{14}$$

which are functionally independent by the relation

$$X_1^2 - 4K^2X_2 + 4HX_2 - 4H^2 = 0.$$

The Darboux space of type three in (1) has the infinitesimal distance

$$ds^2 = (e^{-u} + e^{-2u})(d_u^2 + d_v^2), \tag{15}$$

after making the change of coordinates as $(x, y) \to (u, v)$, where $x = u - iv$, $y = u + iv$. For this system, the corresponding free Hamiltonian has the form

$$H = \frac{1}{4}\frac{e^{2u}(p_u^2 + p_v^2)}{e^u+1}, \tag{16}$$

and the three integrals of motion are [KKMW03]

$$\begin{aligned} K &= p_v, \\ X_1 &= \frac{1}{4}\frac{e^{2u}}{e^u+1}\cos vp_u^2 - \frac{1}{4}\frac{e^u(e^u+2)}{e^u+1}\cos vp_v^2 + \frac{1}{2}e^u \sin vp_u p_v, \\ X_2 &= \frac{1}{4}\frac{e^{2u}}{e^u+1}\sin vp_u^2 - \frac{1}{4}\frac{e^u(e^u+2)}{e^u+1}\sin vp_v^2 - \frac{1}{2}e^u \cos vp_u p_v. \end{aligned} \tag{17}$$

These integrals are functionally independent by the relation

$$X_1^2 + X_2^2 - H^2 - HK^2 = 0.$$

For the last Darboux space that is the Darboux space of type four given in (1), the infinitesimal distance in the coordinates (u,v) where $x = v + iu$, $y = v - iu$ is

$$ds^2 = \frac{(2\cos u + a)(d_u^2 + d_v^2)}{\sin^2 2u}, \tag{18}$$

with the Hamiltonian

$$H = -\frac{\sin^2 2u(p_u^2 + p_v^2)}{2\cos 2u + a} \tag{19}$$

admits three integrals of motion [KKMW03]

$$\begin{aligned} K &= p_v, \\ X_1 &= e^{2v}(-H + \cos 2up_u^2 + \sin 2up_u p_v), \\ X_2 &= e^{-2v}(-H + \cos 2up_v^2 - \sin 2up_u p_v). \end{aligned} \tag{20}$$

which are functionally independent by the relation

$$X_1 X_2 - K^4 - aK^2 H - H^2 = 0.$$

3 The constants of motions for Darboux spaces generated by Killing-Yano tensors

The metric for the Darboux space of type one (8)in the (u,v) coordinates is identified as

$$g_{ij}^{(1)} = \begin{pmatrix} 2u & 0 \\ 0 & 2u \end{pmatrix}. \tag{21}$$

Solving Eqs.(2), the corresponding KY tensor is obtained as $f_{12} = C_1 u$, where C_1 is a constant. Then from this KY tensor, the Killing tensors are calculated by using (4) and the components of the Killing tensor are

$$\begin{aligned} k_{11} &= \frac{-1}{2}(C_1^2 u), \\ k_{12} &= 0, \\ k_{22} &= \frac{-1}{2}(C_1^2 u). \end{aligned} \tag{22}$$

Then by using the method given in (6) we obtained the constant of motion for the metric (21) as

$$I = \frac{-1}{2}(C_1^2 u)(p_u^2 + p_v^2). \tag{23}$$

The Darboux space of type two with the infinitesimal distance (11) in the given (u,v) coordinates is

$$ds^2 = \frac{1}{4}(\frac{-a}{u^2} + b)(d_u^2 + d_v^2). \tag{24}$$

So the corresponding metric has the form

$$g_{ij}^{(2)} = \begin{pmatrix} \frac{1}{4}(\frac{-a}{u^2}+b) & 0 \\ 0 & \frac{1}{4}(\frac{-a}{u^2}+b) \end{pmatrix}. \tag{25}$$

From (2), the KY tensor for the metric (25) is obtained as

$$f_{12} = \frac{C_1(-a+bu^2)}{u^2}, \tag{26}$$

where C_1 is a constant. The Killing tensors are calculated from this KY tensor directly by using (4) and given as follows

$$\begin{aligned} k_{11} &= \frac{-4C_1^2}{u^2}(-a+bu^2), \\ k_{12} &= 0, \\ k_{22} &= \frac{-4C_1^2}{u^2}(-a+bu^2). \end{aligned} \tag{27}$$

The constant of motion we calculated from the KY tensor by the method (6) for the metric (25) is

$$I = \frac{4C_1^2}{u^2}(a-bu^2)(p_u^2+p_v^2). \tag{28}$$

The infinitesimal distance for the Darboux space of type three in(1) has the form (after the given change of coordinates)

$$ds^2 = (ae^{-u}+be^{-2u})(d_u^2+d_v^2), \tag{29}$$

and the metric represents this space is

$$g_{ij}^{(2)} = \begin{pmatrix} ae^{-u}+be^{-2u} & 0 \\ 0 & ae^{-u}+be^{-2u} \end{pmatrix}. \tag{30}$$

The associated KY tensor found as $f_{12} = C_1(ae^{-u}+be^{-2u})$ using (2) where C_1 is a constant. Without solving the Killing equation (5) we calculated the Killing tensors from (4) as

$$\begin{aligned} k_{11} &= -C_1^2(ae^{-u}+be^{-2u}), \\ k_{12} &= 0, \\ k_{22} &= -C_1^2(ae^{-u}+be^{-2u}), \end{aligned} \tag{31}$$

and by (6) the corresponding constant of motion is obtained as

$$I = -C_1^2(ae^{-u}+be^{-2u})(p_u^2+p_v^2). \tag{32}$$

The Darboux space of type four in (1) has the infinitesimal distance in the coordinates (u,v) where $x = v+iu$, $y = v-iu$

$$ds^2 = \frac{(2a\cos u + b)}{-4\sin^2 u}(d_u^2 + d_v^2), \tag{33}$$

and so the identified metric is

$$g_{ij}^{(2)} = \begin{pmatrix} \frac{(2a\cos u+b)}{-4\sin^2 u} & 0 \\ 0 & \frac{(2a\cos u+b)}{-4\sin^2 u} \end{pmatrix}. \tag{34}$$

The KY tensor $f_{12} = C$ (where C is a constant) is calculated by solving (2), then from (4) the components of Killing tensors are obtained as in the following

$$\begin{aligned} k_{11} &= \frac{4C^2\sin^2 u}{(2a\cos u + b)}, \\ k_{12} &= 0, \\ k_{22} &= \frac{4C^2\sin^2 u}{(2a\cos u + b)}, \end{aligned} \tag{35}$$

and (6) gives that the constant of motion calculated from KY tensor for the metric (34) is

$$I = \frac{4C^2\sin^2 u}{(2a\cos u + b)}(p_u^2 + p_v^2). \tag{36}$$

Acknowledgments

One of the authors (D.B.) would like to thank M. Henneaux, B. Edgar and N. Montesinos for their helpful discussions.
Özlem Defterli would like to thank to Prof. Dr. Kenan Taş and the organizers of the MME'06 Symposium for giving her the financial support.

References

[RS99] Ranada, M.F., Santander, M.: Superintegrable systems on two-dimensional sphere S-2 and the hyperbolic plane H-2. J. Math. Phys., **40(10)**, 5026–5057 (1999)

[KMP00] Kalnins, E.G., Miller, W.Jr., Pogosyan, G.S.: Completeness of multiseparability on the complex 2-sphere. J. Phys. A:Math. Gen., **33**, 6791–6806 (2000)

[KKPM01] Kalnins, E.G., Kress, J.M., Pogosyan, G.S., Miller, W.Jr.: Completeness of multiseparability in two-dimensional constant curvature spaces. J. Phys. A:Math. Gen., **34**, 4705–4720 (2001)

[STW01] Sheftel, M.B., Tempesta, P., Winternitz, P.: Superintegrable systems in quantum mechanics and classical Lie theory. J. Math. Phys., **42(2)**, 659–673 (2001)

[TTW01] Tempesta, P., Turbiner, A.V., Winternitz, P.: Exact solvability of superintegrable systems. J. Math. Phys., **42**, 4248–4261 (2001)

[CZ02] Curtright, T.L., Zachos, C.K.: Deformation quantization of superintegrable systems in Nambu mechanics. New J. Phys., **4**, 83–101 (2002)

[BD04a] Baleanu, D., Defterli, Ö.: Killing-Yano tensors and angular momentum. Czech. Journ. Phys., **54**, 157–165 (2004)

[Eva90] Evans, N.W.: Superintegrability in classical mechanics. Phys. Rev. A, **41**, 5666–5676 (1990)

[GPS95] Grosche, G., Pogosyan, G.S., Sissakian, A.N.: Path-integral discussion for Smorodinsky-Winternitz potentials.2. The 2-dimensional and 3-dimensional sphere. Fortschr. Phys., **43(6)**, 523–563 (1995)

[Arn89] Arnold, V.I.: Mathematical Methods of Classical Mechanics. Springer-Verlag, New York (1989)

[Golds80] Goldstein, H.: Classical Mechanics. Addison-Wesley, Massachusetts (1980)

[Wojciech83] Wojciechowski, S.: Superintegrability of the Calogero - Moser system. Phys. Lett. A, **95**, 279–281 (1983)

[FMSUW65] Fris, J., Mandrosov, V., Smorodinsky, Ya.A., Uhlir, M., Winternitz, P.: On higher symmetries in quantum mechanics. Phys. Lett., **16**, 354–360 (1965)

[Hieta84] Hietarinta, J.: Classical versus quantum integrability. J. Math. Phys., **25(6)**, 1833–1840 (1984)

[Hieta87] Hietarinta, J.: Direct methods for the search of the second invariant. Phys. Rep., **147**, 87–154 (1987)

[Hol82] Holt, C.R.: Construction of new integrable Hamiltonians in two degrees of freedom. J. Math. Phys., **23**, 1037–1046 (1982)

[LV95] Létourneau, P., Vinet, L.: Superintegrable systems: polynomial algebras and quasi-exactly solvable Hamiltonians. Ann. Phys., **243**, 144–162 (1995)

[Koen72] Koenigs, G.: Sur les géodésiques a intégrales quadratiques. In: Darboux, G. (ed) Leçons sur la théorie générale des surfaces. Chelsea Publishing, London (1972)

[KKW02] Kalnins, E.G., Kress, J.M., Winternitz, P.: Superintegrability in a two-dimensional space of non-constant curvature. J. Math. Phys., **43**, 970–996 (2002)

[KKMW03] Kalnins, E.G., Kress, J.M., Miller, W.Jr., Winternitz, P.: Superintegrable systems in Darboux spaces. J. Math. Phys., **44(12)**, 5811–5848 (2003)

[KSHM80] Kramer, D., Stephani, H., Herlt, E., MacCallum, M.: Exact Solutions of Einstein's Field Equations. Cambridge University Press, London (1980)

[GRH93] Gibbons, G.W., Rietdijk, R.H., Holten, J.W.van: SUSY in the sky. Nucl. Phys. B, **404**, 42–64 (1993)

[DB04] Defterli, Ö., Baleanu, D.: Killing-Yano tensors and superintegrable systems. Czech. Journ. Phys., **54(11)**, 1215–1221 (2004)

[Def04] Defterli, Ö.: Mathematical Aspects of Superintegrable Systems. MSc Thesis, Çankaya University, Ankara (2004)

[BD04b] Baleanu, D., Defterli, Ö.: Killing-Yano tensors, surface terms and superintegrable systems. AIP Conference Proceedings, **729**, 99–105 (2004)

A fourth order accurate difference-analytical method for solving Laplace's boundary value problem with singularities

A.A. Dosiyev[1] and S. Cival Buranay[2]

[1] Department of Mathematics, Eastern Mediterranean University, Gazimagusa, Mersin 10, Turkey adiguzel.dosiyev@emu.edu.tr

[2] Department of Mathematics, Eastern Mediterranean University, Gazimagusa, Mersin 10, Turkey suzan.buranay@emu.edu.tr

High accurate difference-analytical method of solving the mixed boundary value problem for Laplace's equation on graduated polygons (which can have broken sections and be multiply connected) is described and justified. The uniform estimate for the error of the approximate solution is of order $O(h^4)$, where h is the mesh step, for the errors of derivatives of order p, $p = 1, 2, ...$, in a finite neighbourhood of re-entrant vertices, of order $O(h^4/r_j^{p-\lambda_j})$, where r_j is the distance from the current point to the vertex in question, $\lambda_j = 1/\alpha_j$ or $\lambda_j = 1/2\alpha_j$ depending on the types of boundary conditions, $\alpha_j\pi$ is the value of the angle. The last part of the paper is devoted to illustrate numerical experiments.

1 Introduction

Among many approaches which have been proposed for the solution of the elliptic boundary value problem with singularities, a special emphasis has been placed on the construction of combined methods (see [Li98, Dos04] and references therein).

In [Dos04, Dos92, Dos94] a new combined difference-analytical method called the block-grid method (BGM) in solving the Laplace equation on graduated polygons is introduced. This method is a combination of the exponentially convergent block method (see [Vol80, Vol94]) in "singular" part, and the finite difference method in "nonsingular" part of the polygon. A sixth order gluing operator S^6 is constructed for gluing together the grids and blocks. The uniform estimate for the error of the BGM is of order $O(h^6)$, (h is the mesh step) when the boundary functions on some sides of polygon are given from the Hölder classes $C_{6,\lambda}$, $0 < \lambda < 1$. These requirements cannot be essentially lowered in $C_{k,\lambda}$. Therefore, the problem of construction and justification of

Kenan Taş et al. (eds), Mathematical Methods in Engineering, 167–176.

the methods with simple structure of finite difference equations and gluing operators, when the boundary functions are given from $C_{s,\lambda}$, $s < 6$, gets important. This problem is solved for the classes $C_{2,\lambda}$, $0 < \lambda < 1$, in [DB04], and $C_{1,1}$ in [DB05a] (uniform error is $O(h^2 |\ln h| + 1)$) for the BGM on arbitrary polygons.

In this paper, we justify BGM for the mixed boundary value problem on polygons by using a fourth order gluing operator S^4, which requires 16 functional values at grid nodes (S^6 needs 31). The uniform estimate for the error of the approximate solution is of order $O(h^4)$, whereas it is of order $O(h^4/r_j^{p-\lambda_j})$, for the errors of p-order derivatives ($p = 1, 2, ...$) in a finite neighborhood of the singular vertices; here r_j is the distance from the current point to the vertex in question, and λ_j depends on the magnitude of interior angle and the type of boundary conditions on the sides of considered angle. Finally, we illustrate the effectiveness of the method in solving the cracked-beam problem, which has a singularity due to abrupt changes in the type of boundary conditions.

2 Boundary value problem on polygons

Let G be an open simply connected graduated polygon in the xy plane, let γ_j, $j = 1, 2 \dots N$, be its sides, including the ends, enumerated counterclockwise, ($\gamma_0 = \gamma_N$), and let $\alpha_j \pi$ with $\alpha_j \in \left\{\frac{1}{2}, 1, \frac{2}{3}, 2\right\}$ be the interior angle formed by its sides γ_{j-1} and γ_j. Denote by A_j the vertex of the jth angle, by s the arclength, measured along the boundary of G in the positive direction, and by s_j the value of s at the vertex A_j. Let r_j, θ_j be a polar system of coordinates with pole in A_j, where the angle θ_j is taken counterclockwise from the side γ_j, and let ν_j be a parameter taking the values 0 or 1, and $\overline{\nu}_j = 1 - \nu_j$.

We consider the mixed boundary value problem

$$\Delta u = 0 \quad \text{on } G, \tag{1}$$

$$\nu_j u + \overline{\nu}_j u_n^{(1)} = \nu_j \varphi_j + \overline{\nu}_j \psi_j \quad \text{on } \gamma_j, \quad j = 1, 2, \dots, N, \tag{2}$$

where $\Delta \equiv \partial^2/\partial x^2 + \partial^2/\partial y^2$, $u_n^{(1)}$ is the derivative along the inner normal, φ_j and ψ_j are given functions, and

$$1 \leq \sum_{j=1}^{N} \nu_j \leq N, \tag{3}$$

$$\nu_j \varphi_j + \overline{\nu}_j \psi_j \in C_{4,\lambda}(\gamma_j), \quad 0 < \lambda < 1, \quad 1 \leq j \leq N. \tag{4}$$

At some vertices A_j for $\alpha_j = 1/2$ the following conjugation conditions are fulfilled

$$\nu_j \varphi_j^{(2q+\delta_\tau - 2)} + \overline{\nu}_j \psi_j^{(2q+\delta_\tau)}$$

$$= (-1)^{q+\delta_\tau+\delta_{\tau-1}}(\nu_{j-1}\varphi_{j-1}^{(2q+\delta_{\tau-1})} + \overline{\nu}_{j-1}\psi_{j-1}^{(2q+\delta_\tau)}), \tag{5}$$

where $\tau = \nu_{j-1}+2\nu_j$; $\delta_w = 1$ when $w = 0$; $\delta_w = 0$, when $w \neq 0$, $q = 0, 1, \ldots, \overline{Q}$, $\overline{Q} = [(4 - \delta_{\tau-3} - \delta_{\tau-1} - \delta_{\tau-2})/2] - \delta_\tau$. For the remaining vertices A_j no compatibility conditions are required to hold for the boundary conditions, in particular, the values of φ_{j-1} and φ_j at A_j might be different. In addition, we require that when $\alpha_j \neq 1/2$ or $\alpha_j = 1/2$ but (5) is not satisfied the boundary functions on γ_{j-1} and on γ_j be given as algebraic polynomials of arclength s measured along γ.

Let H be the set of all j for which $\alpha_j \neq 1/2$ or $\alpha_j = 1/2$ but (5) is not fulfilled. In the neighbourhood of A_j, $j \in H$ we construct two fixed block-sectors $T_j^i = T_j(r_{ji}) \subset G$, $i = 1, 2$, where $0 < r_{j2} < r_{j1} < \min\{s_{j+1} - s_j, s_j - s_{j-1}\}$, $T_j(r) = \{(r_j, \theta_j) : 0 < r_j < r,\ 0 < \theta_j < \alpha_j\pi\}$.

We set (see Volkov [Vol80])

$$R(m, m, r, \theta, \eta) = R(r, \theta, \eta) + (-1)^m R(r, \theta, -\eta), \tag{6}$$

$$R(1 - m, m, r, \theta, \eta) = R(m, m, r, \theta, \eta) - (-1)^m R(m, m, r, \theta, \pi - \eta), \tag{7}$$

where

$$R(r, \theta, \eta) = \frac{1 - r^2}{2\pi(1 - 2r\cos(\theta - \eta) + r^2)}, \tag{8}$$

is the kernel of the Poisson integral for a unit circle. We specify the kernel

$$R_j(r_j, \theta_j, \eta) = \lambda_j R(\nu_{j-1}, \nu_j, (\frac{r_j}{r_{j2}})^{\lambda_j}, \lambda_j\theta_j, \lambda_j\eta), j \in H, \tag{9}$$

where

$$\lambda_j = \frac{1}{(2 - \nu_{j-1}\nu_j - \overline{\nu}_{j-1}\overline{\nu}_j)\alpha_j}. \tag{10}$$

The following lemma will be used to construct the proposed method.

Lemma 1. *(Volkov [Vol94]) The solution u of the boundary value problem* (1), (2) *can be represented on $\overline{T}_j^2 \setminus V_j$, $j \in H$ in the form*

$$u(r_j, \theta_j) = Q_j(r_j, \theta_j) + \int_0^{\alpha_j\pi} (u(r_{j2}, \eta) - Q_j(r_{j2}, \eta))R_j(r_j, \theta_j, \eta_j)d\eta, \tag{11}$$

where V_j is the curvilinear part of the boundary of the sector T_j^2, $Q_j(r_j, \theta_j)$ is harmonic on T_j^1 and satisfies the boundary conditions (2) *on $\gamma_{j-1} \cap \overline{T}_j^1$ and $\gamma_j \cap \overline{T}_j^1$, $j \in H$.*

3 The fourth order block-grid method

We describe the procedure of obtaining the algebraic system of equations for the numerical solution of problem (1), (2).

Let us consider in addition to the sectors T_j^1, T_j^2 and T_j^3 (see Sec.2) in the neighborhood each vertex A_j, $j \in H$ of the polygon G a sector T_j^4, where $0 < r_{j4} < r_{j3} < r_{j2}$, and let $G^T = G \setminus (\cup_{j\in H}\overline{T}_j^4)$. Let $\Pi_k \subset G_T$, $k = 1, 2, \ldots M$ $(M < \infty)$ be certain fixed open rectangles with sides a_{1k} and a_{2k}, with a_{1k}/a_{2k} rational and $G \equiv (\cup_{k=1}^M \Pi_k) \cup (\cup_{j\in H} T_j^3)$. We denote by η_k the boundary of the rectangle Π_k and V_j the curvilinear part of the boundary of the sector T_j^2, and let $t_j = (\cup_{k=1}^M \eta_k) \cap \overline{T}_j^3$. The following general requirement (the overlapping condition) is imposed on the arrangement of the rectangles Π_k, $k = 1, 2, \ldots, M$: any point P lying on $\eta_k \bigcap G_T$, $1 \le k \le M$, or located on $V_j \bigcap G$, $j \in H$, falls inside at least one of the rectangles $\Pi_{k(p)}$, $1 \le k(p) \le M$, depending on P, the distance from P to $G_T \bigcap \eta_{k(p)}$ is not less than some constant κ_0 independent of P, which we call a *depth* of a gluing of the rectangles Π_k, $k = 1, 2, ..., M$. We introduce a parameter $h \in (0, \kappa_0/3]$ and define a square grid on Π_k, $1 \le k \le M$, with maximal possible step $h_k \le \min\{h, \min\{a_{1k}, a_{2k}\}/4\}$ such that the boundary η_k lies entirely on the grid lines. Let Π_k^h be the set of grid nodes on Π_k, let η_k^h be the set of nodes on η_k, and let $\overline{\Pi}_k^h = \Pi_k^h \cup \eta_k^h$. We denote the set of nodes in $\overline{\eta_k \cap G^T}$ by η_{k0}^h, the set of remaining nodes on η_k by η_{k1}^h, and the set of nodes on t_j by t_j^h. We also specify a natural number $n \ge [\ln^{1+\bar{\kappa}} h^{-1}] + 1$, where $\bar{\kappa} > 0$ is a fixed number, and the quantities $n(j) = \max\{4, [\alpha_j n]\}$, $\beta_j = \alpha_j \pi / n(j)$, and $\theta_j^m = (m - 1/2)\beta_j$, $j \in H$, $1 \le m \le n(j)$. On the arc V_j we choose points (r_{j2}, θ_j^m), $1 \le m \le n(j)$, and denote the set of these points by V_j^n. Let $\omega^{h,n} = (\cup_{k=1}^M \eta_{k0}^h) \cup (\cup_{j\in H} V_j^n)$, $\overline{G}_*^{h,n} = \omega^{h,n} \cup (\cup_{k=1}^M \overline{\Pi}_k^h)$

Let φ_j and ψ_j be given functions from the boundary conditions (2), and let $\varphi = \{\varphi_j\}_{j=1}^N$ and $\psi = \{\psi_j\}_{j=1}^N$. On the set $\omega^{h,n}$ we introduce the fourth order matching operator S^4 constructed in [Dos02a] (see Case 1). The value of $S^4(u_h, \varphi, \psi)$ at the point $P \in \omega^{h,n}$ is defined linearly in terms of the values of the function u_h at node P_k of the grid constructed on the rectangle $\Pi_{k(p)} \ni P$, part of the boundary $\eta_{k(p),0}$ of which is the maximum distance away from P and in terms of the assigned boundary values of φ (or ψ) at a fixed number of points or in terms of the values of the derivatives $\varphi^{(m)}$, $m = 0, 1, 2, 3$ (or $\psi^{(q)}$, $q = 0, 1, 2$). Moreover, $S^4(u_h, 0, 0)$ has representation

$$S^4(u_h, 0, 0) = \sum_{k\le 16} \xi_k u_{h,k}, \tag{12}$$

where $u_{h,k} = u_h(P_k)$, $\xi_k \ge 0$, $\sum_{k\le 16} \xi_k = 1$, and

$$u - S^4(u, \varphi, \psi) = O(h^4) \tag{13}$$

uniformly on $\omega^{h,n}$. We put

$$R_j^{(q)}(r_j, \theta_j) = \frac{R_j(r_j, \theta_j, \theta_j^q)}{\max\left\{1, \beta_j \sum_{p=1}^{n(j)} R_j(r_j, \theta_j, \theta_j^p)\right\}}, \tag{14}$$

where $R_j(r,\theta,\eta)$ is the kernel defined by (9). Then

$$0 \leq \beta_j \sum_{q=1}^{n(j)} R_j^{(q)}(r_j,\theta_j) \leq 1, \ j \in H. \tag{15}$$

Furthermore, from the estimation (2.29) in [Vol80] follows the existence of the positive constants n_0 and σ such that, for $n \geq n_0$,

$$\max_{\overline{T}_j^3} \beta_j \sum_{q=1}^{n(j)} R_j(r_j,\theta_j,\theta_j^q) \leq \sigma < 1, \tag{16}$$

when $\nu_{j-1}+\nu_j \geq 1$.

Consider the system of linear equations having for all $n \geq n_0$ a unique solution

$$u_h = Bu_h \text{ on } \Pi_k^h, \quad u_h = \bar{\nu}_m B_m u_h + E_{mh}(\varphi,\psi) \text{ on } \eta_{k1}^h \cap \gamma_m, \tag{17}$$

$$u_h = \bar{\nu}_m \bar{\nu}_{m+1} \dot{B}_m u_h + \dot{E}_{mh}(\varphi_m,\varphi_{m+1},\psi_m,\psi_{m+1}) \text{ on } \eta_{k1}^h \cap \gamma_m \cap \gamma_{m+1}, \tag{18}$$

$$u_h(r_j,\theta_j) = Q_j(r_j,\theta_j) + \beta_j \sum_{q=1}^{n(j)} R_j^{(q)}(r_j,\theta_j)(u_h(r_{j2},\theta_j^q) - Q_j(r_{j2},\theta_j^q)),$$

$$(r_j,\theta_j) \in t_j^h \cup \overline{T}_j^3, \tag{19}$$

$$u_h = S^4(u_h,\varphi,\psi) \text{ on } \omega^{h,n}, \tag{20}$$

where $1 \leq m \leq N$, $1 \leq k \leq M$, $j \in H$; B, B_m, $\dot{B}_m$, E_{mh}, and $\dot{E}_{mh}$ are defined in [Vol66] (see also [Dos02a]), $Q_j(r_j,\theta_j)$ is the harmonic polynomial used in Lemma 1.

Definition 1. *The solution of the system* (17) – (20) *is called a numerical solution of the problem* (1), (2) *on* $\overline{G}_*^{h,n} \cup \overline{T}_j^3$.

4 The estimation error

Theorem 1. *Assume that conditions* (3) – (5) *hold. Then, there exists a natural number* n_0 *such that for all* $n = \max\left\{n_0, \left[\ln^{1+\bar{\kappa}} h^{-1}\right] + 1\right\}$, $\bar{\kappa} > 0$ *being a fixed number, the following inequalities are valid:*

$$\max_{\overline{G}_*^{h,n}} |u_h - u| \leq ch^4, \tag{21}$$

$$\left| \frac{\partial^p}{\partial x^{p-q} \partial y^q} (U_h(r_j,\theta_j) - u(r_j,\theta_j)) \right| \leq c_p h^4 \quad on \ \overline{T}_j^3, \tag{22}$$

first, for integer λ_j and any ν_{j-1} and ν_j when $p \geq \lambda_j$, second, for $\nu_{j-1} = \nu_j = 0$ and any λ_j when $p = 0$;

$$\left| \frac{\partial^p}{\partial x^{p-q} \partial y^q} (U_h(r_j, \theta_j) - u(r_j, \theta_j)) \right| \leq c_p h^4 / r^{p-\lambda_j} \quad on\ \overline{T}_j^3, \tag{23}$$

for any λ_j, if $\nu_{j-1}+ \nu_j \geq 1$, $0 \leq p < \lambda_j$ or $\nu_{j-1} = \nu_j = 0$, $1 \leq p < \lambda_j$;

$$\left| \frac{\partial^p}{\partial x^{p-q} \partial y^q} (U_h(r_j, \theta_j) - u(r_j, \theta_j)) \right| \leq c_p h^4 / r^{p-\lambda_j} \quad on\ \overline{T}_j^3 \backslash A_j, \tag{24}$$

for noninteger λ_j, and any ν_{j-1} and ν_j when $p > \lambda_j$. Everywhere $0 \leq q \leq p$, λ_j is the quantity (10), ν_{j-1} and ν_j are parameters entering into the boundary conditions (2), u is a solution of the problem (1), (2), c_p, $p = 0, 1, ...$, are constants independent of $r_j = r_j(x, y)$, $\theta_j = \theta_j(x, y)$, and h.

Proof. Let

$$\varepsilon_h = u_h - u, \tag{25}$$

where u_h is a solution of system (17)-(20) on $\overline{G}_*^{h,n}$, and u is the trace on $\overline{G}_*^{h,n}$ of the solution of (1),(2). On the basis of (1), (2), (17)-(20), and (25) the error ε_h satisfies the system of difference equations

$$\varepsilon_h = B\varepsilon_h + r_h^1 \text{ on } \Pi_k^h, \tag{26}$$

$$\varepsilon_h = \overline{\nu}_m B_m \varepsilon_h + r_h^2 \text{ on } \eta_{k1}^h \cap \gamma_m, \tag{27}$$

$$\varepsilon_h = \overline{\nu}_m \overline{\nu}_{m+1} \dot{B}_m \varepsilon_h + r_h^3 \text{ on } \eta_{k1}^h \cap \gamma_m \cap \gamma_{m+1}, \tag{28}$$

$$\varepsilon_h(r_j, \theta_j) = \beta_j \sum_{q=1}^{n(j)} R_j^{(q)}(r_j, \theta_j) \varepsilon_h(r_{j2}, \theta_j^q) + r_{jh}^4, (r_j, \theta_j) \in t_j^h, \tag{29}$$

$$\varepsilon_h = S^4 \varepsilon_h + r_h^5 \text{ on } \omega^{h,n}, \tag{30}$$

where $1 \leq m \leq N$, $1 \leq k \leq M$, $j \in H$,

$$r_h^1 = Bu - u \quad \text{on } \cup_{k=1}^M \Pi_k^h, \tag{31}$$

$$r_h^2 = \overline{\nu}_m B_m u - u + E_{mh}(\varphi_m, \psi_m) \quad \text{on } \gamma_m \cap (\cup_{k=1}^M \eta_{k1}^h), \tag{32}$$

$$r_h^3 = \overline{\nu}_m \overline{\nu}_{m+1} \dot{B}_m u - u + \dot{E}_{mh}(\varphi_m, \varphi_{m+1}, \psi_m, \psi_{m+1}),$$
$$\text{on } \gamma_m \cap \gamma_{m+1} \cap (\cup_{k=1}^M \eta_{k1}^h), \tag{33}$$

$$r_{jh}^4 = \beta_j \sum_{q=1}^{n(j)} R_j^{(q)}(r_j, \theta_j)(u(r_{j2}, \theta_j^q) - Q_j(r_{j2}, \theta_j^q)) - (u(r_j, \theta_j) - Q_j(r_j, \theta_j))$$
$$\text{on } \cup_{j \in H} t_j^h, \tag{34}$$

$$r_h^5 = S^4 u - u \quad \text{on } \omega^{h,n}. \tag{35}$$

On the basis of Lemma 6.2 from [Dos04] there exists a natural number n_0 such that, for all $n = \max\left\{n_0, \left[\ln^{1+\bar{\kappa}} h^{-1}\right] + 1\right\}$, where $\bar{\kappa} > 0$ is a fixed number,

$$\max_{j\in H} \left|r_{jh}^4\right| \leq ch^4. \tag{36}$$

Since, the set of points $\omega^{h,n}$ located from the vertices of the polygon G at the distance exceeding some positive quantity independent of h, then by virtue of (4), (5), from (13) we obtain

$$\max_{\omega^{h,n}} \left|r_h^5\right| \leq ch^4. \tag{37}$$

On the basis of (31)-(36), and Remark 15 in [Dos03] by analogy with the proof of Theorem 6.3 from [Dos04] we obtain

$$\max_{\overline{G}_*^{h,n}} |\varepsilon_h| \leq ch^4. \tag{38}$$

Let $T_j^* = T_j(r_j^*)$, $r_j^* = (r_{j2} + r_{j3})/r_{j2}$. By virtue of Lemma 6.5 from [Vol94], and Lemma 6.2 from [Dos04], we have

$$\Bigg|\beta_j \textstyle\sum_{q=1}^{n(j)} R_j(r_j,\theta_j,\theta_j^q)(u(r_{j2},\theta_j^q) - Q_j(r_j,\theta_j^q))$$

$$- \int_0^{\alpha_j\pi} R_j(r_j,\theta_j,\eta)(u(r_{j2},\eta) - Q_j(r_{j2},\eta))d\eta\Bigg| \leq ch^4, \quad \text{on } \overline{T}_j^*, j \in H, \tag{39}$$

for all $n = \max\left\{n_0, \left[\ln^{1+\bar{\kappa}} h^{-1}\right] + 1\right\}$, $\bar{\kappa} > 0$ is a fixed number. From (21) and the boundedness of $\beta_j \sum_{q=1}^{n(j)} R_j(r_j,\theta_j,\theta_j^q)$ for all $n = \max\left\{n_0, \left[\ln^{1+\bar{\kappa}} h^{-1}\right] + 1\right\}$, we obtain

$$\left|\beta_j \sum_{q=1}^{n(j)} R_j(r_j,\theta_j,\theta_j^q)(u_h(r_{j2},\theta_j^q) - u(r_j,\theta_j^q))\right| \leq ch^4, \quad \text{on } \overline{T}_j^*, j \in H. \tag{40}$$

From (19), (39), (40), and Lemma 1 for all $n = \max\left\{n_0, \left[\ln^{1+\bar{\kappa}} h^{-1}\right] + 1\right\}$, we have

$$|U_h(r_j,\theta_j) - u(r_j,\theta_j)| \leq ch^4, \quad \text{on } \overline{T}_j^*, \ j \in H. \tag{41}$$

Since $T_j^3 \subset \overline{T}_j^*$, $j \in H$, then from the inequality (41) follows a proof of (21) when $p = 0$. To establish the validity of remainder inequalities of Theorem 1 we put

$$\varepsilon_h(r_j,\theta_j) = U_h(r_j,\theta_j) - u(r_j,\theta_j) \quad \text{on } \overline{T}_j^*, \ j \in H. \tag{42}$$

From (19), (42) follows that the function $\varepsilon_h(r_j,\theta_j)$ is continuous on $\overline{T}_j^*$, and is a solution of the boundary value problem

$$\Delta\varepsilon = 0 \quad \text{on } T_j^*, \tag{43}$$

$$\nu_m \varepsilon_h + \overline{\nu}_m (\varepsilon_h)'_n = 0 \text{ on } \gamma_m \cap \overline{T}^*_j, m = j-1, j, \tag{44}$$

$$\varepsilon_h(r^*_j, \theta_j) = U_h(r^*_j, \theta_j) - u(r^*_j, \theta_j), 0 \le \theta_j \le \alpha_j \pi. \tag{45}$$

Taking into account (41)-(44), from the Lemma 6.12 given by Volkov [Vol94] follows all remainder inequalities of Theorem 1.

Theorem 2. *The system* (17) – (20) *can be solved by Schwarz's alternating method with any accuracy* $\varepsilon > 0$ *in a uniform metric with the number of iterations* $O(\ln \varepsilon^{-1})$, *independent of* h *and* n.

Proof. The proof is obtained by analogy with the proof of Theorem 7.1 from [Dos04].

5 The cracked-beam problem

Let $G = \{(x, y) : -1/2 < x < 1/2, 0 < y < 1/2\}$, and γ be its boundary. We consider the following problem:

$$\Delta u = 0 \text{ in } G, \tag{46}$$

$$u = 0 \text{ on } y = 0, \ -1/2 \le x \le 0, \tag{47}$$

$$u = 0.125 \text{ on } y = 1/2, \ -1/2 \le x \le 1/2 \tag{48}$$

$$\frac{\partial u}{\partial n} = 0 \text{ on the other boundary segments of } \gamma. \tag{49}$$

In the original problem, $\Delta v = -1$ and $v = 0$ along $y = 1/2$. The transformation $u = v + y^2/2$ leads to the problem considered above (see [FGW73, Wig88, OGS91]).

The cracked-beam problem has singularity at the origin due to abrupt change in the type of boundary conditions. In this problem, from an engineering standpoint, the most interesting quantity is the constant (called the stress intensity factor)

$$\sigma^0 = \lim_{r \to 0} r^{-\frac{1}{2}} \left[u(r, 0) - u(0, 0) \right], \tag{50}$$

which gives a measure of "the amount of torsion the beam can endure before fracture occurs" [FGW73]. The quantity σ^0 can be approximated by using the formula (19) as

$$\sigma^0(h^{-1}, n) = \frac{2}{n\sqrt{r_{1,2}}} \sum_{q=1}^{n} u_h(r_{1,2}, \theta^q) \cos \frac{\theta^q}{2}. \tag{51}$$

In Table 1 the values at some points of the stress function v, and in Table 2 the value of stress intensity factor are given. In both tables for the BGM $r_{1,2} = 0.43$, $(h^{-1}, n) = (32, 60)$, are taken. In Table 3 the extremely accurate values (accurate $\approx 10^{-19}$) of the function v obtained in [Dos05b] are shown. The

Table 1. The solution obtained by BGM, of the cracked beam problem at various points compared with results from the literature.

(x, y)	Fix et al.[FGW73]	Wigley[Wig88]	Olson et al.[OGS91]	BGM
$(0, 1/24)$	0.027425	0.027428	0.027429	0.027427895509
$(-11/24, 1/4)$	0.032877	0.032878	0.032879	0.032877886408
$(11/24, 1/4)$	0.070844	0.070844	0.070844	0.07084351329431

Table 2. Value of the stress intensity factor

Fix et al.[FGW73]	Wigley[Wig88]	Olson et al.[OGS91]	BGM
0.1917	0.19112	0.191119	0.19111863199

Table 3. The extremely accurate values of the function v

(x, y)	Dosiyev and Cival [Dos05b]
$(0, 1/24)$	0.02742789550524769 12
$(-11/24, 1/4)$	0.0328778863976422215
$(11/24, 1/4)$	0.0708435132220747256

extremely accurate result for the value of stress intensity factor in [Dos05b] is 0.191118631971872089. By comparing the results in Table 1 and in Table 2 with the extremely accurate results in [Dos05b] follows that the accuracy of the block grid solution is $\approx 10^{-10}$.

Remark 1. The fourth order BGM can be generalized for the problems on arbitrary polygons by using a unique square grid on "nonsingular" part(see [Dos02b]).

Remark 2. The method and results are valid for multiply connected polygons.

References

[Li98] Li, Z.C.: Combined Methods for Elliptic problems with Singularities, Interfaces and Infinities. Kluwer Academic Publishers, Dordrech Boston London (1998)

[Dos04] Dosiyev, A.A.: The high accurate block-grid method for solving Laplace's boundary value problem with singularities. SIAM J. Numer. Anal., **42(1)**, 153–178 (2004)

[Dos92] Dosiyev, A.A.: A block-grid method for increasing accuracy in the solution of the Laplace equation on polygons. Russian Acad. Sci. Dokl.Math., **45(2)**, 396–399 (1992)

[Dos94] Dosiyev, A.A.: A block-grid method of increased accuracy for solving Dirichlet's problem for Laplace's equation on polygons. Comp. Maths Math. Phys., **34(5)**, 591–604 (1994)

[Vol80] Volkov, E.A.: An exponentially converging method for solving Laplace's equation on polygons. Math. USSR Sb., **37(3)**, 295–325 (1980)

[Vol94] Volkov, E.A.: Block method for solving the Laplace equation and constructing conformal mappings. CRC Press, USA (1994)

[DB04] Dosiyev, A.A., Buranay Cival, S.: A difference-analytical method for solving Laplace's boundary value problem with singularities. In: Akca, H., Boucherif, A., Covachev, V. (ed) 2004-Dynamical Systems and Applications. GBS Publishers and Distributers, India (2004)

[DB05a] Dosiyev, A.A., Buranay Cival S.: A combined method for solving Laplace's boundary value problem with singularities. Inter. Journal of Pure and Appl. Math., **21(3)**, 353–367 (2005)

[Dos02a] Dosiyev, A.A.: A fourth order accurate composite grids method for solving Laplace's boundary value problems with singularities. Comp. Maths Math. Phys., **42(6)**, 832–849 (2002)

[Vol66] Volkov, E.A.: Effective error estimates for grid method solutions of boundary-value problems for Laplace's and Poisson's equations on rectangle and certain triangles. Tr. Mat. Inst. Akad. Nauk SSSR., **74**, 55–85 (1966)

[Dos03] Dosiyev, A.A.: On the maximum error in the solution of Laplace equation by finite difference method. Intern. Journal of Pure and Appl. Math., **7(2)**, 229–241 (2003)

[FGW73] Fix, G.J., Gulati, S., Wakoff, G.I.: On the use of singular functions with finite element approximations. J.Comput. Phys., **13**, 209–228(1973)

[Wig88] Wigley, N.M.: An efficient method for subtracting off singularities at corners for Laplace's equation. J. Comput. Phys., **78**, 369–377 (1988)

[OGS91] Olson, L.G., Georgiou, G.C., Schults, W.W.: An efficient finite element method for treating singularities in Laplace's equation. J. Comput. Phys., **96**, 391–410 (1991)

[Dos05b] Dosiyev, A.A., Buranay Cival, S.: On solving the cracked beam problem by a block method. In: Georgiou, G.,Papannastasiou, P., and Papadrakakis, M. (ed) 5th GRACM International Congress on Computational Mechanics. Kantzilaris Publication, Nicosia (2005)

[Dos02b] Dosiyev, A.A.: A high accuracy difference-analytical method for solving Laplace's boundary value problem with singularities. Proceedings of the International Conference on Computational Mathematics, Part II, Novosibirsk, 402–407 (2002)

Modeling of PDE processes with finite dimensional non-autonomous ODE systems*

Mehmet Önder Efe

Department of Electrical and Electronics Engineering, TOBB Economics and Technology University, TR-06560, Ankara, Turkey onderefe@etu.edu.tr

Processes governed by Partial Differential Equations (PDE) display very rich dynamical behavior, which is continuous spatially. Influencing the behavior of PDE systems through boundaries is an interesting research as it is involves the handling of infinite dimensionality, due to which the traditional tools of control theory do not apply directly. This study demonstrates how a nonlinear PDE is converted into a reasonably descriptive Ordinary Differential Equation (ODE) model. The approach is based on Proper Orthogonal Decomposition (POD), which separates the temporal and spatial components of the dynamics. The finite term expansion of the solution results in an autonomous ODE and this paper demonstrates how the external excitations are made explicit in the dynamical model. 2D Burgers equation is used to illustrate the effectiveness of the approach and a finite dimensional dynamical model is shown to be capable of capturing the essential response.

1 Introduction

Systems and control theory is a well-founded framework and the research on the discovery and understanding of system dynamics is an every growing subset of the paradigm. Defining $\alpha \in \Re^n$ and $\gamma \in \Re^m$ as the state vector and the control input vector, respectively, one branch of the control research focuses on the affine nonlinear models having the representation $\dot{\alpha} = f(\alpha) + g(\alpha)\gamma$, [Kha02]. Upon suitably defining the functions $f(\cdot)$ and $g(\cdot)$, the linear state space systems of the form $\dot{\alpha} = A\alpha + B\gamma$ are obtained and these systems constitute a subset for the aforementioned nonlinear models, [Oga97]. Unsurprisingly, in both representations above, we have the control input explicitly. The problem studied in this paper is to obtain dynamic models for processes governed by PDEs. We particularly focus on the 2D Burgers equation

* This work is supported by TOBB ETÜ BAP Program (Contract No: ETÜ BAP 2006/04)

Kenan Taş et al. (eds), Mathematical Methods in Engineering, 177–187.

$w_t = cw_{xx} - \mu w(w_x + w_y)$, [Ef06]. The problem is interesting not only because of its spatial continuity but also because of its nonlinearity. POD method is utilized to perform the modal decomposition and unfortunately an autonomous dynamical model lacking the control input(s) is obtained, [LT01]. The paper demonstrates how the boundary conditions are made explicit in the resulting finite dimensional ODE model having the structure $\dot{\alpha} = f(\alpha) + g(\alpha)\gamma$.

Various kinds of Burgers equation have been studied in the past. In [DH03, McDH04, Blen91, BE03, Hata98, NMT01], a simplified version of Navier-Stokes equations given by the Partial Differential Equation (PDE) set $\underline{w}_t + \epsilon(\underline{w} \cdot \nabla)\underline{w} = \mu\nabla^2\underline{w}$ with $\underline{w}$ being 2-by-1 vector function is described as the 2D Burgers equation. The 2D Burgers equation is therefore considered as a turbulence free cartoon for Navier-Stokes equations and has been studied in the past for modelling traffic flows, shock waves and acoustic transmission. Blender postulates a method to obtain the solution of the above mentioned PDE set iteratively, [Blen91]. In [BE03], Boules and Eick perform the model reduction with Fourier expansions. In [Sire99, Hie00, Zhu96], some other variants of 2D Burgers equation have been considered with the goal of finding exact solutions under certain circumstances. These types are $(w_t + ww_x - w_{xx})_x + w_{yy} = 0$ in [Sire99, Hie00], $w_t + uu_x + w_{xx} + w_{xxx} = 0$ in [Zhu96]. In [NMT01], Nishinari et al. focus on cellular automaton, which is extensively studied for developing models of traffic flow, fluids and immune systems, and therefore a good model to work on is a variant of Burgers equation. In [Hata98], the dynamics that arises upon discretization of 2D Burgers equation is analyzed. The effects of chosen time step (Δt) for getting physically reasonable numerical solutions are elaborated. Wescott et al. present a computational technique to obtain the numerical solutions of PDEs having nonlinear convection terms like 2D Burgers equation and Navier-Stokes equations, [Wes01], the goal in which is to reduce the computation time without giving concessions from the accuracy. Boules and Eick obtain the solution of Burgers equation for a specific boundary regime and initial conditions, [BE03]. Using a truncated Fourier series expansion yields an autonomous ODE set, the solution of which approximates the numerical solution, and the derived model rebuilds the situation implied by the chosen initial and boundary conditions. When the 1D version given by $w_t = -ww_x + w_{xx}$ is taken into consideration, it is seen that a significant amount of research outcome has been reported on modelling and control system design. A majority of the works on 1D Burgers equation emphasize the similar difficulties as the motivating factors and focus on the solutions and solvability issues. The current paper, on the other hand, presents an approach to low dimensional (LD) modeling of PDE processes which are to be controlled through boundaries. The contribution of this paper is to demonstrate that a LD nonlinear model can easily be obtained to represent the essential dynamics of a 2D Burgers equation excited continuously through the boundaries of a square domain. The POD algorithm is presented in the second section. The third section demonstrates how the autonomous ODE set is made non-autonomous. The fourth section is devoted to the justification of the devised LD model and the concluding remarks are given at the end of the paper.

2 Proper orthogonal decomposition

Consider the ensemble $W_i(x,y)$, $i = 1,2,\ldots,N_s$, where N_s is the number of elements. Every element of this set corresponds to a snapshot observed from a process, say for example the flow governed by 2D Burgers equation with initial and boundary conditions,

$$\begin{aligned}
&w_t(x,y,t) = c\left(w_{xx}(x,y,t) + w_{yy}(x,y,t)\right) - \mu w(x,y,t)\left(w_x(x,y,t) + w_y(x,y,t)\right)\\
&w(x,0,t) = f_1(x)\gamma_1(t), w(1,y,t) = f_2(y)\gamma_2(t)\\
&w(x,1,t) = f_3(x)\gamma_3(t), w(0,y,t) = f_4(y)\gamma_4(t),\\
&w(x,y,0) = 0 \quad \forall(x,y),
\end{aligned} \tag{1}$$

where, c and μ are the known constant parameters, and the subscripts $_x$, $_y$ and $_t$ refer to the partial differentiation with respect to x, y and time, respectively. The continuous time process takes place over the physical domain $\Omega := \{(x,y)|(x,y) \in [0,1] \times [0,1]\}$ and the solution is obtained on a spatial grid denoted by Ω_d, which describes the coordinates of the pixels of every snapshot in the ensemble. The entities described over Ω_d are matrices in $\mathbb{R}^{N_y \times N_x}$. Note that in (1), $f_i(\cdot)$ for each i is a function that describes how $\gamma_i(t)$ influences the behavior along the corresponding edge of Ω. $f_i(\cdot)$s can be selected arbitrarily yet for every i, $f_i(0) = f_i(1) = 0$ so that the problem description is consistent at the corners of Ω, and $\gamma_i(t)$ becomes independent from $\gamma_j(t)$ for $i \neq j$ and the external excitations can be selected arbitrarily.

With this problem description, the goal of applying POD is to find an orthonormal basis set letting us to write the solution as

$$w(x,y,t) = \sum_{i=1}^{R_L} \alpha_i(t)\Phi_i(x,y), \tag{2}$$

where $\alpha_i(t)$ is the ith temporal mode, $\Phi_i(x,y)$ is the ith spatial function (basis function or the eigenfunction), R_L is the number of independent basis functions that can be synthesized from the given ensemble, or equivalently that spans the space described by the ensemble. It will later be clear that if the basis set $\{\Phi_i(x,y)\}_{i=1}^{R_L}$ is an orthonormal set, Galerkin projection yields the autonomous set of ODEs directly. Let us summarize the POD procedure.

Step 1. Calculate the $N_s \times N_s$ dimensional correlation matrix L, the $(ij)^{\text{th}}$ entry of which is $L_{ij} := \langle W_i, W_j\rangle_{\Omega_d}$, where $\langle \cdot,\cdot\rangle_{\Omega_d}$ is the inner product operator defined over $\mathbb{R}^{N_y \times N_x}$.

Step 2. Find the eigenvectors (denoted by v_i) and the associated eigenvalues (λ_i) of the matrix L. Sort them in a descending order in terms of the magnitudes of λ_i. Note that every v_i is an $N_s \times 1$ dimensional vector satisfying $v_i^{\mathrm{T}} v_i = \frac{1}{\lambda_i}$, here, for simplicity of the exposition, we assume that the eigenvalues are distinct.

Step 3. Construct the basis set by using

$$\Phi_i(x,y) = \sum_{j=1}^{N_s} v_{ij} W_j(x,y), \tag{3}$$

where v_{ij} is the jth entry of the eigenvector $v_i = (v_{i1}\ v_{i2}\ \dots\ v_{iN_s})^{\mathrm{T}}$, and $i = 1, 2, ..., R_L$, with $R_L = \mathrm{rank}(L)$. It can be shown that $\langle \Phi_i(x,y), \Phi_j(x,y)\rangle_{\Omega} = \delta_{ij}$ with δ_{ij} being the Kronecker delta function. Notice that the basis functions are admixtures of the snapshots, [LT01].

Step 4. Calculate the temporal coefficients. Taking the inner product of both sides of (2) with $\Phi_i(x,y)$, the orthonormality property leads to

$$\begin{aligned}
\alpha_i(t_0) &= \langle \Phi_i(x,y), w(x,y,t_0)\rangle_{\Omega} \\
&= \langle \phi_i, W_{t_0}\rangle_{\Omega_d} \\
&:= \frac{1}{N_s}\sum_{l=1}^{N_x}\sum_{j=1}^{N_y} \phi_i(x_l, y_j) W_{t_0}(x_l, y_j) \\
&:= \phi_i(x,y) \odot W_{t_0}(x,y),
\end{aligned} \tag{4}$$

where $\phi_i \in \mathbb{R}^{N_y \times N_x}$ is a sampled form of the basis function Φ_i defined over Ω. The operator denoted by $\odot$ computes a real number that is the sum of all elements of a matrix obtained through the elementwise multiplication of the two matrices that $\odot$ lies in between. Without loss of generality, an element of the ensemble $\{W_i(x,y)\}_{i=1}^{N_s}$ may be $W(x,y,t_0)$. Therefore, in order to generate the temporal gain, $\alpha_k(t)$, of the spatial eigenfunction ϕ_k, one would take the inner product of ϕ_k with the elements of the ensemble as given below,

$$\langle W_i, \phi_k\rangle_{\Omega_d} \approx \alpha_k(t_i) \qquad i = 1, 2, \dots, N_s \tag{5}$$

The above computation is important for making a comparison between the quantities obtained from the decomposition (See (5)) and the quantities obtained from the model. Note that the temporal coefficients satisfy orthogonality properties over the discrete set $t \in \{t_1, t_2, \dots, t_{N_s}\}$ (See (6)).

$$\sum_{i=1}^{N_s} \langle W_i(x,y), \Phi_k(x,y)\rangle_{\Omega_d}^2 \approx \sum_{i=1}^{N_s} \alpha_i^2(t_i) = \lambda_k. \tag{6}$$

For a more detailed discussion on the POD method, the reader is referred to [LT01, Lum67, RCM04] and the references therein.

Fundamental assumption: The majority of works dealing with POD and model reduction applications presume that the flow is dominated by coherent modes, which means that the flow can be decomposed into distinguishable components in the order of dominance. Because of the dominance of coherent modes, the typical spread of the eigenvalues of the correlation matrix turns out to be logarithmic and the terms decay very rapidly in magnitude. This fact enables us to assume that a reduced order representation, say with M modes ($M < R_L$) can also be written as an equality

$$W(x,y,t) = \sum_{i=1}^{M} \alpha_i(t)\Phi_i(x,y), \tag{7}$$

and the reduced order model is derived under the assumption that (7) satisfies the governing PDE in (1), [LT01, RCM04, Ravi00]. Unsurprisingly, such an assumption results in a model having uncertainties, however, one should keep in mind that the goal is to find a model, which matches the infinite dimensional system in some sense of approximation with typically $M \ll R_L \leq N_s$. To represent how good such an expansion is, a percent energy measure is defined as follows

$$E = \frac{\sum_{i=1}^{M} \lambda_i}{\sum_{i=1}^{R_L} \lambda_i} \times 100\,\%, \tag{8}$$

where the tendency of $E \to 100\%$ means that the model captures the dynamical information contained in the snapshots well. Conversely, an insufficient model will be obtained if E is far below 100%. Clearly, POD lets us reduce the dimensionality of the problem from infinity to R_L, and the fundamental assumption further enables us to reduce the LD model order to M. In the next section, we demonstrate how the boundary conditions are transformed into explicit control terms in the corresponding set of ODEs.

3 Low dimensional modeling

In the order reduction phase, we need to obtain the autonomous ODE model first. Towards this goal, if (7) is a solution to the PDE in (1), then it has to satisfy the PDE. Substituting (7) into (1) with the fundamental assumption yields

$$\sum_{i=1}^{M} \dot{\alpha}_i(t)\Phi_i(x,y) = c\sum_{i=1}^{M} \alpha_i(t)\Psi_i(x,y) - \mu\sum_{i=1}^{M}\sum_{j=1}^{M} \alpha_i(t)\alpha_j(t)\Phi_i(x,y)\Upsilon_j(x,y), \tag{9}$$

where $\Psi_i(x,y) = \frac{\partial^2\Phi_i(x,y)}{\partial x^2} + \frac{\partial^2\Phi_i(x,y)}{\partial y^2}$ and $\Upsilon_j(x,y) = \frac{\partial\Phi_j(x,y)}{\partial x} + \frac{\partial\Phi_j(x,y)}{\partial y}$. Taking the inner product of both sides with $\Phi_k(x,y)$ and remembering $\langle\Phi_i(x,y), \Phi_k(x,y)\rangle_\Omega = \delta_{ik}$ with δ_{ik} being Kronecker delta results in

$$\dot{\alpha}_k = c\sum_{i=1}^{M} \alpha_i \langle \Phi_k, \Psi_i \rangle_{\Omega} - \mu \sum_{i=1}^{M}\sum_{j=1}^{M} \alpha_i \alpha_j \langle \Phi_k, \Phi_i \Upsilon_j \rangle_{\Omega}, \tag{10}$$

where we have dropped the arguments x, y and t for simplicity. Defining ζ_k and β_j as the entities in Ω_d corresponding to the entities Ψ_k and Υ_j in Ω, respectively, one could rewrite (10) as

$$\dot{\alpha}_k(t) = c\sum_{i=1}^{M} \alpha_i(t) \left(\phi_k(x,y) \odot \zeta_i(x,y)\right) - \mu \sum_{i=1}^{M}\sum_{j=1}^{M} \alpha_i \alpha_j \left(\phi_k \odot (\phi_i \otimes \beta_j)\right)_{\Omega_d} \tag{11}$$

where $\otimes$ stands for the elementwise multiplication of the two matrices that it lies in between. For the first term above, notice that $\odot$ operator can be applied individually over $\Omega_d^1, \ldots, \Omega_d^n$ which are n nonoverlapping subdomains of Ω_d such that $\Omega_d^1 \bigcup \ldots \bigcup \Omega_d^n = \Omega_d$. This lets us separate the entries corresponding to boundaries without modifying the value of $\langle \phi_k, \zeta_i \rangle_{\Omega_d}$, i.e. $\phi_k(x,y) \odot \zeta_i(x,y)$ as seen in (12),

$$\begin{aligned}
\dot{\alpha}_k(t) = {} & c\sum_{i=1}^{M} \alpha_i(\phi_k(x,0) \odot \zeta_i(x,0) + c\sum_{i=1}^{M} \alpha_i(\phi_k(1,y) \odot \zeta_i(1,y)) + \\
& c\sum_{i=1}^{M} \alpha_i(\phi_k(x,1) \odot \zeta_i(x,1)) + c\sum_{i=1}^{M} \alpha_i(\phi_k(0,y) \odot \zeta_i(0,y)) + \\
& c\sum_{i=1}^{M} \alpha_i \left(\phi_k^{\circ}(x,y) \odot \zeta_i^{\circ}(x,y)\right) - \mu \sum_{i=1}^{M}\sum_{j=1}^{M} \alpha_i \alpha_j \left(\phi_k^{\circ} \odot (\phi_i^{\circ} \otimes \beta_j^{\circ})\right) \\
& -\mu \sum_{i=1}^{M} \alpha_i \phi_i(x,0) \odot \sum_{j=1}^{M} \alpha_j \left(\phi_k(x,0) \otimes \beta_j(x,0)\right) \\
& -\mu \sum_{i=1}^{M} \alpha_i \phi_i(1,y) \odot \sum_{j=1}^{M} \alpha_j \left(\phi_k(1,y) \otimes \beta_j(1,y)\right) \\
& -\mu \sum_{i=1}^{M} \alpha_i \phi_i(x,1) \odot \sum_{j=1}^{M} \alpha_j \left(\phi_k(x,1) \otimes \beta_j(x,1)\right) \\
& -\mu \sum_{i=1}^{M} \alpha_i \phi_i(0,y) \odot \sum_{j=1}^{M} \alpha_j \left(\phi_k(0,y) \otimes \beta_j(0,y)\right)
\end{aligned} \tag{12}$$

In above, $\phi_k^{\circ}(x,y)$ denotes a matrix which is obtained when the boundary elements of $\phi_k(x,y)$ are removed, i.e. the first and the last rows, and columns. Similarly, in the computation of terms like $\phi_k(x,0) \odot \zeta_i(x,0)$, the terms $\phi_k(x,0)$ and $\zeta_i(x,0)$ correspond to the first rows of the matrices $\phi_k(x,y)$ and $\zeta_i(x,y)$, respectively.

Due to the lengthy nature of the expression above, we will demonstrate how the terms $T_1 := \sum_{i=1}^{M} \alpha_i(\phi_k(x,0) \odot \zeta_i(x,0)$, which is responsible for the linear diffusion type term, and $T_2 := \sum_{i=1}^{M} \alpha_i \phi_i(x,0) \odot \sum_{j=1}^{M} \alpha_j \left(\phi_k(x,0) \otimes \beta_j(x,0)\right)$, which is responsible for the nonlinear term, are manipulated to postulate the model. Notice that the boundary condition along $y=0$ edge of Ω is given by

$$\sum_{i=1}^{M} \alpha_i(t)\phi_i(x,0) = f_1(x)\gamma_1(t), \tag{13}$$

which states that if (7) is a solution, then is must be satisfied at the boundaries as well. Considering this fact constitutes the crux of the LD modeling effort. The boundary condition in (13) can be paraphrased as

$$\alpha_k(t)\phi_k(x,0) = f_1(x)\gamma_1(t) - \sum_{i=1}^{M} (1-\delta_{ik})\,\alpha_i(t)\phi_i(x,0). \tag{14}$$

Separating the kth component of the term T_1, which is obtained when $i=k$, lets us embed the boundary conditions in (14) into the expression of T_1 as given below,

$$T_1 := (f_1(x) \odot \zeta_k(x,0))\gamma_1(t) + \sum_{i=1}^{M} \alpha_i \left(\phi_k(x,0) \odot \zeta_i(x,0) - \phi_i(x,0) \odot \zeta_k(x,0)\right) \tag{15}$$

Similarly, for the term T_2, we have rather simple arrangements to see the excitation terms explicitly,

$$T_2 = \gamma_1(t) \sum_{j=1}^{M} \alpha_j \left(f_1(x) \odot (\phi_k(x,0) \otimes \beta_j(x,0))\right). \tag{16}$$

The representations in (15) and (16) indicate that the terms seen in (12) can be concatenated and the following low dimensional dynamical model is obtained,

$$\dot{\alpha}(t) = A\alpha(t) - C(\alpha(t)) + \sum_{i=1}^{4} (B_i - D_i\alpha)\,\gamma_i(t), \tag{17}$$

where $\alpha(t) = (\alpha_1(t)\ \alpha_2(t)\ \ldots\ \alpha_M(t))^{\mathrm{T}}$ is the state vector and

$$C(\alpha) = (\alpha^{\mathrm{T}} C_1 \alpha \quad \alpha^{\mathrm{T}} C_2 \alpha \quad \ldots \quad \alpha^{\mathrm{T}} C_M \alpha)^{\mathrm{T}}.$$

The $(ki)^{\text{th}}$ entry of matrix A is computed as

$$\begin{aligned}(A)_{ki} = {} & c(\phi_k(x,y) \odot \zeta_i(x,y) - \phi_i(x,0) \odot \zeta_k(x,0) - \phi_i(1,y) \odot \zeta_k(1,y) \\ & -\phi_i(x,1) \odot \zeta_k(x,1) - \phi_i(0,y) \odot \zeta_k(0,y)),\end{aligned} \tag{18}$$

and the kth entries of the column vectors $B_1, \ldots, B_4$ are

$$(B_1)_k = cf_1(x) \odot \zeta_k(x,0)\,, \qquad (B_2)_k = cf_2(y) \odot \zeta_k(1,y),$$
$$(B_3)_k = cf_3(x) \odot \zeta_k(x,1)\,, \qquad (B_4)_k = cf_4(y) \odot \zeta_k(0,y). \tag{19}$$

Likewise, we have the components $C(\alpha)$ and matrices $D_1, \ldots, D_4$ contributed by the nonlinear term $-\mu w(w_x + w_y)$ of the PDE in (1);

$$(C_k)_{ij} = \mu(\phi_k^\circ \odot (\phi_i^\circ \otimes \beta_j^\circ)),$$

where C_k is an $M \times M$ matrix, and

$$(D_1)_{kj} = \mu f_1(x) \odot (\phi_k(x,0) \otimes \beta_j(x,0)), (D_2)_{kj} = \mu f_2(y) \odot (\phi_k(1,y) \otimes \beta_j(1,y)),$$
$$(D_3)_{kj} = \mu f_3(x) \odot (\phi_k(x,1) \otimes \beta_j(x,1)), (D_4)_{kj} = \mu f_4(y) \odot (\phi_k(0,y) \otimes \beta_j(0,y)) \tag{20}$$

This result practically lets us have the representative nonlinear dynamical model in (17) for the infinite dimensional process in (1), which needs to be validated. The next section presents to what extent the modelling strategy discussed here could be successful.

4 Validation of the nonlinear dynamical model

According to the described procedure, several tests have been done. Due to the numerical advantages, the PDE has been solved by using Crank-Nicholson method (See [Far93] for details), with a step size of 1 msec. The initial distribution is taken zero everywhere and we have chosen $c = 2$ and $\mu = 1$. In order to form the solution, a linear grid having $N_x = N_y = 40$ points in x-direction and y-direction respectively. According to the above parameter values, a set of 501 snapshots embodies the entire numerical solution, among which a linearly sampled $N = 251$ snapshots have been used for the POD scheme. Although one may use the entire set of snapshots, it has been shown by Sirovich, [Siro87], that a reasonably descriptive subset of them can be used for the same purpose. In the literature, this approach is called *method of snapshots*, which significantly reduces the computational intensity of the overall scheme, (See also [Ravi00, LT01]). Once the modes have been obtained, we truncate the solution at $M = 12$, which represents %99.9832 of the total energy described in the denominator of the expression in (8).

In order to demonstrate the performance of the dynamic model, we choose the functions that are effective along the boundaries as $f_1(x) = \sin(2\pi x)$, $f_2(y) = \sin(2\pi y)$, $f_3(x) = -\sin(2\pi x)$ and $f_4(y) = -\sin(2\pi y)$. As the temporal excitations we chose the following input signals, $\gamma_1(t) = \sin(2\pi 50t(T-t))$, $\gamma_2(t) =$

$\sin(2\pi 8t(T-\sin(4\pi t))$, $\gamma_3(t) = \sin(2\pi 65t(T-t))$ and $\gamma_4(t) = \sin(2\pi 8t(T-\cos(4\pi t))$, where $T = 0.5$ seconds. The choice of the above set of excitations signals is deliberate as they are spectrally rich enough, i.e, $\alpha_k(t)$'s will undergo regimes that change sometimes slowly and sometimes fast depending on the spectral composition of the external inputs. Under these conditions, the numerical content of the dynamical model is computed and a dynamic model is obtained. It is observed that the temporal variables obtained form the POD algorithm are very close to those obtained from the LD model and this observation indicates that the LD model is a good representative for the chosen conditions. Undoubtedly, one would expect a good match between the state variables obtained from the POD algorithm and the state variables obtained through the numerical solution of the ODE set in (17). One might question whether the model is specific to the boundary conditions above. Remedying this is accomplished by choosing another set of boundary conditions and obtaining the response of the model without modifying the model parameters. For this purpose, we set $\gamma_1(t) = \sin(2\pi 55t(T-t))$, $\gamma_2(t) = \sin(2\pi 9t(T-\sin(2\pi t))$, $\gamma_3(t) = \sin(2\pi 75t(T-t))$ and $\gamma_4(t) = \sin(2\pi 7t(T-\cos(5\pi t))$. The choice of this set of excitation signals is due to the spectral richness and aperiodicity within the selected time. With these excitation signals, without modifying the basis and the model contents, we have obtained the results illustrated in Figure 1, where every subplot contains two curves. It is seen that the state variables are obtained precisely when the relevant signal changes slowly. During the regions where the signals change quickly, there is some small discrepancy due to the neglected modes, chosen excitation signals, effects of numerical differentiation and so on. This result practically tells us that POD is a powerful technique for developing ODE models for PDE systems.

5 Conclusions

This paper considers POD based LD modeling of the flow governed by 2D Burgers equation. The studied problem is interesting due to its nonlinearity and the manner in which the boundary signals excite the process. The 2D nature of the problem makes it further appealing to dwell on. The paper validates the model and emphasizes that the model is useful over a set of operating conditions. The simulation results have shown that the model produces the temporal content of the dynamics precisely, indicating that the POD algorithm associated with the presented separation scheme are successful in deriving a representative LD model. Altholection of M is absolutely a matter of the problem in hand and the auxiliary conditions it is subjectugh not discussed here, it is clear from (8) that the model performance is strictly dependent upon the number of modes chosen, i.e. increasing M yields better yet mode complicated models, however, as M decreases, the similarity of the modes from the LD model to those from the POD algorithm disappears according to the energy expression in (8). In short, the se to.

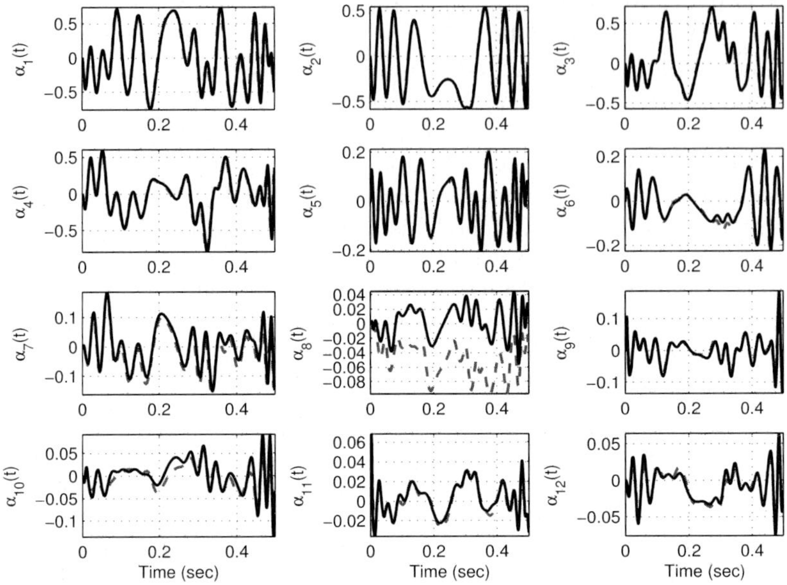

Fig. 1. Temporal variables; solid curves are form the dynamic model in (17) whereas dashed curves are the desired ones obtained from the POD algorithm (See (5))

As a result, POD is a powerful technique but its usefulness depends upon the PDE in hand, problem settings and the associated operating conditions.

References

[Kha02] Khalil, H.K.: Nonlinear systems. Prentice-Hall, 3rd Ed., New Jersey (2002)

[Oga97] Ogata, K.: Modern control engineering. Prentice-Hall, New Jersey (1997)

[Ef06] Efe, M.Ö.: Observer based boundary control for 2D Burgers equation. Trans. of the Institute of Measurement and Control, **28** 177–185 (2006)

[LT01] Ly, H.V., Tran, H.T.: Modeling and control of physical processes using proper orthogonal decomposition. Mathematical and Computer Modelling of Dynamical Systems, **33**, 223-236 (2001)

[DH03] Donea, J., Huerta, A.: Finite element methods for flow problems. John Wiley & Sons, West Sussex (2003)

[McDH04] McDonough, J.M., Huang, M.T.: A 'poor man's NavierStokes equation': derivation and numerical experiments-the 2-D case. Int. J. for Numerical Methods in Fluids, **44**, 545-578 (2004)

[Blen91] Blender, R.: Iterative solution of nonlinear partial-differential equations. Journal of Physics A: Mathematical and General, **24**, L509-L512 (1991)

[BE03] Boules, A.N., Eick, I.J.: A spectral approximation of the two-dimensional Burgers equation. Indian Journal of Pure & Applied Mathematics, **34**, 299-309 (2003)

[Hata98] Hataue, I.: Mathematical and numerical analyses of dynamical structure of numerical solutions of two-dimensional fluid equations. Journal of the Physical Society of Japan, **67**, 1895-1911 (1998)

[NMT01] Nishinari, K., Matsukidaira, J., Takahashi, D.: Two-dimensional Burgers cellular automaton. Journal of the Physical Society of Japan, **70**, 2267-2272 (2001)

[Sire99] Sirendaoreji, S.J.: Exact solutions of the two-dimensional Burgers equation. Journal of Physics A: Mathematical and General, **32**, 6897-6900 (1999)

[Hie00] Hietarinta, L.: Comments on 'Exact solutions of the two-dimensional Burgers equation'. Journal of Physics A: Mathematical and General, **33**, 5157-5158 (2000)

[Zhu96] Zhu, Z.: Exact solutions for a two-dimensional KdV-Burgers-type equation. Chinese Journal of Physics, **34**, 1101-1105 (1996)

[Wes01] Wescott, B.L.: An efficient formulation of the modified nodal integral method and application to the two-dimensional Burgers equation. Nuclear Science and Eng., **139**, 293-305 (2001)

[Lum67] Lumley, J.: The structure of inhomogeneous turbulent flows. In: Yaglom, A., Tatarsky, V. (ed) Atmospheric Turbulence and Wave Propagation. Vol.3, Nauca, Moscow (1967)

[RCM04] Rowley, C.W., Colonius, T., Murray, R.M.: Model reduction for compressible flows using POD and Galerkin projection. Physica D-Nonlinear Phenomena, **189**, 115-129 (2004)

[Ravi00] Ravindran, S.S.: A reduced order approach for optimal control of fluids using proper orthogonal decomposition. Int. Journal for Numerical Methods in Fluids, **34**, 425-488 (2000)

[Far93] Farlow, S.J.: Partial differential equations for scientists and engineers. Dover Publications Inc., New York (1993)

[Siro87] Sirovich, L.: Turbulence and the dynamics of coherent structures. Quarterly of Applied Mathematics, **XLV**, 561-590 (1987)

On solutions of discrete nonlinear elliptic boundary value problems

Gusein SH. Guseinov

Department of Mathematics, Atilim University, 06836 Incek, Ankara, Turkey
guseinov@atilim.edu.tr

In this paper, we consider a boundary value problem (BVP) for second order nonlinear partial difference equations on the lattice rectangles. Some explicit conditions are established that ensure existence and uniqueness or solely existence of solution to the BVP under consideration.

1 Introduction

Let $\mathbf{Z}$ the set of integers. A lattice point $z = (i, j)$ in the plane is defined as a point with integer coordinates i and j. The set of all lattice points is denoted by $\mathbf{Z^2} = \mathbf{Z} \times \mathbf{Z}$. Two lattice points are said to be 4-neighbors or, simply, neighbors if their Euclidean distance is one. The four neighbors of $z = (i, j)$ are $(i-1, j), (i+1, j), (i, j-1)$, and $(i, j+1)$. The lattice points $z_1, z_2, \ldots, z_n$ are said to form a path with terminals z_1 and z_n if z_1 is a neighbor of z_2, z_2 is a neighbor of z_3, etc. A set of lattice points is said to be connected if any two of its points are terminals of a path of points contained in the set. A nonempty connected set of lattice points is called a lattice domain.

Given a lattice domain Ω, a lattice point is a boundary point of Ω if it does not belong to Ω but has at least one neighbor in Ω. The set of boundary points of Ω is denoted by $\partial\Omega$. Let Ω be a finite lattice domain. In [Che03]-Sec.7.8, the nonlinear second order partial difference equation

$$u_{i+1,j} + u_{i-1,j} + u_{i,j+1} + u_{i,j-1} - 4u_{ij} + f(i, j, u_{ij}) = 0, \qquad (i, j) \in \Omega, \quad (1)$$

is considered subject to the Dirichlet boundary condition

$$u_{ij} = 0, \qquad (i, j) \in \partial\Omega, \quad (2)$$

and the following result is established by using the contraction mapping theorem (Banach fixed point theorem).

Kenan Taş et al. (eds), Mathematical Methods in Engineering, 189–196.

Theorem 1. *Suppose* $f : \mathbf{Z^2} \times \boldsymbol{R} \to \boldsymbol{R}$ *satisfies the Lipschitz condition*

$$|f(i,j,t) - f(i,j,s)| \le q_{ij}\,|t-s|\,, \qquad (i,j) \in \Omega, \qquad t,s \in \boldsymbol{R},$$

where (q_{ij}) *is a nontrivial and nonnegative bivariate sequence,* $\boldsymbol{R}$ *denotes the set of real numbers. Suppose further that the least positive eigenvalue* λ^* *of the problem*

$$u_{i+1,j} + u_{i-1,j} + u_{i,j+1} + u_{i,j-1} - 4u_{ij} + \lambda q_{ij} u_{ij} = 0, \qquad (i,j) \in \Omega,$$

$$u_{ij} = 0, \qquad (i,j) \in \partial\Omega,$$

satisfies $\lambda^* > 1$. *Then the boundary value problem (1),(2) has a unique solution.*

In the present paper we replace the condition $\lambda^* > 1$ of Theorem 1 by more explicit conditions in the case where Ω is a lattice rectangle and $q_{ij} = L$ for all $(i,j) \in \Omega$. We consider also the case where $f(i,j,t)$ does not satisfy a Lipschitz condition and apply the Brouwer fixed point theorem in this case to get an existence theorem without uniqueness of solution.

2 The difference operators

Let (f_k) be a given complex sequence, where $k \in \mathbf{Z}$. The forward and backward difference operators Δ and ∇ are defined by

$$\Delta f_k = f_{k+1} - f_k \qquad \text{and} \qquad \nabla f_k = f_k - f_{k-1},$$

respectively. We easily see that

$$\nabla f_k = \Delta f_{k-1},$$

$$\Delta^2 f_k = \Delta(\Delta f_k) = f_{k+2} - 2f_{k+1} + f_k,$$

$$\Delta\nabla f_k = f_{k+1} - 2f_k + f_{k-1} = \nabla\Delta f_k = \Delta^2 f_{k-1} = \nabla^2 f_{k+1}.$$

For any integers $a, b \in \mathbf{Z}$ with $a < b$ we have the summation by parts formulas

$$\sum_{k=a}^{b} (\Delta f_k) g_k = f_{b+1} g_b - f_a g_{a-1} - \sum_{k=a}^{b} f_k (\nabla g_k), \tag{3}$$

$$\sum_{k=a}^{b} (\nabla f_k) g_k = f_b g_{b+1} - f_{a-1} g_a - \sum_{k=a}^{b} f_k (\Delta g_k). \tag{4}$$

The first order partial differences of a bivariate sequence (u_{ij}) are defined by

$$\Delta_1 u_{ij} = u_{i+1,j} - u_{ij}, \qquad \Delta_2 u_{ij} = u_{i,j+1} - u_{ij},$$

$$\nabla_1 u_{ij} = u_{ij} - u_{i-1,j}, \qquad \nabla_2 u_{ij} = u_{ij} - u_{i,j-1}.$$

The discrete Laplacian of u_{ij} is defined by

$$\begin{aligned} \Delta_1\nabla_1 u_{ij} + \Delta_2\nabla_2 u_{ij} &= u_{i+1,j} + u_{i-1,j} + u_{i,j+1} + u_{i,j-1} - 4u_{ij} \\ &= \Delta_1^2 u_{i-1,j} + \Delta_2^2 u_{i,j-1} = \nabla_1^2 u_{i+1,j} + \nabla_2^2 u_{i,j+1}. \end{aligned}$$

3 Application of the Banach fixed point theorem

Let $a \geq 2$ and $b \geq 2$ be fixed integers and

$$\Omega = \{(i,j) : 1 \leq i \leq a-1, 1 \leq j \leq b-1\}$$

be a lattice rectangle with the boundary

$$\begin{aligned}\partial\Omega = \{(0,j) : 1 \leq j \leq b-1\} \cup \{(a,j) : 1 \leq j \leq b-1\} \\ \cup \{(i,0) : 1 \leq i \leq a-1\} \cup \{(i,b) : 1 \leq i \leq a-1\}.\end{aligned}$$

Consider the boundary value problem (BVP)

$$u_{i+1,j} + u_{i-1,j} + u_{i,j+1} + u_{i,j-1} - 4u_{ij} + f(i,j,u_{ij}) = 0, \qquad (i,j) \in \Omega, \quad (5)$$

$$u_{ij} = 0, \qquad (i,j) \in \partial\Omega, \quad (6)$$

where (u_{ij}) for $(i,j) \in \Omega \cup \partial\Omega$, is a desired solution.

Denote by $\mathcal{B}$ the real Banach space of all (finite) bivariate sequences $u = (u_{ij})$, where $(i,j) \in \Omega$ and $u_{ij} \in \mathbf{R}$, with the norm

$$u = \max_{(i,j)\in\Omega} |u_{ij}|.$$

Obviously, $\mathcal{B}$ is an $(a-1)(b-1)$ dimensional real linear space. Next, we define the operators $A : \mathcal{B} \to \mathcal{B}$ and $F : \mathcal{B} \to \mathcal{B}$ as follows. For any $u \in \mathcal{B}$ we set

$$(Au)_{ij} = -(u_{i+1,j} + u_{i-1,j} + u_{i,j+1} + u_{i,j-1} - 4u_{ij}),$$

$$(Fu)_{ij} = f(i,j,u_{ij}),$$

for $(i,j) \in \Omega$, taking into account that when we calculate $(Au)_{ij}$ for $i = 1$ or $i = a-1$ and also for $j = 1$ or $j = b-1$ we use the boundary condition (6).

A proof of the following lemma can be found in [Smi85]- Chap.5.

Lemma 1. *Linear operator A is invertible and*

$$\|A^{-1}u\| \leq \frac{1}{4}(a^2+b^2)\,\|u\| \qquad forall \qquad u \in \mathcal{B}.$$

Note that the operator F is nonlinear in general.

The (BVP) (5),(6) is equivalent to the vector equation $Au = Fu$ for $u \in \mathcal{B}$, which can be written in the form $u = A^{-1}Fu$. Let us set $T = A^{-1}F$. Then we get that the (BVP) (5),(6) is equivalent to the equation $u = Tu(u \in \mathcal{B})$. The last equation is a fixed point problem for the operator T.

Below in the proof of Theorem 2 we will use the following well-known contraction mapping theorem named also as the Banach fixed point theorem: *Let $\mathcal{B}$ be a nonempty complete metric space with a metric d. Assume $T : \mathcal{B} \to \mathcal{B}$ is a contraction mapping, i.e., there is an α, $0 < \alpha < 1$, such that $d(Tu,Tv) \leq \alpha d(u,v)$ for all $u, v \in \mathcal{B}$. Then T has a unique fixed point in $\mathcal{B}$.*

Theorem 2. *Suppose* $f : \boldsymbol{Z}^2 \times \boldsymbol{R} \to \boldsymbol{R}$ *satisfies the Lipschitz condition*

$$f(i,j,t) - f(i,j,s) \leq L\,|t-s|\,, \qquad (i,j) \in \Omega, \tag{7}$$

where L *is a positive constant not depending on* $(i,j) \in \Omega$ *and* $t, s \in \boldsymbol{R}$. *Suppose further that*

$$\frac{1}{4}(a^2+b^2)L < 1 \tag{8}$$

Then the BVP (5),(6) has a unique solution.

Proof. It will be sufficient to show that the operator $T = A^{-1}F$ is a contraction mapping on the space $\mathcal{B}$. For $u, v \in \mathcal{B}$ we have, using Lemma 1,

$$\|Tu - Tv\| = \left\|A^{-1}Fu - A^{-1}Fv\right\| = \left\|A^{-1}(Fu - Fv)\right\|$$

$$\leq \frac{1}{4}(a^2+b^2)\,\|Fu - Fv\|\,.$$

On the other hand, using the Lipschitz condition (7) we have

$$\|Fu - Fv\| = \max_{(i,j)\in\Omega} |f(i,j,u_{ij}) - f(i,j,v_{ij})| \leq L \max_{(i,j)\in\Omega} |u_{ij} - v_{ij}| = L\,\|u - v\|\,.$$

Therefore

$$\|Tu - Tv\| \leq \frac{1}{4}(a^2+b^2)L\,\|u - v\|\,.$$

So we see that under the condition (8), T is a contraction mapping and thus has a unique fixed point by the Banach fixed point theorem.

4 Application of the Brouwer fixed point theorem

To get an existence result without uniqueness of solution, we will apply below the following Brouwer fixed point theorem: *Let* $\mathcal{B}$ *be a finite dimensional Banach space and* S *a nonempty closed, bounded, and convex subset of* $\mathcal{B}$. *Assume* $T : \mathcal{B} \to \mathcal{B}$ *is a continuous operator. If the operator* T *leaves the set* S *invariant, i.e., if* $T(S) \subset S$, *then* T *has at least one fixed point in* S.

Theorem 3. *Let the function* $f : \boldsymbol{Z}^2 \times \boldsymbol{R} \to \boldsymbol{R}$ *be continuous with respect to its third argument. Suppose further that for some* $R > 0$

$$\frac{1}{4}(a^2+b^2) \max_{(i,j)\in\Omega, |t|\leq R} |f(i,j,t)| \leq R. \tag{9}$$

Then the BVP (5),(6) has a solution (u_{ij}) *such that* $|u_{ij}| \leq R$ *for all* $(i,j) \in \Omega$.

Proof. From the continuity of $f(i,j,t)$ with respect to t it follows that the operator F is continuous. Hence the operator $T = A^{-1}F$ is continuous. Next, consider the set (ball) $S = \{u \in \mathcal{B} : \|u\| \leq R\}$. Obviously, S is a nonempty closed, bounded, and convex set in $\mathcal{B}$. Let us show that T maps S into itself. For u in S we have $\|u\| \leq R$ and hence $|u_{ij}| \leq R$ for all $(i,j) \in \Omega$. Therefore using Lemma 1 and the condition (9) we have for $u \in S$

$$\|Tu\| = \left\|A^{-1}Fu\right\| \leq \frac{1}{4}(a^2+b^2)\,\|Fu\| = \frac{1}{4}(a^2+b^2)\max_{(i,j)\in\Omega}|f(i,j,u_{ij})| \leq R.$$

Thus $T : S \to S$. Now the Brouwer fixed point theorem can be applied to obtain a fixed point of T in S. The proof is complete.

Remark 1. The condition (9) is satisfied for a sufficiently large positive number R if $f(i,j,t)$ is a bounded function (for example, of type $\sin t$ or $\cos t$). This condition is satisfied for a sufficiently large positive number R also if

$$|f(i,j,t)| \leq C\,|t|^{\beta} \qquad \textit{for} \qquad (i,j) \in \Omega \qquad \textit{and} \qquad t \in \mathbf{R},$$

where C is a positive constant and $0 \leq \beta < 1$.

5 Conditions in terms of least eigenvalue

In this section we will make use of the Hilbert space technique.

Denote by $\mathcal{H}$ the real finite dimensional Hilbert space of all real (finite) bivariate sequences $u = (u_{ij})$, where $(i,j) \in \Omega$ and $u_{ij} \in \mathbf{R}$, with the inner product (scalar product)

$$\langle u, v\rangle = \sum_{(i,j)\in\Omega} u_{ij}v_{ij} = \sum_{i=1}^{a-1}\sum_{j=1}^{b-1} u_{ij}v_{ij} \qquad \textit{for} \qquad u, v \in \mathcal{H}.$$

Using summation by parts formulas (3),(4), and remembering that, according to the boundary condition (6), we put

$$u_{0j} = u_{aj} = 0(1 \leq j \leq b-1), \qquad u_{i0} = u_{ib} = 0(1 \leq i \leq a-1),$$

for all $u \in \mathcal{H}$, we find that

$$\langle Au, v\rangle = -\sum_{i=1}^{a-1}\sum_{j=1}^{b-1}(\Delta_1\nabla_1 u_{ij} + \Delta_2\nabla_2 u_{ij})v_{ij} = \langle u, Av\rangle\,, \tag{10}$$

$$\langle Au, u\rangle = -\sum_{i=1}^{a-1}\sum_{j=1}^{b-1}\left[(\nabla_1 u_{ij})^2 + (\nabla_2 u_{ij})^2\right], \tag{11}$$

for all $u, v \in \mathcal{H}$. Relation (10) shows that the operator $A$is self-adjoint, while (11) shows that it is positive:

$$\langle Au, u\rangle > 0 \quad forall \quad u \in \mathcal{H},\ u \neq 0.$$

Therefore it follows from Linear Algebra that the operator A has exactly $m = \dim \mathcal{H} = (a-1)(b-1)$ orthonormal eigenvectors $\varphi^{(k)}$, $1 \le k \le m$, and the corresponding eigenvalues λ_k, $1 \le k \le m$, are positive:

$$A\varphi^{(k)} = \lambda_k \varphi^{(k)}, \qquad 1 \le k \le m,$$

$$\left\langle \varphi^{(k)}, \varphi^{(l)} \right\rangle = 0 \quad if \quad k \neq l, and = 1 \quad if \quad k = l,$$

$$0 < \lambda_1 \le \lambda_2 \le ... \le \lambda_m.$$

For arbitrary $u \in \mathcal{H}$ we have

$$u = \sum_{k=1}^{m} c_k \varphi^{(k)}, \qquad c_k = \left\langle u, \varphi^{(k)} \right\rangle,$$

$$\|u\|^2 = \langle u, u\rangle = \sum_{k=1}^{m} c_k^2.$$

Since the operator A is positive, it is invertible. We have

$$Au = \sum_{k=1}^{m} c_k \lambda_k \varphi^{(k)}, \qquad A^{-1}u = \sum_{k=1}^{m} \frac{c_k}{\lambda_k} \varphi^{(k)}. \tag{12}$$

From (12) we get that

$$\left\|A^{-1}u\right\|^2 = \sum_{k=1}^{m} \frac{c_k^2}{\lambda_k^2} \le \frac{1}{\lambda_1^2} \sum_{k=1}^{m} c_k^2 = \frac{1}{\lambda_1^2} \|u\|^2.$$

Thus, we have established the following result.

Lemma 2. *The operator A is invertible and*

$$\left\|A^{-1}u\right\| \le \frac{1}{\lambda_1} \|u\| \qquad forall \qquad u \in \mathcal{H},$$

where λ_1 is the least (i.e., the first) positive eigenvalue of the operator A.

We can compute the eigenvalues of the operator A explicitly. The eigenvalue equation $Au = \lambda u$, $u \in \mathcal{H}$ can be written in the form

$$-(u_{i+1,j} + u_{i-1,j} + u_{i,j+1} + u_{i,j-1} - 4u_{ij}) = \lambda u_{ij} (1 \le i \le a-1, 1 \le j \le b-1), \tag{13}$$

$$u_{0j} = u_{aj} = 0(1 \leq j \leq b-1), u_{i0} = u_{ib} = 0(1 \leq i \leq a-1). \tag{14}$$

The problem (13),(14) can be solved by the method of separation of variables, setting

$$u_{ij} = x_i y_j. \tag{15}$$

Substituting of this into (13) gives

$$-(x_{i+1} - 2x_i + x_{i-1})y_j - x_i(y_{j+1} - 2y_j + y_{j-1}) = \lambda x_i y_j.$$

To separate the variables, we divide both sides by $x_i y_j$:

$$-\frac{x_{i+1} - 2x_i + x_{i-1}}{x_i} = \frac{y_{j+1} - 2y_j + y_{j-1}}{y_j} + \lambda.$$

Both sides must be equal to a constant, by the usual argument. Thus

$$-\frac{x_{i+1} - 2x_i + x_{i-1}}{x_i} = \frac{y_{j+1} - 2y_j + y_{j-1}}{y_j} + \lambda = -\mu.$$

This yields two ordinary linear difference equations for $(x_i)$and (y_j):

$$x_{i+1} - 2x_i + x_{i-1} = \mu x_i, \qquad 1 \leq i \leq a-1 \qquad \textit{and} \tag{16}$$

$$y_{j+1} - 2y_j + y_{j-1} = -(\lambda + \mu)y_j, \qquad 1 \leq j \leq b-1. \tag{17}$$

From (14) and (15) we get the boundary conditions

$$x_0 = x_a = 0 \qquad \textit{and} \tag{18}$$

$$y_0 = y_b = 0. \tag{19}$$

Next, the one-dimensional problems (16),(18) and (17),(19) are discrete Sturm-Liouville problems (see [KP91]-Chap.7). The eigenvalues of (16),(18) are

$$\mu_p = -4\sin^2\frac{\pi p}{2a} \qquad (1 \leq p \leq a-1),$$

and the corresponding orthonormal eigenvectors $x^{(p)} = (x_i^{(p)})$ have the form

$$x_i^{(p)} = \sqrt{\frac{2}{a}}\sin\frac{\pi p}{a}, \qquad i = 1, 2, ..., a-1(1 \leq p \leq a-1).$$

In view of the similar result for the problem (17),(19) we find that

$$\lambda^{(p,q)} = 4(\sin^2\frac{\pi p}{2a} + \sin^2\frac{\pi q}{2b}) \qquad (1 \leq p \leq a-1, 1 \leq q \leq b-1) \tag{20}$$

are eigenvalues of (13),(14) and the corresponding orthonormal eigenvectors $u^{(p,q)} = (u_{ij}^{(p,q)})_{(i,j)\in\Omega}$ are of the form

$$u_{ij}^{(p,q)} = \sqrt{\frac{4}{ab}} \sin\frac{\pi pi}{a} \sin\frac{\pi qj}{b} \qquad (1 \le p \le a-1, 1 \le q \le b-1).$$

Since the vectors $u^{(p,q)}$ $(1 \le p \le a-1, 1 \le q \le b-1)$ form a basis for $\mathcal{H}$, (20) contains all eigenvalues of the problem (13),(14).

From (20) we see that the first (positive) eigenvalue of the operator A is

$$\lambda_1 = 4(\sin^2\frac{\pi}{2a} + \sin^2\frac{\pi}{2b}).$$

Now using Lemma 2 and reasoning as in the proofs of Theorem 2 and Theorem 3 we get the following results, respectively.

Theorem 4. *Suppose $f : \mathbf{Z^2} \times \boldsymbol{R} \to \boldsymbol{R}$ satisfies the Lipschitz condition*

$$|f(i,j,t) - f(i,j,s)| \le L\,|t-s|\,, \qquad (i,j) \in \Omega,$$

where L is a positive constant not depending on $(i,j) \in \Omega$ and $t,s \in \boldsymbol{R}$. If

$$\frac{L}{4(\sin^2\frac{\pi}{2a} + \sin^2\frac{\pi}{2b})} < 1, \tag{21}$$

then the BVP (5),(6) has a unique solution.

Theorem 5. *Let the function $f : \boldsymbol{Z}^2 \times \boldsymbol{R} \to \boldsymbol{R}$ be continuous with respect to its third argument. Suppose further that for some $R > 0$*

$$\frac{1}{4(\sin^2\frac{\pi}{2a} + \sin^2\frac{\pi}{2b})} \max_{(i,j)\in\Omega,|t|\le R} |f(i,j,t)| \le R. \tag{22}$$

Then the BVP (5),(6) has a solution (u_{ij})such that $|u_{ij}| \le R$ for all $(i,j) \in \Omega$.

Remark 2. Since $a \ge 2$ and $b \ge 2$, using the inequality

$$\sin x \ge \frac{2\sqrt{2}}{\pi}x \qquad for \qquad 0 \le x \le \frac{\pi}{4},$$

we have that

$$\sin^2\frac{\pi}{2a} + \sin^2\frac{\pi}{2b} \ge 2(\frac{1}{a^2} + \frac{1}{b^2}) > \frac{1}{a^2+b^2}$$

and, therefore, conditions (21) and (22) improve conditions (8) and (9), respectively.

References

[Che03] Cheng, S.S.: Partial Difference Equations. Taylor and Francis, London New York (2003)

[Smi85] Smith, G.D.: Numerical Solution of Partial Differential Equations, 3rd. ed. Clarendon Press, Oxford (1985)

[KP91] Kelley, W.G., Peterson, A.C.: Difference Equations: An Introduction with Applications. Academic Press, New York (1991)

Some exact solutions of the $(2 + 1)$-dimensional Kadomtsev-Petviashvili equation

E. V. Krishnan

Department of Mathematics and Statistics, Sultan Qaboos University, P.O.Box 36, Al Khod 123, Muscat, Sultanate of Oman `krish@squ.edu.om`

The mapping and modified mapping methods, with a new mapping relation, have been developed to derive some new exact doubly periodic solutions of the (2+1)-D Kadomtsev-Petviashvili equation in terms of squares of Jacobian elliptic functions. The corresponding limit solutions such as triangular solutions, solitary wave solutions, and singular solutions in the case of the modulus of the elliptic function approaching 0 and 1 have also been derived.

1 Introduction

In the theory of nonlinear waves, the study of travelling wave solutions has generated lot of interest among researchers because these waves can be detected easily as they are solutions of constant form moving with a fixed velocity. The three types of travelling waves of our interest are: the localized travelling waves known as solitary waves which are single hump solutions approaching asymptotically to zero at large distances expressed in terms of *sech* functions, the periodic waves expressed in terms of Jacobian elliptic functions which in their infinite period limits reduce to solitary wave solutions and then the kink-antikink waves which rise or descend from one asymptotic state to another which are normally expressed in terms of tanh functions.

Various methods such as Backlund transformations [Lam71], inverse scattering technique [GGKM67], Hirota's direct method [Hir71], tanh method [HK90, Mal92], series method [HKB85, Kri03], Weierstrass elliptic function method [KP05], singular manifold method [PK05a] etc. have been employed for the derivation of these exact solutions of nonlinear evolution equations over the last few decades.

In this paper a mapping method and its extension [Pen03a, Pen03b] have been employed to derive a variety of Jacobian elliptic function solutions [Law89] for the $(2 + 1) - D$ Kadomtsev-Petviashvili equation [PK05b]. The fact that the Jacobian elliptic functions degenerate into hyperbolic functions

Kenan Taş et al. (eds), Mathematical Methods in Engineering, 197–204.

when the modulus approaches 1, has been the key point in generating interest in the investigation of exact solutions.

2 Mapping methods

For a given nonlinear evolution equation, say, in two variables,

$$N(u, u_t, u_x, ...) = 0, \tag{1}$$

we seek a travelling wave solution in the form

$$u(x,t) = u(\xi),\ \xi = k(x - ct), \tag{2}$$

where k, the wave number and c, the wave velocity, are constants to be determined.

We substitute equation (2) into equation (1) which yields an ordinary differential equation.

In the mapping method, $u(\xi)$ is expanded into a polynomial in $f(\xi)$ in the form

$$u(\xi) = \sum_{i=0}^{n} A_i f^i, \tag{3}$$

where A_i are constants to be determined, n is fixed by balancing the linear term of highest order with the nonlinear term in equation (1), and f satisfies the following equations

$$f'' = pf + qf^3,\ f'^2 = pf^2 + \frac{1}{2}qf^4 + r, \tag{4}$$

and p, q and r are constants to be determined.

After equation (3) with equation (4) is substituted into the ordinary differential equation, the coefficients A_i, k, c, p, q and r may be determined. Thus equation (3) establishes an algebraic mapping relation between the solution to equation (4) and that of equation (1).

In the modified mapping method, we assume $u(\xi)$ may be expanded into a polynomial in f with positive and negative powers, that is, in the form

$$u(\xi) = \sum_{i=0}^{n} A_i f^i + \sum_{i=1}^{n} B_i f^{-i}, \tag{5}$$

where n is the same as in equation (3), and f satisfies the equations (4) and (5). When $B_i = 0$, equation (5) degenerates to equation (3).

Due to the presence of the parameters p, q and r, equation (4) has rich structure of solutions.

3 $(2+1)-D$ Kadomtsev-Petviashvili equation

The $(2+1)-D$ Kadomtsev-Petviashvili equation

$$(u_t + 6uu_x + u_{xxx})_x - 3u_{yy} = 0, \tag{6}$$

governs wave propagation in the field of plasma physics, fluid dynamics etc.

We consider the travelling wave solutions of equation (6) in the form

$$u = u(\xi), \ \xi = k_1 x + k_2 y - \omega t, \tag{7}$$

so that equation (6) reduces to

$$k_1(-\omega\, u_\xi + 6k_1\, u\, u_\xi + k_1^3\, u_{\xi\xi\xi})_\xi - 3\, k_2^2\, u_{\xi\xi} = 0. \tag{8}$$

Integrating,

$$-\omega\, k_1\, u_\xi + 6\, k_1^2\, u\, u_\xi + k_1^4\, u_{\xi\xi\xi} - 3\, k_2^2\, u_\xi = 0, \tag{9}$$

where integration constant is taken to be zero. Integrating again, we get

$$A\, u'' + B\, u + C\, u^2 + D = 0, \tag{10}$$

where,

$$A = k_1^4, \ B = -(\omega\, k_1 + 3\, k_2^2), \ C = 3\, k_1^2, \tag{11}$$

and D is the integration constant.

4 Solution using mapping method

We assume equation (10) has the solution of the form

$$u = A_0 + A_1\, f, \tag{12}$$

where f satisfies the equations

$$f'' = p + q\, f + r\, f^2, \ f'^2 = 2\, p\, f + q\, f^2 + \frac{2}{3}\, r\, f^3 \tag{13}$$

and the coefficients A_i, p, q and r will be determined. Equation (12) gives an algebraic mapping relation between solutions to equations (6) and (13).

Substituting equations (12) and (13) in equation (10) and equating the coefficients of f^2, f^1 and f^0, we obtain $A_1 = -\dfrac{Ar}{C}$, $A_0 = -\dfrac{Aq+B}{2C}$, and the constraint relation

$$k_1^8(q^2 - 4pr) - (\omega k_1 + 3k_2^2)^2 + 12\, k_1^2\, D = 0. \tag{14}$$

Using equations (11) and (12), we obtain the new exact solution of equation (6) as

$$u(x,y,t) = \frac{\omega k_1 + 3k_2^2 - k_1^4 q}{6k_1^2} - \frac{1}{3}k_1^2 r f(k_1 x + k_2 y - \omega t). \quad (15)$$

Case 1. $p = 2,\ q = -4(1+m^2),\ r = 6m^2$

Now, equation (13) has two solutions $f(\xi) = sn^2(\xi)$ and $f(\xi) = cd^2(\xi)$. So, we obtain the periodic wave solutions of equation (6) as

$$u(x,y,t) = \frac{\omega k_1 + 3k_2^2 + 4k_1^4(1+m^2)}{6k_1^2} - 2m^2k_1^2 sn^2(k_1 x + k_2 y - \omega t). \quad (16)$$

and

$$u(x,y,t) = \frac{\omega k_1 + 3k_2^2 + 4k_1^4(1+m^2)}{6k_1^2} - 2m^2k_1^2 cd^2(k_1 x + k_2 y - \omega t). \quad (17)$$

When $m \to 1$, equation (16) reduces to the solitary wave solution

$$u(x,y,t) = \frac{\omega k_1 + 3k_2^2 + 8k_1^4}{6k_1^2} - 2k_1^2 \tanh^2(k_1 x + k_2 y - \omega t). \quad (18)$$

Case 2. $p = 2(1-m^2),\ q = 4(2m^2-1),\ r = -6m^2$

So, the equation (13) has solution $f(\xi) = cn^2(\xi)$. Thus the periodic wave solution of equation (6) is

$$u(x,y,t) = \frac{\omega k_1 + 3k_2^2 - 4k_1^4(2m^2-1)}{6k_1^2} + 2m^2k_1^2 cn^2(k_1 x + k_2 y - \omega t). \quad (19)$$

and the corresponding solitary wave solution when $m \to 1$ is

$$u(x,y,t) = \frac{\omega k_1 + 3k_2^2 - 4k_1^4}{6k_1^2} + 2k_1^2 sech^2(k_1 x + k_2 y - \omega t). \quad (20)$$

Case 3. $p = -2(1-m^2),\ q = 4(2-m^2),\ r = -6$

Thus the equation (13) has solution $f(\xi) = dn^2(\xi)$ and we obtain the periodic wave solution of equation (6) as

$$u(x,y,t) = \frac{\omega k_1 + 3k_2^2 - 4k_1^4(2-m^2)}{6k_1^2} + 2k_1^2 dn^2(k_1 x + k_2 y - \omega t). \quad (21)$$

When $m \to 1$, we get the same solitary wave solution (20).

Case 4. $p = 2m^2,\ q = -4(1+m^2),\ r = 6$

In this case, the equation (13) has two solutions $f(\xi) = ns^2(\xi)$ and $f(\xi) = dc^2(\xi)$. Thus we get the periodic wave solutions of equation (6) as

$$u(x,y,t) = \frac{\omega k_1 + 3k_2^2 + 4k_1^4(1+m^2)}{6k_1^2} - 2k_1^2 ns^2(k_1 x + k_2 y - \omega t), \quad (22)$$

and

$$u(x,y,t) = \frac{\omega k_1 + 3k_2^2 + 4k_1^4(1+m^2)}{6k_1^2} - 2k_1^2 dc^2(k_1x + k_2y - \omega t). \quad (23)$$

When $m \to 1$, equation (22) gives rise to the singular solution

$$u(x,y,t) = \frac{\omega k_1 + 3k_2^2 + 8k_1^4}{6k_1^2} - 2k_1^2 \coth^2(k_1x + k_2y - \omega t). \quad (24)$$

Case 5. $p = -2m^2,\ q = 4(2m^2-1),\ r = 6(1-m^2)$

So, the equation (13) has the solution $f(\xi) = nc^2(\xi)$. Thus the periodic wave solution of equation (6) is

$$u(x,y,t) = \frac{\omega k_1 + 3k_2^2 - 4k_1^4(2m^2-1)}{6k_1^2} - 2(1-m^2)k_1^2 nc^2(k_1x + k_2y - \omega t). \quad (25)$$

Case 6. $p = -2,\ q = 4(2-m^2),\ r = -6(1-m^2)$

Now, the equation (13) has the solution $f(\xi) = nd^2(\xi)$. So, we have the periodic solution of equation (6) as

$$u(x,y,t) = \frac{\omega k_1 + 3k_2^2 - 4k_1^4(2-m^2)}{6k_1^2} + 2(1-m^2)k_1^2 nd^2(k_1x + k_2y - \omega t). \quad (26)$$

Case 7. $p = 2,\ q = 4(2-m^2),\ r = 6(1-m^2)$

In this case, the equation (13) has the solution $f(\xi) = sc^2(\xi)$. Thus the periodic wave solution of equation (6) is,

$$u(x,y,t) = \frac{\omega k_1 + 3k_2^2 - 4k_1^4(2-m^2)}{6k_1^2} - 2(1-m^2)k_1^2 sc^2(k_1x + k_2y - \omega t). \quad (27)$$

Case 8. $p = 2,\ q = 4(2m^2-1),\ r = -6m^2(1-m^2)$

In this case, the equation (10) has the solution $f(\xi) = sd^2(\xi)$. So, our corresponding periodic wave solution of equation (6) is,

$$u(x,y,t) = \frac{\omega k_1 + 3k_2^2 - 4k_1^4(2m^2-1)}{6k_1^2} + 2m^2(1-m^2)k_1^2 sd^2(k_1x + k_2y - \omega t). \quad (28)$$

Case 9. $p = 2(1-m^2),\ q = 4(2-m^2),\ r = 6$

So, the equation (13) has the solution $f(\xi) = cs^2(\xi)$. Thus the periodic wave solution of equation (6) is

$$u(x,y,t) = \frac{\omega k_1 + 3k_2^2 - 4k_1^4(2-m^2)}{6k_1^2} - 2k_1^2 cs^2(k_1x + k_2y - \omega t). \quad (29)$$

and when $m \to 1$, equation (29) degenerates to the singular solitary wave solution

$$u(x,y,t) = \frac{\omega k_1 + 3k_2^2 - 4k_1^4}{6k_1^2} - 2k_1^2 cosech^2(k_1 x + k_2 y - \omega t). \tag{30}$$

Case 10. $p = -2m^2(1-m^2)$, $q = 4(2m^2-1)$, $r = 6$

In this case, the equation (13) has the solution $f(\xi) = ds^2(\xi)$. So, we obtain the periodic wave solution of equation (6) as

$$u(x,y,t) = \frac{\omega k_1 + 3k_2^2 - 4k_1^4(2m^2-1)}{6k_1^2} - 2k_1^2 ds^2(k_1 x + k_2 y - \omega t). \tag{31}$$

When $m \to 1$, equation (31) degenerates to the singular solitary wave solution (30).

It may be noted that when the constant D in equation (10) is assumed to be zero, the constraint relation (14) reduces to

$$k_1^8 (q^2 - 4pr) = (\omega k_1 + 3 k_2^2)^2. \tag{32}$$

Now, the expression $q^2 - 4pr$ in all cases is $16m^4 - 16m^2 + 16$ which is always positive for $0 \leq m \leq 1$. Thus all our solutions are valid with the constraint relation (32).

We assume that equation (10) has the solution of the form

$$u = A_0 + A_1 f + B_1 f^{-1}, \tag{33}$$

where f satisfies the equation (13) and the coefficients A_i, B_i, p, q and r will be determined. Equation (33) is an algebraic mapping relation between solutions to equations (6) and (13).

Substituting equations (33) and (13) in equation (10) and equating the coefficients of like powers of f, we obtain $A_0 = -\dfrac{Aq+B}{2C}$, $A_1 = -\dfrac{Ar}{C}$ and $B_1 = -\dfrac{3pA}{C}$ and the constraint relation

$$k_1^8(q^2 + 16pr) - (\omega k_1 + 3k_2^2)^2 + 12k_1^2 D = 0. \tag{34}$$

Using equations (11) and (33), we obtain the new exact solution of equation (6) as

$$u(x,y,t) = \frac{\omega k_1 + 3k_2^2 - k_1^4 q}{6k_1^2} - \frac{1}{3}k_1^2 r f(k_1 x + k_2 y - \omega t) - k_1^2 p f^{-1}(k_1 x + k_2 y - \omega t). \tag{35}$$

Case 1. $p = 2$, $q = -4(1+m^2)$, $r = 6m^2$

Now, equation (13) has two solutions $f(\xi) = sn^2(\xi)$ and $f(\xi) = cd^2(\xi)$. So, we obtain the periodic wave solutions of equation (6) as

$$u(x,y,t) = \frac{\omega k_1 + 3k_2^2 + 4k_1^4(1+m^2)}{6k_1^2}$$

$$-2k_1^2\left\{m^2sn^2(k_1x+k_2y-\omega t)+ns^2(k_1x+k_2y-\omega t)\right\}, \tag{36}$$

and

$$u(x,y,t) = \frac{\omega k_1+3k_2^2+4k_1^4(1+m^2)}{6k_1^2}$$
$$-2k_1^2\left\{m^2cd^2(k_1x+k_2y-\omega t)+dc^2(k_1x+k_2y-\omega t)\right\}. \tag{37}$$

When $m\to 1$, equation (36) gives rise to the solution

$$u(x,y,t) = \frac{\omega k_1+3k_2^2+8k_1^4}{6k_1^2}-2k_1^2\left\{\tanh^2(k_1x+k_2y-\omega t)+\coth^2(k_1x+k_2y-\omega t)\right\}. \tag{38}$$

Case 2. $p = 2(1-m^2),\ q = 4(2m^2-1),\ r = -6m^2$

So, the equation (13) has the solution $f(\xi) = cn^2(\xi)$. Thus the periodic wave solutions of equation (6) is

$$u(x,y,t) = \frac{\omega k_1+3k_2^2-4k_1^4(2m^2-1)}{6k_1^2}$$
$$+2k_1^2\left\{m^2cn^2(k_1x+k_2y-\omega t)-(1-m^2)nc^2(k_1x+k_2y-\omega t)\right\}. \tag{39}$$

When $m\to 1$, we obtain the solitary wave solution (20).

Case 3. $p = -2(1-m^2),\ q = 4(2-m^2),\ r = -6$

Thus the equation (13) has solution $f(\xi) = dn^2(\xi)$ and we obtain the periodic wave solution of equation (6) as

$$u(x,y,t) = \frac{\omega k_1+3k_2^2-4k_1^4(2-m^2)}{6k_1^2}$$
$$+2k_1^2\left\{dn^2(k_1x+k_2y-\omega t)+(1-m^2)nd^2(k_1x+k_2y-\omega t)\right\}. \tag{40}$$

When $m\to 1$, we get the same solitary wave solution (20).

Case 4. $p = 2,\ q = 4(2-m^2),\ r = 6(1-m^2)$

In this case, the equation (13) has the solution $f(\xi) = sc^2(\xi)$. Thus the periodic wave solution of equation (6) is

$$u(x,y,t) = \frac{\omega k_1+3k_2^2-4k_1^4(2-m^2)}{6k_1^2}$$
$$-2k_1^2\left\{(1-m^2)sc^2(k_1x+k_2y-\omega t)+cs^2(k_1x+k_2y-\omega t)\right\}. \tag{41}$$

When $m\to 1$, the equation (41) reduces to the singular solution (30).

Case 5. $p = 2,\ q = 4(2m^2-1),\ r = -6m^2(1-m^2)$

In this case, the equation (13) has the solution $f(\xi) = sd^2(\xi)$. So, our corresponding periodic wave solution of equation (6) is

$$u(x,y,t) = \frac{\omega k_1+3k_2^2-4k_1^4(2m^2-1)}{6k_1^2}$$

$$+2k_1^2\left\{m^2(1-m^2)sd^2(k_1x+k_2y-\omega t)-ds^2(k_1x+k_2y-\omega t)\right\}. \quad (42)$$

When $m \to 1$, it reduces to the singular solution (30).

It may be noted again that when the constant D in equation (10) is assumed to be zero, the constraint relation (34) reduces to

$$k_1^8\,(q^2+16pr) = (\omega k_1+3k_2^2)^2. \quad (43)$$

Now, the expression q^2+16pr reduces to $16m^4+224m^2+16$ in cases 1, and to $16m^4-256m^2+256$ in cases 3 and 4 which are always positive for $0 \le m \le 1$. But q^2+16pr becomes $256m^4-256m^2+16$ in cases 2 and 5 which is not positive for all values in the interval $0 \le m \le 1$. Thus in the case of $D = 0$, the solutions $f(\xi) = cn^2(\xi)$ and $f(\xi) = sd^2(\xi)$, are not valid since the constraint relation is not satisfied.

References

[Lam71] Lamb, G.L.: Analytical description of ultra short pulse propagation in a resonant medium. Rev. Mod. Phys., **43**, 99-124 (1971)

[GGKM67] Gardner, C.S., Greene, J.M., Kruskal, M.D., Miura, R.M.: Method for solving the Korteweg-de Vries equation. Phys. Rev. Lett., **19**, 1095-1097 (1967)

[Hir71] Hirota, R.: Exact solution of the Kortweg-de Vries equation for multiple collision of solitons. Phys. Rev. Lett., **27**, 1192-1194 (1971)

[HK90] Huibin, L., Kelin, W.: Exact solutions for two nonlinear equations. I, J. Phys. A, **23**, 3923-3928 (1990)

[Mal92] Malfliet, M.: Solitary wave solutions of nonlinear wave equations. Am. J. Phys., **60**, 650-655 (1992)

[HKB85] Hereman, W., Korpel, A., Banerjee, P.P.: A general physical approach to solitary wave construction from linear solutions. Wave motion, **7**, 283-290 (1985)

[Kri03] Krishnan, E.V.: Series solutions for a coupled wave equation. International J. of Diff. Eqs. and Applics., **8**, 13-22 (2003)

[KP05] Krishnan, E.V., Peng, Y: A new solitary wave solution for the new Hamiltonian amplitude equation. J. Phys. Soc. Japan, **74**, 896-897 (2005)

[PK05a] Peng, Y., Krishnan, E.V.: The singular manifold method and exact periodic wave solutions to a restricted BLP dispersive long wave system. Reports on Math. Phys., **56**, 367-378 (2005)

[Pen03a] Peng, Y.: Exact periodic wave solutions to a new Hamiltonian amplitude equation. J. Phys. Soc. Japan, **72**, 1356-1359 (2003)

[Pen03b] Peng, Y.: New exact solutions to a new Hamiltonian amplitude equation. J. Phys. Soc. Japan, **72**, 1889-1890 (2003)

[Law89] Lawden, D.W.: Elliptic Functions and Applications. Springer-Verlag, New York (1989)

[PK05b] Peng, Y., Krishnan, E.V.: Exact travelling wave solutions to the (3+1)-D Kadomtsev-Petviashvili equation. Acta Physica Polonica, **108**, 421-428 (2005)

Varadhan estimates without probability: lower bound

Rémi Léandre

Institut de Mathématiques, Université de Bourgogne, 21000, Dijon, France
Remi.leandre@u-bourgogne.fr

We translate in semi-group theory Varadhan estimates, lower bound, got by ourself by using the Malliavin Calculus for hypoelliptic heat-kernels.

1 Introduction

Let us consider $m+1$ vector fields on R^d with derivatives at each order bounded. Let us consider the vector spaces defined inductively by:

$$E_0(x) = (X_1(x), .., X_m(x)) \tag{1}$$

$$E_{l+1}(x) = E_l(x) \cup [E_l, (X_1, .., X_m)](x) \tag{2}$$

We suppose that in the starting point x there exists an l such that $E_l(x)$ spans R^d (Strong Hoermander's hypothesis).

We consider an Hoermander's type operator:

$$L = X_0 + 1/2 \sum_{i \geq 1} X_i^2 \tag{3}$$

The heat semi-group associated to L has a smooth density $p_t(x, y)$. It is the purposed of Hoermander's theorem [Hoer67, Koh69, Mall78]. We are concerned in this paper by the behaviour of $p_t(x, y)$ when $t \to 0$. We introduce as it is classical the Carnot-Caratheory distance defined as follows: let $t \to h_t^i$ m elements of $L^2([0, 1])$. We consider the horizontal curve:

$$dx_t(h) = \sum_{i \geq 1} X_i(x_t(h)) h_t^i dt \tag{4}$$

We define $d^2(x, y)$ as the infimum of $\sum \int_0^1 |h_t^i|^2 dt = \|h\|^2$ when $x_0(h) = x$ and $x_1(h) = y$. In the sequel we will do the following assumption: $d(x, y) < \infty$ for all y in R^d.

Kenan Taş et al. (eds), Mathematical Methods in Engineering, 205–215.

Our result is the following:
Main theorem: When $t \to 0$

$$\underline{\lim} 2t \log p_t(x,y) \geq -d^2(x,y) \tag{5}$$

There is now a huge literature about estimate of hypoelliptic heat-kernels: we refer to [Davi90, VSC92] for analytical approaches, to the two surveys of Léandre [Lean88a, Lean03] for a probabilistic approach as well as the survey of Kusuoka [Kus92] and Watanabe [Wata92] about this topic.

There are two versions of Malliavin Calculus:

One uses the apparatus of Sobolev Spaces on the Wiener space. Let us recall namely that the main originallity of Malliavin Calculus with respect of his prelimary versions (See works of Fomin, Hida, Albeverio, Elworthy, Berezanskii...) is to complete the classical differential operations on the Wiener space in all the L^p. Diffusions, although they are only almost surely defined, belong to all the Sobolev spaces on the Wiener space, because there is no Sobolev imbedding theoerem. It is the original approach of Malliavin [Mall78, IW81, Stro83].

The second one uses the stochastic flow theorem. It is the approach used by Bismut [Bis81a], which avoids the previous apparatus of functional analysis.

In [Lean], we have translated in semi-group theory the approach of Bismut. Our goal is to prove again Main Theorem, originally proved by Léandre [Lean85, Lean87] by using Malliavin Calculus, by using the translation of Malliavin Calculus of [Lean].

2 Algebraic scheme of the proof

In order to do the asymptotic expansion of $p_t(x,y)$ when $t \to 0$, we will put $t = \epsilon^2$ according an old trick of Molchanov [Mol75] and study the asymptotic in time 1 of the heat-kernel of the generator

$$L_\epsilon = 1/2\epsilon^2 \sum_{i>0} X_i^2 + \epsilon^2 X_0 \tag{6}$$

$\exp[L_\epsilon]$ has an heat-kernel $q_\epsilon(x,y)$ provided the first-condition is checked.

Let us consider the Hilbert space H of L^2 maps from $[0,1]$ into R^d. We can consider according Bismut [Bis84] the elements h where $h \to x_1(h)$ is a submersion.

The main remark of Léandre [Lean85, Lean87] is that

$$d^2(x,y) = d_R^2(x,y) \tag{7}$$

where in $d^2_R(x,y)$ (so-called Bismutian distance) we take the infimum of $\|h\|^2$ where $x_1(h)(x) = y$, $x_0(h) = x$ and $h \to x_1(h)$ is a submersion in h. Let us introduce an h satisfying the three previous conditions and such that $\|h\|^2 \leq d^2(x,y) + \eta$ for a small η.

First step: introduction of Molchanov's translation.

Let us introduce the generator $L_\epsilon(h)$:

$$L_\epsilon(h) = L_\epsilon + \epsilon \sum_{i>0} 1/\epsilon h^i_t X_i \tag{8}$$

Let us introduce the vector fields on R^{d+1}

$$\tilde{X}_i(\epsilon) = (\epsilon X_i, -1/\epsilon h^i_t u) \tag{9}$$

for $i > 0$ and the generator

$$\tilde{L}_\epsilon(h)(\tilde{f}) = \epsilon^2 < X_0, \tilde{D}\tilde{f} > + \sum_{i>0} h^i_t < X_i, \tilde{D}\tilde{f} >$$

$$+1/2\epsilon^2 \sum_{i>0} < DX_i X_i, \tilde{D}\tilde{f} > +1/2 \sum_{i>0} < \tilde{X}_i(\epsilon), \tilde{D}^2\tilde{f}, \tilde{X}_i(\epsilon) > \tag{10}$$

According of the quasi-invariance formula of [Lean], translation in semi-group theory of the traditional quasi-invariance formulas of stochastic-analysis, we have:

$$\exp[L_\epsilon](f)(x) = \exp[\tilde{L}_\epsilon(h)][uf](x,1) \tag{11}$$

We consider the generator:

$$\overline{L}_\epsilon(h) = \epsilon^2 X_0 + \sum_{i>0} h^i_t X_i + 1/2 \sum \tilde{X}^2_i(\epsilon) \tag{12}$$

It differs from $\tilde{L}_\epsilon(h)$ by $-1/2 \sum |h^i_t|^2 \|^2 u \frac{\partial}{\partial u}$. This last vector field commutes with $\tilde{L}_\epsilon(h)$. We deduce that:

$$\exp[\tilde{L}_\epsilon(h)][uf](x,1) = \exp[-\|h\|^2/2\epsilon^2] \exp[\overline{L}_\epsilon(h)][uf](x,1) \tag{13}$$

Let us consider the vector field for $i > 0$

$$\overline{Y}_i(\epsilon) = (\epsilon X_i, -h^i_t) \tag{14}$$

and the generator $\overline{Q}_\epsilon(h)$:

$$\overline{Q}_\epsilon(h) = \epsilon^2 X_0 + \sum_{i>0} h^i_t X_i + 1/2 \sum_{i>0} \overline{Y}^2_i(\epsilon) \tag{15}$$

We have clearly

$$\exp[\overline{L}_\epsilon(h)][uf](x,1) = \exp[\overline{Q}_\epsilon(h)][\exp[u/\epsilon]f](x,0) \tag{16}$$

Let g be a function from R into $[0,1]$ equals to 1 in a small neighborhood of 0 and equals to 0 outside a small neighborhood of 0. We have, if f is positive, the fundamental inequality:

$$\exp[L_\epsilon]f(x) \geq \exp[-\frac{d^2(x,y)+2\eta}{2\epsilon^2}]\exp[\overline{Q}_\epsilon(h)][g(\epsilon u)f](x,0) \tag{17}$$

Second step: introduction of Molchanov's rescaling.

Let us consider the flot ϕ_t associated to the ordinary equation:

$$dx_t(h) = \sum X_i(x_t(h))dt \tag{18}$$

Let us introduce the vector fields for $i > 0$

$$Y_i(\epsilon) = (\epsilon\phi_t^{*-1}X_i, -h_t^i) \tag{19}$$

and the generator

$$Q_\epsilon(h) = \epsilon^2\phi_t^{*-1}X_0 + 1/2\sum_{i>0} Y_i^2(\epsilon) \tag{20}$$

We have the main formula:

$$\exp[t\overline{Q}_\epsilon(h)][g(\epsilon u)h](x,0) = \exp[tQ_\epsilon(h)][g(\epsilon u)f_t](x),0) \tag{21}$$

f_t is the map which to z associates $f(\phi_t(z))$. (21) arises from the Itô-Stratonovitch formula of Bismut [Bis81b]. We can show (21) by using semi-group theory: we can suppose by a density result that the semi-group associated to $Q_\epsilon(h)$ satisfies in the starting point strong Hoermander's hypothesis, such that it has a heat-kernel satisfying a parabolic equation. Since the generator is intrinsic, when we do the change of variable $z \to \phi_t(z)$, the got heat-kernel satisfies the transformed parabolic equation by this transformation, which is the parabolic equation associated to $\overline{L}_\epsilon(h)$ because the change of variable depends on time t.

By this result, we have only to estimate the density $r_\epsilon^1(.)$ of the measure

$$f \to \exp[Q_\epsilon(h)][g(\epsilon u)f](y,0) \tag{22}$$

at y since $\phi_1(x) = y$. We will do the rescaling, after supposing $y = 0$ (it is always possible) $y' \to y'/\epsilon$. This means we have to consider the vector fields for $i > 0$,

$$Z_i(\epsilon) = (\phi_t^{*-1}X_i(\epsilon.), -h_t^i) \tag{23}$$

and the generator

$$R_\epsilon(h) = \epsilon\phi_t^{*-1}X_0(\epsilon.) + \sum_{i>0} Z_i^2(\epsilon) \tag{24}$$

We consider the density $r_\epsilon^2(.)$ of the measure associated to

$$f \to \exp[R_\epsilon(h)](g(\epsilon u)f](0,0) \tag{25}$$

(Let us recall, we have supposed that $y = 0$). We have clearly the relation:

$$r_\epsilon^1(0) = \epsilon^{-d} r_\epsilon^2(0) \tag{26}$$

Moreover $r_\epsilon^2(.)$ is the density of a non-degenerated Gaussian measure on R^m by (7). The result holds by Malliavin Calculus depending on a parameter of the next part.

3 Malliavin calculus without probability depending on a parameter

This part is mainly the translation in semi-group theory of the work of Léandre [Lean85, Lean87] (See the work of Watanabe [Wata87] for the counterpart of this theorem in Malliavin's approach).

Let us recall the formalism of [Lean]: We consider $X_i(s,y)$, $i = 0,..,m$ some vector fields on R^d bounded with bounded derivatives at each order dependly smoothly for a parameter $s \in [0,1]$ with the same boundedness assumptions. We consider some vector fields on $R^d \times Gl(R^d) \times M^d$ where $Gl(R^d)$ is the space of invertible matrices on R^d and M^d the space of symmetric matrices on R^d:

$$\hat{X}_i(s) = (X_i(s), DX_i(s)U, 0) \qquad \hat{X}(s) = (0, 0, \sum_{i>0} < U^{-1}X_i(s), . >^2) \tag{27}$$

We consider the semi group $P_t(s)$ associated to

$$L_s = X_0(s) + 1/2 \sum_{i>0} X_i(s)^2 \tag{28}$$

and the semi-group $\hat{P}_t(s)$ associated to

$$\hat{L}_s = \hat{X}(s) + \hat{X}_0(s) + 1/2 \sum_{i>0} \hat{X}_i^2(s) \tag{29}$$

We get:

Theorem 1. *If* $\sup_s \hat{P}_t(s)[V^{-p}](x, I, 0) < \infty$ *for all* $p > 1$, *we have* $P_t(s)f(x) = \int_{R^d} p_t(s,x,y)f(y)dy$ *where* $(s,y) \to p_t(s,x,y)$ *is continuous in* s *and smooth in* y.

Proof. We remark that:

$$\frac{\partial}{\partial t}\frac{\partial}{\partial s} P_t(s)f(x) = L_s \frac{\partial}{\partial s} P_t(s)f(x) + (\frac{\partial}{\partial s} L_s) P_t(s)f(x) \tag{30}$$

such that:

$$\frac{\partial}{\partial s}P_t(s)f = \int_0^t P_{t-u}(s)(\frac{\partial}{\partial s}L_s)P_u(s)f du \tag{31}$$

Let us compute $D_x P_t(s)f(x)$. It is the subject of Lemma III.2 of [Lean]: let be the vector fields on $R^d \times Gl(R^d)$

$$\overline{X}_i(s) = (X_i(s), D_x X_i(s)U) \tag{32}$$

and the generator

$$\overline{L}_s = \overline{X}_0(s) + 1/2\sum_{i>0}\overline{X}_i^2(s) \tag{33}$$

To $\overline{L}_s$ is associated a semi-group $\overline{P}_t(s)$ and we have:

$$D_x P_t(s)f(x) = \overline{P}_t(s)[DfU](x, I) \tag{34}$$

We write

$$\frac{\partial}{\partial s}L_s = \frac{\partial}{\partial s}X_0(s) + 1/2\sum_{i>0}(\frac{\partial}{\partial s}X_i(s)X_i(s) + X_i(s)\frac{\partial}{\partial s}X_i(s)) \tag{35}$$

We deduce that

$$(\frac{\partial}{\partial s}L_s)P_t(s)f =< \overline{P}_t(s)[Dfu](x,I), \frac{\partial}{\partial s}X_0(s) >$$

$$+\frac{1}{2}\sum_{i>0}\frac{\partial}{\partial s}X_i(s) < \overline{P}_t(s)[DfU](x,I), X_i(s) > +\frac{1}{2}\sum_{i>0}X_i(s) < \overline{P}_t(s)[DfU](x,I), \frac{\partial}{\partial s}X_i(s) > \tag{36}$$

In the various brackets which appear in the previous formula, the vector fields are not considered as differential operators but as vectors.

On the other hand, we consider the vector fields on $R^d \times R^d$

$$Y_i(s) = (X_i(s), D_x X_i(s)U + \frac{\partial}{\partial s}X_i(s)) = \overline{X}_i(s) + Z_i(s) \tag{37}$$

and we consider the semi-group $R_t(s)$ spanned by the generator $Y_i(s) + 1/2\sum_{i>0} Y_i^2(s)$. We have:

Lemma 1. *We have*

$$\frac{\partial}{\partial s}P_t(s)f(x) = R_t(s)[DfU](x,0) \tag{38}$$

Proof. We decompose the generator of $R_t(s)$ into

$$\overline{L}(s) + \xi(s) \tag{39}$$

We remark by using Volterra expansion that

$$R_t(s)[DfU](x,0] = \sum \int_{\Delta_n(t)} \overline{P}_{u_1}(s)\xi(s)...\xi(s)\overline{P}_{t-u_n}(s)[DfU](x,0) \quad (40)$$

where $\Delta_n(t)$ is the simplex $0 < u_1 < .. < u_n < t$ provided the series converges for the smooth topology. The main remark is that $\overline{P}_t(s)[DfU](x, U_0)$ is linear in U_0 such that:

$$R_t(s)[DfU](x,0) = \int_{o<u<t} P_u(s)\xi(s)\overline{P}_{t-u}(s)[DfU](x,0) \quad (41)$$

But $Z_i^2(s)\overline{P}_t(s)[DfU](x,0) = 0$ and we have the following identities:

$$Z_0(s)\overline{P}_t(s)[DfU](x,0) =< \overline{P}_t(s)[DfU](x,I), \frac{\partial}{\partial s}X_0(s) > \quad (42)$$

$$\overline{X}_i(s)Z_i(s)\overline{P}_t(s)[Df](x,0) = X_i(s) < \overline{P}_t(s)[DfU](x,I), \frac{\partial}{\partial s}X_i(s) > \quad (43)$$

$$Z_i(s)\overline{X}_i(s)\overline{P}_t(s)[DfU](x,0) = \frac{\partial}{\partial s}X_i(s) < \overline{P}_t(s)[DfU](x,I), X_i(s) > \quad (44)$$

In the brackets which appear in the 3 previous formulas, the vector fields are considered as vector and not as first order differential operators. We deduce our remark by using (36) and (31). ◊

Let (α) be a multi-index on R^d. We have

$$\frac{\partial}{\partial s}P_t(s)[\frac{\partial^\alpha}{\partial x^\alpha}f](x) = R_t(s)[D_x\frac{\partial^\alpha}{\partial x^\alpha}fU](x,0) \quad (45)$$

By using the integration by parts formulas analoguous to the proof of Theorem III.1 of [Lean], we deduce that

$$|\frac{\partial}{\partial s}P_t(s)[\frac{\partial^\alpha}{\partial x^\alpha}f](x)| \leq C\|f\|_\infty \quad (46)$$

where $\|.\|_\infty$ is the uniform norm. Therefore the result. ◊

Proof of the main theorem: We consider the semi-group $\exp[t\hat{R}_\epsilon(h)]$ enlarged to the semi-group associated to $R_\epsilon(h)$ with the Malliavin matrix involved with the component in R^d. Let v be a function on R with values in $[0,1]$ equals to 0 in a neighborhood of 0 and equals to 1 outside a small neighborhood of 0. The density r_ϵ^2 is larger than the density r_ϵ^3 of the measure on R^d:

$$f \to \exp[\hat{R}_\epsilon(h)](g(\epsilon u)v(det(V_\epsilon(h)))f](0,0,I,0) \quad (47)$$

$\epsilon \to r_\epsilon^3(0)$ is continuous in ϵ by a small adaptation of Theorem 1 and moreover $V_0(h) = \sum_{i>0}\int_0^1 < \phi_u^{*-1}X_i, >^2 du$ which is a non-degenerated symmetric positive matrix. Therefore r_0^3 is the strictly positive density of a non-degenerated Gaussian measure on R^d. ◊

Remark 1. We can follow this proof in order to show without using probabilities the following result which was shown by Ben Arous-Léandre [BL91] by using probabilities. Let L_ϵ be the generator $X_0 + \frac{\epsilon^2}{2}\sum_{i>0} X_i^2$ where the involved vector fields satisfy still the strong Hoemander's hypothesis. The associated semi-group under assumption similar to the assumptions of the introduction has an heat-kernel $p_t^\epsilon(x,y)$. Instead of considering (4), we consider:

$$dx_t(h) = \sum_{i>0} X_i(x_t(h))h_t^i dt + X_0(x_t(h))dt \tag{48}$$

We introduce the quantity $d_R^2(x,y)$ (so called pseudo Bismutian distance) which is the infimum of $\|h\|^2$ such that $x_0(h) = x$, $x_1(h) = y$ and $h' \to x_1(h')$ is a submersion in h. We get when $\epsilon \to 0$

$$\underline{\lim} 2\epsilon^2 \log p_1^\epsilon(x,y) \geq -d_R^2(x,y) \tag{49}$$

4 Conclusion

We have shown that the algebraic sheme of our proof of the Varadhan estimates, lower bound, where the Girsanov formula and the Itô-Stratonovitch formula play a big role, can be interpreted in semi-group theory. Namely, in [Lean85] and [Lean87], all the quantities considered are solutions of convenient stochastic differential equations, which can be interpreted by using suitable semi-groups. The role of the Bismut condition in this proof can be interpreted by using the Malliavin Calculus without probability of [Lean].

5 Appendix

Estimates on the Malliavin matrix without probability

We follow the method of [Lean84] and [Lean88b] to invert Malliavin's matrix, which can be easily adapted in semi-group theory as it was already remarked in [Lean].

Let ξ be a element of R^d of norm 1. Let U_0 a matrix with norm bounded by C as well as its inverse. We suppose that x_0 lies in a small neighborood O where the strong Hoermander's assumption is still checked. Let

$$F_l(x,U,\xi) = \sum_{E_l(x)} < U^{-1}Y(x), \xi >^2 \tag{50}$$

Lemma 2. *Let us suppose that for arbitrary small t_0, we have:*

$$\overline{P}_{t_0}[F_l(.,.,\xi) > t_0^\alpha](x_0,U_0) > C > 0 \tag{51}$$

Then (51) remains true for an interval starting from t_0 and of length t_0^β for β depending only from α.

Proof. We introduce a function g from R^+ into $[0,1]$ with bounded derivatives equals to 1 on $[1,\infty[$ and equal to 0 in 0. We consider the auxiliary function:

$$t \to \overline{P}_t[g(\frac{F_l(.,.,\xi)}{t_0^\alpha})](x_0, U_0) \tag{52}$$

It has a first derivative bounded by $t_0^{2\alpha}$. Therefore the result. ♢

We recall the result of [Lean]:

$$P_t[y \in O^c](x) + \overline{P}_t[|U^{-1}| > C](x, I) + \overline{P}_t[|U| > C](x, I) = O(t^\infty) \tag{53}$$

where $O(t^\infty)$ means that this quantity is smaller than $C_p t^p$ for all p

Lemma 3. *Let us suppose that*

$$\overline{P}_t[F_l(..,\xi) > t^\alpha](x_0, U_0) > C > 0 \tag{54}$$

on an interval $I(x_0, U_0)$ *starting from* t_0*, where* t_0 *can be choosed arbitrarily small, and of length* t_0^β*. Then there exists an* α' *depending only from* α *and* β *and a* $t_1(x_0, U_0)$ *in this interval, such that:*

$$\overline{P}_{t_1(x_0,U_0)}[F_{l-1}(.,.,\xi) > t_1^{\alpha'}(x_0, U_0)](x_0, U_0) > C > 0 \tag{55}$$

Proof. Let us consider the probability law $\overline{P}_{t_0}(x_0, U_0)$. Either $F_{l-1}(.,.,\xi) > t_0^{\alpha_1}$ for a big probability for this law, and the result is proved. Or it is not the case. In this case, we consider

$$G_{l-1}(x", U", \xi) = \sum_{E_{l-1}} (< (U")^{-1}Y(x"), \xi > - < U^{-1}Y(x), \xi >)^2 \tag{56}$$

and we consider a function from R^+ into $[0,1]$, strictly increasing such that: $g(0) = g"(0) = 0$ and $g'(0) > 0$. We consider the auxiliary function:

$$t \to \overline{P}_t[g(\frac{G_{l-1}}{t_0^{\alpha"}})](x, U) \tag{57}$$

for $\alpha"$ depending only from α and x and U being chosen according the law $\overline{P}_{t_0}(x_0, U_0)$. It is equals to 0 in $t = 0$, has a first derivative larger than $C > 0$ and a second derivative bounded by $Ct_{t_0}^{-2\alpha"}$. The result arises clearly. ♢

From the strong Hoermander's hypothesis, we deduce the following lemma:

Lemma 4. *For any* (x_0, U_0) *satisfying the previous conditions, there exists an interval* $I(x_0, U_0)$ *starting from* t_0*, which can be chosen arbitrarily small, and of length* t_0^β*, and a* α *such that on* $I(x_0, U_0)$

$$\overline{P}_t[F_0(.,.,\xi) > t^\alpha](x_0, U_0) > C > 0 \tag{58}$$

Lemma 5. *For any (x_0, U_0) satisfying the previous conditions, there exists an interval $I'(x_0, U_0)$ starting from t_0, which can be choosed arbitrarily small, and of length t_0^β such that on $I(x_0, U_0)$, we have*

$$\hat{P}[V(\xi) < t^\alpha](x_0, U_0, 0) < C < 1 \tag{59}$$

Moreover α does not depend on (x_0, U_0) as well as β.

Proof. We consider the interval $I(x_0, U_0)$ of the previous lemma starting from t_0. We consider the auxilary function

$$t \to \hat{P}_s[g(\frac{V(\xi)}{t_0^\alpha})](x_0, U_0, 0) \tag{60}$$

If $V(\xi)$ at the departure time t_0 is small, the auxilary function has a derivative in t_0 strictly smaller than $C < 0$ and a second derivative bounded by $t^{-2\alpha}$. The result if we choose g strictly decreasing, convex from R^+ into $[0, 1]$ equals to 0 in 1 and tending to 0 at the infinity. ♢

Theorem 2. $\hat{P}_t[V^{-p}](x, I, 0) < \infty$ *for all* p.

Proof. We choose ϵ^{-r} points ξ_i on the the unit sphere of R^d such that:

$$\hat{P}_t(|V^{-1}| > \epsilon](x, I, 0) \leq \sum \hat{P}_t[V(\xi_i) < \epsilon](x, I, 0) + \hat{P}_t[|V| > \epsilon^{-\gamma}](x, I, 0) \tag{61}$$

The last quantity is bounded by all ϵ^p for all p [Lean]. It remains to show that uniformly $\hat{P}_t[|V\xi| < \epsilon] = O(\epsilon^\infty)$.

We slice the time interval in $\epsilon^{-\delta}$ small intervals. By using (53), (59) and the semi-group property, we deduce that:

$$\hat{P}_t[V\xi < \epsilon](x, I, 0) \leq \{ \sup_{x_0 \in O, |U_0| + |U_0^{-1}| < C} \hat{P}_{\epsilon^\delta}[V\xi < \epsilon](x_0, U_0, 0)\}^{\epsilon^{-\delta}} \tag{62}$$

But by Lemma 5, if δ is small enough,

$$\hat{P}_{\epsilon^\delta}[V\xi < \epsilon](x_0, U_0, 0) < C < 1 \tag{63}$$

The result arises. ♢

As a consequence, we deduce the existence of the considered heat-kernels in this paper.

References

[Hoer67] Hoermander, L.: Hypoelliptic second order differential equation. Acta. Math., **119**, 147-171 (1967)

[Koh69] Kohn, J.J.: Pseudo-differential operators and hypoellipticity. Proc. Symp. Pure. Maths., **23**, 61-69 (1969)

[Mall78] Malliavin, P.: Stochastic Calculus of variations and hypoelliptic operators. In: Itô, K. (ed.) Stochastic analysis. Kinokuniya, Tokyo (1978)

[Davi90] Davies, E.B.: Heat kernel and spectral theory. Cambridge Tracts in Math., **92**, Cambridge (1990)

[VSC92] Varopoulos, N., Saloff-Coste, L., Coulhon, T.: Analysis and geometry on groups. Cambridge Tracts of Maths., **100**, Cambridge (1992)

[Lean88a] Léandre, R.: Applications quantitatives et géométriques du Calcul de Malliavin. In: Séminaire Franco-Japonais. Métivier, M., Watanabe, S. (ed.) Lect. Notes. Math., **1322**, Springer, Heidelberg (1988). English translation: In: Geometry of random motion. Durrett, R., Pinsky, M. (ed.) Cont. Maths., **73**, A.M.S., Providence (1988)

[Lean03] Léandre, R.: Malliavin Calculus for a general manifold. In: Séminaire X-E.D.P. Bony, J.M., Chemin, J.Y., Sjoestrand, J. (ed.) Exposé XXIV. Publi. Ecole Polytechnique, Palaiseau (2003)

[Kus92] Kusuoka, S.: More recent theory of Malliavin Calculus. Sugaku, **5**, 155-173 (1992)

[Wata92] Watanabe, S.: Stochastic analysis ans its application. Sugaku, **5**, 51-72 (1992)

[IW81] Ikeda, N., Watanabe, S.: Stochastic differential equations and diffusion processes. North-Holland, Amsterdam (1981)

[Stro83] Stroock, D.: Some applications of stochastic Calculus to partial differential equations. In: Ecole d'Été de Saint-Flour X. Hennequin, P. (ed.) Lect. Notes Math., **876**, Springer, Heidelberg (1983)

[Bis81a] Bismut, J.M.: Martingales, the Malliavin Calculus and hypoellipticity under general Hoermander's conditions. Z.W., **56**, 469-505 (1981)

[Lean] Léandre, R.: Malliavin Calculus of Bismut type without probability. To appear Festchrift in honour of K. Sinha. A.M. Boutet de Monvel edt.

[Lean85] Léandre, R.: Estimation en temps petit de la densité d'une diffusion hypoelliptique. C.R.A.S., **301**, Série I, 801-804 (1985)

[Lean87] Léandre, R.: Minoration en temps petit de la densité d'une diffusion dégénérée. J.F.A., **74**, 399-414 (1987)

[Mol75] Molchanov, S.: Diffusion processes and Riemannian geometry. Rus. Math. Surveys, **30**, 1-63 (1975)

[Bis84] Bismut, J.M.: Large deviations and the Malliavin Calculus. Prog. Math., **45**, Birkhauser, Berlin (1984)

[Bis81b] Bismut, J.M.: Mécanique aléatoire. Lect. Notes. Math., **866**, Springer, Heidelberg (1981)

[Wata87] Watanabes, S.: Analysis of Wiener functionals (Malliavin Calculus) and its applications to heat-kernels. Ann. Probab., **15**, 1-39 (1987)

[BL91] Ben Arous, G., Léandre, R.: Décroissance exponentielle du noyau de la chaleur sur la diagonale (II). P.T.R.F., **90**, 372-402 (1991)

[Lean84] Léandre, R.: Extension du théorème de Hoermander à divers processus de sauts. Thèse. Université de Franche-Comté, Besançon (1984)

[Lean88b] Léandre, R.: Régularité de processus de sauts dégénérés (II). Ann. Inst. Henri Poincaré, **24**, 209-236 (1988)

Dirichlet problem for orthotropic bounded cylinder with combined boundary conditions

Raid Al-Momani[1] and Khalid Al-Momani[2]

[1] Department of Mathematics and Physics, Qatar University, Qatar
raid@yu.edu.jo
[2] King Abdullah University Hospital, Irbid, Jordan momanki@just.edu.jo

We solve Dirichlet problem of the two dimensional nonstationary heat conduction problem for orthotropic bounded cylinder with boundary conditions of the first and second kinds of circular discontinuity of temperature density of heat flow. The solution of such problem was derived with the help of Laplace and Hankel transformations.

1 Introduction

In the last years a valuable attention were given to the study of heat transfer in different solids with local sources of heat of different geometric form on planes [KMY01, Man02, Man01]. The solution of stationary heat conduction problems were well studied in the fundamental research of Kozlov, Mandrik, Popov and others, see for example [Koz86, LUS79]. The increasing interest to the nonstationary heat conduction problems of bounded and semi-bounded solids is because the solution of such problems can be applied to different theoretical and practical applications.

A very interesting case of temperature fields distributions in cylindrical axi-symmetrical domains is when the boundary conditions on the plane ($z = 0$) are defined as Hankel and Dirac delta functions integrals:

$$H_R(\delta_{r_0}) = \lim_{\epsilon \to 0} \int_{r_0}^{r_0+\epsilon} r J_0(p_r) q_\epsilon(r) dr = \frac{J_0(pr_0)}{2\pi} \tag{1}$$

where the density of heat flow equals

$$q_\epsilon(r) = \left\{ \begin{matrix} \frac{1}{\pi(2r+\epsilon)}, r_0 \le r \le r_0 + \epsilon \\ 0, r \notin [r_0, r_0 + \epsilon] \end{matrix} \right\}.$$

Kenan Taş et al. (eds), Mathematical Methods in Engineering, 217–222.

2 Statement of the problem

It's need to solve the following heat conduction problem

$$K_a \frac{1}{r}\frac{\partial}{\partial r}\left[r\frac{\partial T(r,z,\tau)}{\partial r}\right] + \frac{\partial^2 T(r,z,\tau)}{\partial z^2} = \frac{1}{\alpha_z}\frac{\partial T(r,z,\tau)}{\partial \tau},$$
$$0 < r < R, 0 < z < h, \tau > 0 \quad (2)$$

with the initial conditions

$$T(r,z,0) = T_0 = constant \quad (3)$$

and the following boundary conditions

$$T(r,0,\tau) = T_v(r,\tau), \qquad 0 \le r \le R, z = 0, \tau > 0, \quad (4)$$
$$\frac{\partial T}{\partial z}\,|_{z=h} = -\frac{\alpha_h}{\lambda_z}\left[T(r,h,\tau) - T_0\right], 0 \le r \le R, z = h, \tau > 0, \quad (5)$$
$$\frac{\partial T}{\partial r}\,|_{r=R} = -\frac{\alpha_R}{\lambda_r}\left[T(R,z,\tau) - T_0\right], r = R, 0 < z < h, \tau > 0, \quad (6)$$
$$\frac{\partial T}{\partial r}\,|_{r=0}\;, r = 0, 0 < z < h, \tau > 0, \quad (7)$$

where R and h are the radius and half of the height of the bounded orthotropic cylinder respectively, T_0 is the initial temperature, α_R and α_h are the heat transfer coefficients on lateral and one of front surfaces of the cylinder, $K_a = \frac{a_r}{a_z}$ is the coefficients relation of temperature conductivity in the direction of cylindrical coordinates r and z.

The solution of the two dimensional nonstationary heat conduction problem (1)-(7) for bounded orthotropic cylinder derived with the help of Laplace and Hankel transforms for $\overline{T_H}(p,z,s)$ has the following form

$$\overline{T_H}(P,z,s) - \frac{T_0 R J_1(PR)}{Ps} = \int_0^R rJ_0(Pr)\left[\overline{T_v}(r,s) - \frac{T_0}{s}\right]dr$$

$$\left\{\frac{\sqrt{K_aP2\frac{s}{a_z}}\cosh(\sqrt{K_aP2\frac{s}{a_z}}(h-z)) + \left(\frac{\alpha_h}{\lambda_z}\right)\sinh(\sqrt{K_aP2\frac{s}{a_z}}(h-z))}{\sqrt{K_aP2\frac{s}{a_z}}\cosh(\sqrt{K_aP2\frac{s}{a_z}}(h-z)) + \left(\frac{\alpha_h}{\lambda_z}\right)\sinh(\sqrt{K_aP2\frac{s}{a_z}}h)}\right\} \quad (8)$$

where

$$\overline{T_H}(P,z,s) = \int_0^R rJ_0(Pr)\int_0^\infty T(r,z,\tau)e^{-s\tau}d\tau dr,$$

$$\overline{T_v}(r,s) = \int_0^\infty T_v(r,\tau)e^{-s\tau}\,d\tau.$$

The inverse Hankel transform for the solution (8) has the following form

$$H^{-1}\left[\overline{T_H}(P,z,s)\right] = \overline{T}(r,z,s) = \frac{2}{R^2}\sum_{n=1}^{\infty}\frac{J_0(P_n r)\overline{T_H}(P_n,z,s)}{J_0^2(P_n R)+J_1^2(P_n R)},$$

where $\mu_n = P_n R$ are the roots of the characteristic equation

$$\mu_n J_1(\mu_n) - B_{i_R} J_0(\mu_n) = 0$$

where $B_{i_R} = \frac{\alpha_R R}{\lambda_r}$ is Bio criteria on the lateral surface of the cylinder $r = R, 0 < z < h$.

When the thermophysical characteristics in the corresponding directions are equal, that is, $\lambda_r = \lambda_z = \lambda$ and $a_r = a_z = a$ ($K_a = 1$), the expression (8) is a generalized solution of the two dimensional nonstationary heat conduction problem for isotropic cylinder with the given boundary conditions (4)-(7).

As an example of a particular application of (8), we find the solution of the two dimensional nonstationary heat conduction problem for bounded orthotropic cylinder with the following boundary conditions: the initial temperature of the bounded orthotropic cylinder is T_0. On the surfaces ($z = h, 0 < r < R$ and $r = R$, $0 < z < h$) we assign a constant initial temperature, that is, $T(r,h,\tau) = T(R,z,\tau) = T_0$. On the surface ($z = 0, 0 < r < R$) we assign a function

$$T(r,0,\tau) = T_v(r,\tau).$$

For $T_v(r,\tau)$ the Laplace transform always exist.

It is necessary to define the two dimensional temperature field $T(r,z,\tau)$. Applying (8) for the case when $\alpha_h, \alpha_R \to \infty$ ($B_{i_h}, B_{i_R} \to \infty$), we get

$$\overline{T_H}(P,z,s) - \frac{T_0 R J_1(PR)}{Ps} = \frac{\sinh(\sqrt{K_a P^2 \frac{s}{a_z}} z(h-z))}{\sinh(\sqrt{K_a P^2 \frac{s}{a_z}} h)} \cdot \int_0^R r J_0(\mathrm{Pr})\left[\overline{T_v}(r,s) - \frac{T_0}{s}\right] dr. \tag{9}$$

Here, the inverse Hankel transform of (9) has the form

$$H^{-1}\left[\overline{T_H}(P,z,s)\right] = \overline{T_H}(r,z,s) = \frac{2}{R}\sum_{n=1}^{\infty}\frac{J_0(P_n r)}{J_1^2(P_n R)}\overline{T_H}(P_n,z,s), \tag{10}$$

where $P_n R = \mu_n$ are the roots of $J_0(\mu_n) = 0$.

Using the inverse Laplace transform, the solution (9) for the original of $T(r,z,\tau)$ can be written in the following form

$$T(r,z,\tau)-T_0=\frac{a_z}{h}\sum_{n=1}^{\infty}\frac{J_0(\mu_n\frac{r}{R})}{J_1^2(\mu_n)}\int_0^R e^{-\mu_n^2\frac{a_r\xi}{R^2}}$$

$$\cdot\frac{\partial}{\partial z}\theta_0(\frac{h-z}{h}\Big|i\pi\frac{\alpha_z\xi}{h^2})\frac{2}{R^2}\int_0^R xJ_0(\mu_n\frac{x}{R})\left[T_0-T_v(x,\tau-\xi)\right]dxd\xi, \tag{11}$$

where $\theta_0(\frac{v}{\tau})$ is theta function [GR80].

If the boundary conditions of first kind are given in discontinuous form (in the ring domain)

$$\overline{T}(r,z,s)\left|_{z=0}-\frac{T_0}{s}=\begin{cases}\overline{T_v}(r,s)-\frac{T_0}{s}, & r_1>r>r_0;\\ 0, \quad r_1<r<R & and \quad 0\le r<r_0\end{cases}\right\} \tag{12}$$

then the solution (11) has the form

$$T(r,z,\tau)-T_0=\frac{a_z}{h}\sum_{n=1}^{\infty}\frac{J_0(\mu_n\frac{r}{R})}{J_1^2(\mu_n)}\int_0^\tau e^{-\mu_n^2\frac{a_r\xi}{R^2}}$$

$$\cdot\frac{\partial}{\partial z}\theta_0(\frac{h-z}{h}(i\pi\frac{\alpha_z\xi}{h^2})\frac{z}{R^2}\int_{r_0}^{r} xJ_0(\mu_n\frac{x}{R})\left[T_0-T_v(x,\tau-\xi)\right]dxd\xi \tag{13}$$

We can get the solution of the unbounded orthotropic plate with conditions (12) on one of the plate surfaces at $R\to\infty$ from (13)

$$\lim_{R\to\infty}\left[T(r,z,\tau)-T_0\right]=T_1(r,z,\tau)-T_0$$

$$=\frac{\alpha_z}{2h\alpha_r}\int_0^r\frac{1}{\xi}e^{\frac{-r^2}{4a_r\xi}}\frac{\partial}{\partial z}\theta_0(\frac{h-z}{h}\Big|i\pi\frac{a_z\xi}{h^2})\cdot e^{\frac{-x^2}{4a_r\xi}}I_0(\frac{rx}{2a_r\xi})\left[T_0-T_v(x,\tau-\xi)\right]dxd\xi. \tag{14}$$

The solution (14) in the domain of Laplace transform $\overline{T}_1(r,z,s)$ can be written in the following form (s is Laplace transform parameter)

$$\overline{T}_1(r,z,s) - \frac{T_0}{s} = \int_0^\infty PJ_0(\mathrm{Pr}) \sinh\left[\frac{(h-z)}{\sqrt{a_z}}\sqrt{s+a_r P^2}\right] \cdot \int_{r_0}^{r_1} xJ_0(Px)\left[\overline{T}_v(x,s) - \frac{T_0}{s}\right] dxdp. \tag{15}$$

We can get the solution of the two dimensional nonstationary solution $\overline{T}_2(r,z,s) = r \to \infty \lim \overline{T}_1(r,z,s)$ for half semi-infinite orthotropic half-space with the initial conditions (3) and boundary conditions (12)

$$\overline{T}_2(r,z,s) - \frac{T_0}{s} = \int_0^\infty PJ_0(\mathrm{Pr}) e^{-\frac{z}{\sqrt{a_z}}\sqrt{s+a_r P^2}} \cdot \int_{r_0}^{r_1} xJ_0(Px)\left[\overline{T}_v(x,s) - \frac{T_0}{s}\right] dxdp \tag{16}$$

Applying the inverse Laplace transform to (16), we get

$$T_2(r,z,\tau) - T_0 = \frac{z}{4\sqrt{\pi a_r}\sqrt{a_z}} \int_0^\tau \frac{1}{\xi^{\frac{s}{2}}} e^{\left(\frac{-r^2}{4a_r\xi} - \frac{z^2}{4a_z\xi}\right)} \cdot \int_{r_0}^{r_1} xe^{\left(\frac{-x^2}{4a_r\xi}\right)} I_0\left(\frac{rx}{2a_r\xi}\right) \left[T_v(x,\tau-\xi) - T_0\right] dxd\xi. \tag{17}$$

If the inner radius r_0 of the ring domain goes to zero ($r_0 \to 0$), we have from (17) the two dimensional nonstationary solution for semi-infinite orthotropic solid with initial conditions (3) and boundary conditions (12) in the circular domain $0 \le r < r_1$, $z = 0$

$$T_2(r,z,\tau) - T_0 = \frac{z}{4\sqrt{\pi \alpha_z}} \int_0^\tau \frac{1}{\xi^{\frac{s}{2}}} e^{\left(-\left(\frac{r^2}{a_r} + \frac{z^2}{a_z}\right)\frac{1}{4\xi}\right)} \cdot \int_{r_0}^{r_1} xe^{\left(\frac{-x^2}{4a_r\xi}\right)} I_0\left(\frac{rx}{2a_r\xi}\right) \left[T_v(x,\tau-\xi) - T_0\right] dxd\xi.$$

Suppose that the redundant temperature in the circular domain ($0 \le r \le r_1$) on the surface ($z = 0$) of semi-bounded orthotropic solid is constant, that is,

$T_v(r,z) - T_0 = T_v - T_0 = const\ (T_0 \neq T_v)$.

Then it is not hard to get an expression for the temperature difference $T_2(r,z,\tau) - T_0$ in the form

$$\frac{T_2(r,z,\tau) - T_0}{T_v - T_0} = \frac{1}{2}\int_0^\infty J_0(\frac{rx}{r_1})J_1(x)e^{-\frac{z\sqrt{K_a}}{r_1}x}$$

$$\{erfc(\frac{z}{2\sqrt{a_z\tau}} - \frac{\sqrt{a_r\tau}}{r_1}x) + e^{-(\frac{z\sqrt{K_a}}{r_1}x)}erfc(\frac{z}{2\sqrt{a_z\tau}} + \frac{\sqrt{a_r\tau}}{r_1}x)\}dx.$$

At $r = 0$, we get the solution $T_2(0,z,\tau)$ on the axis $z \geq 0$

$$\frac{T_2(0,z,\tau) - Tv}{T_0 - T_v} = erfc(\frac{z}{2\sqrt{a_z\tau}}) + \frac{1}{\sqrt{1 + \frac{r_1^2}{z^2K_a}}}$$

$$erfc(\frac{r_1}{2\sqrt{a_z\tau}} + \sqrt{1 + \frac{z^2K_a}{r_1^2}}.$$

References

[KMY01] Kozlov, V.P., Mandrik, P.A., Yurchuk, N.I.: Method for Solving Nonstionary Heat Problems with Mixed Discontinuous Boundary Conditions on the Boundary of a Half-Space. Differential Equations, **37(2)**, 257-261 (2001)

[Man02] Mandrik, P.A.: Application of Laplace and Hankel Transforms to Solution of Mixed Nonstationary Boundary Value Problems. Integral Transforms and Special Functions, **13(3)**, 277-283 (2002)

[Man01] Mandrik, P.A.: Solution of the Heat Equation with Mixed Boundary Conditions on the Surface of an Isotropic Half-Space. Differential Equations, **37(2)**, 257-261 (2001)

[Koz86] Kozlov, V.P.: Two Dimensional Nonstationary Axial Symmetric Heat Conduction Problems. Nauka and Technika, Minsk (1986) (Russian)

[LUS79] Lebedev, N.N., Uflyand, Y.S., Skalskaya, I.P.: Worked Problems in Applied Mathematics. Dover Publications, New York (1979)

[GR80] Gradshteyn, I.S., Ryzhik, I.M.: Tables of Integrals. Series and Products, Academic Press, New York (1980)

A numerical analysis of variational finite difference schemes for steady state heat conduction problems with discontinuous coefficients

Ebru Ozbilge

Department of Mathematics, Applied Mathematical Sciences Research Center, Kocaeli University, 41300 Anitpark, Izmit - Kocaeli, Turkey
ebruozbilge@gmail.com

A class of monotone conservative schemes is derived for the boundary value problem for second order differential equation with discontinuous coefficient. The necessary condition for conservativeness of the finite difference scheme is obtained. The examples are presented for different discontinuous coefficients and the theoretical statements for the conservativeness conditions are supported by the results of numerical experiments.

1 Conservative finite difference schemes

We consider the steady-state heat conduction model in the case $q(x) = 0$

$$-(k(x)u'(x))' = f(x), \quad x \in (0,1). \tag{1}$$

Integrating equation (1), we at once obtain

$$\varphi_2 - \varphi_1 = \int_{x_1}^{x_2} f(x)dx, \;\; \varphi_i = -(k(x)u^{'}(x))_{x=x_i}, x_i \in (0,1), i = 1,2, \tag{2}$$

where $\varphi(x) := -k(x)u^{'}(x)$ denotes the flux. The number φ_i is the value of the flux at the point x_i, $i = 1,2$. The left hand side of this equation represents differences of flux at the end points and the right hand side of the equation represents the heat given to the rod. Equality at (2) implies the conservation law of heat, that is, the heat given to the rod is equal to the flux differences at the end points.(See in [LW60]) We will study the difference analogy of this property for the solution of difference schemes. Note that finite difference schemes, satisfying conservation laws, are called conservative finite difference schemes.(See in [Sama01]) Let us make this clear by an example. The finite difference scheme of problem (1) is given as:

Kenan Taş et al. (eds), Mathematical Methods in Engineering, 223–232.

$$-\tfrac{1}{h}\left[b_i\tfrac{y_{i+1}-y_i}{h} - a_i\tfrac{y_i-y_{i-1}}{h}\right] = f_i, \quad i=\overline{1,N-1}, Nh=1. \tag{3}$$

Let us write the finite difference scheme (3) in the following form

$$-\tfrac{1}{h}\left[a_{i+1}\tfrac{y_{i+1}-y_i}{h} - a_i\tfrac{y_i-y_{i-1}}{h}\right] = f_i + \tfrac{(b_i-a_{i+1})}{h}\tfrac{y_{i+1}-y_i}{h}, i=\overline{1,N-1}. \tag{4}$$

It is equivalent to the difference equation

$$\varphi_{i+1}^h - \varphi_i^h = hf_i + (b_i - a_{i+1})\frac{y_{i+1}-y_i}{h}, i=\overline{1,N-1}, \tag{5}$$

where

$$\varphi_{i+1}^h = -a_{i+1}\frac{y_{i+1}-y_i}{h}, \quad \varphi_i^h = -a_i\frac{y_i-y_{i-1}}{h}, i=\overline{1,N-1}. \tag{6}$$

This is similar to the conservation laws. It is obvious that if

$$b_i \neq a_{i+1}, i=\overline{1,N-1} \tag{7}$$

then conservation law can not be satisfied. If we write (5) in the whole mesh then the following equation can be obtained:

$$\varphi_N^h - \varphi_1^h = \sum_{i=1}^{N-1} hf_i + \sum_{i=1}^{N-1}(b_i - a_{i+1})\frac{y_{i+1}-y_i}{h}. \tag{8}$$

The conservation law is satisfied in the whole mesh if and only if the condition

$$b_i = a_{i+1}, i=\overline{1,N-1} \tag{9}$$

holds.

2 Conservativeness of difference schemes in the class of discontinuous coefficients: necessary and sufficient conditions

It is well known that the classical finite difference schemes are based on the notion of classical solution[SA76]. Because of this reason the coefficients of the considered equation are assumed to be continuous. But in many physical problems, the coefficients, which explain the properties of material, may not be continuous.

Let us consider the following one dimensional steady-state heat conduction model

$$\begin{cases} -(k(x)u'(x))' + q(x)u(x) = f(x), & x \in (0,1), \\ u(0)=1, \; u(1)=0. \end{cases} \tag{10}$$

where $c_1 \geq k(x) \geq c_0 > 0$, $q(x) \geq 0$. In the case $f(x) = 0$ and $q(x) = 0$, the finite difference approximation of problem (10) is given as follows:

$$\begin{cases} -\frac{1}{h}\left[b_i \frac{y_{i+1}-y_i}{h} - a_i \frac{y_i - y_{i-1}}{h}\right] = 0, \quad i = \overline{1, N-1}, \\ y_0 = 1, \ y_N = 0. \end{cases} \tag{11}$$

Assume that the coefficient $k(x)$ is discontinuous at $\xi \in (0,1)$, piecewise constant function:

$$k(x) = \begin{cases} k_1, \ 0 < x \leq \xi \\ k_2, \ \xi < x < 1, \quad \xi \in (0,1), \ k_i = const, i = 1,2, \ k_2 > k_1 > 0. \end{cases} \tag{12}$$

This situation models the steady state heat conduction in the one dimensional rod, which consists of two different homogeneous materials. Taking into account the continuity of the weak solution[Ad75] $u \in H^1[0,1]$ of problem (10) and the continuity of the flux $-k(x)u'(x)$, we can write

$$\begin{cases} (u)_{x=\xi} := u(\xi+0) - u(\xi-0) = 0, \\ (ku')_{x=\xi} := k(\xi+0)u'(\xi+0) - k(\xi-0)u'(\xi-0) = 0. \end{cases} \tag{13}$$

Then the analytical solution of the boundary value problem (10) is represented as follows:

$$u(x) = \begin{cases} 1 - \alpha x, \quad 0 \leq x \leq \xi, \\ \beta(1-x), \ \xi \leq x \leq 1. \end{cases} \tag{14}$$

The continuity conditions (13) permit one to find the unknown parameters $\alpha, \beta \in R$ [HMSS02]:

$$\alpha = \frac{k_2}{(k_2-k_1)\xi + k_1}, \quad \beta = \frac{k_1}{(k_2-k_1)\xi + k_1}, \quad \xi \in (0,1). \tag{15}$$

Now assuming that $\xi \in (x_m, x_{m+1})$, we consider the monotone finite difference (11)(see in [SMMM02, SZ81]) in the case of discontinuous coefficients (12). We have

$$\begin{cases} y_{i-1} - 2y_i + y_{i+1} = 0, & i \neq m, m+1, \\ b_m(y_{m+1} - y_m) - a_m(y_m - y_{m-1}) = 0, & i = m, \\ b_{m+1}(y_{m+2} - y_{m+1}) - a_{m+1}(y_{m+1} - y_m) = 0, & i = m+1 \end{cases} \tag{16}$$

In order to obtain the approximate solution $y^h = (y_0, ..., y_N)$ of the discrete problem (16) in the following form:

$$y_i = \begin{cases} 1 - \alpha_h x_i, \quad 0 \leq i \leq m, \\ \beta_h(1 - x_i), \ m+1 \leq i \leq N, \end{cases} \tag{17}$$

we need to determine the unknown parameters α_h and β_h from the second and third equations of (16). Eliminating the difference $y_{m+1} - y_m$ from these equations and finding the differences $y_{m+2} - y_{m+1}$ and $y_m - y_{m-1}$ from the second and first parts of (17), respectively, we get the following relationship:

$$\beta_h = \frac{a_m a_{m+1}}{b_m b_{m+1}} \alpha_h \tag{18}$$

between the parameters β_h and α_h, via the coefficients a_i and b_i of the finite difference scheme (16). Let us define now the values of the approximate solution (17) at the mesh points $i = m$ and $i = m+1$:

$$y_m = 1 - \alpha_h x_m, \quad y_{m+1} = \beta_h(1 - x_{m+1}),$$

to find the difference

$$y_{m+1} - y_m = \beta_h(1 - x_{m+1}) - 1 + \alpha_h x_m.$$

Substituting this, with the difference $y_m - y_{m-1} = -\alpha_h h$, into the second equation of (16), we obtain

$$b_m[\beta_h(1 - x_{m+1}) - 1 + \alpha_h x_m] + a_m \alpha_h h = 0.$$

Using (18) here, we can find the unknown parameter α_h:

$$\alpha_h = \left[\frac{a_m}{b_m} h + x_m + (1 - x_{m+1}) \frac{a_m a_{m+1}}{b_m b_{m+1}} \right]^{-1}. \tag{19}$$

Thus, formulas (18) and (19) define one-to-one the approximate solution (17), corresponding to the discontinuous coefficient (12). Let us take approximate solution as follows[Vol88]:

$$y_h(x) = \begin{cases} 1 - \alpha_h x, & x \in [0, \xi] \\ \beta_h(1 - x), & x \in [\xi, 1] \end{cases}$$

and suppose that approximate solution approaches to exact solution $u(x)$ in $C - norm$. That is $\|u(x) - y_h(x)\|_{C[0,1]} \to 0, \quad h \to 0$. This is equivalent to $\alpha_h \to \alpha$ and $\beta_h \to \beta$, as $h \to 0$. Comparing the coefficients α and α_h, we will get the following condition:

$$\Delta_h := \frac{a_m a_{m+1}}{k_1} - \frac{b_m b_{m+1}}{k_2} \to 0, \quad h \to 0.$$

Lemma 1. *The necessary and sufficient condition for the convergence in C-norm of the monotone finite difference scheme (11) in the class of the discontinuous coeficients (12) is given as:*

$$\Delta_h := \frac{a_m a_{m+1}}{k_1} - \frac{b_m b_{m+1}}{k_2} \to 0, \quad h \to 0. \tag{20}$$

(20) can be rewritten in the following form:

$$\frac{k_1}{k_2} = \frac{a_m}{b_m}\frac{a_{m+1}}{b_{m+1}}. \tag{21}$$

Therefore, homogeneous finite difference scheme is convergent when condition (20) is satisfied. By starting out this, various conservative finite difference schemes can be constructed. Now by considering the family of homogeneous finite difference schemes in the class of discontinuous coefficients, it can be proved that convergent finite difference schemes are conservative finite difference schemes [Shis92]. In the following section let us prove this.

3 A family of conservative schemes in the class of discontinuous coefficients

Taking into account the tri-diagonality of the schemes, corresponding to the second order differential operators (with the order of approximation $O(h^2)$, on a uniform mesh, in general), we can obtain the coefficients a_i, b_i of the monotone scheme (11), via the coefficients $k = k(x)$ as follows:

$$\begin{cases} a_m = \alpha_{-1}k_{m-1} + \alpha_0 k_m + \alpha_{+1}k_{m+1}, \\ b_m = \beta_{-1}k_{m-1} + \beta_0 k_m + \beta_{+1}k_{m+1}, \end{cases} \tag{22}$$

where the unknown parameters $\alpha \pm i$, $\beta \pm i$, $i = 0, 1$, satisfy the conditions

$$\alpha_{-1} + \alpha_0 + \alpha_1 = 1, \quad \beta_{-1} + \beta_0 + \beta_1 = 1. \tag{23}$$

Now we are going to show, on the basic model (10), (12), how the family of conservative schemes can be derived by using criterion (21). Since $\xi \in (x_m, x_{m+1})$ is the discontinuity point of the function $k(x)$, given by (12), for the mesh points $i = m-1,\ m,\ m+1$ we have

$$k_{m-1} = k_m = k_1; \quad k_{m+1} = k_2.$$

Hence, by (22)-(23), the coefficient a_m,b_m,a_{m+1} and b_{m+1} can be determined as follows:

$$a_m = (1-\alpha_1)k_1 + \alpha_1 k_2, \quad b_m = (1-\beta_1)k_1 + \beta_1 k_2, \tag{24}$$

$$a_{m+1} = \alpha_{-1}k_1 + (1-\alpha_{-1})k_2, \quad b_{m+1} = \beta_{-1}k_1 + (1-\beta_{-1})k_2. \tag{25}$$

Thus, in the interval $[x_m, x_{m+1}]$, including the discontinuity point $\xi \in [x_m, x_{m+1}]$, of the uniform mesh w_h, the coefficients a_i, b_i of the monotone finite difference scheme (11) satisfies conditions (24)-(25). Now we need to add these conditions to the convergence condition (21) to obtain the required family of conservative schemes. For this aim, we introduce the parameter $t = k_1/k_2 > 0$ and use conditions (24) and (25) in (21):

$$\frac{(1-\alpha_1)k_1+\alpha_1 k_2}{(1-\beta_1)k_1+\beta_1 k_2}\cdot\frac{\alpha_{-1}k_1+(1-\alpha_{-1})k_2}{\beta_{-1}k_1+(1-\beta_{-1})k_2}-\frac{k_1}{k_2}=0$$

This condition implies

$$\begin{aligned}&-\alpha_{-1}(1-\alpha_1)t^2-[(1-\alpha_1)(1-\alpha_{-1})+\alpha_1\alpha_{-1}]t-\alpha_1(1-\alpha_{-1})\\ &+\beta_{-1}(1-\beta_1)t^3+[(1-\beta_1)(1-\beta_{-1})+\beta_1\beta_{-1}]t^2+\beta_1(1-\beta_{-1})t=0,\quad \forall t>0.\end{aligned}$$

Due to the arbitrarity of the parameter $t>0$, we obtain that all coefficients of this polynomial are equal to zero:

$$\begin{cases}\alpha_1(1-\alpha_{-1})=0,\\ \beta_{-1}(1-\beta_1)=0,\\ \beta_1\beta_{-1}+(1-\beta_1)(1-\beta_{-1})-\alpha_{-1}(1-\alpha_1)=0,\\ \beta_1(1-\beta_{-1})-(1-\alpha_1)(1-\alpha_{-1})-\alpha_1\alpha_{-1}=0.\end{cases}$$

Eliminating free parameters and taking into the approximation conditions

$$\alpha_{-1}+\alpha_0+\alpha_1=1,\quad \beta_{-1}+\beta_0+\beta_1=1,\quad \beta_1-\beta_{-1}=1+\alpha_1-\alpha_{-1},$$

we obtain

$$\beta_{-1}=\alpha_1=0,\quad \alpha_{-1}=\beta_0=1-\alpha_0,\quad \beta_1=\alpha_0,\quad \alpha_0\in[0,1].$$

Substituting this in (22), we obtain the following one parametric family of coefficients corresponding to the nodal point x_m:

$$\begin{cases}a_m=(1-\alpha_0)k_{m-1}+\alpha_0 k_m,\\ b_m=(1-\alpha_0)k_m+\alpha_0 k_{m+1},\quad \alpha_0\in[0,1].\end{cases}\tag{26}$$

The conservativeness condition $a_{m+1}=b_m$, for the case of discontinuous coefficients, evidently follows from (26). From (26) if we select different α_0 then we construct various conservative schemes.

Theorem 1. *The solution of the finite difference scheme (11) is convergent in the class of discontinuous coefficients if and only if the finite difference scheme (11) is conservative. That is:* $a_{i+1}=b_i$ $i=\overline{1,N-1}$.

4 Special cases and numerical examples

As a first example, let us give the analysis the nonconservative case for the weak solution $u(x)\in\overset{0}{H}{}^1[0,1]$ of the boundary value problem (10) with $q(x)=0$, $f(x)=0$, corresponding to the discontinuous coefficient $k=k(x)$ at $\xi=1/3$, given by (12), with $k_1=1$, $k_2=2$. The analytical solution of this problem, given by (14)-(15), is plotted in Figure 1 (solid line). The results presented in this figure correspond to the two finite difference schemes with the following coefficients

$$a_m^{(1)} = k_1,\ \ b_m^{(1)} = k_1,\ \ a_{m+1}^{(1)} = (k_1+k_2)/2,\ \ b_{m+1}^{(1)} = k_2;$$
$$a_m^{(2)} = k_1,\ \ b_m^{(2)} = k_2,\ \ a_{m+1}^{(2)} = (k_1+k_2)/2,\ \ b_{m+1}^{(2)} = k_2.$$

Evidently in both cases, conservativeness condition $b_m = a_{m+1}$ is not satisfied.

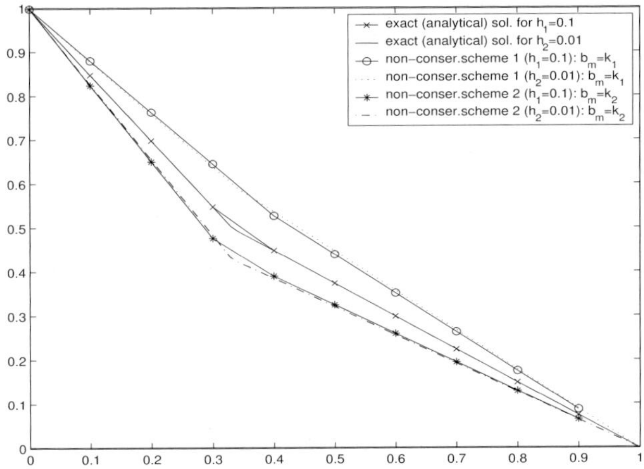

Fig. 1. Behaviour of the non-conservative schemes on the uniform meshes

Figure 1 shows that in the both cases the relative errors $\varepsilon_u^{(j)}$ are high enough, especially near the discontinuity point $x = \xi$. Moreover, by decreasing the mesh step from $h = 0.1$ to $h = 0.01$ these errors doesn't decrease. This shows that in the class of discontinuous coefficients, the monotone scheme does not converge. Formula (26) shows that having the values of the parameter $\alpha_0 \in [0,1]$ one can construct various conservative finite difference schemes, in particular, well-known in literature schemes. However, as will be shown below, accuracy of these convergent schemes will be different, depending on the choice of the parameter $\alpha_0 \in (0,1)$, as well as on the jump $[k]_{x=\xi}$ of the discontinuous function $k(x)$ at $x = \xi$. Next consider the class of conservative schemes given by formula (26), for the weak solution $u(x) \in \overset{0}{H}{}^1[0,1]$ of the boundary value problem (10) with $q(x) = 0$, $f(x) = 1$, corresponding to the following discontinuous coefficients at the point $\xi = 1/2$.

$$k(x) = \begin{cases} x^2+1, & x \in [0,\xi), \\ \frac{7}{2}x + \frac{1}{2}, & x \in [\xi,1] \end{cases} \tag{27}$$

$$k(x) = \begin{cases} x^2+1, \ x \in [0,\xi), \\ 4x+6, \ x \in [\xi,1] \end{cases} \tag{28}$$

respectively. Take the weak solution $u(x)$ as a second order polynomial in the space $\overset{0}{H}{}^1[0,1] = \{u \in H^1[0,1] : u(0) = u(1) = 0\}$.

$$u(x) = \begin{cases} x + bx^2, & x \in [0,\xi), \\ (1-x) + d(1-x)^2, & x \in [\xi, 1] \end{cases} \tag{29}$$

From continuity conditions (13) for the function $u(x) \in H^1[a,b]$ at the discontinuity point $x = \xi = 1/2$, the coefficients b and d can be obtained as:

$$\begin{cases} b = -\frac{1}{2}\frac{(2\xi^3-23\xi^2+6\xi-1)}{\xi}\frac{1}{(2\xi^3-9\xi^2+\xi-2)}, \\ d = \frac{1}{2}\frac{(6\xi^3-11\xi^2+5\xi-4)}{(10\xi^2-11\xi^3+2\xi^4+2-3\xi)}. \end{cases} \tag{30}$$

$$\begin{cases} b = -\frac{1}{2}\frac{(\xi^3-13\xi^2-13\xi+5)}{\xi}\frac{1}{(\xi^3-5\xi^2-\xi-1)}, \\ d = \frac{1}{2}\frac{(-6\xi^2+3\xi^3-2-3\xi)}{(-6\xi^3+\xi^4+1+4\xi)}. \end{cases} \tag{31}$$

Table 1. Relative errors for the conservative schemes: $[k]_{\xi=1/2} = 1$

$\alpha_0^{(j)}$	0	1/3	2/3	1/4	3/4	1/5	4/5
a_{m+1} $b_m = a_{m+1}$	k_1	$2k_1/3$ $+k_2/3$	$k_1/3$ $+2k_2/3$	$3k_1/4$ $+k_2/4$	$k_1/4$ $+3k_2/4$	$4k_1/5$ $+k_2/5$	$k_1/5$ $+4k_2/5$
ε_u $(h = 10^{-2})$	0.1114	0.3437	0.6299	0.2717	0.7010	0.2283	0.7436
ε_u $(h = 10^{-3})$	0.0116	0.0344	0.0627	0.0273	0.0698	0.0230	0.0741

The analytical solution (29) is shown in Figure 2 with solid line. In Figure 2, $k_1 = x^2 + 1, k_2 = \frac{7}{2}x + \frac{1}{2}, h = 0.01, \alpha_0^{(1)} = 0$, and $\xi = 1/2$ is taken. In addition to this, the relative errors are given in Table 1 and Table 2 with respect to the parameters $\alpha_0^{(1)} = 0, \alpha_0^{(2)} = 1/3, \alpha_0^{(3)} = 2/3, \alpha_0^{(4)} = 1/4, \alpha_0^{(5)} = 3/4, \alpha_0^{(6)} = 1/5, \alpha_0^{(7)} = 4/5$, for the mesh steps $h = 0.01$ and $h = 0.001$ with jumps $k_1 = x^2 + 1, k_2 = \frac{7}{2}x + \frac{1}{2}, [k]_{\xi=1/2} = 1$ and $k_1 = x^2 + 1, k_2 = 4x + 6, [k]_{\xi=1/2} = 27/4$, respectively. As it seen from the Table 1 and Table 2, in both cases the minimal relative error achieves at the parameter $\alpha_0^{(1)} = 0$.

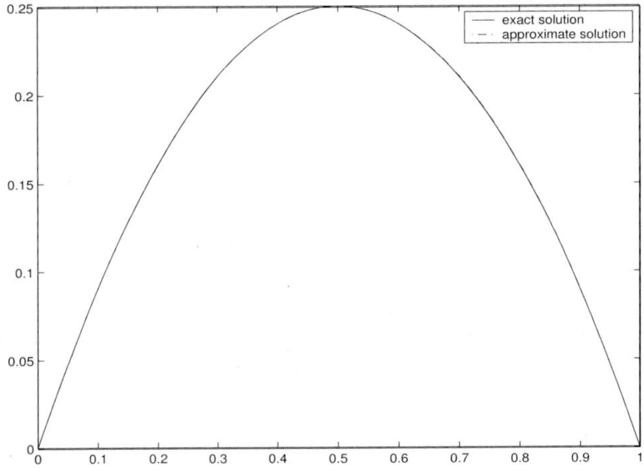

Fig. 2. Convergence of conservative schemes on the uniform meshes

Table 2. Relative errors for the conservative schemes: $[k]_{\xi=1/2} = 27/4$

$\alpha_0^{(j)}$	0	1/3	2/3	1/4	3/4	1/5	4/5
a_{m+1} $b_m = a_{m+1}$	k_1	$2k_1/3$ $+k_2/3$	$k_1/3$ $+2k_2/3$	$3k_1/4$ $+k_2/4$	$k_1/4$ $+3k_2/4$	$4k_1/5$ $+k_2/5$	$k_1/5$ $+4k_2/5$
ε_u $(h = 10^{-2})$	1.1173	1.2682	1.4129	1.2301	1.4482	1.2065	1.4693
ε_u $(h = 10^{-3})$	0.1134	0.1274	0.1414	0.1239	0.1449	0.1218	0.1469

If we compare Table 1 and Table 2, we see that as jump of the discontinuity point increases, the relative error increases. Note that the proposed approach to the construction of the family of conservative finite difference schemes can also be applied to the multidimensional problems.

5 Acknowledgement

The author thanks A. Hasanoglu (Hasanov) for the statement of this problem and helpful discussions.

References

[LW60] Lax, P.D., Wendroff, B.: System of conservation laws. Comm.Pure Appl. Math., **13(2)**, 217–237 (1960)

[Sama01] Samarskii, A.A.: The theory of difference schemes. Marcel Dekker, New York (2001)

[SA76] Samarskii, A.A., Andreev, V.: Difference methods for elliptic equations. Nauka, Moscow (1976)

[Ad75] Adams, R.A.: Sobolev Spaces. Academic Press, New York (1975)

[HMSS02] Hyman, J., Morel, J., Shashkov, M., Steinberg, S.: Mimetic finite difference methos for diffusion equations. Computational Geosciences, **6**, 333–352 (2002)

[SMMM02] Samarskii, A.A., Matus, P., Mazhukin, V., Mozolevski, I.: Monotone difference schemes for equations with mixed derivatives. Computers and Mathematics with Applications, **44**, 501–510 (2002)

[SZ81] Schneider, G., Zedan, M.: A modifed strongly implicit procedure for the numerical solution of field problems. Numerical Heat Transfer, **4**, 1–19 (1981)

[Vol88] Voligt, W.: Finite-difference schemes for parabolic problems with first and mixed second derivatives. Z. Angew. Math. und Mech., **68(7)**, 281–288 (1988)

[Shis92] Shishkin, G.: Grid approximation of the singularly perturbed boundary-value problem for quasilinear parabolic equations in the case of total degeneracy with respect to space variables. In: Kuznetsov, Yu.A. (ed) Numerical methods and mathematical modeling. Russian Academy of Sciences, Institute of Computational Mathematics, Moscow (1992)

On the solution of a mathematical model of a viscoelastic bar

Arpad Takači[1] and Djurdjica Takači[2]

[1] Department of Mathematics and Informatics, Faculty of Science, University of Novi Sad, Trg D. Obradovića 4, 21000 Novi Sad, Serbia takaci@im.ns.ac.yu
[2] Department of Mathematics and Informatics, Faculty of Science, University of Novi Sad, Trg D. Obradovića 4, 21000 Novi Sad, Serbia djtak@im.ns.ac.yu

A hyperbolic type equation with certain initial and boundary conditions, appropriate for application of the Mikusiński calculus, is considered. Similar problems appeared as mathematical models of the shock between a solid body and a viscoelastic bar.

The exact solution of the corresponding problem in the field of Mikusiński operators is constructed, and the character and regularity of the operational function solution of the problem is analyzed. Then the solution of the starting problem is obtained as a finite sum of continuous functions. An algorithm for constructing an approximate solution is given, and an example is presented.

1 Introduction

We consider the following partial differential equation:

$$\frac{\partial^2 u(x,t)}{\partial t^2} - \mu \frac{\partial^2 u(x,t)}{\partial x^2} + K u(x,t) + \lambda \frac{\partial u(x,t)}{\partial t} = f(x,t), \tag{1}$$

for $0 < x < 1$, and $0 < t < T$, with the following conditions:

$$u(0,t) = 0, \tag{2}$$

$$\mu \frac{\partial u(1,t)}{\partial x} + K_1 u(1,t) + \lambda_1 \frac{\partial u(1,t)}{\partial t} = g(t) + \int_0^t g_1(t-\tau) u(1,\tau) d\tau, \tag{3}$$

$$u(x,0) = u_0(x), \quad \frac{\partial u(x,0)}{\partial t} = u_1(x), \tag{4}$$

where μ, K, λ, K_1 and λ_1 are given positive constants, u_0, u_1, g, g_1 and f are given functions ($g_1(t) > 0$ for $t \in [0,T]$), and $u = u(x,t)$ is the unknown function.

Kenan Taş et al. (eds), Mathematical Methods in Engineering, 233–242.

In [BLD01], the equation (1), for $\mu = 1$, with the condition (4) and with the conditions:

$$\frac{\partial u(0,t)}{\partial x} = P(t), \tag{5}$$

$$\frac{\partial u(1,t)}{\partial x} + K_1 u(1,t) + \lambda_1 \frac{\partial u(1,t)}{\partial t} = 0, \tag{6}$$

was considered. The, initially unknown, function P satisfies is the solution of the equation

$$P'' + \omega^2 P(t) = h\frac{\partial^2 u(0,t)}{\partial t^2}, \qquad 0 < t < T, \tag{7}$$

with the conditions

$$P(0) = P_0, \qquad P'(0) = P_1, \tag{8}$$

where $\omega > 0$, $h > 0$ and P_0, P_1 are given constants.

The solution of the problem (7), (8) has the form

$$P(t) = g(t) + hu(0,t) - h\omega \int_0^t \sin(\omega(t-\tau))u(0,\tau)d\tau, \tag{9}$$

where

$$g(t) = (P_0 - hu_0(0))\cos\omega t + \frac{1}{\omega}(P_1 - hu_1(0))\sin\omega t. \tag{10}$$

This mathematical model expresses the shock between a solid body and a linear viscoelastic bar, resting on a viscoelastic base with linear elastic constrains at the side with a viscous frictional resistance. In [TT06], the same problem as in [BLD01] was considered within the frames of the Mikusiński calculus. The definition of the Mikusiński operator field, $\mathcal{F}$, and some of its properties are exposed in Section 2. For a complete survey on the Mikusiński calculus, see [Mi83] and [MB87].

In [LUT01], the problem (1), (2), (3), (4), (μ a given function) was considered. In this paper, we assume that μ is a positive constant and construct the solution of the corresponding problem in the field $\mathcal{F}$. We analyze the character of the obtained solution, $u(x)$, in the field $\mathcal{F}$, and, in particular, we give conditions ensuring that $u(x)$ represents a continuous function. The form of the function $u(x,t)$, corresponding to the operational function $u(x)$, together with sufficient conditions enabling it to become the solution of the starting problem. An algorithm for constructing an approximate solution is given.

We apply our procedure on a numerical example in order to present the obtained results. The difference between the exact and the approximate solution is presented with two plots from the *Scientific Workplace* program, version 5.5.

Note that in the paper [TT04] the authors considered certain classes of partial differential equations, that correspond to ordinary differential equations in $\mathcal{F}$. In fact, we analyzed the existence and character of the solution, constructed an approximate one and estimated the error of approximation.

2 Some elements of the Mikusiński calculus

The set of continuous functions $\mathcal{C}_+$ with supports in $[0, \infty)$, with the usual addition and the multiplication given by the convolution

$$(f * g)(t) = \int_0^t f(\tau)g(t-\tau)\,d\tau, \qquad t \geq 0,$$

is a commutative ring without unit element. By the Titchmarsh theorem, $\mathcal{C}_+$ has no divisors of zero, hence its quotient field, called the Mikusiński operator field, and denoted by $\mathcal{F}$, can be defined. Its elements are called *operators*; they are quotients of the form

$$\frac{f}{g}, \quad f \in \mathcal{C}_+, \; 0 \not\equiv g \in \mathcal{C}_+,$$

where the last division is observed in the sense of convolution (see [Mi83]). Clearly, every continuous function $a = a(t)$ with support in $[0, \infty)$ can be observed as a (unique) operator of the form $(a * g)/g$ (where g is an arbitrary nonzero element from $\mathcal{C}_+$); we shall simply denote this operator by a. Then we say that the operator a represents the continuous function $a(t)$ and write $a = \{a(t)\}$. In view of these remarks, the multiplication in $\mathcal{F}$ of two continuous functions $a = a(t)$ and $b = b(t)$ from $\mathcal{C}_+$ will be simply denoted by ab; this product is thus the operator c representing the continuous function $c(t) = a * b(t)$, $t \geq 0$. We shall denote by $\mathcal{F}_c$ the proper subset of $\mathcal{F}$ consisting of the operators representing continuous functions.

For examples of operators, we have the integral operator $\ell \in \mathcal{F}_c$ representing the constant function 1 on $[0, \infty)$, and the α powers of ℓ, ℓ^α :

$$\ell = \{1\}, \qquad \ell^\alpha = \left\{ \frac{t^{\alpha-1}}{\Gamma(\alpha)} \right\}, \quad \alpha \geq 1.$$

Also, among the most important operators are the inverse operator to ℓ, the differential operator s, and the identity operator, I, i.e., $\ell s = I$. Neither s nor I are operators from $\mathcal{F}_c$.

Note that for $a > 0$ the operator e^{-as} becomes the shift operator (see [Mi83]), given by

$$e^{-as}g = \left\{ \begin{array}{ll} 0, & 0 \leq t \leq a, \\ g(t-a), & 0 < a < t \end{array} \right\}, \tag{11}$$

where g is any operator representing a continuous function. In our consideration the condition $a > 0$ is essential for the existence of operator e^{-as}.

For the theory of differential equations, the following relation, connecting the operator representing the nth derivative of an $n$times derivable function $x = x(t)$ with the operator x is essential:

$$\{x^{(n)}(t)\} = s^n x - x(0)s^{n-1} - \cdots - x^{(n-1)}(0)I.$$

3 The exact solution of an operational differential equation

In the field of Mikusiński operators, $\mathcal{F}$, the problem

$$s^2u(x) - \mu u''(x) + Ku(x) + s\lambda u(x) = f_1(x) + su_0(x) + u_1(x) + \lambda u_0(x), \quad (12)$$

$$u(0) = 0, \quad (13)$$

$$\mu u'(1) + K_1 u(1) + \lambda_1(su(1) - u_0(1)) = G + G_1 u(1), \quad (14)$$

corresponds to the problem (1), (2), (3), (4). In (12), (13), (14), s is the differential operator, $u(x)$ is the unknown operational function corresponding to $u(x,t)$, while f_1 is the operational function corresponding to the right-hand side function f. The operators G and G_1 correspond to the functions g and g_1.

Equation (12) is a nonhomogeneous ordinary differential equation in the field of Mikusiński operators and its general solution can be found by using a well known procedure from classical analysis. Namely, the solution is a sum of the solution of the corresponding homogeneous equation and the particular one.

The characteristic equation of (12), in the field $\mathcal{F}$, has the form:

$$\mu v^2 = s^2 + K + s\lambda.$$

Hence, putting

$$W = \frac{I}{\sqrt{\mu}}\sqrt{s^2 + K + s\lambda}, \quad (15)$$

its solutions are:

$$v = \pm W = \pm\frac{I}{\sqrt{\mu}}\sqrt{s^2 + K + s\lambda}.$$

Thus the exact solution of the homogeneous differential equation has the form

$$u_h(x) = C_1 \exp(xW) + C_2 \exp(-xW),$$

where C_1 and C_2 are operators to be determined by using the conditions (13) and (14). Now, the solution of (12) has the form

$$u(x) = C_1 \exp(xW) + C_2 \exp(-xW) + u_p(x), \quad (16)$$

where the particular solution $u_p(x)$ can be determined similarly as in the classical case.

From (16) we have

$$\begin{aligned} u(0) &= C_1 + C_2 + u_p(0), \\ u'(1) &= C_1 W e^W - C_2 W e^{-W} + u_p'(1). \end{aligned} \quad (17)$$

Using (13), we get

$$C_1 = -C_2 - u_p(0).$$

From (14) and (17), we obtain

$$G + G_1 u(1) = \mu(C_1 W e^W - C_2 W e^{-W} + u_p'(1)) + K_1 \left(C_1 e^W + C_2 e^{-W} + u_p(1)\right)$$
$$+ \lambda_1 (s \left(C_1 e^W + C_2 e^{-W} + u_p(1)\right) - u_0(1)).$$

Substituting C_1 into this equation, and solving the obtained equation in C_1, we obtain

$$C_2 = \frac{G + G_1 u(1) - K_1 u_p(1) - \lambda_1 (s u_p(1) - u_0(1)) - \mu u_p'(1)}{(-\mu W - K_1 - \lambda_1 s)(I + \frac{-\mu W + K_1 + \lambda_1 s}{-\mu W - K_1 - \lambda_1 s} e^{-2W})} e^{-W} + \frac{-u_p(0)}{I + \frac{-\mu W + K_1 + \lambda_1 s}{-\mu W - K_1 - \lambda_1 s} e^{-2W}}. \tag{18}$$

4 The character of the solution

In order to analyze the character of the obtained exact solution, u, given by (16), we have to analyze the character of the operator C_2 from (18). To that end, we first transform the operator μW, W from (15), as follows:

$$\begin{aligned} \mu W &= \sqrt{\mu}\sqrt{s^2 + K + s\lambda} \;=\; \sqrt{\mu} s \sqrt{I + K\ell^2 + \ell\lambda} \\ &= s\sqrt{\mu} \sum_{i=0}^{\infty} \binom{1/2}{i} (K\ell^2 + \ell\lambda)^i \\ &= s(\sqrt{\mu} + \phi) \;=\; \sqrt{\mu} s + \frac{\lambda\sqrt{\mu}}{2} I + \phi_1 \end{aligned} \tag{19}$$

(see [Mi83], p. 57), where ϕ and ϕ_1 are operators from $\mathcal{F}_c$. Next, in order to simplify the forthcoming formulas, we introduce the operators $\mathcal{A}$ and $\mathcal{B}$ as follows:

$$\mathcal{A} = \frac{I}{\mu W + K_1 + \lambda_1 s} = \frac{I}{s(\sqrt{\mu} + \phi) + K_1 + \lambda_1 s}, \tag{20}$$

$$\mathcal{B} = \mathcal{A}(-\mu W + K_1 + \lambda_1 s) - I = \frac{-\mu W + K_1 + \lambda_1 s}{\mu W + K_1 + \lambda_1 s} - I. \tag{21}$$

One easily checks that both $\mathcal{A}$ and $\mathcal{B}$ are from $\mathcal{F}_c$.

The character of the operator $\mathcal{D}$, given by

$$\mathcal{D} = \mathcal{A} \cdot (G + G_1 u(1) - K_1 u_p(1) - \lambda_1 (s u_p(1) - u_0(1)) - \mu u_p'(1)), \tag{22}$$

depends on the character of the particular solution, because G and G_1 are from $\mathcal{F}_c$. Clearly, if the operational function $u_p(x)$ represents a continuous function, then the operator $\mathcal{D}$ also represents one. Further on, using (21), we can write

$$\frac{I}{I+\dfrac{-\mu W+K_1+\lambda_1 s}{-\mu W-K_1-\lambda_1 s}e^{-2W}}=\frac{I}{I-(I+\mathcal{B})e^{-2W}}=\sum_{i=0}^{\infty}(I+\mathcal{B})^i e^{-2Wi}$$
$$=I+(I+\mathcal{B})e^{-2W}+(I+\mathcal{B})^2e^{-4W}+\cdots \tag{23}$$

Finally, combining (20), (21) and (23), we obtain that the coefficient C_2 from (18) can be written in a form

$$C_2=\mathcal{D}\left(\sum_{i=0}^{\infty}(-1)^i(I+\mathcal{B})^i e^{-2Wi}\right)e^{-W}-u_p(0)\sum_{i=0}^{\infty}(-1)^i(I+\mathcal{B})^i e^{-2Wi}. \tag{24}$$

Similarly, the other coefficient appearing in the solution (16), C_1, has the form

$$C_1=\mathcal{D}\left(\sum_{i=0}^{\infty}(-1)^i(I+\mathcal{B})^i e^{-2Wi}\right)e^{-W}-u_p(0)\sum_{i=1}^{\infty}(-1)^i(I+\mathcal{B})^i e^{-2Wi}. \tag{25}$$

This analysis shows that if $\mathcal{D}$ is from $\mathcal{F}_c$, then so are the coefficients C_1 and C_2. As noted above, this is certainly true if $u_p(x)$ represents a continuous function.

Using (19), the operators e^{-rW}, $r\in\mathbb{N}$, appearing in (25) and (24), can be written as

$$e^{-rW}=e^{-\frac{r}{\sqrt{\mu}}s-\frac{r\lambda}{2\sqrt{\mu}}-\frac{r}{\sqrt{\mu}}\phi_1}=e^{-\frac{r}{\sqrt{\mu}}s}(I+W_r),$$

where W_r are operators from $\mathcal{F}_c$. Then the operators C_2 and C_1 can be conveniently written in following forms:

$$\begin{aligned}C_2&=-u_p(0)+D_1e^{-s/\sqrt{\mu}}+D_2e^{-2s/\sqrt{\mu}}+D_3e^{-3s/\sqrt{\mu}}+D_4e^{-4s/\sqrt{\mu}}+\cdots,\\ C_1&=D_1e^{-s/\sqrt{\mu}}+D_2e^{-2s/\sqrt{\mu}}+D_3e^{-3s/\sqrt{\mu}}+D_4e^{-4s/\sqrt{\mu}}+\cdots,\end{aligned} \tag{26}$$

where D_1, D_2, ..., are operators from $\mathcal{F}_c$.

Then the solution of the considered problem can be written as

$$\begin{aligned}u(x)=&\left(D_1e^{-s/\sqrt{\mu}}+D_2e^{-2s/\sqrt{\mu}}+D_3e^{-3s/\sqrt{\mu}}+D_4e^{-4s/\sqrt{\mu}}+\cdots\right)\\ &\cdot e^{(xs+x\frac{\lambda}{2}+x\phi_1)/\sqrt{\mu}}\\ &+\left(-u_p(0)+D_1e^{-s/\sqrt{\mu}}+D_2e^{-2s/\sqrt{\mu}}+D_3e^{-3s/\sqrt{\mu}}\right.\\ &\left.+D_4e^{-4s/\sqrt{\mu}}+\cdots\right)\cdot e^{(-xs-x\frac{\lambda}{2}-x\phi_1)/\sqrt{\mu}}+u_p(x).\end{aligned} \tag{27}$$

Analogously as in [TT06], we denote by $\delta_i(x)$ and $\gamma_i(x)$, $i=0,1,\ldots$, the following functions:

$$\delta_0 = -u_0 \exp(-x(\lambda/2) - x\phi_1)/\sqrt{\mu}),$$

$$\gamma_i(x) = E_i(x) \exp(x(\lambda/2) + x\phi_1)/\sqrt{\mu}), \tag{28}$$

$$\delta_i(x) = D_i(x) \exp(-x(\lambda/2) - x\phi_1)/\sqrt{\mu}),$$

for $i = 1, 2, \ldots$ and ϕ_1 from (19). Then the solution of equation (12) with conditions (13) and (14) can be written as

$$\begin{aligned} u(x) &= \gamma_1(x)e^{-(1-x)s/\sqrt{\mu}} + \gamma_2(x)e^{-(2-x)s/\sqrt{\mu}} + \gamma_3(x)e^{-(3-x)s/\sqrt{\mu}} + \cdots \\ &+ \delta_0(x)e^{-xs/\sqrt{\mu}} + \delta_1(x)e^{-(x+1)s/\sqrt{\mu}} + \delta_2(x)e^{-(x+2)s/\sqrt{\mu}} \\ &+ \delta_3(x)e^{-(x+3)s/\sqrt{\mu}} + \cdots + u_p(x), \end{aligned} \tag{29}$$

where $\gamma_1, \gamma_2, \ldots, \delta_0, \delta_1, \delta_2, \ldots,$ represent continuous functions.

5 The solution of the problem (1), (2), (3), (4)

The operational function $u(x)$ from (29) is, in the field $\mathcal{F}$, the solution of the problem (12), (13), (14). Now, our task is to show that the function $u(x,t)$, corresponding to this $u(x)$, is in fact the solution of the problem (1), (2), (3), (4). Using (28), we can determine the functions $\xi_{1,k}$, $k = 1, 2, 3, \ldots$, and $\xi_{2,j}$, $j = 0, 1, 2, \ldots$, corresponding to the operational functions

$$e^{-(k-x)s/\sqrt{\mu}}\gamma_k(x), \quad k = 1, 2, 3, \ldots, \qquad \text{and} \qquad e^{-(j+x)s/\sqrt{\mu}}\delta_j(x), \quad j = 0, 1, 2, \ldots,$$

respectively. In fact, they correspond to the addends of the exact solution $u(x)$ in (29), as follows:

$$e^{-(k-x)s/\sqrt{\mu}}\gamma_k(x) = \begin{cases} \{0\}, & t < \frac{k-x}{\sqrt{\mu}}, \\ \{\xi_{1,k}(x, t - \frac{k-x}{\sqrt{\mu}})\}, & t \geq \frac{k-x}{\sqrt{\mu}}, \end{cases} \qquad k = 1, 2, 3, \ldots, \tag{30}$$

$$e^{-(j+x)s/\sqrt{\mu}}\delta_j(x) = \begin{cases} \{0\}, & t < \frac{j+x}{\sqrt{\mu}}, \\ \{\xi_{2,j}(x, t - \frac{j+x}{\sqrt{\mu}})\}, & t \geq \frac{j+x}{\sqrt{\mu}}, \end{cases} \qquad j = 0, 1, 2, 3, \ldots. \tag{31}$$

Note that the interval $[0, T]$ has been divided into subintervals $[0, \frac{1}{\sqrt{\mu}}]$, $(\frac{1}{\sqrt{\mu}}, \frac{2}{\sqrt{\mu}}]$, $(\frac{2}{\sqrt{\mu}}, \frac{3}{\sqrt{\mu}}]$, and so on. The last of these subintervals is of the form $(\frac{n-1}{\sqrt{\mu}}, \frac{n}{\sqrt{\mu}}]$, where $\frac{n-1}{\sqrt{\mu}} < T \leq \frac{n}{\sqrt{\mu}}$.

The function $u(x,t)$, corresponding to the operational function $u(x)$ given by (29), has the form:

$$\begin{aligned} u(x,t) = u_p(x,t) + \xi_{2,0}(x, t - \tfrac{x}{\sqrt{\mu}}) + \xi_{1,1}(x, t - \tfrac{1+x}{\sqrt{\mu}}) \\ + \cdots + \xi_{2,n-1}(x, t - \tfrac{n-1+x}{\sqrt{\mu}}) + \xi_{1,n}(x, t - \tfrac{n+x}{\sqrt{\mu}}), \end{aligned} \tag{32}$$

where the functions $\xi_{1,k}$, $k = 1, 2, 3, \ldots$, and $\xi_{2,j}$, $j = 0, 1, 2, \ldots$, are given by (30) and (31), respectively.

Let us remark that on the right-hand side of the last equality there appears only a finite number of addends, i.e., $u(x,t)$ is expressed as a finite sum, in view of the shift operator property, see (11). Therefore, it is neither necessary to analyze the convergence of the infinite series appearing in (29), nor in previous relations, namely in (24), (25), (26) and (27).

If the function $u(x,t)$ from (32) satisfies the conditions (4) of (1), i.e.,

$$u(x,0) = u_0(x), \quad \frac{\partial u(x,0)}{\partial t} = u_1(x),$$

then it is the solution of the problem (1) with (2), (3), (4).

6 The approximate solution

In order to obtain the exact solution of the considered problem in the field $\mathcal{F}$, we used infinite sums in expressing the exact solution in relations (19), (20), (21). If instead of infinite sums in these relations the finite ones are considered, then an approximate solution in the field $\mathcal{F}$ is obtained.

The approximate solution of the characteristic equation can be written as:

$$\begin{aligned} W_n &= s\frac{1}{\sqrt{\mu}} \sum_{i=0}^{n} \binom{1/2}{i} (K\ell^2 + \ell\lambda)^i \\ &= s\frac{1}{\sqrt{\mu}}(I + \phi_n) \ = \ \frac{1}{\sqrt{\mu}} s + \frac{\lambda}{2} I + \phi_{1,n}. \end{aligned}$$

Analogously, we can obtain the approximations for the operators $\mathcal{A}$, $\mathcal{B}$, $\mathcal{D}$, given by (20), (21), (22), respectively. Using these approximations and also the analogously obtained approximation for D_i, $i = 1, 2, \ldots$ we can get $C_{1,n}$ and $C_{2,n}$, as the approximations for C_1 and C_2, from (26).

The approximate particular solution can be obtained by taking a finite sum instead of an infinite one.

The approximate solution of the considered problem in the field of Mikusiński operators, $u_n(x)$, can be obtain from (29) by taking the approximations for $\delta_i(x)$ and $\gamma_i(x)$, $i = 0, 1, \ldots,$ from (28).

The error of approximation can be obtained by determining the difference between the finite and infinite sums wherever the approximation are made.

7 Numerical example

For an example, let us consider the following equation:

$$\frac{\partial^2 u(x,t)}{\partial t^2} - 2\frac{\partial^2 u(x,t)}{\partial x^2} + 3u(x,t) + 2\frac{\partial u(x,t)}{\partial t} = 0,$$

$$u(0,t) = 0, \quad u(x,0) = e^x - e^{-x}, \quad \frac{\partial u(x,0)}{\partial t} = e^{-x} - e^x.$$

This is the problem (1), with (2), (3), and (4) for $\mu = 2$, $K = 3, \lambda = 2$.

In the field of Mikusiński operators, this problem corresponds to the problem

$$s^2 u(x) - 2u''(x) + 2su(x) + 3u(x) = (s+1)(e^x - e^{-x}),$$

$$u(0) = 0,$$

$$G + G_1 u(1) - K_1 u_p(1) - \lambda_1(su_p(1) - u_0(1)) - 2u_p'(1) = 0. \tag{33}$$

Let us remark that the numerical constants λ_1, g, and g_1 are chosen such that the second condition is satisfied.

Using (16) we obtain the exact solution of the problem (33) as

$$u(x) = C_1 \exp(x\sqrt{s^2 + I + s}) + C_2 \exp(-x\sqrt{s^2 + I + s}) + u_p(x),$$

and $C_1 = C_2 = 0$.

The particular solution of this problem is

$$u_p(x) = \frac{(s+1)\,(e^x - e^{-x})}{s^2 + 2s + 1} = \frac{\ell\,(e^x - e^{-x})}{I + \ell}.$$

The exact solution of the considered problem in $\mathcal{F}$ is

$$u(x) = u_p(x) = \frac{\ell(e^x - e^{-x})}{I + \ell} = (e^x - e^{-x}) \sum_{i=0}^{\infty} (-1)^i \ell^{i+1}$$
$$= \left\{ (e^x - e^{-x}) \sum_{i=0}^{\infty} (-1)^i \frac{t^i}{i!} \right\}.$$

The approximate solution in $\mathcal{F}$ is

$$u_n(x) = (e^x - e^{-x}) \sum_{i=0}^{n} (-1)^i \ell^{i+1} = \left\{ (e^x - e^{-x}) \sum_{i=0}^{n} (-1)^i \frac{t^i}{i!} \right\}.$$

In the Figure 1, both in a) and b) the exact solutions are given in Wire Frame, while the approximate ones are given in Color Patch. In a), $(x,t) \in [0,1] \times [0,1]$, while in b) and $(x,t) \in [0,1] \times [0,0.5]$.

As one could expect, these figures confirm that by decreasing the interval for t, the approximate solution becomes a better approximation of the exact one.

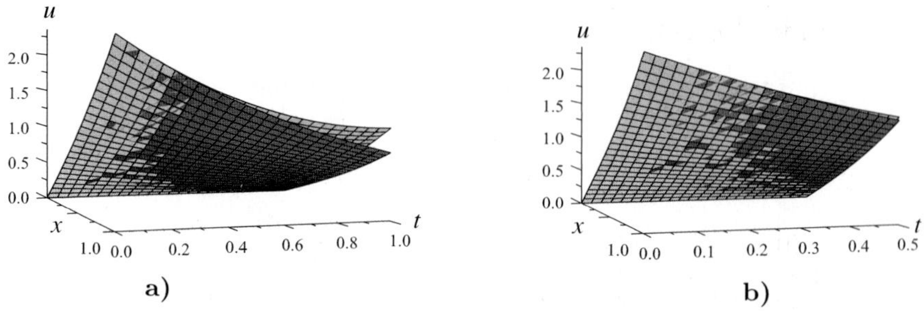

Fig. 1. Exact and approximate solutions of the example from Section 7.

Acknowledgement.

The authors want to thank for the financial support of the Ministry of Science and Environmental Protection of Serbia.

References

[BLD01] Bergounioux, M., Long, N.T., Dinh, A.P.N.: Mathematical model for a shock problem involving a linear viscoelastic bar. Nonlinear Analysis, Theory, Methods & Applications, **43**, 547–561 (2001)

[TT06] Takači, Dj., Takači, A.: On the approximate solution of mathematical model of a viscoelastic bar. Nonlinear Analysis, Theory, Methods & Applications (in print).

[Mi83] Mikusiński, J.: Operational Calculus. Vol.I, PWN – Polish Scientific Publishers, Warszawa and Pergamon Press, Oxford (1983)

[MB87] Mikusiński, J., Boehme, T.K.: Operational Calculus. Vol.II, PWN – Polish Scientific Publishers, Warszawa and Pergamon Press, Oxford (1987)

[LUT01] Long, N.T., Ut, L.V., Truc, N.T.T.: On a shock problem involving a linear viscoelastic bar. Nonlinear Analysis, Theory, Methods & Applications, **63**, 198–224 (2005)

[TT04] Takači, Dj., Takači, A.: The approximate solution of nonlinear equations. Publicationes Mathematicae Debrecen, **65(1–2)**, 65–79 (2004)

Dynamics of wall bounded flow

Ozan Tuğluk and Hakan I. Tarman

Department of Engineering Sciences, Middle East Technical University, Ankara, Turkey {tugluk, tarman}@metu.edu.tr

There are various scenarios proposed in literature for transition in plane channel (Poiseuille) flow. In this work, one of these scenarios, namely, streak breakdown, is tested numerically using a Karhunen-Loeve (K-L) based model. The K-L basis was empirically generated earlier using a numerical database representing the flow. This basis is modified in this work to include the mean flow. A K-L basis provides an optimal parametrization of the underlying flow in energy norm. Since it is specific to the flow, each basis element carries an independent characteristic of the flow and has physical interpretation. A system of model amplitude equations is then obtained by Galerkin projection of the governing equations onto the space spanned by the K-L basis. The physical interpretation of the basis elements is used to truncate the resulting system to obtain a low dimensional model.

1 Introduction

Despite much research and progress, there is still a lack of complete picture of transition to turbulence in the wall-bounded flows. This is also true in the case of plane channel (Poiseuille) flow which is a very attractive flow to study the dynamics of transition due to its simple geometry. It started with the conflict between the linear stability analysis leading to famous Orr-Sommerfeld equation predicting stability for Reynolds number $Re < 5572$ [OP83] and experimental observations showing transition to turbulence for Re as low as 1000 [PH69]. This conflict motivated much research on nonlinear analysis and various transition scenarios [RSBH98].

A recent theory on transition through transient growth of disturbances before decaying is supported by experiments and numerical simulations and is to play a fundamental role in the initial stages of transition [Kli92, RH93]. In channel flows, the disturbances that yield the greatest transient growth, called optimals, are independent, or nearly independent, of the streamwise coordinate. Streamwise streaks periodic in the spanwise direction and their

Kenan Taş et al. (eds), Mathematical Methods in Engineering, 243–253.

subsequent breakdown leading to transition have been proposed as a transition scenario and investigated in [RSBH98].

In this preliminary study, the breakdown of streamwise streaks is numerically simulated using a low dimensional model reduced from the governing Navier-Stokes (NS) equations through Karhunen-Loéve (KL) procedure. This procedure is useful in studying complicated physical phenomena by decomposing an experimental or numerical realization of flow into structures (KL modes) each of which carries an independent character of the flow. This may be a static analysis in which the evolution of the flow is studied by projecting an existing database representing a spatial realization of the flow at discrete times onto the subspace spanned by the KL modes (KL subspace), [WHS97] or a dynamic analysis in which the evolution of the flow is studied by numerically integrating a relatively low dimensional model of the governing equations obtained by Galerkin projection onto the KL subspace [SZ92, Tar03]. Contrary to the static analysis being limited to the control parameter (such as Reynolds number) values at which the database is obtained, the dynamic analysis allows a study in a range of the parameter values.

2 Governing equations

In contrast to the boundary layer slowly growing in the streamwise directon over a flat plate, the channel flow between two parallel plates is homogeneous in both streamwise and spanwise directions after an entrance region in which both upper and lower boundary layers are growing. That is why channel flow has a special place in the experimental or numerical study of the transitional mechanisms. Channel flow is driven by a constant pressure gradient k in the x-direction in a geometry, $-\infty < x, z < \infty$; $-h < y < h$. Here, x is the streamwise direction, z the spanwise direction and y the direction between the parallel plates. For velocity and spatial coordinates, the symbols, $\mathbf{u} = (u_1, u_2, u_3) = (u, v, w)$ and $\mathbf{x} = (x_1, x_2, x_3) = (x, y, z)$ will be used interchangibly.

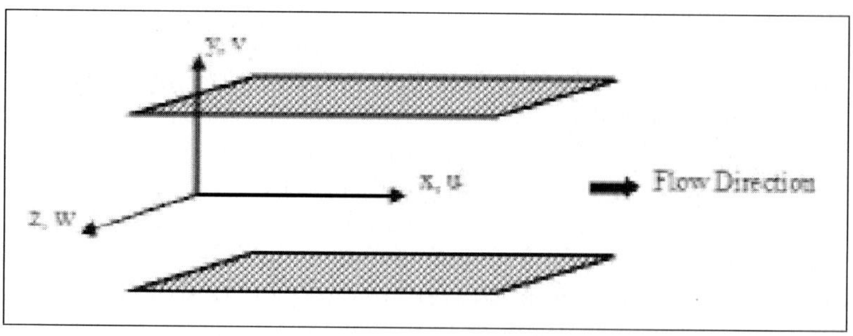

Fig. 1. Flow geometry and the notation

The governing Navier-Stokes (NS) equations and the boundary conditions are:

$$\frac{\partial u_j}{\partial x_j} = 0, \tag{1a}$$

$$\frac{\partial u_i}{\partial t} + \frac{\partial}{\partial x_j} u_i u_j + \frac{\partial p}{\partial x_i} = \delta_{i1} + \frac{1}{R_\tau} \nabla^2 u_i \tag{1b}$$

for $i = 1, 2, 3$, and

$$\mathbf{u}(x, -1, z, t) = \mathbf{u}(x, 1, z, t) = 0 \tag{2}$$

with periodic boundary conditions in the x and y directions. Here, the standard normalization is used, based on the friction velocity $u_\tau = \sqrt{kh/\rho}$ and $R_\tau = \frac{u_\tau h}{\nu}$ is the Reynolds number [TL72].

3 KL procedure

KL modes are computed as eigenfunctions, $\{\Phi^{(n)}\}$, of the integral equation :

$$\int_V K_{ij}(\mathbf{x}, \mathbf{x'})\, U_j^{(n)}(\mathbf{x'})\, d\mathbf{x'} = \lambda_n U_i^{(n)}(\mathbf{x}) \tag{3}$$

whose kernel is the two-point correlation tensor

$$K_{ij}(\mathbf{x}, \mathbf{x'}) = \frac{1}{K} \sum_{k=1}^{K} u_i^{(k)}(\mathbf{x})\, u_j^{(k)}(\mathbf{x'}) \tag{4}$$

obtained using flow database

$$\mathbf{u}_i^{(k)}(x, y, z) = \mathbf{u}_i(x, y, z, k\Delta t) \qquad k = 1, 2 \cdots K, \quad i = 1, 2, 3. \tag{5}$$

Since the flow is taken as periodic in the x and z directions, KL modes take the form

$$\mathbf{U}^{(\mathbf{m})}(\mathbf{x}) = e^{2\pi i \left(\frac{mx}{L_x} + \frac{nz}{L_z}\right)} \mathbf{V}^{(q)}(m, n; y) \tag{6}$$

where $\mathbf{m} = (m, n, q)$ is an index vector. KL modes form an orthogonal basis in an inner product space

$$\left(\mathbf{U}^{(n)}, \mathbf{U}^{(m)}\right)_{\mathbf{x}} \equiv \int_V \sum_{j=1}^{3} U_j^{(\mathbf{n})}(\mathbf{x}) U_j^{(\mathbf{m})}(\mathbf{x}) dV = \delta_{\mathbf{nm}}, \tag{7}$$

and represent the flow in an expansion

$$\mathbf{u}(\mathbf{x}, t) = \sum_{\mathbf{m}} a_{\mathbf{m}}(t)\, \mathbf{U}^{(\mathbf{m})}(\mathbf{x}), \tag{8}$$

where the expansion coefficients are statistically independent in time,

$$(a_{\mathbf{m}}, a_{\mathbf{n}})_t \equiv \frac{1}{K} \sum_{k=1}^{K} a_{\mathbf{m}}(k\Delta t)\, a_{\mathbf{n}}(k\Delta t) = \lambda_{\mathbf{m}} \delta_{\mathbf{nm}}. \tag{9}$$

Due to the symmetries in flow geometry [WHS97], KL modes come out degenerate as such

$$\lambda_{(m,n,q)} = \lambda_{(m,-n,q)} = \lambda_{(-m,n,q)} = \lambda_{(-m,-n,q)}. \tag{10}$$

Consequently, KL modes need to be computed only for the positive index pair (m, n), $m, n > 0$, the other members of the index family $\{m, n, q\}$, where

$$\{m, n, q\} \equiv \{\underbrace{(m, n, q), (-m, -n, q)}_{\text{conjugate pairs}}; \underbrace{(m, -n, q), (-m, n, q)}_{\text{conjugate pairs}}\} \tag{11}$$

are obtained using the symmetry relations. KL modes satisfy all the spatial constraints, such as continuity equation, boundary conditions and form the building blocks of the flow

$$\begin{aligned}\mathbf{v}^{(\mathbf{m})}(\mathbf{x}, t) \equiv {} & \underbrace{a_{(m,n,q)}(t)\mathbf{U}^{(m,n,q)}(\mathbf{x}) + a_{(-m,-n,q)}(t)\mathbf{U}^{(-m,-n,q)}(\mathbf{x})}_{\text{complex conjugate pairs}} + \\ & \underbrace{a_{(-m,n,q)}(t)\mathbf{U}^{(-m,n,q)}(\mathbf{x}) + a_{(m,-n,q)}(t)\mathbf{U}^{(m,-n,q)}(\mathbf{x})}_{\text{complex conjugate pairs}}\end{aligned} \tag{12}$$

for each index family $\{m, n, q\}$.

These computed KL modes, as they satisfy spatial constraints and carry independent features of the flow, form a convenient basis for reducing NS equations to a relatively low dimensional dynamical system via Galerkin projection. Further, their divergence-free nature causes the gradiant term to drop during projection. In order to construct a relatively low dimensional KL based dynamical model of the governing equations, first, an approximation of the flow

$$\mathbf{u} \approx \mathbf{u}_s = \sum_{\mathbf{n} \in S} \mathbf{v}^{(\mathbf{n})}(\mathbf{x}, t) \tag{13}$$

in terms of a set of KL modes selected based on their physical importance is to be obtained. Here, S denotes index set of the selected KL modes. This approximation is, in turn, forced to satisfy the governing NS equations ($\mathrm{NS}(\mathbf{u}) = 0$) using Galerkin projection

$$\left(\mathbf{U}^{(\mathbf{m})}, \mathrm{NS}(\mathbf{u}_s)\right)_{\mathbf{x}} = 0 \tag{14}$$

resulting in a dynamical system

$$\frac{d}{dt}a_{\mathbf{m}} = PROD(\mathbf{m}) + \frac{1}{R_\tau}DISS(\mathbf{m};\mathbf{n})a_{\mathbf{n}} + NLIN(\mathbf{m};\mathbf{n},\mathbf{r})a_{\mathbf{n}}a_{\mathbf{r}}, \tag{15}$$

where, using the index vector representations $\mathbf{m} = (m_1, m_3, q_m)$, $\mathbf{n} = (n_1, n_3, q_n)$ and $\mathbf{r} = (r_1, r_3, q_r)$, the coefficients are

$$PROD(\mathbf{m}) = \left(\mathbf{U}^{(\mathbf{m})}, \delta_{i1}\right)_{\mathbf{x}}, \tag{16}$$

for $m_1 = m_3 = 0$,

$$DISS(\mathbf{m};\mathbf{n}) = \left(\mathbf{U}^{(\mathbf{m})}, \Delta\mathbf{u}_s\right)_{\mathbf{x}}, \tag{17}$$

for $m_1 = n_1$, $m_3 = n_3$, and

$$NLIN(\mathbf{m};\mathbf{n},\mathbf{r}) = \left(\mathbf{U}^{(\mathbf{m})}, \mathbf{u}_s \times (\nabla \times \mathbf{u}_s)\right)_{\mathbf{x}} \tag{18}$$

for $m_1 = n_1 + r_1$, $m_3 = n_3 + r_3$. In the computation of the coefficients, $NLIN(\mathbf{m};\mathbf{n},\mathbf{r})$, the use of the equivalent form

$$\mathbf{u} \cdot \nabla\mathbf{u} = \frac{1}{2}(\mathbf{u} \cdot \mathbf{u}) - \mathbf{u} \times (\nabla \times \mathbf{u}) \tag{19}$$

of the nonlinear term of NS equations is observed to improve the computation as verified by the satisfaction of the triad relation

$$NLIN(\mathbf{m};\mathbf{n},\mathbf{r}) + NLIN(-\mathbf{n};-\mathbf{m},\mathbf{r}) + NLIN(-\mathbf{r};\mathbf{n},-\mathbf{m}) = 0 \tag{20}$$

where $-\mathbf{m} = (-m_1, -m_3, q_m)$.

4 Numerical procedure

NS equations are numerically integrated in [WHS97] by using Fourier-Chebyshev spectral method in which the flow is taken periodic in x and z directions. The flow variables are expanded in the form

$$\mathbf{u}(\mathbf{x}, t) = \sum_{m=-\frac{M}{2}}^{\frac{M}{2}-1} \sum_{n=-\frac{N}{2}}^{\frac{N}{2}-1} \sum_{p=0}^{P} \hat{\mathbf{u}}(m, n, p; t)\, T_p(x_2)e^{ik_1x_1}e^{ik_3x_3} \tag{21}$$

where $k_1 = 2\pi m/L_x$, $k_3 = 2\pi n/L_3$ and $T_p(y)$ are the Chebyshev polynomials. The values $L_x = \pi$, $L_z = 0.3\pi$, $P = 128$, $M = 48$, $N = 24$, $R_\tau = 135.5$ are selected based on the work [JM90] which corresponds to the minimal channel geometry to obtain turbulent flow at both upper and lower boundary layers.

Table 1. KL modes as computed in [WHS97]

KL mode $\mathbf{m} = (m, n, q)$	(0,1,1)	(0,1,2)	(1,1,1)	(1,1,2)	(0,1,3)	(0,1,4)
Energy content % $\lambda_{\mathbf{m}} / \sum_{\mathbf{n}} \lambda_{\mathbf{n}}$	13	10	5	5	3	2

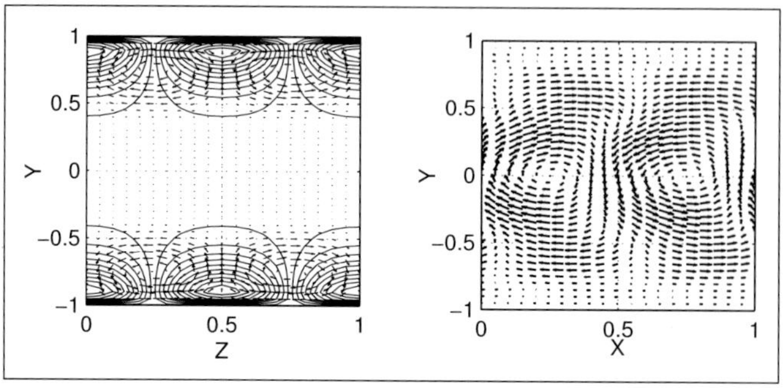

Fig. 2. Flow structure of KL modes (0,1,1) and (1,0,1)

The computed flow field after the removal of the mean flow is used to compute the KL modes as shown in Table 1. Each KL mode carries an independent feature of the flow.
For example, the most energetic KL mode $\mathbf{m} = (0, 1, 1)$, along which carries 13% of the flow energy, physically represents a pair of counter-rotating rolls extending in the streamwise direction in the upper and lower sections of the channel as shown in Figure 2. This rolling motion carries the fluid from the center towards the wall regions. KL modes are classified in [WHS97] as roll modes $(m = 0, n \neq 0, q)$, core or propagating modes $(m \neq 0, n, q)$, and net-flux modes $(m = 0, n = 0, q)$ depending on their physical roles. They vary with respect to the quantum number q in a way of increasing complexity in y-direction as shown in Figure 3. As q increases KL modes exhibit smaller scale structures and form symmetric and anti-symmetric pairs in the plane $y = 0$. This form necessitates the inclusion of KL modes in pairs in the set S.

In this work, the KL modes are re-computed in a way to include the mean flow which was removed in [WHS97]. This eliminates the need to model the mean flow, which would result in a cubic nonlinear term in the dynamical system. The inclusion of the mean flow in the extraction process of the KL modes effects only the net-flux modes $\mathbf{m} = (0, 0, q)$, especially the most energetic mode $\mathbf{m} = (0, 0, 1)$. As shown in Figure 3, $\mathbf{m} = (0, 0, 1)$ mode is in a form very similar to the mean profile.

The resulting dynamical system is numerically integrated in time for various forcing parameter R_τ values and for a selected set S of KL modes. Numerical solution is in turn used for a nonlinear stability analysis in two stages. In the first stage, the stability of the laminar flow, $U(y) = \frac{1}{2}R_\tau(1-y^2)$, is numerically tested againts disturbances having no streamwise variation $(m = 0)$ at

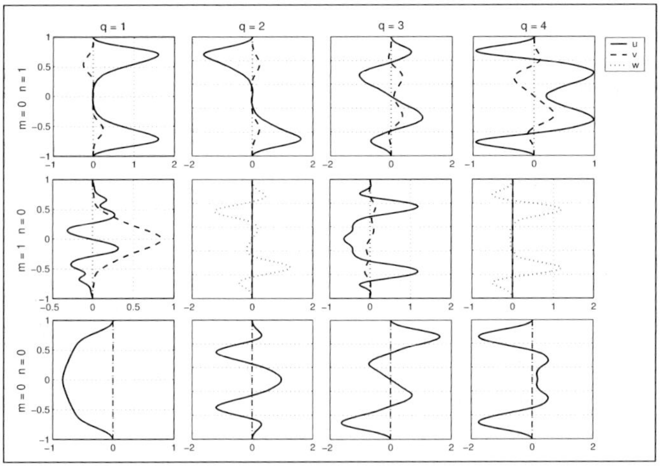

Fig. 3. Vertical profiles of KL modes $\mathbf{V}^q(m,n;y)$ (real part)

various R_τ values and the secondary flow , to which the laminar flow looses its stability, is obtained. In the second stage, the stability of the secondary flow is tested against disturbances that have streamwise variation ($m \neq 0$) through a numerical perturbation procedure. In this study, the (kinetic) energy of the flow

$$E(\mathbf{u}) = \frac{1}{2}\int_0^{L_z} dz \int_0^{L_x} dx \int_{-1}^{1} dy\, \mathbf{u}\cdot\mathbf{u} \tag{22}$$

relative to that of laminar flow ($E(U(y)\delta_{i1}) = 2L_xL_zR_\tau/15$), relative energy,

$$\varepsilon(\mathbf{u}) = \frac{E(\mathbf{u})}{2L_xL_zR_\tau/15} \tag{23}$$

is used as a measure.

In this work, KL modes that are thought to play the most important role in the transition process, are selected as:

$$S = \{(m,n,q) \mid (0,0,1\cdots 25); (0,1,1\cdots 4); (1,0,1\cdots 4); (1,1,1\cdots 4)\}.$$

Here, the main idea in the the inclusion of the $(0,0,q)$ modes with $1 \leq q \leq 25$ is to have as complete representation of the modes carrying the flow energy as possible in the dynamical model system. Moreover, the parabolic laminar flow profile $U(y)$ is included in the system through the projection onto these modes

$$a_{\mathbf{m}}(t) = (\mathbf{U}^{\mathbf{m}}, U(y)\delta_{i1})_{\mathbf{x}} \quad \text{for} \quad \mathbf{m} = (0,0,q) \in S. \tag{24}$$

Other modes are included in pairs $(q = 1; 2), (q = 3; 4)$ due to the appearance of the modes in pairwise symmetric and antisymmetric form around $y = 0$ plane (see Figure 3). Furthermore, the number of modes, $1 \leq q \leq 4$, is included to provide the necessary degrees of freedom for the highly resolved KL modes to adjust to the low resolution requirements at low R_τ values. Recall that KL modes carry high vertical resolution due to their construction from data at high R_τ values ($R_\tau = 135.5$ in this case).

The dynamical model system is also an initial value problem. In the first stage of the two stage nonlinear stability analysis, an initial value of

$$a_{(m,n,q)}(t=0) = \begin{cases} A_0 & \text{for} \quad (m=0, n \neq 0, q) \in S \\ 0 & \quad\quad \text{otherwise} \end{cases} \tag{25}$$

is taken for the streamwise independent components $(m = 0, n \neq 0, q) \in S$ of the flow which is superimposed onto the laminar flow $(m = 0, n = 0, q) \in S$ of Eq. 24. Here, A_0 is selected for the relative energy $\varepsilon\left(\mathbf{u}(t=0)\right) = \varepsilon_0^0$ to take a particular value ε_0^0 . In the second stage, the initial value of

$$a_{(m,n,q)}(t=0) = \begin{cases} A_0 & \text{for} \quad (m=0, n \neq 0, q) \in S \\ A_1 & \text{for} \quad (m \neq 0, n, q) \in S \\ 0 & \quad\quad \text{otherwise} \end{cases} \tag{26}$$

is taken for the secondary flow arising from the first stage, $(m = 0, n, q) \in S$, and for the superimposed streamwise dependent component of the flow, $(m \neq 0, n, q) \in S$. Here, A_0 and A_1 are selected for the relative energy, $\varepsilon\left(\{\mathbf{u}(t=0) \mid A_1 = 0\}\right) = \varepsilon_0^0$ and $\varepsilon(\mathbf{u}(t = 0) \mid A_0 = 0) = \varepsilon_1^0$ take separate particular values, ε_0^0 and ε_0^1 .

5 Results

In Figure 4, a typical solution obtained at $R_\tau = 20$ is shown. This solution, obtained at the particular initial relative energy value of $\varepsilon_0^0 = 10^{-2}$, shows that the laminar flow modeled by the system is stable at $R_\tau = 20$. The solution obtained at various other particular values of ε_0^0 supports the same conclusion. In another numerical experiment (Figure 5) at $R_\tau = 40$ and $\varepsilon_0^0 = 10^{-2}$, the laminar flow loses its stability to give rise to a secondary flow. The response of the system to the perturbation is shown as a burst in energy initially. Later, the flow reaches an equilibrium as a result of the dissipative effects of the nonlinear terms activated by the initial burst.

The secondary flow is characterized by the transfer of high momentum at the centre towards the wall layers. This can be seen in Figure 6(a) in the relative flattness of the mean secondary flow profile due to this momentum transfer in comparison to the laminar flow profile. A detailed look at the secondary flow profile in Figure 6(b) shows the streamwise high and low velocity streaks extending in the streamwise direction in the wall region. This is in accordance with the streamwise streak formation in literature.

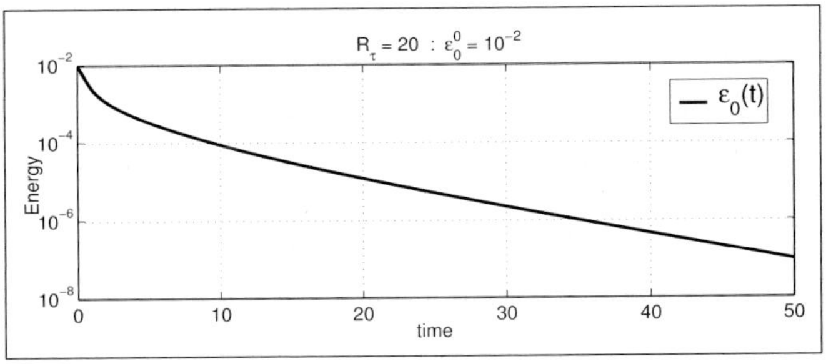

Fig. 4. The solution obtained at the particular R_τ and ε_0^0 values indicated.

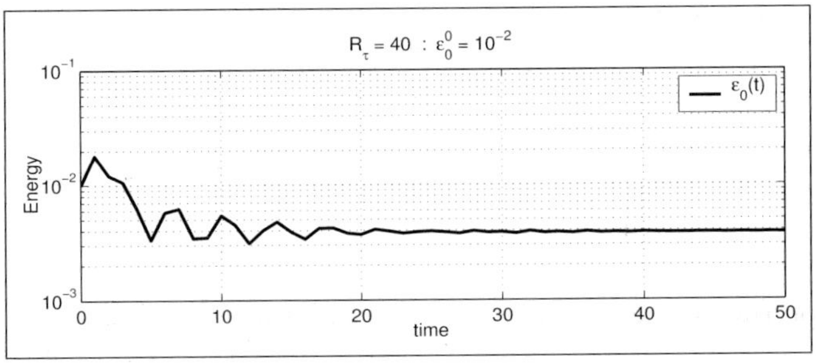

Fig. 5. The solution obtained at the particular R_τ and ε_0^0 values indicated.

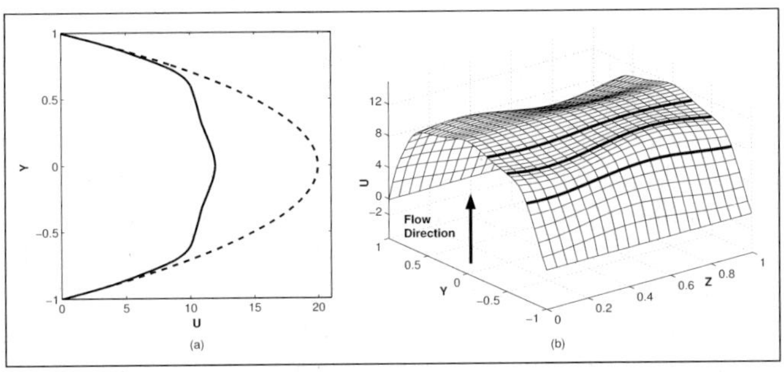

Fig. 6. (a) The comparison between the laminar flow and the mean secondary flow profiles obtained by the perturbation of the laminar flow at $R_\tau = 40$. (b) The highlighted curves indicate the spanwise variation in the streamwise velocity component of the flow. At each point on these curves, the value of the streamwise velocity component extend in the streamwise direction wihout varying, thus forming the streamwise high and low velocity streaks.

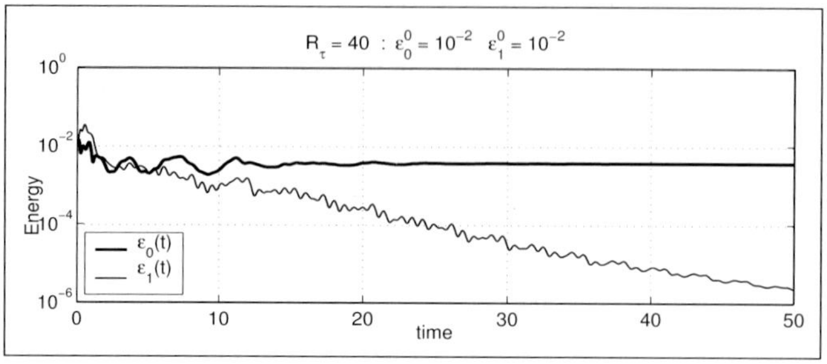

Fig. 7. The solution obtained at the particular R_τ and ε_0^0, ε_1^0 values indicated.

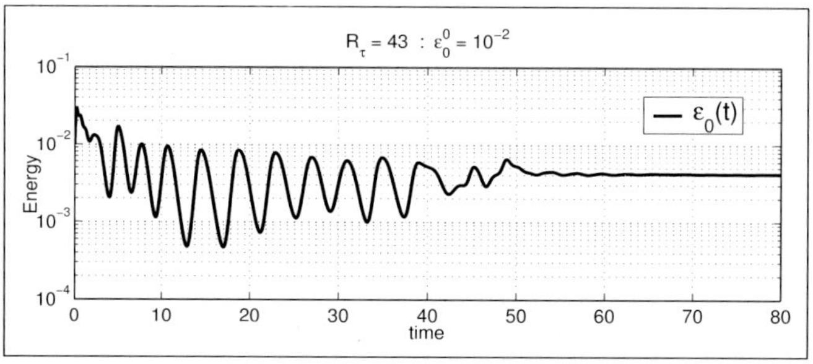

Fig. 8. The solution obtained at the particular R_τ and ε_0^0 values indicated.

The perturbation of the secondary flow obtained at $R_\tau = 40$ by three dimensional components of the flow shows the stability of the secondary flow (Figure 7). The numerical experiments at various other ε_1^0 values supports the same conclusion. Laminar flow, when perturbed at $R_\tau = 43$ and at various relative energy values ε_0^0 similarly loses its stability and gives rise to secondary flow. A typical solution for $\varepsilon_0^0 = 10^{-2}$ is shown in Figure 8. The secondary flow at $R_\tau = 43$ loses its stability to perturbations by three dimensional components of the flow (Figure 9). The numerical experiments at various other ε_1^0 values supports the same conclusion. This result is in agreement with the scenario of transition in literature through the breakdown of high and low velocity streamwise streaks.

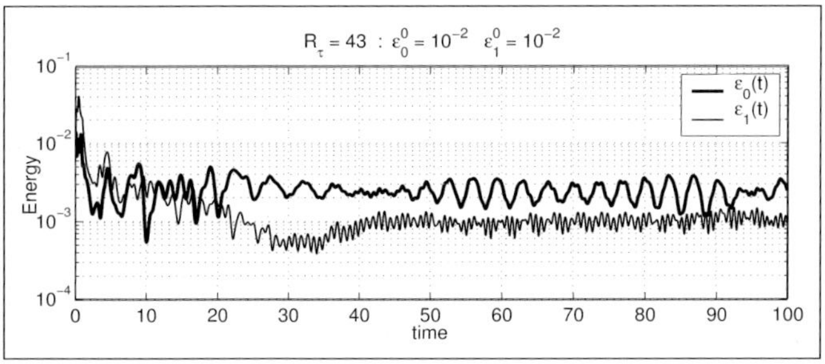

Fig. 9. The solution obtained at the particular R_τ and ε_0^0, ε_1^0 values indicated.

References

[OP83] Orszag, S.A., Patera, A.T.: Secondary instability of wall-bounded shear flows. Journal of Fluid Mechanics, **128**, 347-360 (1983)

[PH69] Patel, V.C., Head, R.: Some observations on skin friction and velocity profiles in fully developed pipe and channel flows. Journal of Fluid Mechanics, **38**, 181-201 (1969)

[RSBH98] Reddy, S.C., Schmid, P.J., Bagget, J.S., Henningson, D.S.: On stability of streamwise streaks and transition tresholds in plane channel flows, Journal of Fluid Mechanics, **365**, 269-303 (1998)

[Kli92] Klingmann, B.G.B.: On transition due to three-dimensional disturbances in plane Poiseuille flow. Journal of Fluid Mechanics, **240**, 167-195 (1992)

[RH93] Reddy, S.C., Henningson, D.S.: Energy growth in viscous channel flows. Journal of Fluid Mechanics, **252**, 209-238 (1993)

[WHS97] Webber, G.A., Handler, R.A., Sirovich, L.: Karhunen Loéve decomposition of minimal channel flow. Physics of Fluids, **9**, 1054-1066 (1997)

[SZ92] Sirovich, L., Zhou, X.: Coherence and chaos in a model of turbulent boundary layer. Physics of Fluids, **4**, 2855-2870 (1992)

[Tar03] Tarman, I.H.: A Karhunen-Loéve-based approach to numerical simulation of transition in Rayleigh-Benard convection. Num. Heat Transfer B., **43**, 567-586 (2003)

[TL72] Tennekes, H., Lumley, J.L.: A First Course in Turbulence. The MIT Press, Massachusetts (1972)

[JM90] Jiménez, J., Moin, P.: The minimal flow unit in near-wall turbulence. Journal of Fluid Mechanics, **225**, 213-240 (1990)

Applications of Wavelets

Wavelet transform for the simultaneous prediction of the colorants in food product

Hakan A. Aktaş[1], Erdal Dinç[2], Güzide Pekcan[1], Özgür Üstündag[2] and Aysegül Taş[3]

[1] Department of Chemistry, Faculty of Science and Art, Suleyman Demirel University, 32260 Isparta, Turkey ahakan@fef.sdu.edu.tr
[2] Department of Analytical Chemistry, Faculty of Pharmacy, Ankara University,06100, Tandogan, Ankara, Turkey dinc@pharmacy.ankara.edu.tr
[3] Department of Computer Engineering, Faculty of Engineering and Architecture Çankaya University, 06530 Balgat, Ankara, Turkey aysegul@cankaya.edu.tr

Wavelet transform method was successfully applied to the multicomponent analysis of the binary mixtures containing different colorants in commercial food product. In this application, wavelet transform method is suitable for the quantitative resolution of the mixtures of these colorants and this hybrid approach doesn't require any separation and extraction steps. The method was tested by using various synthetic ternary mixtures and applied to the sample and successfully results were obtained.

1 Introduction

As it is known, many foods and drinks contain the synthetic and natural colorants. In some cases, the synthetic colorants can make the pathogenic effect for the human health, in particular if they are excessively consumed. For this reason the use of the additive synthetic colorants in foods and drinks has been controlled according to the laws and regulations. As a result, the quantitative analysis of the artificial colorants in foods and drinks are very important for human health. Quantitative determination of these colorants in foods and drinks is very important task for the analytical chemistry. The quantitative determination of some colorants in samples has been performed by various analytical methods such as spectrophotometric methods [NFLF99]-[OAK02], HPLC [LL99], capillary zone electrophoresis [Gera99], polarography and chemometric calibration [LLD02]. In some cases, these analytical methods may not give better results in analytical applications. In the other hand, chromatography and capillary electrophoresis, which have complex components, require a chemical separation procedure. In this context the wavelet technique

Kenan Taş et al. (eds), Mathematical Methods in Engineering, 257–264.

[Daud92]-[DB4] is a promising tool for solving the above drawbacks. In this study, the continuous wavelet transform (CWT) was applied to the multicomponent analysis of soft drink powders (food product) containing tartrazine (T) and allura red (A) in the presence of the overlapping absorption spectra. In this application, Haar CWT approach in combination with zero crossing technique was found to be suitable for the quantitative evaluation of two colorants in samples. This combined method was successfully applied to the synthetic mixtures and real samples.

2 Experimental

2.1 Apparatus

Shimadzu UV-1601 double beam UV-Visible spectrophotometer with a fixed slit width (2 nm) was connected to a computer loaded with Shimadzu UVPC soft-ware, equipped with a Lexmark 320 printer. This instrument was used for the registration of all the absorption spectra. All data treatments in the application of the method were carried out by using Matlab 7.0 and EXCEL softwares.

2.2 Reagents and standard solutions

All chemicals were of analytical reagent grade. The A and T colorants were used from Aldrich products. Stock solutions of T and A (50 mg/50 mL) were prepared in 0.1 M HCl for the following procedure. A series of standard solutions for each colorant in the concentration range of 5-30 g/ml for T and 4-24 g/ml for A was prepared by using the stock solutions. A validation set of 12 synthetic mixture solutions containing T and A colorants in the working range was prepared from stock solutions.

3 Results and discussion

3.1 Continuous wavelet transform

Wavelet is expressed as a series of functions $\Psi_{a,b}(\lambda)$ having the following forms

$$\Psi_{a,b}(\lambda) = \frac{1}{\sqrt{\mid a \mid}}\Psi(\frac{\lambda - b}{a}), a \neq 0, a, b \in R \tag{1}$$

Here a represents the scale parameter controlling the dilatation and b is the translation parameter controlling the shift and R is the domain of real numbers In this study, various wavelet families was tested for the resolution of the over-lapping absorption signals and Haar CWT approach was found to be appropriate family for the determination of T and A colorants in food product.

The application of the CWT method in combination with zero crossing will be explained below. The absorption spectra of T and A colorants in concentration range of 5-30 $\mu g/ml$ for T and 4-24 $\mu g/ml$ for A were recorded in the spectral range of 200.0 -609.5 nm as shown in Figure 1A. The absorption spectra of two colorants and their samples were transferred as vectors into wavelet domain. The absorption data vec-tors consisting of 2046 data points were transformed by using Haar CWT approach with a=525 (scale parameter). After that, the CWT spectra were obtained by plot-ting CWT-coefficients against wavelengths. The CWT spectra were presented in Figure 1a for the standard series for both colorants.

3.2 CWT application to quantitative evaluation

For the quantitative evaluation of T and A colorants in samples, the linear regression functions were obtained by measuring the transformed amplitude at 294.7 nm for T corresponding to the zero crossing point for A and at 355 nm for A corresponding to the zero crossing points for T in the CWT spectra (see Fig. 1b). The calibration functions and their statistical results were presented in Table 1.

Table 1. Linear regression analysis and its statistical results

Parameter	T	A
λ (nm)	294.7	555.0
Concentration range (μg/mL)	5 - 30	4 -24
Slope (m)	0.2062	0.1740
Intercept (n)	-0.1335	0.3724
r	0.9991	0.9996
SE(m)	0.0044	0.0026
SE(n)	0.0855	0.0404
SE(r)	0.0918	0.0434
LOD (μg/mL)	0.62	0.29
LOQ (μg/mL)	2.08	0.96

r= correlation coefficient of the calibration function,
SE(m)= Standard error of slope, SE(n)= Standard error of intercept,
SE(r)= Standard error of correlation coefficient

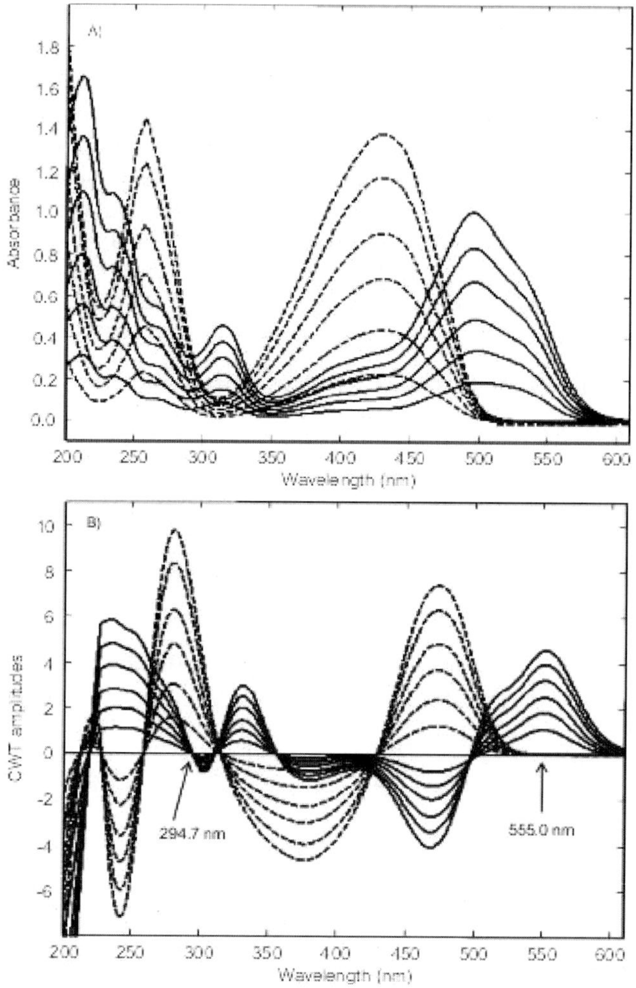

Fig. 1. Absorption spectra (a) and their CWT spectra of 4-24 $\mu g/mL$ for A colorant (b) and 5-30 $\mu g/mL$ for T colorant (—-).

The calibration range, according to each compound concentration existing in the soft drinking powder product, was designated in the practical range, to give accurate, precise and linear response. The correlation coefficients of calibration curves were found to be higher than 0.9990 for all selected zero crossing points and proposed wavelet family.

4 Validation of the method

Haar CWT method was tested by analyzing different mixtures of the studies compounds at various concentrations within the linearity range. Good accuracy and precision were reported for the proposed CWT method. The recoveries, their means and their relative standard deviations were presented in Table 2. The obtained results gave good accuracy for the investigated CWT method. It was observed that the mixture results indicated good agreement with each other.

Table 2. Recovery data obtained by application of Haar CWT method to the synthetic binary mixtures.

Actual Concentration (µg/mL)		Predicted Concentration (µg/mL)		Recovery (%)	
T	A	T	A	T	A
10	4	9.77	3.97	97.7	99.2
10	8	9.60	7.99	96.0	99.9
10	12	9.50	11.70	95.0	97.5
10	16	9.92	15.85	99.2	99.1
10	20	9.58	20.40	95.8	102.0
10	24	9.46	24.02	94.6	100.1
5	12	4.89	12.09	97.9	100.8
10	12	9.84	12.62	98.4	105.2
15	12	14.41	12.09	96.1	100.8
20	12	19.69	12.35	98.5	102.9
25	12	24.39	12.32	97.6	102.7
30	12	28.93	12.55	96.4	104.6
			Mean	96.9	101.2
			SD	1.47	2.31
			RSD	1.52	2.28

RSD = Relative standard deviation

5 Food product analysis

The proposed CWT family was applied to the simultaneous prediction of T and A colorants in a soft drinking powder product. Eight replicates determinations for this commercial soft drinking powder product were made. Experimental results were presented in Table 3. The applied wavelet family gave a good coincidence for the experimental results of the soft drinking powder product.

Table 3. Results of the quantitative prediction of T and A colorants in commercial by Haar CWT method

	g/g (%)	
No.	T	A
1	8.85×10^{-2}	6.26×10^{-2}
2	8.83×10^{-2}	6.39×10^{-2}
3	8.86×10^{-2}	6.49×10^{-2}
4	8.86×10^{-2}	6.46×10^{-2}
5	8.99×10^{-2}	6.54×10^{-2}
6	8.80×10^{-2}	6.25×10^{-2}
7	8.34×10^{-2}	6.15×10^{-2}
8	8.47×10^{-2}	6.18×10^{-2}
Mean	8.75×10^{-2}	6.34×10^{-2}
SD	2.22×10^{-3}	1.50×10^{-3}
RSD	2.54	2.37
SE	7.84×10^{-4}	5.31×10^{-4}
CL (p=0.05)	1.54×10^{-3}	1.04×10^{-3}

CL = confidence limit, SE = Standard error

6 Conclusions

The main aim of our study was to investigate the application of CWT to the binary mixture analysis. We combined, in an optimal way, Haar method and zero-crossing technique to obtain a high resolution of a binary mixture containing T and A colorants. We believe that the application of the wavelet family represents an alternative way and offers new possibilities for the mixture analysis, quality control and routine analysis of T and A colorants.

References

[NFLF99] Nevado, J.J.B., Flores , J.R., Llerena, M.J.V., Farinas, N.R.: Simultaneous spectrophotometric determination of tartrazine, patent blue V, and indigo carmine in commercial products by partial least squares and principal component regression methods. Talanta, **48(4)**, 895–903 (1999)

[LLGE99] Lopez-de-Allba, P.L., Lopez-Martinez, L., Gonzalez-Leal, M., Estrada-Hernandez, Y.: Extraction of sunset yellow and allura red by ion-pair formation with aliquat-336 in IBMK. Simultaneous spectrophotometric determination by bivariate calibration and derivative spectrophotometry. Quim. Anal., **18(4)**, 291–297 (1999)

[JLMBD00] Joseph-Charles J., Langlois M.H., Montagut M., Boyer C., Dubost, J.P.: Simultaneous determination of two synthetic dyes erytrosine and sunset yellow in a pharmaceutical syrup by first derivative visible spectrophotometry. Anal. Lett., **33(8)**, 1567-1575 (2000)

[NQK01] Ni, Y.N., Qi, M., Kokot, K.S.: Simultaneous spectrophotometric determination of ternary mixtures of Tartrazine, Sunset Yellow, and Ponceau 4R by H-point standard addition method. Anal. Lett., **34(14)**, 2585-2596 (2001)

[OK03] Ozgur, M.U., Koyuncu, I.: The Simultaneous Determination of Quinoline Yellow (E-104) and Sunset Yellow (E-110) in Syrups and Tablets by Second Derivative Spectrophotometry. Turk. J. Chem. **26(4)** 501-508 (2003)

[LLCD01] Lopez-de-Alba, P.L. , Lopez-Martinez, L., Cerda, V., De-Leon- Rodriguez, L.M.: Simultaneous determination of tartrazine, sunset yellow and allura red in commercial soft drinks by multivariate spectral analysis. Quim. Anal., **20(2)**, 63–72 (2001)

[BOK00] Bozdogan, A., Ozgur, M.U., Koyuncu, I.: Simultaneous determination of sunset yellow and Ponceau 4R in gelatin powder by derivative spectrophotometry and partial least-squares multivariate spectrophotometric calibration. Anal. Lett., **33(14)**, 2975-2982 (2000)

[OAK02] Ozgur, M.U., Alpdogan, G., Koyuncu I.: Zero-crossing derivative spectrophotometric determination of mixtures of sunset yellow and erythrosine in pure form and in tablets. Anal. Lett., **35(4)**, 721–732 (2002)

[LL99] Lancaster, F.E., Lawrence, J.F.: Determination of benzidine in the food colours tartrazine and sunset yellow FCF, by reduction and derivatization followed by high-performance liquid chromatography, Food Addit. Contam. **16(9)**, 381–390 (1999)

[Gera99] Gerasimov, A.V.: Qualitative and quantitative treatment of thin-layer chromatograms of incompletely separated synthetic food colors. J. Anal. Chem., **55(12)**, 1161–1165 (2000)

[LLD02] Lopez-de-Alba, P.L., Lopez-Martinez, L., De-Leon-Rodriquez, L.M.: Simultaneous determination of synthetic dyes tartrazine, allura red and sunset yellow by differential pulse polarography and partial least squares. A multivariate calibration method. Electroanal., **14(3)**, 197–205 (2002)

[Daud92] Daubechies, I.: Ten Lectures on Wavelets. Society for Industrial and Applied Mathematics, Philadelphia (1992)

[Walc92] Walczak, B.: Wavelets in Chemistry. Elsevier Press, Amsterdam, The Netherland (2000)

[MS04] Ma, C.X., Shao, X.G.: Continuous wavelet transform applied to removing the fluctuating background in near-infrared spectra. J. Chem. Inf. Comp. Sci., **44(3)**, 907-911 (2004).

[DB04a] Dinc, E., Baleanu, D: Multicomponent quantitative resolution of binary mixtures by using continuous wavelet transform. J. AOAC Int., **87(2)**, 360-365 (2004)

[DB04b] Dinc, E., Baleanu, D: Application of the wavelet method for the simultaneous quantitative determination of benazepril and hydrochlorothiazide in their mixtures. J.AOAC Int., **87(4)**, 834–841 (2004)

[DBU03] Dinc, E., Baleanu, D., Ustundag, O.: An approach to quantitative two-component analysis of a mixture containing hydrochlorothiazide and spironolactone in tablets by one-dimensional continuous Daubechies and biorthogonal wavelet analysis of UV-spectra. Spectr. Lett., **36**, 341–355 (2003)

[DB03a] Dinc, E., Baleanu, D.: A zero-crossing technique for the multidetermination of thiamine HCl and pyridoxine HCl in their mixture by using one-dimensional wavelet transform. J. Pharm. and Biomed. Anal., **31**, 969–978 (2003)

[DB4] Dinc, E., Baleanu, D.: One-dimension continuous wavelet resolution for the simultaneous analysis of binary mixture of benazepril and hydrochlorothiazide in tablets using spectrophotometric absorbance data. Rom. Journ. Chem., **49(11)**, 917–925 (2004)

A review on the wavelet transform applications in analytical chemistry

Erdal Dinç[1] and Dumitru Baleanu[2,3]

[1] Department of Analytical Chemistry, Faculty of Pharmacy, Ankara University, 06100, Tandogan, Ankara, Turkey dinc@pharmacy.ankara.edu.tr

[2] Çankaya University, Faculty of Arts and Science, Department of Mathematics and Computer Sciences, Ankara, Turkey dumitru@cankaya.edu.tr

[3] Institute of Space Sciences, P.O. BOX, MG-23,R 76900 Magurele-Bucharest, Romania baleanu@venus.nipne.ro

In the spectral analysis, the continuous wavelet transform or very recently developed fractional wavelet transform are powerful tools for the data reduction, de-noising, compressing and baseline correction of the analytical signals and resolution of multicomponent overlapping signals. Recently, continuous wavelet transform in combination of zero-crossing approach and spectral ratio treatment has been used for the quantitative resolution and the prediction of multi-mixtures in the presence of the original overlapping signals. This combined approach provides a short time analysis, accurate, precision, rapid and low cost for the quality control and routine analysis of the commercial products containing active compounds. This hybrid approach indicates that this technique is perfectly suitable for the multicomponent analysis of the overlapping analytical signals in the various fields of the analytical chemistry. In addition, the wavelet transform method are an alternative and promising signal analysis approach for the elimination or reduction of the disadvantageous of the classical spectral derivative methods for the analytical purposes. This review presents briefly the theoretical basis of the applications of continuous wavelet transform and fractional wavelet transform with the classical analytical approaches and reports some of their analytical applications.

1 Introduction

The new advancements in computer and information science, statistics and applied mathematics in recent years have caused the major changes in the content of the analytical chemistry. All these developments together with new combined analytical instrumentation devices offer a new possibility to chemists and pharmacists for their researches and analytical applications. As combined analytical instrumentations, a new sophisticated technology called

Kenan Taş et al. (eds), Mathematical Methods in Engineering, 265–284.

hyphenated instrumentation such as the liquid chromatography-diode array detector system (LC-DAD), gas chromatography-mass spectrometry (GC-MS) and liquid chromatography-mass spectrometry (LC-MS), etc., has been used for the multicomponent determination. The above sophisticated analytical methods require chemical separation step such as derivation, extraction and other chemical processes during analysis. In some cases, these analysis methods may not give always better results for some of the multicomponent determination of active compounds in samples. At the same time, the related techniques having complex components bring high cost and time consumption. We can clearly say that there is a need of new analytical approaches, techniques and methods to solve the above drawbacks and to provide alternative resolutions for the complex problems of analytical chemistry.

In recent years, the developments of wavelet transform (WT) method and its applications in the analytical chemistry have significantly amplified the potential power of various spectral techniques [Da92, Ma98, Le98, Wa00]. In the spectral analytical studies, continuous wavelet transform (CWT) is great useful for the simultaneous determination of active compounds in samples. In this context, CWT approach is a powerful signal processing tool for data reduction, de-noising, baseline correction and resolution of multi-component overlapping spectra.

In our previous studies, CWT method with zero-crossing technique and ratio spectra procedure were directly applied to the multicomponent determination of compounds in samples in presence of the strongly overlapping absorption signals, without using any priory chemical treatment such as derivation and extraction,(see Refs. [Wa00]-[AN05] for more details) without using any chemical separation procedure, and successful results were obtained.

It was observed that CWT method in the combination with zero crossing technique and spectral ratio procedure is a new hybrid analytical approach with very simple application for the higher resolution of the overlapping signals and ratio signals, despite the difficult mathematical theory of these wavelet families.

Very recently, fractional wavelet transform (FWT) method, which is a new wavelet transform based on the fractional B-splines [UB99]-[BU02] was successfully applied to image and signal analysis [UB00]. The presence of a continuous order parameter makes FWT more powerful than other methods, especially when fractal signals or fractional Brownian motions are present.

In the spectral analysis studies, the usage of the usual derivative spectrophotometry to the original absorption spectra give us several disadvantages such as: peak intensity diminished with higher order derivation, it required the additional smooth function mode as well as the additional scaling factor processes. In the spectral derivation, when the above mentioned parameters were used for obtaining the derivative spectrum, the form of obtained derivative spectrum may suffer some deformations from the original one. As a result the usual derivative method produces several errors in the quantitative analysis. Therefore, the drawbacks can be eliminated by applying FWT approach to the original absorption spectra. In the cases of low concentration and corresponding undetectable signals in the analytical chemistry, the combined use of FWT and CWT approaches give the better results than of the classical analytical approaches.

The aim of this paper is to demonstrate the analytical applications of the CWT and FWT to the multicomponent determination in the analytical chemistry.

The plan of this study is as follows:
In section two we briefly present some basic definitions of CWT. Section three illustrates the basic definitions of FWT. Section four contains briefly the theoretical information of the chemical applications of CWT and FWT in analytical chemistry. Finally, section five is dedicated to our conclusions.

2 Continuous wavelet transform

Nowadays wavelets are popular as signal and image processing methods for various fields of science and engineering. Wavelets are mathematical functions generated from one basic function by the dilatation (scale parameter) $(W(x) \longrightarrow W(2x))$ and a translation (shift parameter) $(W(x) \longrightarrow W(x+1))$. Projection of a signal onto wavelet basic functions is called the wavelet transform (WT). Given a mother wavelet [Da92, Wa00] $\Psi(\lambda)$ by scaling and shifting of $\Psi(\lambda)$ a set of functions denoted by $\Psi_{a,b}$ is obtained as indicated below

$$\Psi_{a,b}(\lambda) = \frac{1}{\sqrt{(|a|)}}\Psi(\frac{\lambda - b}{a}), a \neq 0, a, b \in R. \tag{1}$$

Here a represents the scale parameter, which is a variable, used to control the scaling, b represents the translation parameter controlling the translation and R is the domain of real numbers.

The action of a given CWT on a function $f(\lambda)$ is given below

$$CWT\{f(\lambda); a, b\} = \int_{-\infty}^{\infty} f(\lambda)\psi^{*}_{a,b}(\lambda)d\lambda = \langle f(\lambda), \psi_{a,b}\rangle, \tag{2}$$

where the superscript $*$ denotes the complex conjugate and $\langle f(\lambda), \psi_{a,b}\rangle$ denotes the inner product of function $f(\lambda)$ onto the wavelet function $\Psi_{a,b}(\lambda)$.

3 Fractional wavelet transform

Recently, a new wavelet transform based on the fractional B-splines was initiated [UB99],[UB00],[BU02]. The mathematical idea of fractional derivatives has represented the subject of interest for various branches of science [Po99]. As it is already known the splines play a significant role on the early development of the theory of the wavelet transform. The generalization of the spline constructions was proposed in [UB99],[UB00],[Po99], namely new wavelet bases with a continuous order parameter was obtained. The new fractional splines have all properties of the polynomial splines with the exception of compact support when the order is non-integer. The main advantage of this construction is that we can build the wavelet bases parameterized by the continuously-varying regularity parameter α.

In the following we define the notion of B-spline. As it is already known a B-spline is a generalization of the Bezier curve. Let a vector known as the knot is defined $T = \{t_0, t_1, \cdots t_m\}$ where T is a non-decreasing sequence with $t_i \in [0,1]$, and define control points P_0, P_n. Let us define degree as $p = m - n - 1$. The knots $t_{p+1}, \cdots, t_{m-p+1}$ are called internal knots. If we define the basis functional as

$$N_{i,0}(t) = 1 \ , if t_i \leq t \leq t_{i+1}, \tag{3}$$

$$0 \ , otherwise \tag{4}$$

and $N_{i,p}(t) = \frac{t-t_i}{t_{i+p}-t_i} N_{i,p-1}(t) + \frac{i+p+l-t}{t_{i+p+l}-t_{i+l}} N_{i+l,p-l}(t)$, then the curve defined by

$$C(t) = \sum_{i=0}^{n} P_i N_{i,p}(t) \tag{5}$$

is a B-spline.

Fractional B-spline

As it was proved in the literature this family interpolates between the integer degrees of polynomial B-splines and that they allow a fractional order of approximation [UB99],[UB00],[BU02]. The fractional B-spline is defined as

$$\beta_+^\alpha = \frac{\sum_{k=0}^{\infty} (-1)^k \binom{\alpha+1}{k} (x-k)_+^\alpha}{\Gamma(\alpha+1)}, \tag{6}$$

where

$$\Gamma(\alpha+1) = \int_0^{+\infty} x^\alpha e^{-x} dx \tag{7}$$

and

$$(x-k)_+^\alpha = max(x-k, 0)^\alpha. \tag{8}$$

The forward fractional finite difference operator of order α is defined as

$$\Delta_+^\alpha f(x) = \sum_{k=0}^{\infty} (-1)^k \binom{\alpha}{k} f(x-k), | \begin{matrix} \alpha+1 \\ k \end{matrix} || \ x-k \ |_*^\alpha, \tag{9}$$

where

$$\binom{\alpha}{k} = -\frac{\Gamma(\alpha+1)}{\Gamma(k+1)\Gamma(1-k)}. \tag{10}$$

The above defined B-splines fulfill the convolution property

$$\beta_+^{\alpha_1} * \beta_+^{\alpha_2} = \beta_+^{\alpha_1+\alpha_2}. \tag{11}$$

The centered fractional B-splines of degree α are given by

$$\beta_*^{\alpha}(x) = \frac{1}{\Gamma(\alpha+1)} \sum_{k \in Z} (-1)^k | \begin{matrix} \alpha+1 \\ k \end{matrix} | x-k|_\star^{\alpha}, \tag{12}$$

where $| x |_*^{\alpha}$ has the following form

$$| x |_*^{\alpha} = \begin{matrix} \frac{|x|^{\alpha}}{-2\sin(\frac{\pi\alpha}{2})}, & \alpha \ not \ even \\ \frac{x^{2n} logx}{(-1)^{1+n}}, & \alpha \ even \end{matrix} . \tag{13}$$

Fractional B-spline wavelets

The definition of the fractional B-spline wavelets is given as follows

$$\psi_+^{\alpha}(\frac{x}{2}) = \sum_{k \in Z} \frac{(-1)^k}{2^{\alpha}} \sum_{l \in Z} (\begin{matrix} \alpha+1 \\ l \end{matrix}) \beta_\star^{2*\alpha+1}(l+k-1)\beta_+^{\alpha}(x-k). \tag{14}$$

The fractional splines wavelets obey

$$\int_{-\infty}^{+\infty} x^n \Psi_+^{\alpha}(x) dx = 0 \tag{15}$$

and the Fourier transform fulfills the following relations

$$\hat{\psi}_+^{\alpha}(\omega) = C(j\omega)^{\alpha+1}, \ as \ \omega \longrightarrow 0 \tag{16}$$

and

$$\hat{\psi}_{+*}^{\alpha}(\omega) = C(j\omega)^{\alpha+1}, \ as \ \omega \longrightarrow 0. \tag{17}$$

Here $\hat{\psi}_{+*}^{\alpha}(\omega)$ is symmetric. The last formula indicates that the fractional spline wavelets behave like fractional derivative operator.

4 Analytical applications of continuous wavelet transform and fractional wavelet transform

Especially, the transformation of original absorption spectra is widely used in the analytical chemistry for the quantitative resolution of mixtures containing active compounds. This method, which is called derivative spectrophotometry, is based on the derivation of absorption spectra. However, one of the main problems of the derivative method is that the signal-to-noise ratio (S/N) becomes progressively worse for higher order. This influences the accuracy and precision of the analysis results.

To eliminate the above mentioned analytical problem, CWT and FWT methods are used as a powerful signal processing tool for analytical purpose. In addition, CWT and FWT signal processing methods are simultaneously used in the resolution of the complex analytical problems [DB06, DRIB06]. The application of the FWT and CWT in this review is classified as follows:

-CWT-zero crossing method

-Ratio spectra-CWT method

-Ratio spectra-CWT method-zero crossing method

-FWT-zero-crossing method

-FWT- CWT method.

CWT and FWT methods as well as their above mentioned analytical applications will be explained in the following sections.

4.1 Continuous wavelet transform-zero crossing approach

If a mixture of two analytes, M and N is considered and if the absorbance value of this binary mixture is measured at λ_i, the following equation can be written as

$$A_{mix_{\lambda_i}} = \alpha_{M,\lambda_i} C_M + \beta_{N,\lambda_i} C_N, \tag{18}$$

where $A_{mix_{\lambda_i}}$ is the absorbance of the binary mixture at wavelength λ_i, and the coefficients α_{λ_i} are β_{λ_i} absorptivities of M and N analytes. C_M and C_N represent the concentrations of analytes.

If CWT is applied to Eq. (18), the following equation is obtained as follows

$$CWT(A_{mix_{\lambda_i}}) = CWT(\alpha_{M,\lambda_i} C_M) + CWT(\beta_{N,\lambda_i} C_N). \tag{19}$$

If $CWT(\alpha_{M,\lambda_i} C_M) = 0$, then we obtain the following result

$$CWT(A_{mix_{\lambda_i}}) = CWT(\beta_{N,\lambda_i} C_N). \tag{20}$$

In Eq. (20), the CWT amplitudes of the N compounds in mixture depend only on C_N and are independent of the concentration of M in mixture. After that, the calibration graphs can be obtained by measuring the CWT amplitudes corresponding to zero-crossing points. This procedure is repeated for M compound. Calibration functions obtained are used for the quantitative determination of the compounds in their mixtures.

Application of continuous wavelet transform-zero crossing approach

An example of the application of this CWT-zero crossing approach to binary mixtures is the quantitative resolution of the binary mixture of diminazene aceturate (D) and phenazone (P) in the veterinary granules for injection without any chemical separation [DBK05].

In this study, the absorption spectra of two standard series of D and P components in the concentration range 4-20 $\mu g/mL$ were plotted and stored in the range of 200-310 nm as shown in Figure 1.

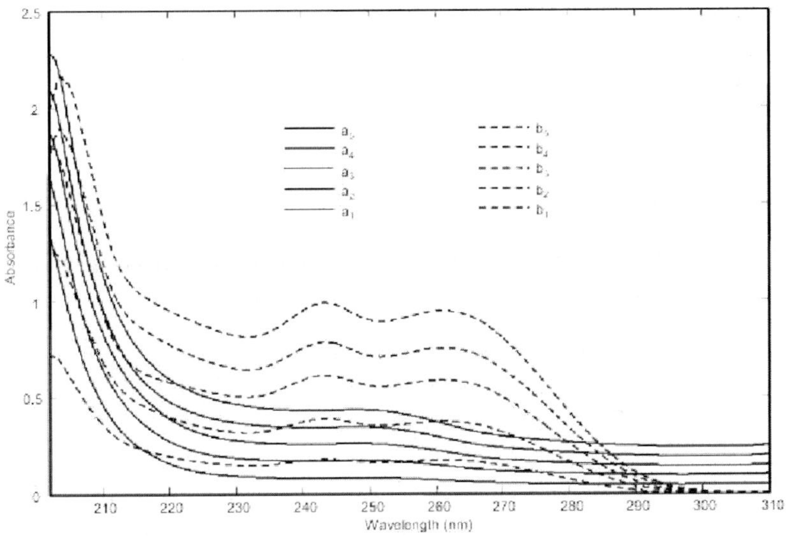

Fig. 1. Absorption spectra of D (—) a1) 4$\mu g/mL$, a2) 8 $\mu g/mL$, a3) 12 $\mu g/mL$,a4) 16 $\mu g/mL$, a5) 20 $\mu g/mL$ and P (- - - -) b1) 4 $\mu g/mL$, b2) 8 $\mu g/mL$, b3) 12 $\mu g/mL$, b4) 16 $\mu g/mL$, b5) 20$\mu g/mL$ in 0.01 M NaOH and methanol (50:50, V/V).

As it can be seen from Figure 1, the quantitative determination of D and P compounds is not possible by using the direct absorbance measurements due to their overlapping spectra in the same spectral region. To solve this problem, several wavelet families with the various values of the scale parameter (a) were tested and Daubechies (db) wavelet approach was found to be the optimal.

In this study, the original spectral data vectors were processed by db4 wavelet transform. A linear regression function for P was obtained by measuring the CWT signal amplitudes at 279.5 nm corresponding to a zero-crossing point for D. In the similar way, the calibration function for D was constructed by measuring the CWT signal amplitude at the 286.1 nm which corresponds to the zero-crossing point for P as shown in Figure 2.

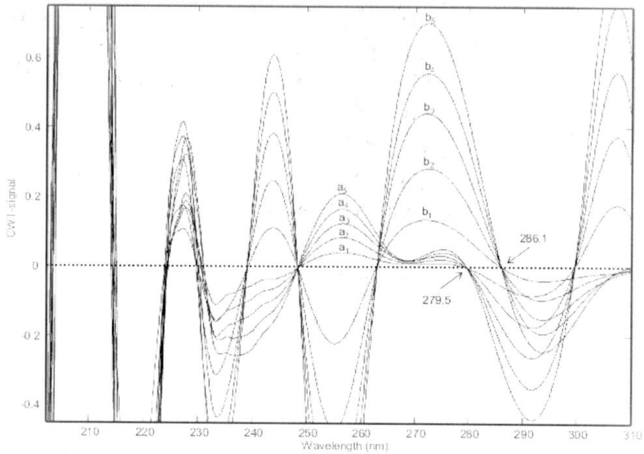

Fig. 2. The CWT spectra of D a1) 4 $\mu g/mL$, a2) 8 $\mu g/mL$, a3) 12 $\mu g/mL$, a4) 16 $\mu g/mL$, a5) 20 $\mu g/mL$ and P b1) 4 $\mu g/mL$, b2) 8 $\mu g/mL$, b3) 12 $\mu g/mL$, b4) 16 $\mu g/mL$, b5) 20 $\mu g/mL$.

The linear regression functions and their statistical parameters for both analytical characteristics were given in Table 1.

Table 1. Linear regression functions and their statistical results

Method	λ (nm)	Regression function	r	SE(a)	SE(b)	SE(r)	LOD	LOQ
CWT-zero crossing	286.1	$S=0.0141C_D+0.0006$	0.9999	0.0012	0.0001	0.0011	0.50	1.67
	279.5	$S=0.0232C_P-0.0028$	0.9992	0.0072	0.0005	0.0068	0.64	2.14

SE(b) : Standard error of slop , SE(a) : Standard error of intercept
SE(r) : Standard error of regression constant, C : Concentration (µg/ mL)
S : Peak amplitude of derivative transform, r : Regression coefficient
LOD : Limit of detection, LOQ : Limit of quantitation

When method was applied to the real veterinary samples, the obtained experimental results can be seen in Table 2.

Examples including the applications of the CWT-zero crossing approach can be find for example in Ref. [DB03a].

Table 2. Experimental results obtained by applying the proposed approach to the commercial veterinary formulation

	CWT-zero crossing db4 (a=180)	
	D	P
	469.9	543.8
	465.2	539.7
	469.0	526.6
	495.6	520.5
	478.9	529.0
Average	475.7	531.9
SD	10.93	8.59
RSD	2.30	1.62

4.2 Ratio spectra- continuous wavelet transform approach

As in the CWT-zero crossing approach, if a mixture of M and N compounds is considered and if the absorbance of this mixture is measured at λ_i, the following equation can be written as

$$A_{mix_{\lambda_i}} = \alpha_{M,\lambda_i} C_M + \beta_{N,\lambda_i} C_N, \tag{21}$$

where $A_{mix_{\lambda_i}}$ denotes the absorbance of the mixture at the wavelength λ_i. The coefficients α_{M,λ_i} and β_{N,λ_i} are absorptivities of M and N analytes. C_M and C_N represent the concentrations of M and N respectively.

The standard spectrum of one of compounds in the same mixture is expressed by the following equation

$$A_{N,\lambda_i} = \beta_{N,\lambda_i} C_N^0. \tag{22}$$

If Eq. (21) is divided by Eq.(22), the ratio spectrum is obtained as

$$\frac{A_{mix_{\lambda_i}}}{\beta_{N,\lambda_i} C_X^0} = \frac{\alpha_{M,\lambda_i} C_M}{\beta_{N,\lambda_i} C_N^0} + \frac{\beta_{N,\lambda_i} C_N}{\beta_{N,\lambda_i} C_N^0}. \tag{23}$$

The equation (8) can be simplified as follows

$$\frac{A_{mix_{\lambda_i}}}{\beta_{N,\lambda_i}} = \frac{\alpha_{M,\lambda_i}}{\beta_{N,\lambda_i}} C_M + C_N. \tag{24}$$

The data vector of the ratio signal, corresponding to Eq.(23) is transferred into wavelet domain and then if CWT is applied on Eq.(23) we obtain

$$CWT(\frac{A_{mix_{\lambda_i}}}{\beta_{N,\lambda_i}}) = CWT(\frac{\alpha_{M,\lambda_i}}{\beta_{N,\lambda_i}}) C_M. \tag{25}$$

In Eq.(25) the CWT signal, $CWT(\frac{A_{mix_{\lambda_i}}}{\beta_{N,\lambda_i}})$, corresponding to minimum and maximum depends only on the values of C_M, and it independent of the C_N value in the binary mixture.

A calibration function for M is obtained by measuring the CWT amplitudes at the minimum or maximum points in the wavelet domain. By using the calibration function, the amount of M in its binary mixture with N compound is determined.

The determination of N compound is carried out by similar procedure.

The modified version of this application is the ratio spectra-CWT-zero-crossing method for the ternary mixture analysis. An example of this application can be find in the literature [DOB05a].

For example, when CWT approach is applied to the ratio spectra of pure B compound (16 $\mu g/mLB(x)$) and its binary mixture (m) containing 16 $\mu g/mLB$ and 20 $\mu g/mL$ H, the transformed CWT spectrum is obtained as shown in Figure 3.

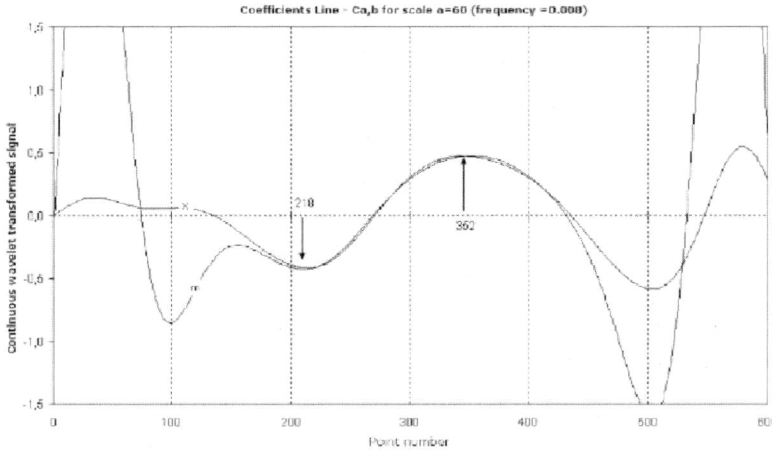

Fig. 3. Continuous wavelet transformed coincident spectrum of 16 $\mu g/mLB(x)$ and its binary mixture (m) with 20 $\mu g/mL$ H (the transformed coincident spectrum of data signals obtained by applying a 6-level one-dimensional db10 compression wavelet and continuous wavelet transform to the ratio spectra data of $\frac{16\mu g/mLBE+20\mu g/mLHCT}{20\mu g/mLHCT}$(m) and $\frac{16\mu g/mLBE}{20\mu g/mLHCT}$(x) in the wavelet domain).

The coincident points corresponding to the maximum and minimum of the wavelength is selected as working wavelength to obtain calibration function [DB04b]. The similar procedure is repeated for the determination of other compound in the binary mixture.

The advantage of this proposed method versus the classical derivative and CWT-zero crossing is that there is no need of any critical point to obtain calibration graph in the qualitative analysis.

Application of ratio spectra-continuous wavelet transform approach

An application of ratio spectra-CWT method to multicomponent analysis of the mixtures containing benazepril (B) and hydrochlorothiazide (H) in tablets can be given as an example [DB04b]. In first step, the absorption spectra of H and B were recorded as indicated in Figure 4.

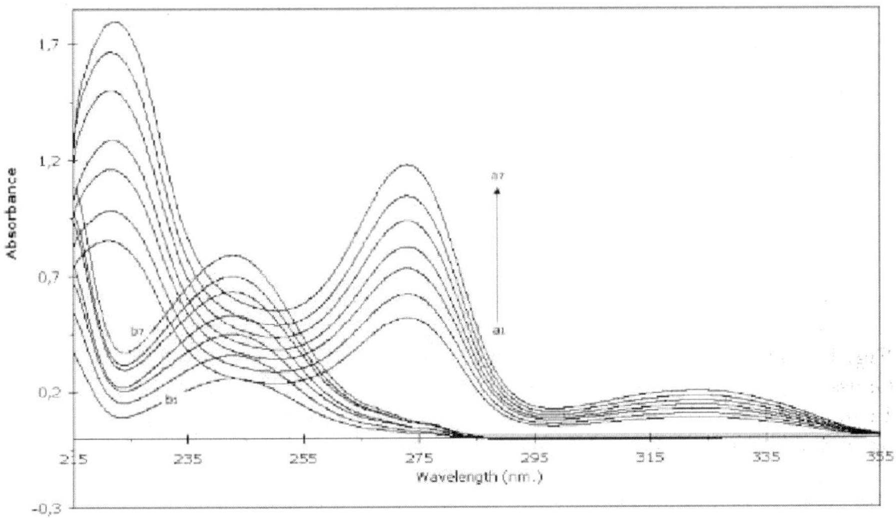

Fig. 4. Absorption spectra of a1) 10 $\mu g/mL$, a2) 12 $\mu g/mL$, a3) 14 $\mu g/mL$, a4) 16 $\mu g/mL$, a5) 18 $\mu g/mL$, a6) 20 $\mu g/mL$, a7) 22 $\mu g/mL$ HTC and b1) 12 $\mu g/mL$, b2) 16 $\mu g/mL$, b3) $\mu 16g/mL$, b4) 16 $\mu g/mL$, b5) 16$\mu g/mL$, b6) 16 $\mu g/mL$, b7) 16 $\mu g/mL$ BE in 0.1 M NaOH.

In the ratio treatment, the absorption spectra of H were divided by the standard spectrum of B (Figure 5A) and the obtained ratio spectra were transferred into the wavelet domain (Figure 5B).

As it can be see the transferred signal (see Figure 5) is not smooth so, we decided to make it smooth by using the discrete wavelet transform (DWT) to reduce the noise and to increase the spectral resolution.

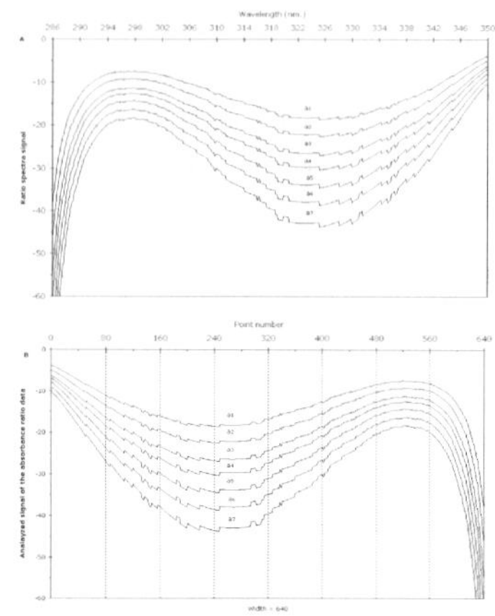

Fig. 5. Ratio spectrum (A) , Analyzed spectrum (B), Compressed spectrum (C) , Gauss4 CWT spectrum (D) and Coif2 CWT spectrum (E) of : a1) 10 $\mu g/mL$, a2) 12 $\mu g/mL$, a3) 14 $\mu g/mL$, a4) 16 $\mu g/mL$, a5) 18 $\mu g/mL$, a6) 20 $\mu g/mL$, a7) 22 $\mu g/mL$ HTC using 16 $\mu g/mL$ BE as a divisor.

In this study, all transferred ratio data signals of H and B compounds were compressed by using a 6-level of Daubechies 10 discrete wavelet (see Figure 6).

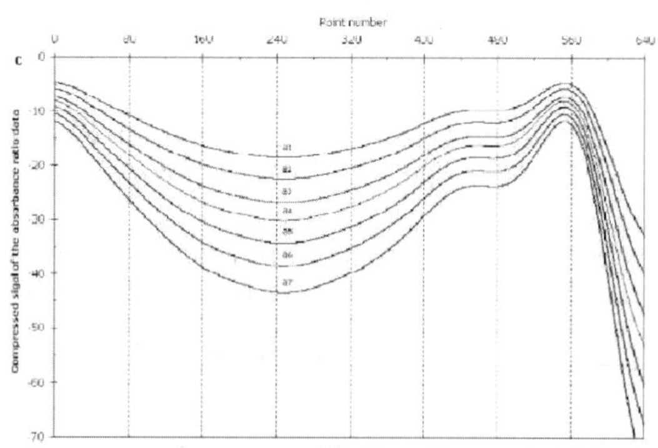

Fig. 6. Compressed spectrum (C) of: a1) 10 $\mu g/mL$, a2) 12 $\mu g/mL$, a3) 14 $\mu g/mL$, a4) 16 $\mu g/mL$, a5) 18 $\mu g/mL$, a6) 20 $\mu g/mL$, a7) 22 $\mu g/mL$ HTC using 16 $\mu g/mL$ **B** as a divisor.

After the above compression processing, to identify the optimal CWT families, various CWT were tested and GAUSS (GAUS) and COIFLETS (COIF) CWT methods were found to be suitable to obtain the coincident CWT signals in the wavelet domain. In these cases, the concentration of H in its mixture were proportional to the transformed amplitudes of the maxima and minima. Other compound (B) in the mixture was processed by using the similar way.

In this study, the calibration graphs were obtained by measuring the CWT amplitudes of the maxima or minima. These calibration graphs were used for the determination of H and B compounds in their binary mixtures and commercial pharmaceutical samples.

The calibration equations and their statistical results were presented in Table 3.

Table 3. Linear regression analysis and its statistical results

Method	Working point	Equation $S = m \times C + n$	r	SE(r)	S(m)	SE(n)
db10 level=6+Gaus4 (a=60)	557max	$S = 7.3318 \times C_{HCT} - 4.6943$	0.9990	1.7634	0.1666	2.7481
	620min	$S = -12.0510 \times C_{HCT} + 9.3864$	0.9990	2.7740	4.3230	0.2621
	218min	$S = -0.0168 \times C_{BE} - 0.0141$	0.9994	0.0056	0.0003	0.0066
	352max	$S = 0.0190 \times C_{BE} + 0.0045$	0.9991	0.0077	0.0004	0.0092
db10 level=6+Coif2 (a=75)	569max	$S = 7.3318 \times C_{HCT} - 4.6943$	0.9990	1.2056	0.1139	1.8788
	622min	$S = -12.0510 \times C_{HCT} + 9.3864$	0.9990	2.5348	0.2395	3.9502
	216min	$S = -0.0068 \times C_{BE} - 0.0037$	0.9996	0.0019	0.0001	0.0023
	349max	$S = 0.0087 \times C_{BE} + 0.0033$	0.9996	0.0024	0.0001	0.0029

S – CWT signal; SE – Standard error
m – Slope of linear regression line; SE(r) – Standard error of correlation coefficient
n – Intercept of linear regression line; SE(m) – Standard error of slope
C – Concentration (µg/mL); SE(n) – Standard error of intercept
r – Correlation coefficient

Two CWT methods were applied to two commercial tablet formulations and satisfactory results were obtained as indicated in Table 4 [DBK05],[DB04b].

Table 4. Commercial tablet results by applying two wavelet transforms

	HCT (mg/tablet)							
	Compress db10 level=6 + gaus4 a=60				Compress db10 level=6 + coif2 a=75			
	557 max		620 min		569 max		622 min	
	I	II	I	II	I	II	I	II
mean	6.34	12.68	6.33	12.63	6.28	12.62	6.52	13.10
SD	0.01	0.26	0.10	0.13	0.07	0.14	0.14	0.47
	BE (mg/tablet)							
	Comprss db10 level=6 + gaus4 a=60				compress db10 level=6 + coif2 a=75			
	218 min		352 max		216 min		349 max	
	I	II	I	II	I	II	I	II
mean	5.22	10.41	4.94	9.93	4.91	10.41	4.94	9.99
SD	0.17	0.36	0.10	0.18	0.24	0.36	0.10	0.07

4.3 Ratio spectra-continuous wavelet transform and zero crossing technique

Another interesting method is the combined application of CWT approach with ratio signal and zero crossing technique to the ternary mixture analysis. This approach is based on the simultaneous use of ratio spectra-CWT and zero crossing technique for the overlapping absorption spectra of ternary mixtures [DOB05a]. In the application of this hybrid approach, the method gave successful results for the quantitative multiresolution of tablets and ternary mixtures consisting of paracetamol, acetylsalicylic acid and caffeine without any separation step.

4.4 Application of continuous wavelet transform in combination with multivariate calibration techniques

The experimental results indicate that the CWT method in combination with multivariate calibration techniques is a promising mathematical too for the multicomponent determination of active compounds in complex mixtures. Some applications of this combined approach can be summarized in the following references [DBUK05],[DOB05b]- [DBK04].

4.5 Fractional wavelet transform-derivative approach

One of the important applications of the CWT methods is the FWT to obtain higher peak amplitude, less noise, and sharper peaks [DB06],[DRIB06]. To compare the proposed FWT approach, first derivative spectrophotometry and chemometric methods (CLS, PCR and PLS) were used for resolving the same samples.

In this application, the spectra of A and S were processed by FWT approach. After that the derivative technique was subjected to the FWT and original spectra. The results obtained from two approaches were compared with each other.

We observed that the FWT method in combination with derivative technique has a higher performance then the classical derivative technique. Figures 7A and 7B indicate the original absorption spectra and their FWT spectra of A and S.

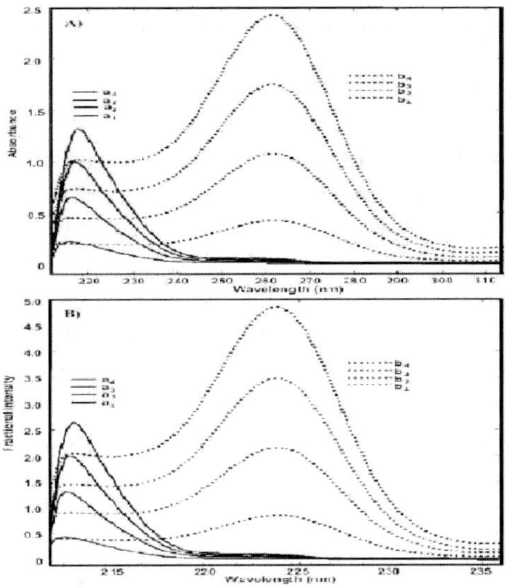

Fig. 7. Compressed spectrum (C) of: a1) 10 $\mu g/mL$, a2) 12 $\mu g/mL$, a3) 14 $\mu g/mL$, a4) 16 $\mu g/mL$, a5) 18 $\mu g/mL$, a6) 20 $\mu g/mL$, a7) 22 $\mu g/mL$ HTC using 16 $\mu g/mL$ **B** as a divisor.

First derivative forms of original absorption spectra and their FWT spectra are presented in Figure 8A and 8B.

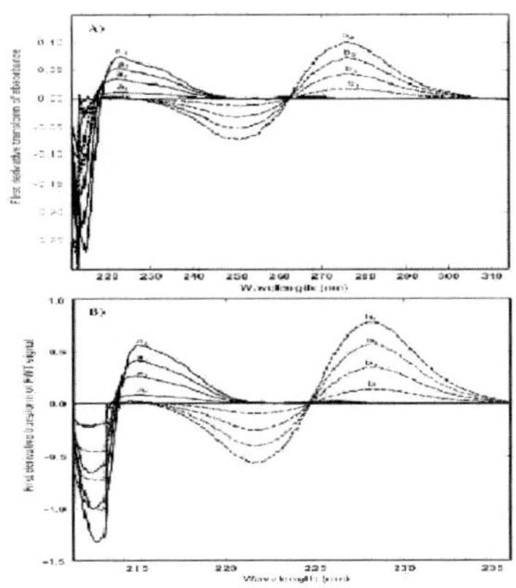

Fig. 8. First derivative transform of absorption spectra (A) and FWT spectra (B) of of 10 $\mu g/mL$, 30$\mu g/mL$, 50$\mu g/mL$, 70$\mu g/mL$ AP (, a1-a4) and of 10 $\mu g/mL$, 25$\mu g/mL$, 40$\mu g/mL$, 55$\mu g/mL$ SB (., b1-b4) in acetonitrile and 0.1 M NaOH (50:50 (v/v)).

4.6 Fractional and continuous wavelet transform

FWT was applied to the original absorption spectra of lacidipine (LAC) and its photo-degradation product (LACD) [DRIB06]. The resulting FWT spectra were processed by CWT and multilinear regression calibration for the simultaneous quantitative resolution of lacidipine and its photo-degradation product in their binary mixtures. These methods don't require any chemical separation step and chemical complex reaction to obtain a detectable signal for the degradation product. By using the Mexican hat function, two calibration functions for LAC and LACD were obtained by measuring the CWT transformed signals at 416.1 nm for LAC and 414.6 nm for LACD, after FWT processing of the original absorption spectra. For a comparison MLRC approach was constructed using the linear regression functions for the individual drug and its degradation product.

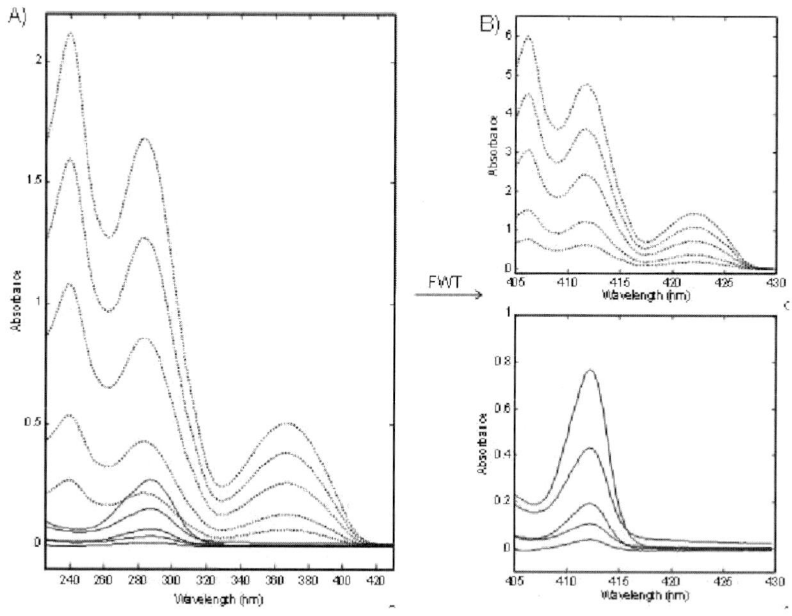

Fig. 9. Original absorption spectra (A) and their corresponding FWT spectra (B) of 5.08, 10.16, 20.32, 30.48 and 40.64 $\mu g/mL$ LAC $(\cdots)$ and 0.51, 1.02, 2.04, 4.08 and 8.16 LACD $(-)$.

The absorption spectra of lacidipine and its degradation product are shown in Figure 9 (A). The examination of this figure shows that the simultaneously determination of LAC and LACD isn't possible by using the classical spectrophotometric method and its modified versions due to the low content of the degradation product. To bypass this difficulty FWT approach was applied to the absorption spectra of LAC and LACD in the spectral range of 225-430 nm.

To determine LAC and LACD in their mixture samples, FWT data vectors were processed by Mexican hat function at the scaling factor a=40. CWT spectra were obtained by plotting the CWT $C_{a,b}$ coefficients versus wavelengths as seen in Figure 10.

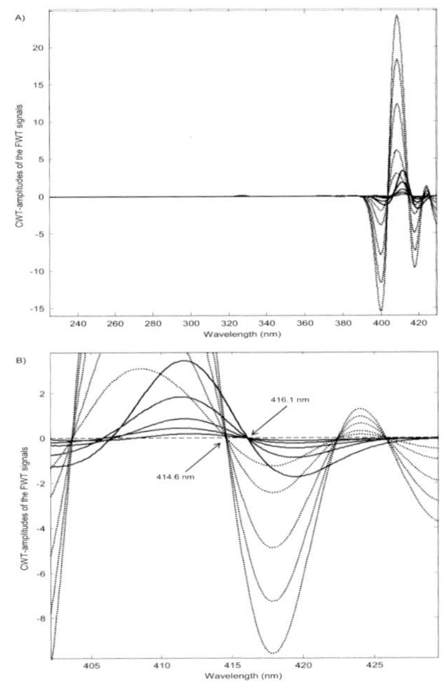

Fig. 10. The CWT spectrum (A) and its detailed form (B) of the FWT signals of 5.08, 10.16, 20.32, 30.48 and 40.64 $\mu g/mL\ LAC(\cdots)$ and 0.51, 1.02, 2.04, 4.08 and 8.16 LACD (−).

4.7 Application of CWT to electrochemical and IR signals

Some of the main problems of analytical studies based on signal analysis are some limitations such as noise, irrelevant information and signal baseline correction to obtain accurate, precise and safe results. A promising tool to solve these mentioned problems is to use wavelets having versatile mathematical properties. In some cases, the nature of electrochemical and IR manipulations can give us the voltammetric and IR signals containing noise, irrelevant information and signal baseline problems. For these reasons, wavelets functions generated from one basic function by dilatation and translation has been used in the areas of data compression, de-noising and data reduction in analytical applications [UB99]-[UB00].

5 Conclusions

Despite of the developments of the analytical instrumentations analytical chemistry needs the advanced new analytical approaches for the resolution of the complex analytical problems. This study presents some of the combined used of the wavelet method and the classical analytical approaches.

The hybrid approaches in this study offer new possibilities and alternative ways for the resolution of mixtures of the active compounds having overlapping spectra.

Contrary to classical derivative spectrophotometry, CWT approaches do not need any optimization such as smoothing function, scaling factor and sampling interval ($\Delta\lambda$). In addition, these combined CWT methods do not produce any problem like diminishing peak intensity in higher order derivative calculation as well as derivative spectrophotometry.

One of the main advantages of CWT approach is the simultaneous data reduction and de-noising for the signal analysis. Beside, CWT and FWT approaches provide higher peak amplitude, less noise, and sharper peaks then derivative spectroscopy.

All wavelet families fulfilling the optimal conditions can be used for the multicomponent determination of active compounds in real samples.

The combination of CWT, FWT and the classical approaches provide safe, reliable, accuracy and rapid analysis of the experimental results for the quality control, routine analysis in drug industry and related branches.

Acknowledgments

One of the authors (E.D.) would like to thank to the organizers of MME06 symposium for giving him the financial support.

References

[Da92] Daubechies, I. : Ten Lectures on Wavelets. Society for Industrial and Applied Mathematics, Philadelphia, (1992).

[Ma98] Mallat S. : A wavelet tour of signal processing. New York, Academic Press, (1998).

[Le98] Leung A.K.M.: A review on applications of wavelet transform techniques in chemical analysis: 1989-1997 Chem. Intell. Lab. Sys. **43** 165– 184 (1998).

[Wa00] Walczak, B. : Wavelets in Chemistry. Elsevier Press, Amsterdam, The Netherland, (2000).

[MS04] Ma, C. X., Shao, X. G.: Continuous wavelet transform applied to removing the fluctuating background in near-infrared spectra. J.Chem. Inf. Compt. Sci., **44(3)**, 907–911 (2004).

[DB03b] Dinc, E., Baleanu, D. : Multidetermination of thiamine HCl and pyridoxine HCl in their mixture using continuous daubechies and biorthogonal wavelet analysis. Talanta, **59**, 707–717 (2003).

[DBU03] Dinc, E., Baleanu, D., Ustundag, O.: An approach to quantitative two-component analysis of a mixture containing hydrochlorothiazide and spironolactone in tablets by one-dimensional continuous Daubechies and biorthogonal wavelet analysis of UV-spectra. Spect. Lett., **36**, 341–355 (2003).

[DB04a] Dinc,E., Baleanu, D. : Multicomponent quantitative resolution of binary mixtures by using continuous wavelet transform. J.AOAC Int., **87(2)**, 360–365 (2004).

[DBH03b] Dinc, E., Baleanu, D., Aboul-Enein, H. : Wavelet analysis for the multi-component determination in a binary mixture of caffeine and propyphenazone in tablets. Il Farmaco, **59**, 335–342 (2004).

[DB04b] Dinc, E., Baleanu, D. : Application of the wavelet method for the simultaneous quantitative determination of benazepril and hydrochlorothiazide in their mixtures. J. AOAC Int., **87(4)**, 834–841 (2004).

[DBOH04] Dinc, E., Baleanu, D., Aboul-Enein, H. : Continuous wavelet transformation applied to the simultaneous quantitative analysis of two-component mixtures. Die Pharmazie, **59**, 618–623 (2004).

[DOB05a] Dinc, E., Ozdemir, A., Baleanu,D.: An application of derivative and continuous wavelet transforms to the overlapping ratio spectra for the quantitative multiresolution of a ternary mixture of paracetamol, acetylsalicylic acid and caffeine in tablets. Talanta, **65**, 36–47 (2005).

[DOB05b] Dinc, E., Ozdemir, A., Baleanu, D.: Comparative study of the continuous wavelet transform, derivative and partial least squares methods applied to the overlapping spectra for the simultaneous quantitative resolution of two-component mixtures. J.Pharm. Biomed. Anal., **37(3)**, 569–575 (2005).

[DBK04] Dinc, E., Baleanu, D., Kanbur, M. : Spectrophotometric multicomponent determination of tetramethrin, propoxur and piperonyl butoxide in insecticide formulation by principal component regression and partial least squares techniques with continuous wavelet transform. Can. J. Anal. Scienc. Spectr., **49(4)**, 218–225 (2004).

[DB4] Dinc, E., Baleanu, D. : One-dimension continuous wavelet resolution for the simultaneous analysis of binary mixture of benazepril and hydrochlorothiazide in tablets using spectrophotometric absorbance data. Rom. Journ. Chem., **49(11)**, 917–925 (2004).

[DBUK05] Dinc, E., Baleanu, D., Ustundag, O., Tas, K.: Chemometric calibration based on the wavelet transform for the quantitative resolution of two-colorants mixtures., Rom. Journ. Chem., 50(4), 283–290 (2005).

[DBK05] Dinc, E., Baleanu, D., Kanbur, M.: A comparative application of wavelet approaches to the absorption and ratio spectra for the simultaneous determination of diminazene aceturate and phenazone in veterinary granules for injection. Die Pharmazie, **60(12)** 892–896 (2005).

[DB06] Dinc, E., Baleanu, D.: A new fractional wavelet approach for the simultaneous determination of ampicillin sodium and sulbactam sodium in a binary mixture. Spectr. Acta Part A, **63(3)**, 631–638 (2006).

[DB05] Dinc, E., Baleanu, D. :Bivariate and multivariate spectral resolution of a mixture of benazepril hydrochloride and hydrochlorothiazide in tablets by using linear regression lines. Rev. Chim., **56(9)**, 937–942 (2005).

[DB03a] Dinc, E., Baleanu, D. : A zero-crossing technique for the multidetermination of thiamine HCl and pyridoxine HCl in their mixture by using one-dimensional wavelet transform. J.Pharm. Biomed. Anal., **31**, 969–978 (2003).

[DBOH046] Dinc, E., Baleanu, D., Ustundag,O., Aboul-Enein, H.:Discrete and continuous wavelet transforms for the multicomponent determination of sunset yellow and tartrazine in their soft drink powders. Rev. Chim., **57(1)**, 29–35 (2006).

[DRIB06] Dinc, E., Ragno, G., Ioele, G., Baleanu, D.:Fractional wavelet analysis for the simultaneous quantitative resolution of lacidipine and its photo-degradation product by continuous wavelet transform and multilinear regression calibration. J.AOAC Int., to appear (2006).

[DBUK05] Dinc, E., Ozdemir, A., Baleanu, D., Tas, K. :Wavelet transform with chemometric techniques for quantitative multiresolution analysis of a ternary mixture consisting of paracetamol, ascorbic acid and acetylsalicylic acid in effervescent tablets, Rev. Chim., **57(5)**, 505–510 (2006).

[DB06] Dinc, E., Baleanu, D.:Continuous wavelet transform applied to the overlapping absorption signals and their ratio signals for the quantitative resolution of mixture of oxfendazole and oxyclozanide in bolus. J.Food Drug Anal., in press, (2006).

[AN05] El-Sayed,A-A. Y., El-Salem, N. A.:Recent Developments of derivative spectroscopy and their analytical applications. Anal.Scien., **21**, 595–614 (2005).

[UB99] Unser, M., Blu, T.:Construction of fractional spline wavelet bases. in SPIE Wavelets Applications in Signal and Image Processing VII, Denver, USA, **3813**, 422–431 (1999).

[UB00] Unser, M., Blu, T.: Fractional splines and wavelets. SIAM Review., **42(1)**, 43–67 (2000).

[BU02] Blu, T., Unser, M.: Wavelets, fractals, and radial basis functions .IEEE Transactions on Signal Processing. **50(3)**, 543–553 (2002).

[Po99] Podlubny, I.: Fractional Differential Equations, Academic Press, (1999).

[NWWZLR01] Nie, L., Wu, S. G., Wang, J. W., Zheng, L. Z., Lin, X. Q., Rui, L. :Continuous wavelet transform and its application to resolving and quantifying the overlapped voltammetric peaks. Anal. Chim. Acta, **450(1-2)**, 185–192 (2001).

[SZ04] Shao, X. G., Zhuang, Y. D. : Determination of chlorogenic acid in plant samples by using near-infrared spectrum with wavelet transform preprocessing. Anal. Sci. **20(3)**, 451–454 (2004).

Continuous wavelet analysis for the ratio signals of the absorption spectra of binary mixtures

Erdal Dinç[1], Dumitru Baleanu[2,3] and Kenan Taş[2]

[1] Department of Analytical Chemistry, Faculty of Pharmacy, Ankara University, 06100, Tandogan, Ankara, Turkey dinc@pharmacy.ankara.edu.tr

[2] Department of Mathematics and Computer Science, Faculty of Arts and Science, Çankaya University, Ankara, Turkey dumitru@cankaya.edu.tr

[3] Institute of Space Sciences, P.O. BOX, MG-23,R 76900 Magurele-Bucharest, Romania baleanu@venus.nipne.ro

Wavelet analysis is successfully applied to the quantitative determination of the components in the binary mixture. This mathematical application is based on the use of the division of the absorption signals by the standard absorption signal and the transformation of the ratio signals. Calibration functions are obtained by measuring the continuous wavelet amplitudes corresponding to the minimum points of the wavelengths. The method is validated and applied to one example of binary mixture analysis.

1 Introduction

In the field of chemistry the analytical signals for de-noising, compressing, image analysis and other chemical purposes were processed by the wavelets transform or classical Fourier transform [Wa00, Da92, MS04].

The chemical instruments produce noises and fluctuations coming from instrumentation recording of a spectrum. Therefore, the cardinal signal of the analyte at lower concentration in samples can interfere with noise peaks.

From this reasons, the above mentioned signal analysis methods should be used for the elimination of the noise effect or diminished signal to noise ratio.

Wavelets are mathematical functions that cut up data into different frequency components, and then study each component with a resolution matched to its scale. They have advantages over traditional Fourier methods in analyzing physical situations where the signal contains discontinuities and sharp spikes. Wavelets, as being a powerful tool of signal processing, have been used in various signal processing domains [Wa00, Da92, MS04, DB04a, DB04b, DBU03]. In the analytical studies various graphical and numerical

Kenan Taş et al. (eds), Mathematical Methods in Engineering, 285–291.

methods have been used for the simultaneous quantitative analysis of the mixtures containing two or more constituents without any separation step. These methods contain some analytical disadvantages such us selectivity and sensitivity, namely these methods don't give correct results in some cases. Continuous wavelet transform (CWT) has flexible and versatile properties for the analytical purposes give a good resolution of the above mentioned inconvenient states.

The main aim of this paper is to present the wavelet method in combination with ratio signals approach for component analysis of binary mixtures of levamizole (LEV) and oxyclozanide (OXC).

2 Theory

2.1 Wavelet method

The most general principle of the wavelet construction is to use both dilations and translations. Commonly used wavelets form a complete orthonormal system of functions with a finite support. Therefore, by changing a scale they can distinct the local characteristics of a signal at various scales, and by translations they cover the whole region in which it is investigated.

CWT method is an important signal processing technique for the overlapping peak resolution and for the significant peak identification. This method is successfully applied to the spectrophotometric multicomponent analysis of relevant compounds in samples. Given a mother wavelet [Wa00, Da92] $\Psi(\lambda)$ by scaling (or dilatation) and shifting (or translation) of $\Psi(\lambda)$ a set of functions denoted by $\Psi_{a,b}$ is obtained as indicated below

$$\Psi_{a,b}(\lambda) = \frac{1}{\sqrt{(|a|)}}\Psi(\frac{\lambda - b}{a}), a \neq 0, a, b \in R. \tag{1}$$

Here, a represents the scale parameter, used to control the scaling, b represents the translation parameter controlling the translation and R is the domain of real numbers.

The action of CWT on a given signal $f(\lambda)$ is defined as follows

$$CWT\{f(\lambda); a, b\} = \int_{-\infty}^{\infty} f(\lambda)\psi^*_{a,b}(\lambda)d\lambda = \langle f(\lambda), \psi_{a,b}\rangle, \tag{2}$$

where the superscript $*$ denotes the complex conjugate and $\langle f(\lambda), \psi_{a,b}\rangle$ represents the inner product of function $f(\lambda)$ onto the wavelet function $\Psi_{a,b}(\lambda)$.

The wavelet Ψ is called invertible if it fulfils the following condition

$$\int_{-\infty}^{\infty} \frac{|\hat{\Psi(\omega)}|^2}{\omega} d\omega \prec \infty. \tag{3}$$

In this study we used Daub5 wavelet family [Da92].

2.2 Ratio spectra CWT approach

Let us consider a binary mixture of two analytes, X and Y and consider the absorbance value of this binary mixture measured at λ_i.

The following equation can be written as

$$A_{m_{\lambda_i}} = \alpha_{\lambda_i} C_X + \beta_{\lambda_i} C_Y, \tag{4}$$

where $A_{m_{\lambda_i}}$ is the absorbance of the binary mixture at wavelength λ_i,and the coefficients α_{λ_i} and β_{λ_i} are absorptivities of X and Y analytes. C_X and C_Y represent the concentrations of analytes.

If Eq. (4) is divided by the spectrum ($A_{m_{\lambda_i}} = \alpha_{\lambda_i} C_X^0$) of a standard solution of one of compounds in binary mixture, the equation (6) becomes:

$$\frac{A_{m_{\lambda_i}}}{\alpha_{\lambda_i} C_X^0} = \frac{\alpha_{\lambda_i} C_X}{\alpha_{\lambda_i} C_X^0} + \frac{\alpha_{\beta_i} C_X}{\alpha_{\lambda_i} C_X^0} \tag{5}$$

If CWT is applied to Eq. (5), the following equation can be obtained as

$$CWT(\frac{A_{m_{\lambda_i}}}{\alpha_{\lambda_i} C_X^0}) = CWT(\frac{\alpha_{\lambda_i}}{\alpha_{\lambda_i}})\frac{C_X}{C_X^0} + CWT(\frac{\beta_{\lambda_i}}{\beta_{\lambda_i}})\frac{C_X}{C_X^0}. \tag{6}$$

If $\frac{\alpha_{\lambda_i}}{\alpha_{\lambda_i}} = 0$, then we obtain the following result

$$CWT(\frac{A_{m_{\lambda_i}}}{\alpha_{\lambda_i} C_X^0}) = CWT(\frac{\beta_{\lambda_i}}{\beta_{\lambda_i}})\frac{C_X}{C_X^0}. \tag{7}$$

Eq. (7) implies that CWT amplitudes of the ratio spectra of binary mixture depend only on C_Y and C_X^0 and not on the concentration of other analyte in the binary mixture. The calibration graphs can be obtained by plotting $CWT(\frac{A_{m_{\lambda_i}}}{\alpha_{\lambda_i} C_X^0})$ signals versus the concentration of C_Y.

This procedure is repeated for X analyte in the binary mixture. The concentration of Y compound is determined by a similar procedure.

2.3 Component analysis

In this section the hybrid method presented above is applying for a binary mixture consists of LEV and OXC. Absorption spectra of two-component, LEV and OXC in the range of 5-25 $\frac{\mu g}{mL}$ were obtained and component signals were shown in Figure 1.

Ratio absorption signals of LEV and OXC were obtained by dividing the standard component signals. Their ratio signals were given in Figure 2(a) and Figure 3(a). The obtained ratio signals were processed by CWT daub5 family and their corresponding transformed signals were indicated in Figure 2(b) and Figure 3(b).

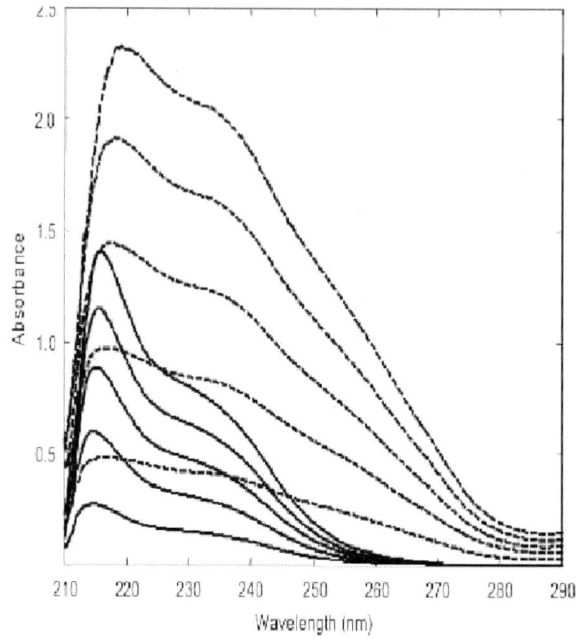

Fig. 1. Absorption spectra of LEV (-) and OXC (—) in the concentration range of 5, 10, 15, 20 and 25 $\mu g/mL$ in 0.1 M NaOH and methanol (50:50,v/v).

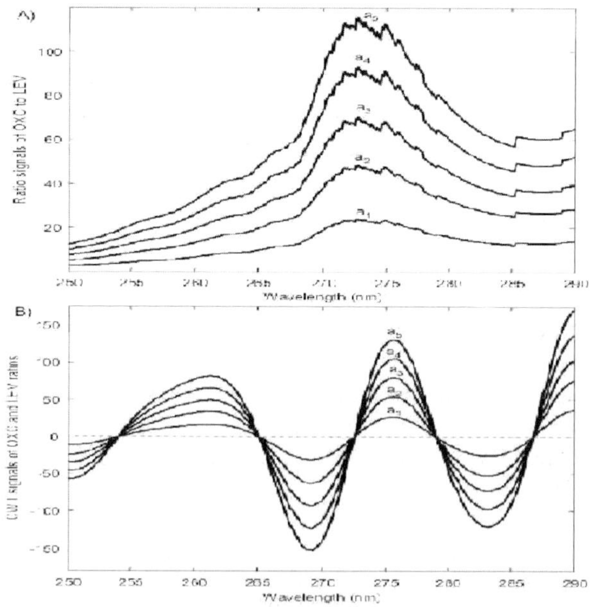

Fig. 2. Ratio spectra (A) and their CWT signals (B) of 5, 10, 15, 20 and 25 $\mu g/mL$ LEV.

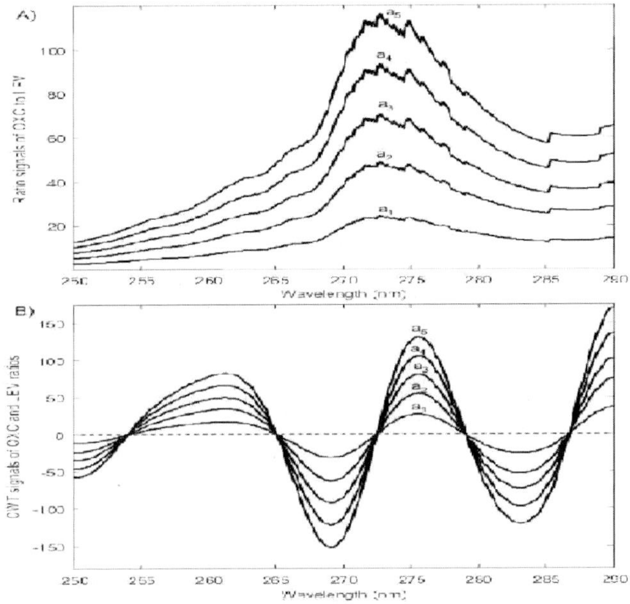

Fig. 3. Ratio spectra (A) and their CWT signals (B) of 5, 10, 15, 20 and 25 $\frac{\mu g}{mL}$ OXC.

A calibration function for each component in samples was computed by measuring the CWT amplitude at 236.1 nm for LEV and at 283 nm for OXC. Calibration functions and their statistical results are illustrated in Table 1.

Table 1. Regression analysis and its results

λ(nm)	Equation	r	s(m)	s(n)
236.1	$S= -3.72\times10^{-3}\ C_{LEV} + 6.13\times10^{-4}$	0.9988	0.0037	0.0017
283	$S= -20.19\times10^{-2}\ C_{OXC} + 10.73\times10^{-3}$	1.0000	0.2018	0.0107

S = CWT signals of the ratio spectra.
r = correlation coefficient of calibration equation.
C = Concentration (μg/mL), s(m) = standard error of slope.
s(n) = standard error of intercept

Both calibration functions obtained in Table 1 were applied to the synthetic samples containing LEV and OXC and percent mean recovery with relative standard deviations are calculated and presented in Table 2. The application results of CWT method to our example of binary mixture containing two components were given in Table 3.

Table 2. Experimental recovery results in the application of CWT method to the synthetic mixtures.

Mixture µg/mL		Recovery (%)	
LEV	OXC	LEV	OXC
5.0	20.0	94.9	95.8
10.0	20.0	100.4	97.6
15.0	20.0	99.1	97.7
20.0	20.0	106.9	99.4
25.0	20.0	97.7	102.0
10.0	5.0	100.5	106.5
10.0	10.0	100.5	99.5
10.0	15.0	100.8	98.8
10.0	20.0	101.0	97.9
10.0	25.0	97.7	98.0
	Mean	99.9	99.3
	SD	2.95	2.84
	RSD	2.95	2.86

Table 3. Results obtained by application of the CWT method to the commercial veterinary product.

	LEV	OXC
	303.70	578.25
	293.21	562.79
	293.24	576.15
	290.71	601.62
	288.04	584.11
Mean	293.78	580.58
SD	5.95	14.11
RSD	2.02	2.43

3 Conclusions

CWT is a powerful tool for the data reduction, denoising, compressing and baseline correction of the analytical signals and resolution of multicomponent overlapping signals. CWT in combination of zero-crossing approach has been used recently for the quantitative resolution and prediction of multimixtures in the presence of the original overlapping signals. CWT transforms a spectra containing noises, as indicated in Figure 3(a), into a smooth one as shown in Figure 3(b). In this way some other techniques can be successfully applied on the transformed signal. In this paper we combine the ratio signal procedure with CWT and it was applied for a binary mixture containing LEV and OXC. This combined approach provides a short analysis time, accurate, precision, rapid and low cost for the quality control and routine analysis of the commercial products. This hybrid method indicates that this technique is perfectly suitable for the multicomponent resolution of the overlapping analytical signals in the various fields of the analytical chemistry.

Acknowledgments

One of the authors (E.D.) would like to thank to the organizers of MME06 symposium for giving him the financial support.

References

[Wa00] Walczak, B.: Wavelets in Chemistry. Elsevier Press, Amsterdam (2000)

[Da92] Daubechies, I.: Ten Lectures on Wavelets. Society for Industrial and Applied Mathematics, Philadelphia (1992)

[MS04] Ma, C.X., Shao, X.G.: Continuous wavelet transform applied to removing the fluctuating background in near-infrared spectra. J.Chem. Inf. Compt. Sci., **44(3)**, 907–911 (2004)

[DB04a] Dinc, E., Baleanu, D.: Multicomponent quantitative resolution of binary mixtures by using continuous wavelet transform. J.AOAC Int., **87(2)**, 360–365 (2004)

[DB04b] Dinc, E., Baleanu, D.: Application of the wavelet method for the simultaneous quantitative determination of benazepril and hydrochlorothiazide in their mixtures. J.AOAC Int., **87(4)**, 834–841 (2004)

[DBU03] Dinc, E., Baleanu, D., Ustundag, O.: An approach to quantitative two-component analysis of a mixture containing hydrochlorothiazide and spironolactone in tablets by one-dimensional continuous Daubechies and biorthogonal wavelet analysis of UV-spectra. Spect. Lett., **36**, 341–355 (2003)

Improved incremental self-organizing map for the segmentation of ultrasound images

Zafer İşcan, Zumray Dokur and Tamer Ölmez

Department of Electronics and Communication Engineering, Istanbul Technical University, Istanbul, Turkey zafer@ehb.itu.edu.tr

This paper presents an Improved Incremental Self-Organizing Map (I2SOM) network that utilizes automatic threshold (AT) value for the segmentation of ultrasound (US) images. I2SOM network has been compared with the well-known unsupervised Kohonen's SOM network (KSOM) and a supervised Grow and Learn (GAL) network in terms of classification accuracy, learning time and number of nodes. For the feature extraction process, two-dimensional discrete cosine transform (2D-DCT) and 2D continuous wavelet transform (2D-CWT) were individually considered and were comparatively investigated to form the feature vectors of US breast and phantom images.

It is observed that the proposed automatic threshold scheme has significantly enhanced the robustness of I2SOM algorithm. Obtained results show that I2SOM can segment US images as good as Kohonen's network.

1 Introduction

Continuous developments and inherently unhazardous structure of ultrasonography preserve its importance in diagnosis of diseases. However, the technique inherently produces speckle noise that restricts diagnosis performance. Since it is an operator-dependent diagnosis technique, obtained results may be interpreted in different manners.

In this study, realization of automatic tissue segmentation was aimed and a diagnosis method which may be useful for especially inexperienced operators is presented. Determination of features which represent the tissues best is still a serious problem which affects the results of segmentation directly. Frequently used feature extraction methods are auto-correlation coefficients [CCH00], gray-level based approaches [LCPIPN02, PKKBSZ00], co-occurrence matrices [KTP02], wavelet [Raj02], discrete Fourier [TMVB03] and discrete cosine [SAD01] transforms.

Main problems of image processing applications include coping up with large number of features, position and scaling changes. Therefore, more clever

Kenan Taş et al. (eds), Mathematical Methods in Engineering, 293–302.

and robust feature extraction methods are needed. It is considered that using some parameters like a priori information or expected noise content will meet these needs [PRH02].

Although the features are determined well, the segmentation algorithm must be chosen well enough to obtain good results. Up to now, various schemes have been introduced in order to accomplish this task. In our study, three different Artificial Neural Networks (ANNs) were compared; GAL [Alp94], Kohonen's SOM [Koh88] and I2SOM network. GAL is an incremental network. It has a fast algorithm but it needs a priori class information in order to achieve segmentation. The Kohonen network is a well-known network which is used in many fields. In this study, I2SOM network is being introduced. I2SOM is an improved version of ISOM [KDO01] network with the addition of an automatic threshold function that produces a suitable threshold value.

A node-coloring scheme that shows the vicinity of nodes visually in multi-dimensional feature space is also presented. 2D-DCT and 2D-CWT were compared to find the best features representing the tissues.

2 Methods

2.1 2D discrete cosine transform

Cosine transform is a sinusoidal based transform like Fourier transform. However, DCT does not contain imaginary coefficients. After transform, most of the energy belonging to the signal (image) accumulates in the initial coefficients. Thus, the signal can be represented only with a few transform coefficients.

Mathematical expression for 2D-DCT transform $C(u, v)$ of an $f(x, y)$ image of MxN size is shown in Eq. 1.

$$C(u,v) = \frac{2}{MN}\alpha(u)\alpha(v)\sum_{x=0}^{M}\sum_{y=0}^{N} f(x,y)\cos[\frac{(2x+1)u\pi}{2M}]\cos[\frac{(2y+1)u\pi}{2N}] \quad (1)$$

$$\alpha(u) = \begin{cases} \frac{1}{\sqrt{2}} & \text{u=0} \\ 1 & 1 \leq u \leq N-1 \end{cases} \qquad u = 0,1,2,...,M-1, v = 0,1,2,...,N-1$$

In this study, 2D-DCT of a 5x5 window of pixel data centered on the pixel of interest is calculated to form the feature vectors of US images. Nine more distinctive low-frequency coefficients were considered as features from a total of 25 coefficients.

2.2 2D continuous wavelet transform

CWT is a convenient transform for non-stationary signals. DCT shows the frequency spectrum of signals but it can not detect the time interval in which any frequency component occurs [Pol96]. By using CWT, time-frequency spectrum of a signal can be obtained which means a higher information level. In this study, 2D-CWT of whole US image was calculated for eight different scale parameters. Thus, eight transformed images were obtained from the original image. For each pixel, the intensity values from the nine images were considered together, and therefore, generating a nine-dimensional feature vector, to be also used as input to the classifiers. Fig. 1 shows the formation of feature vectors.

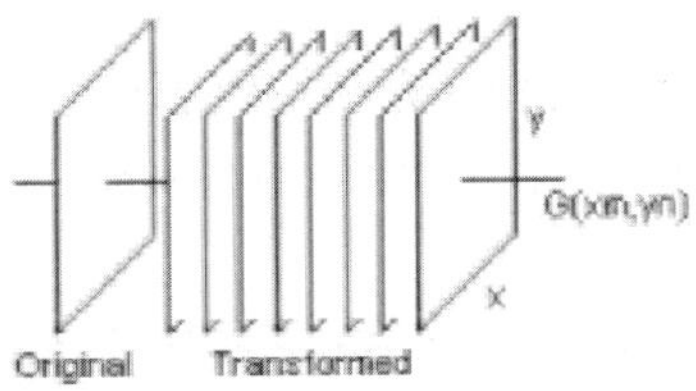

Fig. 1. Formation of nine-dimensional 2D-CWT features

After forming the required feature vectors, self-organization process is accomplished in order to arrange network nodes in multi-dimensional feature space. Fig. 2 depicts the segmentation process.

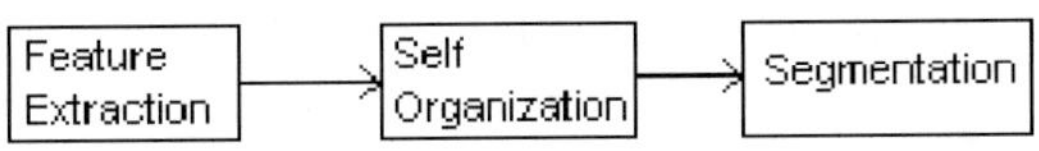

Fig. 2. Processing blocks in segmentation

2.3 Automatic threshold

The main difference of presented network from its former versions is the automatic threshold (AT) value computed by a function before starting to self-organization stage. At past, threshold values were being determined by method of trial and error. When using different US images or different training sets for the same image, recalculation of threshold was required and it was really a time-consuming process. By using AT function (Eq. 2), a standard calculation method is generated for segmentation of US images. Therefore, the robustness of algorithm is provided.

$$AT = \sqrt{\frac{1}{M}\sum_{i=1}^{M}\sum_{j=1}^{N}(x_{ij} - m_j)^2} \tag{2}$$

In Eq. 2, X is the feature vector matrix of size MxN. Each row of the matrix is constituted by the elements of the feature vectors, hence, X holds

N-dimensional M feature vectors. m_j denotes the mean value of the features on column j.

In fact, AT value represents the distribution of feature vectors in multi-dimensional feature space. Although AT function was simply defined, it shows high performance in generating proper (reference) threshold values depending on the number of features (dimension of vectors) and distribution in the feature space.

2.4 Node coloring

In order to visualize the difference between tissue structures, a node-coloring scheme based on interpolation technique was used. The mathematical formulation of the method is expressed in Eq. 3.

$$C(n) = \left[\frac{C(a)}{d(n,a)} + \frac{C(b)}{d(n,b)}\right]\left[\frac{1}{d(n,a)} + \frac{1}{d(n,b)}\right]^{-1} \tag{3}$$

where $C(n)$ denotes the color value of the node number n. $d(n,a)$ and $d(n,b)$ are the Euclidean distances in the feature space between the node n, and the nodes a and b, respectively. In this scheme, first of all, two most distant nodes (a and b) in the network are colored with 0 and 255 gray values. Afterwards, the remaining nodes' colors are assigned according to their Euclidean distances to the formerly colored two nodes. Finally, the segmented image colors are formed according to related nodes' colors.

3 Artificial neural networks

Artificial neural networks (ANNs) have been widely used in many image processing applications. The main advantages of ANNs can be listed as follows [Dok00].

- The solution of the given problem is searched in an iterative scheme by using the vectors in the training set. Hence, there is no need to know or estimate the probability distribution function of the vectors in the feature space.
- Proper results can be achieved for input vectors which do not exist in the training set thanks to ANNs' generalization ability.
- Most of the ANNs in literature can be physically implemented in an easy manner.
- Parallel structure of ANNs produces fast responses.
- ANNs are capable of representing complex class distributions.

3.1 Improved incremental self-organizing map

I2SOM network is a two-layer, self-organizing incremental network (Fig. 3). It is an improved version of formerly developed ISOM [KDO01].

Before the training of I2SOM network, the pixels which will be used in feature extraction process are selected from the image at random. Initially, a feature vector is randomly chosen from the training set, and is assigned as the first node of the network. Then, the learning rate is set.

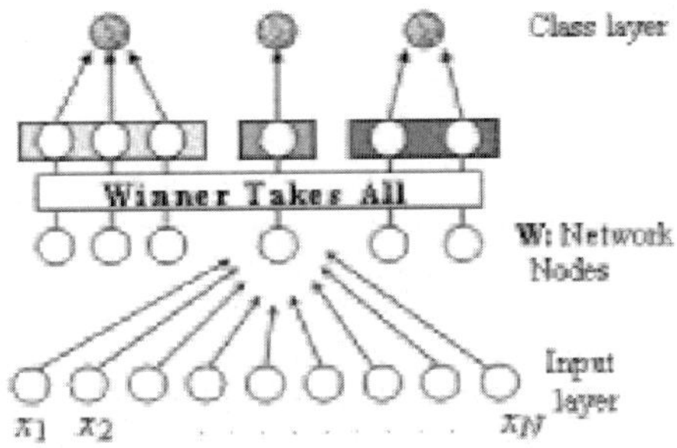

Fig. 3. Structure of I2SOM, N is the feature space dimension

The learning algorithm steps can be summarized as follows:

1. Take a feature vector from the training set.
2. Compute the Euclidean distances between this input feature vector and the nodes in the network, and find the minimum distance.
3. If the minimum distance is higher than the automatic threshold value, include the input vector as a new node of I2SOM. Assign a counter to this new node and set the counter value to one, then go back to the first step. Otherwise, update the weights of the nearest node (winner) as in Eq. 4 Increase the counter value of the winner node by one. Decrease the learning rate.

$$w_{ji}(k+1) = w_{ji}(k) + \eta(t)(x_i(k) - w_{ji}(k)) \quad (4)$$

   Here, w_{ji} is the ith weight of the jth (winner) node nearest to the input vector, x_i is the ith element of the input vector, $\eta(t)$ is the learning rate, and k is the iteration number.
4. Go to step 1 until all feature vectors are exhausted.

After the completion of training period, the counter values of the nodes are plotted to form a node histogram. In Fig. 4a, a sample node histogram can be seen. The bars in the figure show the relative percentage of feature vectors that each node represents in the training set. If the counter value of a node is too low, it means that this node represents a small portion of image pixels, and therefore, it can be removed(e.g. nodes #4, 5 and 6). Without the removal

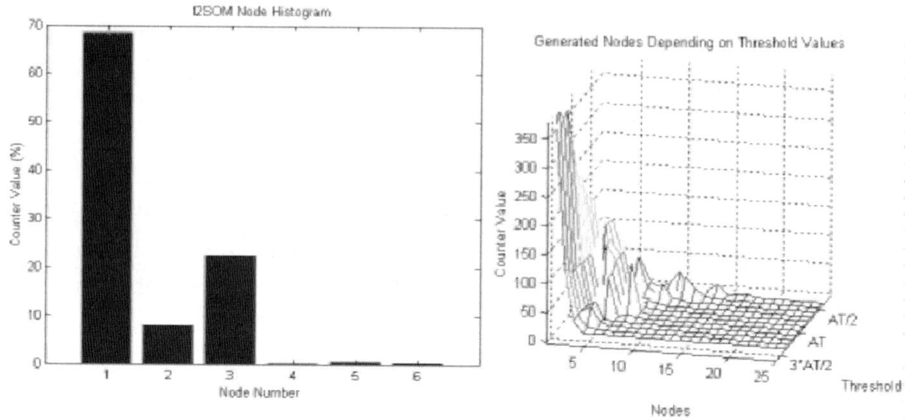

Fig. 4. a)Node histogram, b)Node generation depending on different thresholds

process, segmentation time will become longer. If the operator who will segment the image can predict the number of different classes, threshold value can be adjusted by examining the generated nodes versus threshold values and counter contents. Such an approach is depicted in Fig. 4b.

4 Computer simulations

In the study, a phantom image and an US breast image (Fig. 5) were segmented by using GAL, Kohonen's SOM (KSOM) and I2SOM networks. All simulations were performed on 2 GHz PC by using MATLAB 6.0. Breast image was segmented into three classes. 2D-DCT and 2D-CWT were used in feature extraction processes. Figs. 6-8 show the segmentation results. Related parameters can be seen in Table 1.

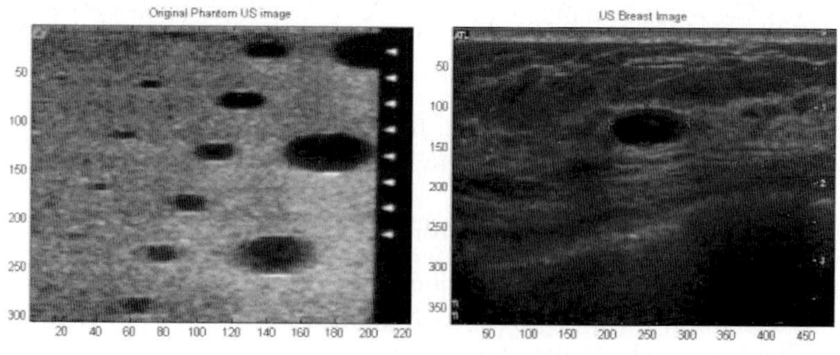

Fig. 5. a) US phantom, b)US breast images

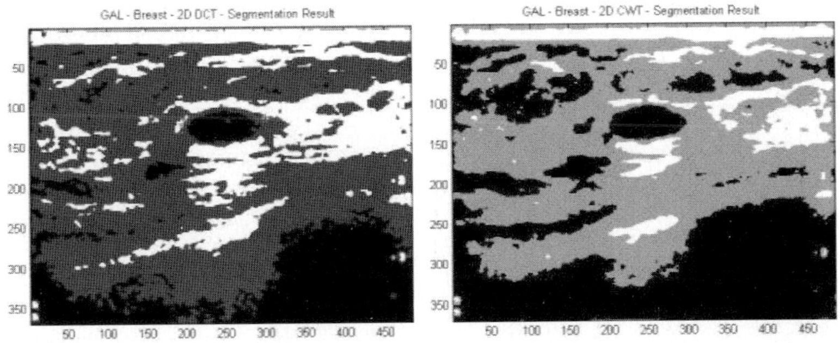

Fig. 6. Segmented breast image by GAL with a) 2D-DCT, b) 2D-CWT features

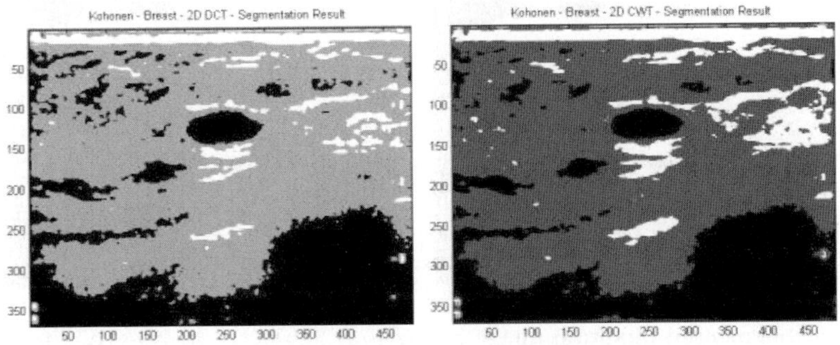

Fig. 7. Segmented breast image by KSOM with a) 2D-DCT, b) 2D-CWT features

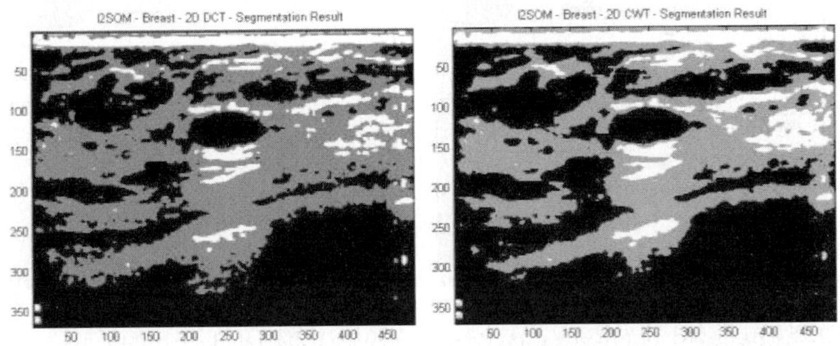

Fig. 8. Segmented breast image by I2SOM with a) 2D-DCT, b) 2D-CWT features

In order to show the performances of the networks comparatively, 100 training feature vectors & 100 test feature vectors (50 for each class) were formed from the US phantom image. Since the characteristics of phantom image [FANXX] are known, it was segmented into two classes. Segmentation results can be seen in Figs. 9-11. Performance comparison was made according to the results presented in Table 2.

Table 1. Performance results of ANNs for the US breast image

ANN	Feature	TT	ST	NoN
GAL	2D-DCT	0	108.17	3
GAL	2D-CWT	0.1	14.51	3
KSOM	2D-DCT	0.01	121.67	3x3
KSOM	2D-CWT	0.01	24.18	3x3
I2SOM	2D-DCT	1.24	109.3	15
I2SOM	2D-CWT	4.33	15.91	6

NoN: Number of generated nodes, TT: Training time (sec.), ST: Segmentation time (sec.)

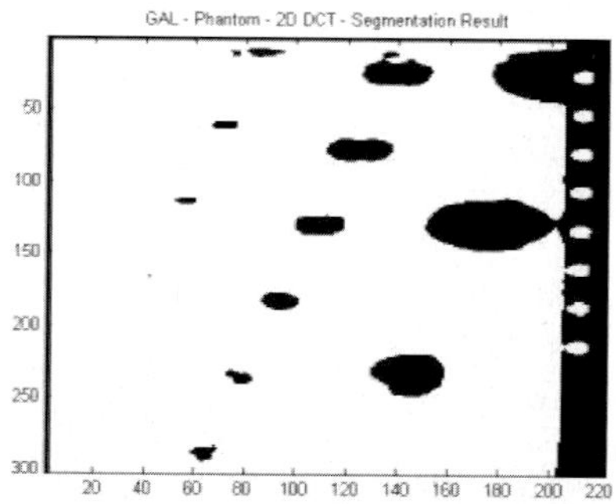

Fig. 9. Segmented phantom image by GAL a) 2D-DCT, b) 2D-CWT features

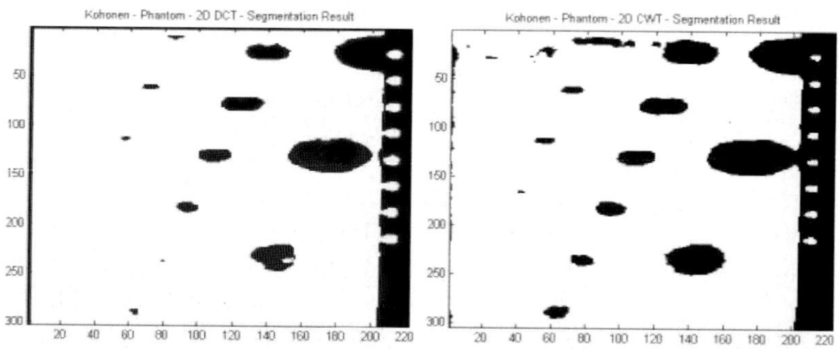

Fig. 10. Segmented phantom image by KSOM a) 2D-DCT, b) 2D-CWT features

5 Conclusions

It is observed that I2SOM network is capable of segmenting the phantom image with a good result. In fact, segmentation performance is too sensitive to the training set. Excessive sensitivity against the training vectors cause GAL network to generate a low performance. Although Kohonen's SOM is a fast algorithm, it is not an incremental network. Besides, the strategy of the learning algorithm of the Kohonen network makes the output nodes locate in the feature space homogenously rather than concentrating on class

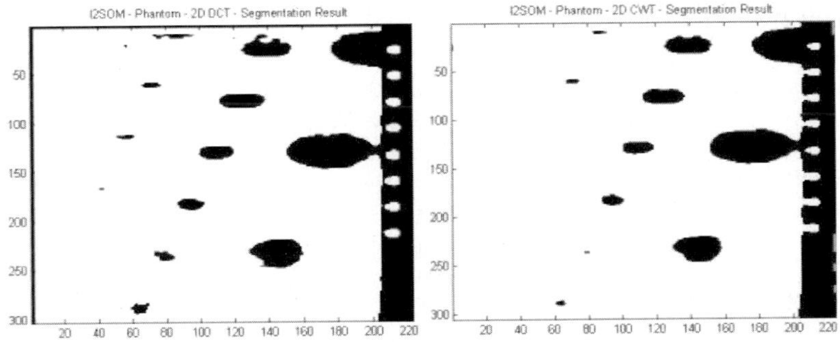

Fig. 11. Segmented phantom image by I2SOM a) 2D-DCT, b) 2D-CWT features

Table 2. Performance results of ANNs for the phantom image

ANN	Feature	MC	RC	SR(%)
GAL	2D-DCT	5	95	95
GAL	2D-CWT	10	90	90
KSOM	2D-DCT	14	86	86
KSOM	2D-CWT	2	98	98
I2SOM	2D-DCT	5	95	95
I2SOM	2D-CWT	11	89	89

MC: Misclassifications, RC: Right Classifications, SR: Success Rate

boundaries. This structure may require excessive number of nodes in the network. Moreover, the problem of determining the optimum number of output nodes and network topology is another disadvantage of the Kohonen network. However, since I2SOM is an incremental network, it automatically determines the proper number of nodes required for the segmentation. Furthermore, AT function significantly eliminated the threshold sensitivity of network's former versions. I2SOM is able to detect different clusters within a given training set by calculating a reference AT value by using statistical distributions of features.

In Table 1, it can be observed that segmentation using 2D-CWT features accomplishes in a shorter time compared to that of 2D-DCT features.

The proposed network was also tested on different medical US images like abdomen, bladder and prostate. Obtained results show that, I2SOM is a highly promising network for the segmentation of US images. Thus, it can be a useful tool for inexperienced operators working in this area.

6 Acknowledgements

Special thanks to Dr. Yi Hong Chou (Chief in US Department of Radiology, Veterans General Hospital, Taipei, Taiwan) for providing us breast images.

References

[CCH00] Chen, D.R., Chang, R.F., Huang, Y.L.: Breast cancer diagnosis using self-organizing map for sonography. World Federation for Ultrasound in Medicine & Biology, **26(3)**, 405-411 (2000)

[LCPIPN02] Loizou, C., Christodoulou, C., Pattichis, C., Istepanian, R., Pantziaris, M., Nicolaides A.: Speckle reduction in ultrasonic images of atherosclerotic carotid plaque. Proceedings of the 14th International IEEE Conference on Digital Signal Processing(Santorini, Greece), **2**, 525-528 (2002)

[PKKBSZ00] Pavlopoulos, S., Kyriacou, E., Koutsouris, D., Blekas, K., Stafylopatis, A., Zoumpoulis, P.: Fuzzy neural network computer assisted characterization of diffused liver diseases using image texture techniques on ultrasonic images. IEEE Engineering in Medicine and Biology Magazine, **19(1)**, 39-47 (2000)

[KTP02] Kadyrov, A., Talepbour, A., Petrou, M.: Texture classification with thousands of features. Proceedings of the 13th British Machine Vision Conference(Cardiff, UK), **2**, 656-665 (2002)

[Raj02] Rajpoot, N.M.: Texture classification using discriminant wavelet packet subbands. Proceedings of the 45th IEEE Midwest Symposium on Circuits and Systems(Oklahoma, USA), **3**, 300-303 (2002)

[TMVB03] Tao, Y., Muthukkumarasamy, V., Verma, B., Blumenstein, M.: A Texture Feature extraction technique using 2D-DFT and Hamming Distance. Proceedings of the 5th International Conference on Computational Intelligence and Multimedia Applications(Xi'an, China), 120-125 (2003)

[SAD01] Sorwar, G., Abraham, A., Dooley, L.S.: Texture classification based on DCT and soft computing. Proceedings of the 10th IEEE International Conference on Fuzzy Systems(Melbourne, Australia), 545-548 (2001)

[PRH02] Petersen, M., Ridder, D., Handels, H.: Image processing with neural networks - A review. Journal of Pattern Recognition, **35**, 2279-2301 (2002)

[Alp94] Alpaydin, E.: GAL: Networks that grow when they learn and shrink when they forget. International Journal of Pattern Recognition and Artificial Intelligence, **8**, 391-414 (1994)

[Koh88] Kohonen T.: Self-Organization and Associative Memory. Springer-Verlag, New York (1988)

[KDO01] Kurnaz, M.N., Dokur, Z., Olmez, T.: Segmentation of ultrasound images by using an incremental self-organized map. Proceedings of the 23rd Annual International Conference of the IEEE Engineering in Medicine and Biology Society(Istanbul, Turkey), **3**, 2638-2640 (2001)

[Pol96] Polikar, R.: Fundamental Concepts & An overview of the wavelet theory. Tutorial, http://users.rowan.edu/ polikar/WAVELETS/WTpart1.html (1996)

[Dok00] Dokur, Z.: Classification of ECG beats by using artificial neural networks and genetic algorithms(in Turkish). Ph.D. Thesis, Istanbul Technical University, Istanbul (2000)

[FANXX] http://www.fantom.suite.dk/571.htm

An application of continuous wavelet transform to electrochemical signals for the quantitative analysis

İncilay Süslü[1], Erdal Dinç[2] and Sacide Altinöz[1]

[1] Department of Analytical Chemistry, Faculty of Pharmacy, Hacettepe University, 06100, Turkey isuslu@hacettepe.edu.tr, saltinoz@hacettepe.edu.tr
[2] Department of Analytical Chemistry, Faculty of Pharmacy, Ankara University, 06100, Tandogan, Ankara, Turkey dinc@pharmacy.ankara.edu.tr

Continuous wavelet transform (CWT) is a new powerful tool for removing noise, irrelevant information and signal baseline correction of voltammograms. In this application, morlet continuous wavelet transforms (ML-CWT) for signal treatments were found to be suitable among the wavelet families. ML-CWT approach was applied to the peak current data vectors consisting of 139 data points in the potential range of (-1004) - (-1556) mV versus Ag/AgCl reference electrode. Peak current data for the calibration and prediction steps in the concentration range of 83.0-375.0 $\mu g/mL$ zafirlukast were obtained by using Osteryoung Square Wave Adsorption Stripping Voltammetry (OSWAdSV).

Three different calibration models namely mean centering calibration (MCC), principal component regression (PCR) and partial least squares (PLS) were constructed by using the relationship between concentration set and CWT-coefficients of the peak current data. The proposed methods were validated by analyzing the synthetic samples and standard addition samples. These methods were successfully applied to the quantitative analysis of zafirlukast in tablets and satisfactory results were reported.

1 Introduction

Some of the main problems of analytical studies based on signal analysis are some limitations such as noise, irrelevant information and signal baseline correction to obtain accurate, precise and safe results. A promising tool to solve these mentioned problems is to use wavelets having versatile mathematical properties [Da92, Wa00, LCG98, DB04]. In some cases, the nature of electrochemical manipulations can give us the voltammetric signals containing noise, irrelevant information and signal baseline problems. For these reasons,

Kenan Taş et al. (eds), Mathematical Methods in Engineering, 303–313.

wavelets functions generated from one basic function by dilatation and translation has been used in the areas of data compression, de-noising and data reduction in analytical applications [RG00, ZM97]. Recently, the CWT techniques have been applied to resolve and quantify voltammetric [NWWZLR01, CHNPSU97] and spectrophotometric [DBK04, DOB05, CSHS05, KKSS05] signals in various fields of analytical chemistry. In addition, the CWT with PCR, PLS and other graphical approaches have been reported for the multicomponent analysis of complex mixtures [DBUAE04, DB00, DB03]. Zafirlukast (ZAF) indicated for the prophylaxis and treatment mild to moderate persistent and chronic asthma. ZAF has been shown effective in the inhibition of allergen-, exercise-, sulphur dioxide- and aspirin induced asthma. The determination of ZAF in pharmaceutical formulation and biological samples has been performed by HPLC [FFTMCFS00, BCC97, RSS02], derivative spectrophotometry [RSS02] and electrochemical methods [SA05a, SA05b]. The aim of the present study is to demonstrate the application of CWT technique to the electrochemical signals to assure the best quantitative determination of ZAF in tablets and to increase the precision and accuracy of analytical results obtained from different calibration models based on the CWT-coefficients. This application of CWT method to the voltammetric study provides a new idea and possibility for the determination of ZAF in samples. The proposed hybrid approaches in this study were tested by using the synthetic sample and standard addition technique. These methods were applied to the real samples and the experimental results were simply compared with each other.

2 Experimental

2.1 Instrument

A BAS 100 B/W (Bioanalytical System, USA) electrochemical analyzer was used to plot voltammograms. The working electrode is a hanging mercury drop electrode (HMDE). The platinum wire serves as an auxiliary electrode, while the reference electrode is Ag/AgCl with saturated 3 M KCI.

2.2 Chemicals and solutions

ZAF was kindly provided from Dr. Reddy's Laboratories (Hyderabad, India). Accolate Tablets (20 mg ZAF per tablet) were kindly supplied by Astra Zeneca Industry. All solvents and chemicals used are of analytical reagent grade. The stock solution of ZAF (1000 $\mu g/mL$) was prepared in acetonitrile. Standard solutions of ZAF were prepared in the working range of 83-375 $\mu g/mL$ using borate buffer (pH = 8) (supporting electrolyte).

3 Results and discussions

3.1 Experimental conditions and optimization

In this study, many supporting electrolytes at different pH were tested for the optimization of experimental conditions to provide successfully electrochemical reduction of our drug. In our case, borate buffer (pH = 8) was found to be optimal for determination of ZAF in samples. Other instrumental conditions were selected as frequency (f) = 70 Hz, scanning increment (E) = 4 mV and pulse amplitude (ESW) = 25 mV with accumulation time of 20 s at accumulation potential of -1000 mV. In practice, voltammograms of ZAF in the concentration range of 83- 375 $\mu g/mL$ within supporting electrolyte were plotted in the voltammetric range from -1004 mV to -1556 mV as shown in Figure 1a.

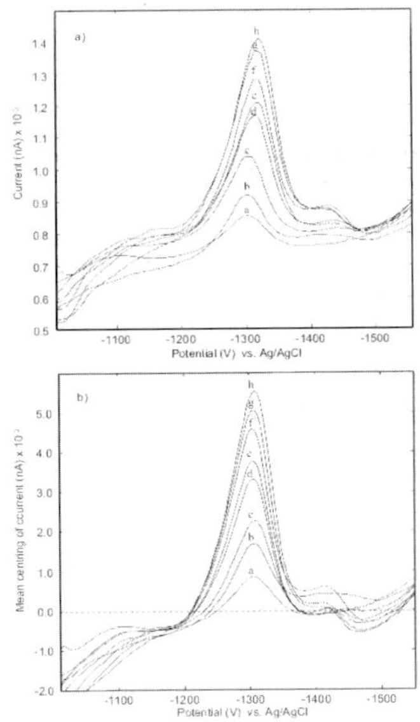

Fig. 1. Voltammograms and their corresponding mean centering forms of a) 83 $\mu g/mL$, b) 115 $\mu g/mL$, c) 167 $\mu g/mL$, d) 231 $\mu g/mL$, e) 259 $\mu g/mL$, f) 286 $\mu g/mL$, g) 355 $\mu g/mL$ and h) 286 zafirlukast by using square wave adsorption stripping voltammetry (OSWAdSV) in buffer solution (pH = 8).

The voltammograms obtained in the above step were processed by mean centering (MC) as a baseline correction procedure (Figure 1b) and the MC voltammograms were transformed by using the continuous wavelet transform.

Both original data (raw data) and transformed data were used to construct the calibrations based on the relationship between concentration and its corresponding signal. In order to increase the precision and accuracy of the analytical calibration models transformed voltammetric signals were used. In first step, a calibration graph was obtained by using the raw data. In the second step, CWT, PCR and PLS calibration were applied to the MC signal data obtained from the original voltammograms of sample solutions. In the next steps of signal analysis, the application of CWT technique to the sample analysis can be summarized in the following sequences:

a) Recording the voltammograms of pure drugs and its sample, b) Transfer the voltammogram data vectors into the wavelet domain, c) Apply CWT to the voltammogram signal data vectors in the wavelet domain.

For CWT signal processing, the type of the wavelet transforms and the scales have to specified. All the wavelet analysis parameters have several values of options. The parameters are closely related and must be considered at the same time. The majority of analyses to date involve environmental and clinical samples, which is the focus of this section.

4 Regression and calibration

Four different calibration approaches for the voltammetric quantitative evaluation of ZAF in the working range of 83- 375 $\mu g/mL$ were discussed in this study. One of them, the calibration equation based on the use of the raw voltammetric data was obtained by measuring the voltammetric peak currents at -1328 mV (Figure 1a). This calibration equation and its statistical results shown in Table 1 was not considered further in this study due to lower correlation coefficient (r) and poor recovery results.

Table 1. Linear regression analysis and its statistical results.

	Potential (mV)	Equation	r	LOD (ng/mL)	LOQ (ng/mL)
Raw data	-1328	$S=1.78x10^{-6}\,C-7.23x10^{-3}$	0.9890	18.90	62.46
MCC	-1304	$S=1.50x10^{-6}\,C-1.92x10^{-5}$	0.9945	17.15	57.17
CWT	-1288	$S=3.48x10^{-6}\,C-5.26x10^{-5}$	0.9991	14.49	48.31

In the linear regression based on the MC voltammetric data, the relationship between drug concentration and peak current at -1304 mV were used to calculate the calibration graph after MC procedure (Figure 2b). Linear regression equation and its statistical results were presented in Table 1. The above mentioned procedure was indicated as MCC. In case of CWT-linear regression, Morl-CWT approach was applied to the MC data of ZAF and its samples. The transformed voltammograms were indicated in Figure 2. The CWT-calibration graph was obtained by using the relation between drug concentration and CWT-peak current at -1288 mV in the transformed voltammograms. The CWT-linear regression equation and its statistical data were summarized in Table 1. It was observed that high correlation coefficients (r) and satisfactory slope, intercept, the limit of detection and the limit of quantitation provided by application of the Morl-CWT with the selected optimal parameters. This hybrid approach was applied to the samples in the above mentioned optimized experimental conditions. According to ICH (22), the calculations of the limit of detection and quantitation (LOD and LOQ, respectively) were achieved by using the standard deviation of the response and the slope of the calibration equation (see Table 1). As alternative methods, principal component regression (PCR) and partial least squares (PLS) were applied to the CWT-coefficients of the peak current data for the quantitative determination of ZAF in tablet. A concentration set of ZAF in the range of 83-375 $\mu g/mL$ was used for the PCR and PLS calibrations. Two chemometric calibrations were constructed by using the relationship between peak current CWT-coefficients and concentration set.

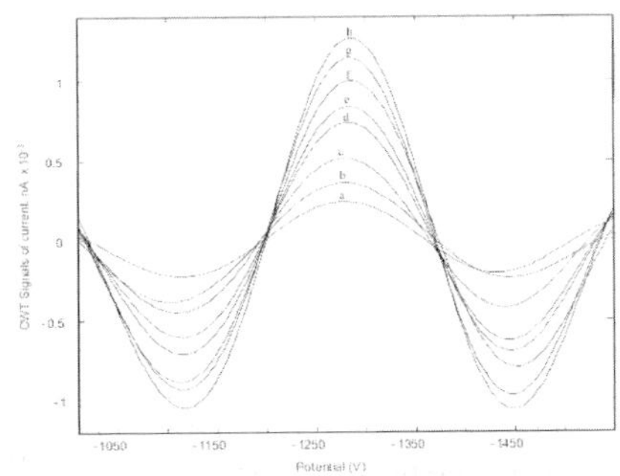

Fig. 2. The CWT signals of voltammograms of a) 83 $\mu g/mL$, b) 115 $\mu g/mL$, c) 167 $\mu g/mL$, d) 231 $\mu g/mL$, e) 259 $\mu g/mL$, f) 286 $\mu g/mL$, g) 355 $\mu g/mL$ zafirlukast

5 Method validation

The proposed approaches in this study were validated by using synthetic samples and standard addition technique. The experiment was repeated three replicates for each concentration. Mean recoveries and their relative standard deviations were calculated and presented in Table 2.

Table 2. Recovery results obtained from synthetic samples by the proposed approaches

Actual Concentration (μg/mL)		Predicted Concentration (μg/mL)		Recovery (%)	
T	A	T	A	T	A
10	4	9.77	3.97	97.7	99.2
10	8	9.60	7.99	96.0	99.9
10	12	9.50	11.70	95.0	97.5
10	16	9.92	15.85	99.2	99.1
10	20	9.58	20.40	95.8	102.0
10	24	9.46	24.02	94.6	100.1
5	12	4.89	12.09	97.9	100.8
10	12	9.84	12.62	98.4	105.2
15	12	14.41	12.09	96.1	100.8
20	12	19.69	12.35	98.5	102.9
25	12	24.39	12.32	97.6	102.7
30	12	28.93	12.55	96.4	104.6
			Mean	96.9	101.2
			SD	1.47	2.31
			RSD	1.52	2.28

RSD = Relative standard deviation

The experimental results indicate that the proposed approaches are suitable for the determinations of ZAF in samples. In addition, the standard addition technique was applied and its results were given in Table 3. No interferences and systematical errors were observed during quantitative analysis procedure.

Table 3. The results obtained by standard addition method.

	MCC	PLS	PCR
Mean recovery (%)	101.2	97.6	100.3
SD	0.70	0.45	1.95
RSD (%)	0.69	0.42	1.94

5.1 Statistical parameters

For the chemometric approaches (PCR and PLS), the ability of a chemometric calibration model can be defined by various ways. The most general expression of the applicability of the calibration is the standard error of calibrations (SEC).

In the chemometric approaches based on the CWT-coefficients, the eight voltammograms obtained from the training set were used to construct the PCR and PLS calibrations for the determination of ZAF. The SEC values for subject drug were computed by the data obtained from the difference between the actual and predicted concentrations in the calibration steps for the analyzed drug. The linear regression analysis and statistical results with SEC values are given in Table 4. According to the cross validation procedure, the first two factors for PCR and PLS gives a good prediction for zafirlukast in samples.

In the chemometric approaches, another important parameter is the standard error of prediction (SEP). The SEP values and their statistical values were computed by using a similar procedure as in the above SEC calculation. The obtained SEP, correlation coefficient (r), slope (m) and intercept (n) are presented in Table 4. The statistical data indicate that the minimum values of SEC and SEP give acceptable results under optimized conditions in the calibration and prediction steps.

Table 4. Statistical parameters in the calibration and prediction steps.

Parameter	PCR	PLS
SEC	1.9593	1.1133
r	1.0000	0.9990
m	1.0000	0.9863
n	0.0024	1.2243
SEP	1.9180	0.0143
r	0.9996	0.9992
m	-0.0393	0.9937
n	0.9987	0.4417

SEC = Standard error of calibration, SEP = Standard error of prediction
r = Correlation coefficient, m = slope of linear regression equation
n = intercept of linear regression equation

5.2 Tablet analysis

The proposed raw data, CWT, PCR and PLS approaches were applied to the determination of ZAF in tablets. The experimental results of commercial preparation are shown in Table 5. A good agreement was observed between the results obtained by the proposed methods.

One-way ANOVA test was applied to the assay results of commercial tablets for the significance of difference between the applied methods. The calculated-F value for tabulated-F value (3.01) are found as 1.25 (P=0.05). The calculated F-value did not exceed the tabulated value, indicating that there was no significant difference among the experimental results of the methods. The numerical values of the statistical test indicate that the proposed approaches are suitable for the determination of ZAF in tablets.

Table 5. Commercial tablets results by the proposed approaches (n=8).

	MCC	CWT	PCR	PLS
Mean	19.82	20.24	19.76	20.17
SD	0.663	0.661	0.069	0.131
RSD	3.343	3.264	0.352	0.652
SE	0.234	0.234	0.025	0.046
CL (0.05)	0.459	0.458	0.048	0.091

Label claim : 20 mg ZAF per tablet , SE = Standard error, CL = confidence limit.

6 Conclusions

Four different calibration methods in this study were developed and applied to the quantitative determination of ZAF in tablets. The aim of this study provides the new alternative and possibilities for the data treatments of voltammogram data to assure the best resolution and results of the voltammetric sample analysis. The proposed MCC, CWT, PCR and PLS approaches can be used for the routine analysis and quality control of ZAF in commercial tablets.

Acknowledgments

The authors would like to thank to the organizers of MME06 symposium for giving him the financial support.

References

[Da92] Daubechies, I.: Ten Lectures on Wavelets. Society for Industrial and Applied Mathematics, Philadelphia (1992)

[Wa00] Walczak, B.: Wavelets in Chemistry. Elsevier Press, Amsterdam, The Netherlands (2000)

[LCG98] Leung, A.K., Chau, F.T., Gao, J.: A review on applications of wavelet transform techniques in chemical analysis: 19891997.Chemom. Intell. Lab. Syst., **43**, 165–184 (1998)

[DB04] Dinc, E., Baleanu, D.: Application of the wavelet method for the simultaneous quantitative determination of benazepril and hydrochlorothiazide in their mixtures. J.A.OAC Int., **87(4)**, 834–841 (2004)

[RG00] Ren, S., Gao, L.: Simultaneous quantitative analysis of overlapping spectrophotometric signals using wavelet multiresolution analysis and partial least squares ARTICLE Pages. Talanta, **50**, 1163–1173 (2000)

[ZM97] Zou, X., Mo, J.: Spline wavelet analysis for voltammetric signals. Anal. Chim. Acta, **340**, 115–121 (1997)

[NWWZLR01] Nie, L., Wu, S., Wang, J., Zheng, L., Lin, X., Rui, L.: Continuous wavelet transform and its application to resolving and quantifying the overlapped voltammetric peaks. Anal.Chim. Acta, **450**, 185–192 (2001)

[CHNPSU97] Cocchi, M., Hidalgo-de-Cisneros, J.L., Naranjo-Rodriquez, I., Palacios-Santander, J.M., Seeber, R., Ulrici, A.: Multicomponent analysis of electrochemical signals in the wavelet domain. Talanta, **59**, 735–749 (2003)

[DBK04] Dinc, E., Baleanu, D., Kanbur, M.: Spectrophotometric multicomponent determination of tetramethrin, propoxur and piperonyl butoxide in insecticide formulation by principal component regression and partial least squares techniques with continuous wavelet transform. Can. J. Anal. Scienc. Spectr., **49(4)**, 218–225 (2004)

[DOB05] Dinc,E., Ozdemir, A., Baleanu,D.: Comparative study of the continuous wavelet transform, derivative and partial least squares methods applied to the overlapping spectra for the simultaneous quantitative resolution of two-component mixtures. J.Pharm. and Biomed. Anal., **37(3)**, 569–575 (2005)

[CSHS05] Chen, D., Shao, X.G., Hu, B., Su, Q.D.: Simultaneous Wavelength Selection and Outlier Detection in Multivariate Regression of Near-Infrared. Spectra. Anal. Scien., **21(2)**, 161–166 (2005)

[KKSS05] Kharintsev, S.S., Kamalova, D.I., Salakhov, M.Kh., Sevastianov, A.A.: Resolution enhancement of composite spectra using wavelet-based derivative spectrometry. Spectr. Acta Part A, **61(1-2)**, 149–156 (2005)

[DBUAE04] Dinc, E., Baleanu, D., Ustundag, O. Aboul-Enein, H.Y.: Continuous wavelet transformation applied to the simultaneous quantitative analysis of two-component mixtures. Pharmazie, **59(8)**, 618–623 (2004)

[DB00] Dinc, E., Baleanu, D.: Multicomponent quantitative resolution of binary mixtures by using continuous wavelet transform. J.AOAC Int., **87(2)**, 360–365 (2000)

[DB03] Dinc, E., Baleanu, D.: Multidetermination of thiamine HCl and pyridoxine HCl in their mixture using continuous daubechies and biorthogonal wavelet analysis. Talanta, **59**, 707–717 (2003)

[FFTMCFS00] Ficarra, R., Ficarra, P., Tommasini, S., Melardi, S., Calabro, M.L., Furlanetto, S., Semreen, M.: Determination of zafirlukast, a selective leukotriene antagonist, human plasma by normal-phase high-performance liquid chromatography with fluorescence detection. J. Pharm. Biomed. Anal., **23(1)**, 169–174 (2000)

[BCC97] Bui, K.H., Coleen, M.K., Connie, T.A.: Determination of zafirlukast, a selective leukotriene antagonist, human plasma by normal-phase high-performance liquid chromatography with fluorescence detection. J. Chromatogr. B., **696(1)**, 131–136 (1997)

[RSS02] Radhakrishna, T., Satyanarayana, J., Satyanarayana, A.: Determination of zafirlukast by stability indicating LC and derivative spectrophotometry. J.Pharm. Biomed. Anal., **30(3)**, 695–703 (2002)

[SA05a] Suslu, I., Altinoz, S.: Electrochemical characteristics of zafirlukast and its determination in pharmaceutical formulations by voltammetric methods. J. Pharm. Biomed. Anal., **39**, 535–542 (2005)

[SA05b] Suslu, I., Altinoz, S.: Differential pulse adsorptive stripping voltammetric determination of zafirlukast in pharmaceutical formulations. Anal. Lett., **38**, 1625–1639 (2005)

[EAEMP96] European Agency for the Evaluation of Medical Products. ICH Topic Q2B Note for Guidance on Validation of Analytical Procedures: Methodology GPMP/ICH/281/95 (1996)

The threshold of compression in wavelet transform with Haar's coefficients - numerical examples

Zlatko Udovicic

Department of Mathematics, Faculty of Sciences, Zmaja od Bosne 33-35, 71000 Sarajevo, Bosnia and Herzegovina zzlatko@pmf.unsa.ba

A couple of numerical examples, which justify one way of determining the threshold in dependence of the allowed relative error, are given. Specificity of the mentioned way is the fact that geometric interpretation of the pyramidal algorithm and basic laws of the theory of probability were used in determining the threshold.

1 Introduction

Definition 1. *Multiresolution analysis is a sequence of closed subspaces of the space $L_2(\mathbb{R})$ which have the following properties:*

1. $\{0\} \subseteq \ldots \subseteq v_2 \subseteq v_1 \subseteq v_0 \subseteq v_{-1} \subseteq \ldots \subseteq L_2(\mathbb{R})$,
2. $\bigcap_{k\in\mathbb{Z}} v_k = \{0\}$ *and* $cl(\bigcup_{k\in\mathbb{Z}} v_k) = L_2(\mathbb{R})$,
3. $(\forall k \in \mathbb{Z})(f(\cdot) \in v_k \Leftrightarrow f(2\cdot) \in v_{k-1})$,
4. $f(\cdot) \in v_0 \Leftrightarrow (\forall k \in \mathbb{Z}) f(\cdot - k) \in v_0$ *and*
5. *there exist the function* $\varphi(\cdot) \in v_0$ *(so-called scaling function) such that the set* $\{\varphi(\cdot - k) \big| k \in \mathbb{Z}\}$ *is the orthonormal base of the space* v_0.

Definition 2. *Wavelet spaces, denoted by* $w_k, k \in \mathbb{Z}$, *are the orthogonal complements of the spaces* $v_k, k \in \mathbb{Z}$, *in relation to the spaces* $v_{k-1}, k \in \mathbb{Z}$. *So,* $v_{k-1} = v_k \oplus w_k$.

As in the case of the space v_0, in the space w_0 there also exists the function $\psi(\cdot)$ (so-called mother wavelet) such that the set $\{\psi(\cdot - k)\big|k \in \mathbb{Z}\}$ is the orthonormal base of the space w_0.

From the definition of the wavelet spaces it can be seen that

$$v_{j-1} = v_j \oplus w_j = \ldots = v_J \oplus w_J \oplus w_{J-1} \oplus \ldots \oplus w_j,$$

Kenan Taş et al. (eds), Mathematical Methods in Engineering, 315–322.

for some $j, J \in \mathbb{Z}, j < J$. So, if $f_{j-1} \in v_{j-1}$, then the function f_{j-1} can be represented as a linear combination of one function from the space v_J ("average value" of the function f_{j-1}, its approximation in the space with the coarsest resolution) and $J-j+1$ functions from the spaces $w_l, l \in \{j, j+1, \ldots, J\}$, ("details" about the function f_{j-1}, its approximations in the spaces at all levels of resolution). In other words,

$$f_{j-1}(x) = \sum_k a_{J,k}\varphi_{J,k}(x) + \sum_{l=j}^{J}\sum_k b_{l,k}\psi_{l,k}(x), \tag{1}$$

where $\varphi_{J,k}(x) = 2^{-J/2}\varphi(2^{-J}x-k), k \in \mathbb{Z}$, and $\psi_{l,k}(x) = 2^{-l/2}\psi(2^{-l}x-k)$, $l \in \{j, j+1, \ldots, J\}, k \in \mathbb{Z}$. It can be proved that the sets $\{\varphi_{J,k}(\cdot)|k \in \mathbb{Z}\}$ and $\{\psi_{l,k}(\cdot)|k \in \mathbb{Z}\}$ are the orthonormal bases of the spaces v_J, and w_l, $l \in \{j, j+1, \ldots, J\}$, respectively. Much more details about multiresolution concept can be found in classical monographs [Daub92] or [Mall98].

The basic example of the multiresolution analysis is so-called Haar's multiresolution analysis. Approximation spaces are the spaces of the piecewise constant functions, i.e. $v_j = \{f(\cdot) \in L_2(\mathbb{R})|(\forall k \in \mathbb{Z})x \in [2^{-j}k, 2^{-j}(k+1)) \Rightarrow f(x) = f(2^{-j}k)\}$. The corresponding scaling function and mother wavelet are $\varphi(x) = \begin{cases} 1, x \in [0,1) \\ 0, \text{othervise} \end{cases}$ and $\psi(x) = \begin{cases} 1, x \in [0,\frac{1}{2}) \\ \text{-1}, x \in [\frac{1}{2},1) \\ 0, \text{othervise} \end{cases}$, respectively.

It is trivial to check that the coefficients of the linear combination (1) satisfy the following relations

$$a_{j,k} = \frac{1}{\sqrt{2}}(a_{j-1,2k} + a_{j-1,2k+1}) \quad \text{and} \quad b_{j,k} = \frac{1}{\sqrt{2}}(a_{j-1,2k} - a_{j-1,2k+1}).$$

This is the so-called direct pyramidal algorithm. The coefficients of the approximation at the coarser level of resolution are computed by using the coefficients of the approximation at the finer level of resolution. If the inequality $|b_{j,k}| < \varepsilon$, where $\varepsilon > 0$ is given constant (so-called threshold), holds, then the coefficient $b_{j,k}$ can be replaced by the zero. The described procedure is the so-called thresholding technics.

The inverse procedure of the direct pyramidal algorithm is the so-called inverse pyramidal algorithm. Now, the coefficients of the approximation at the finer level of resolution are computed by using the coefficients of the approximation at the coarser level of resolution. The inverse pyramidal algorithm can be described by the formulas

$$a_{j-1,2k} = \frac{1}{\sqrt{2}}(a_{j,k} + b_{j,k}) \quad \text{and} \quad a_{j-1,2k+1} = \frac{1}{\sqrt{2}}(a_{j,k} - b_{j,k}).$$

If the thresholding technics were used, instead of the coefficients $a_{j-1,2k}$ and $a_{j-1,2k+1}$, the coefficients $\widetilde{a}_{j-1,2k}$ and $\widetilde{a}_{j-1,2k+1}$ will be obtained.

Let the numbers $a_{0,k}, 0 \leq k \leq 2^J - 1$ be the input data for the direct pyramidal algorithm and let the numbers $a_{J,0}$ and $b_{j,k}, J \geq j \geq 1, 0 \leq k \leq 2^{J-j} - 1$, be obtained after J steps of the direct pyramidal algorithm. It can be proved by the induction that the numbers

$$a_{0,m} = 2^{-J/2}\big[a_{J,0} + \sum_{k=0}^{J-1}(\pm 1)_m 2^{k/2} b_{J-k,n_m}\big], 0 \leq m \leq 2^J - 1,$$

or

$$\widetilde{a}_{0,m} = 2^{-J/2}\big[a_{J,0} + \sum_{k=0}^{J-1}(\pm 1)_m 2^{k/2} \widetilde{b}_{J-k,n_m}\big], 0 \leq m \leq 2^J - 1,$$

will be obtained after J steps of the inverse pyramidal algorithm.

In the last equality was put $\widetilde{b}_{J-j,n} = \begin{cases} 0, & |b_{J-j,n}| < \varepsilon \\ b_{J-j,n}, & |b_{J-j,n}| \geq \varepsilon \end{cases}, j \in \{1, 2, \ldots, J\}$, while the symbol $(\pm 1)_m$ means that some coefficients were added and that some were subtracted depending on which coefficient $a_{0,m}$ was calculated.

If the coefficients $a_{0,k}$ and $\widetilde{a}_{0,k}, 0 \leq k \leq 2^J - 1$, were treated like vectors from $\mathbb{R}^{2^J}$ it is natural to ask about the difference between vectors a_0 and $\widetilde{a}_0$. If all coefficients $b_{j,k}$ were replaced by zeroes then the following inequalities holds

$$\|a_0 - \widetilde{a}_0\|_p \leq \varepsilon \cdot \frac{2^{J/2} - 1}{2^{1/2} - 1} \cdot \begin{cases} 2^{J/2}, & p = 1, \\ 1, & p = 2, \\ 2^{-J/2}, & p = \infty. \end{cases}$$

Hence, if $\rho \in (0, 1)$ is the allowed relative error, then the threshold should be determined in the following way

$$\varepsilon = \rho \cdot \|a_0\|_p \frac{2^{1/2} - 1}{2^{J/2} - 1} \cdot \begin{cases} 2^{-J/2}, & p = 1, \\ 1, & p = 2, \\ 2^{J/2}, & p = \infty. \end{cases} \tag{2}$$

It is obvious that this way of determining the threshold has no justification in the theory of probability.

Geometrical interpretation of the pyramidal algorithm makes possible a more realistic estimation of the error. Geometrically, the direct pyramidal algorithm maps the point $(a_{j-1,2k}, a_{j-1,2k+1})$ into the point $(a_{j,k}, b_{j,k})$ by the symmetry in relation to the line $y = \tan \frac{\pi}{8} x$. The analogous geometrical interpretation has the inverse pyramidal algorithm. It can be checked that if the thresholding technics were used, instead of the original points, the points from the line $y = x$ will be obtained (original points will not be perfectly recovered) by the inverse pyramidal algorithm.

Let the set $S_j = \{(a_{j,2k}, a_{j,2k+1}) \big| k \in \{0, 1, \ldots, 2^{J-j} - 1\}\}$, where the numbers $a_{j,k}$ are the input data for the $j+1^{st}$ step of the direct pyramidal algorithm, be considered. Obviously, all the points from the set S_j belong to the sphere

$L_j = \{(x,y) \in \mathbb{R}^2 \big| \|(x,y)\|_p \leq r_j\}, p \in \{1,2,\infty\}$, where $r_j = \max\{\|(a_{j,2k}, a_{j,2k+1})\|_p \big| k \in \{0,1,\dots,2^{J-j}-1\}, p \in \{1,2,\infty\}\}$. From the mentioned geometrical interpretation it is easy to see that the points from the set S_j, which belong to the sector $V = \{(x,y) \in \mathbb{R}^2 \big| |x-y| < \varepsilon \wedge x \neq y\}$, will not be perfectly recovered. If k_j is the number of those points and if those points are uniformly distributed inside the sphere L_j, then the number k_j can be estimated in the following way

$$\frac{k_j}{2^{J-j-1}} = \frac{m(V \cap L_j)}{m(L_j)} = c_j(\varepsilon),$$

where $m(A)$ denotes the measure of the set $A \subseteq \mathbb{R}^2$. By using the previous relation, the following estimations can be given

$$\|a_0 - \widetilde{a}_0\|_p \leq \varepsilon \cdot \frac{2^{J/2}-1}{2^{1/2}-1} \cdot \begin{cases} \frac{\varepsilon}{r} \cdot 2^{J/2}, \ p = 1, \\ \sqrt{c(\varepsilon)}, \ p = 2, \\ 2^{-J/2}, \ p = \infty, \end{cases}$$

where $r = \min\{r_j \big| 0 \leq j \leq J-1\}$ and $c(\varepsilon) = \max\{c_j(\varepsilon) \big| 0 \leq j \leq J-1\}$. Finally, if $\rho \in (0,1)$ is the allowed relative error, then the threshold should be determined as (unique) solution of the corresponding equation

$$\frac{\rho \cdot \|a_0\|_p}{\varepsilon} \cdot \frac{2^{1/2}-1}{2^{J/2}-1} = \begin{cases} \frac{\varepsilon}{r} \cdot 2^{J/2}, \ p = 1, \\ \sqrt{c(\varepsilon)}, \ p = 2, \\ 2^{-J/2}, \ p = \infty. \end{cases} \tag{3}$$

2 Numerical examples

In the text below, a couple of examples are given. Input data were obtained by discretization of a given function at the gird with step 2^{-12} on the corresponding interval. Abbreviations in the tables have the following meanings:

- ARE-the allowed relative error (in percents)
- RDT-the threshold determined by using the formulas (2)
- PDT-the threshold determined by using the formulas (3)
- RRE-realized relative error (in percents)
- RRL-realized reduction of the length (ratio between the non zero coefficients in linear combination (1) and the number of the input data in direct pyramidal algorithm, in percents).

In every row of the presented tables, the first subrow corresponds to the absolute, the second one to the euclidian and the third one to the uniform norm. The numbers in the square brackets denote the power of number 10.

Example 1. Discretization of the function
`Random[Real,{-0.05,0.05}]` at the interval $[-1, 1]$.

Table 1. Example 1

ARE	RDT	RRE	RRL	ARE/RRE	PDT	RRE	RRL	ARE/RRE
1.0	.1044[−3]	.16[−2]	1.0[2]	.64[3]	.1663[−2]	.36	95	2.8
	.1206[−3]	.14[−1]	1.0[2]	70	.6382[−3]	.17	98	5.8
	.2104[−3]	.28	99	3.5	.2104[−3]	.28	99	3.5
2.5	.2610[−3]	.70[−2]	99	.36[3]	.2630[−2]	.79	93	3.2
	.3016[−3]	.58[−1]	99	43	.1176[−2]	.43	97	5.8
	.5259[−3]	.69	99	3.6	.5259[−3]	.69	99	3.6
5.0	.5221[−3]	.33[−1]	99	.15[3]	.3719[−2]	1.5	90	3.4
	.6031[−3]	.16	98	31	.1867[−2]	.85	95	5.9
	.1052[−2]	2.0	97	2.5	.1052[−2]	2.0	97	2.5
7.5	.7831[−3]	.12	98	63	.4555[−2]	2.3	87	3.2
	.9047[−3]	.27	98	28	.2447[−2]	1.2	93	6.1
	.1578[−2]	2.7	96	2.8	.1578[−2]	2.7	96	2.8
10.0	.1044[−2]	.17	97	60	.5260[−2]	2.9	85	3.5
	.1206[−2]	.46	97	22	.2965[−2]	1.7	92	6.0
	.2104[−2]	4.6	94	2.2	.2104[−2]	4.6	94	2.2
20.0	.2088[−2]	.53	94	37	.7439[−2]	5.1	80	3.9
	.2413[−2]	1.2	93	17	.4714[−2]	3.5	87	5.8
	.4208[−2]	8.1	88	2.5	.4208[−2]	8.1	88	2.5

Example 2. Discretization of the function
`If[Abs[x]<=0.005,1,0]` at the interval $[-1, 1]$.

Table 2. Example 2

ARE	RDT	RRE	RRL	ARE/RRE	PDT	RRE	RRL	ARE/RRE
1.0	.2055[−4]	.85[−13]	.51	.12[14]	.2534[−2]	.85[−13]	.51	.12[14]
	.2940[−3]	.64[−13]	.51	.16[14]	.2769[−2]	.64[−13]	.51	.16[14]
	.4208[−2]	.67[−13]	.51	.15[14]	.4208[−2]	.67[−13]	.51	.15[14]
2.5	.5137[−4]	.85[−13]	.51	.29[14]	.4006[−2]	.85[−13]	.51	.29[14]
	.7351[−3]	.64[−13]	.51	.39[14]	.5100[−2]	.64[−13]	.51	.39[14]
	.1052[−1]	.67[−13]	.51	.38[14]	.1052[−1]	.67[−13]	.51	.38[14]
5.0	.1027[−3]	.85[−13]	.51	.59[14]	.5666[−2]	.85[−13]	.51	.59[14]
	.1470[−2]	.64[−13]	.51	.78[14]	.8095[−2]	.64[−13]	.51	.78[14]
	.2104[−1]	.67[−13]	.51	.75[14]	.2104[−1]	.67[−13]	.51	.75[14]
7.5	.1541[−3]	.85[−13]	.51	.88[14]	.6939[−2]	.85[−13]	.51	.88[14]
	.2205[−2]	.64[−13]	.51	.12[15]	.1061[−1]	.64[−13]	.51	.12[15]
	.3156[−1]	.67[−13]	.51	.11[15]	.3156[−1]	.67[−13]	.51	.11[15]
10.0	.2055[−3]	.85[−13]	.51	.12[15]	.8013[−2]	.85[−13]	.51	.12[15]
	.2940[−2]	.64[−13]	.51	.16[15]	.1285[−1]	.64[−13]	.51	.16[15]
	.4208[−1]	.67[−13]	.51	.15[15]	.4208[−1]	.67[−13]	.51	.15[15]
20.0	.4109[−3]	.85[−13]	.51	.24[15]	.1133[−1]	.85[−13]	.51	.24[15]
	.5881[−2]	.64[−13]	.51	.31[15]	.2040[−1]	.64[−13]	.51	.31[15]
	.8416[−1]	.67[−13]	.51	.30[15]	.8416[−1]	.67[−13]	.51	.30[15]

Example 3. Discretization of the function `Sin[x]` at the interval $[0, 2\pi]$.

Table 3. Example 3

ARE	RDT	RRE	RRL	ARE/RRE	PDT	RRE	RRL	ARE/RRE
1.0	.2677[−2]	.12	31	8.1	.7317[−1]	1.4	2.7	.73
	.2974[−2]	.15	27	6.6	.2142[−1]	.54	8.0	1.9
	.4208[−2]	.29	21	3.5	.4208[−2]	.29	21	3.5
2.5	.6692[−2]	.22	17	11	.1157	1.5	2.5	1.7
	.7436[−2]	.26	16	9.7	.3946[−1]	.81	5.1	3.1
	.1052[−1]	1.1	11	2.2	.1052[−1]	1.1	11	2.2
5.0	.1338[−1]	.37	10	14	.1636	2.0	2.0	2.5
	.1487[−1]	.41	9.8	12	.6264[−1]	1.2	3.8	4.3
	.2104[−1]	1.6	8.2	3.1	.2104[−1]	1.6	8.2	3.1
7.5	.2008[−1]	.46	8.5	16	.2004	2.7	1.4	2.8
	.2231[−1]	.57	7.6	13	.8208[−1]	1.5	2.7	4.9
	.3156[−1]	2.7	5.5	2.8	.3156[−1]	2.7	5.5	2.8
10.0	.2677[−1]	.66	5.7	15	.2314	2.7	1.4	3.8
	.2974[−1]	.76	5.5	13	.9944[−1]	1.6	2.6	6.3
	.4208[−1]	5.8	5.0	1.7	.4208[−1]	5.8	5.0	1.7
20.0	.5354[−1]	.90	4.3	22	.3272	3.2	1.2	6.3
	.5949[−1]	1.1	3.9	18	.1579	1.9	2.2	10
	.8416[−1]	7.0	2.7	2.8	.8416[−1]	7.0	2.7	2.8

Example 4. Discretization of the function `If[Abs[x]<1,x,1]` at the interval $[0, 2]$.

Table 4. Example 4

ARE	RDT	RRE	RRL	ARE/RRE	PDT	RRE	RRL	ARE/RRE
1.0	.3156[−2]	.65[−1]	6.3	15	.7944[−1]	.52	.81	1.9
	.3436[−2]	.97[−1]	6.3	10	.2358[−1]	.39	1.6	2.6
	.4208[−2]	.17	6.3	5.8	.4208[−2]	.17	6.3	5.8
2.5	.7889[−2]	.65[−1]	6.3	38	.1256	.52	.81	4.8
	.8589[−2]	.19	3.1	13	.4344[−1]	.39	1.6	6.4
	.1052[−1]	.37	3.1	6.8	.1052[−1]	.37	3.1	6.8
5.0	.1578[−1]	.13	3.1	38	.1776	.52	.81	9.6
	.1718[−1]	.19	3.1	26	.6896[−1]	.78	.81	6.4
	.2104[−1]	.37	3.1	14	.2104[−1]	.37	3.1	14
7.5	.2367[−1]	.26	1.6	29	.2176	1.0	.42	7.2
	.2577[−1]	.39	1.6	19	.9037[−1]	.78	.81	9.6
	.3156[−1]	.76	1.6	9.9	.3156[−1]	.76	1.6	9.9
10.0	.3156[−1]	.26	1.6	38	.2512	1.0	.42	9.6
	.3436[−1]	.39	1.6	26	.1095	.78	.81	13
	.4208[−1]	.76	1.6	13	.4208[−1]	.76	1.6	13
20.0	.6311[−1]	.26	1.6	77	.3553	1.0	.42	19
	.6871[−1]	.78	.81	26	.1739	.78	.81	26
	.8416[−1]	1.5	.81	13	.8416[−1]	1.5	.81	13

Example 5. Discretization of the function
`If[Abs[x]<=0.005,1,0]+Random[Real,{-0.05,0.05}]` at the interval $[-1,1]$.

Table 5. Example 5

ARE	RDT	RRE	RRL	ARE/RRE	PDT	RRE	RRL	ARE/RRE
1.0	1253.[−3]	.19[−2]	1.0[2]	.52[3]	.6546[−2]	3.4	82	.30
	.3211[−3]	.26[−1]	99	39	.3022[−2]	.66	91	1.5
	.4418[−2]	.45	87	2.2	.4418[−2]	.45	87	2.2
2.5	.3133[−3]	.15[−1]	99	.16[3]	.1035[−1]	7.1	73	.35
	.8027[−3]	.10	97	24	.5566[−2]	1.6	84	1.5
	.1104[−1]	1.5	71	1.7	.1104[−1]	1.5	71	1.7
5.0	.6266[−3]	.47[−1]	98	.11[3]	.1464[−1]	12	64	.41
	.1605[−2]	.26	95	19	.8836[−2]	3.1	76	1.6
	.2209[−1]	3.5	48	1.4	.2209[−1]	3.5	48	1.4
7.5	.9400[−3]	.12	97	61	.1793[−1]	17	57	.45
	.2408[−2]	.47	93	16	.1158[−1]	4.5	70	1.7
	.3313[−1]	5.2	29	1.5	.3313[−1]	5.2	29	1.5
10.0	.1253[−2]	.20	96	49	.2070[−1]	20	51	.49
	.3211[−2]	.73	90	14	.1403[−1]	5.9	65	1.7
	.4418[−1]	5.5	14	1.8	.4418[−1]	5.5	14	1.8
20.0	.2507[−2]	.60	92	33	.2927[−1]	33	35	.60
	.6421[−2]	1.9	82	10	.2227[−1]	11	48	1.8
	.8836[−1]	6.0	.54	3.3	.8836[−1]	6.0	.54	3.3

Example 6. Discretization of the function
`Sin[x]` + `Random[Real,{-0.05,0.05}]` at the interval $[0,2\pi]$.

Table 6. Example 6

ARE	RDT	RRE	RRL	ARE/RRE	PDT	RRE	RRL	ARE/RRE
1.0	.2319[−1]	.13	50	7.8	2.868	1.3	1.6	.75
	.5579	.42[−1]	2.2	24	4.326	0.12	1.5	8.2
	18.84	.96[−1]	1.4	10	18.84	.96[−1]	1.4	10
2.5	.5799[−1]	.38	9.5	6.5	4.535	2.2	1.5	1.2
	1.395	.62[−1]	1.7	40	7.969	.15	1.5	17
	47.10	.91	1.2	2.7	47.10	.91	1.2	2.7
5.0	.1160	.45	4.5	11	6.413	2.2	1.5	2.3
	2.789	.74[−1]	1.6	67	12.65	.21	1.4	24
	94.19	1.9	1.1	2.7	94.19	1.9	1.1	2.7
7.5	.1740	.49	3.8	15	7.855	2.5	1.5	3.0
	4.184	.12	1.5	62	16.58	.32	1.4	24
	141.3	1.7	1.1	4.4	141.3	1.7	1.1	4.4
10.0	.2319	.52	3.3	19	9.070	3.1	1.5	3.2
	5.579	.12	1.5	82	20.08	.39	1.4	26
	188.4	1.7	1.0	5.9	188.4	1.7	1.0	5.9
20.0	.4639	.63	2.6	32	12.83	3.3	1.4	6.1
	11.16	.21	1.4	96	31.89	.67	1.3	30
	376.8	3.4	.83	5.9	376.8	3.4	.83	5.9

Example 7. Discretization of the function
`If[Abs[x]<1,x,1] + Random[Real,{-0.05,0.05}]` at the interval $[0,2]$.

Table 7. Example 7

ARE	RDT	RRE	RRL	ARE/RRE	PDT	RRE	RRL	ARE/RRE
1.0	.4675[−1]	.15	18	6.6	6.435	.74	3.7	1.4
	.7406	.44[−1]	3.9	23	6.582	.72[−1]	3.7	14
	20.99	.15	3.5	6.8	20.99	.15	3.5	6.8
2.5	.1169	.22	6.9	11	10.18	1.0	3.6	2.4
	1.852	.52[−1]	3.7	48	12.13	.22	3.6	11
	52.48	1.3	3.3	1.9	52.48	1.3	3.3	1.9
5.0	.2338	.25	5.9	20	14.39	1.7	3.6	2.9
	3.703	.52[−1]	3.7	97	19.25	.31	3.5	16
	105.0	1.3	3.1	3.8	105.0	1.3	3.1	3.8
7.5	.3506	.27	5.4	28	17.62	1.8	3.5	4.2
	5.555	.72[−1]	3.7	.10[3]	25.22	.43	3.5	18
	157.4	1.4	2.9	5.5	157.4	1.4	2.9	5.5
10.0	.4675	.29	5.0	35	20.35	1.8	3.5	5.5
	7.406	.72[−1]	3.7	.14[3]	30.56	.43	3.5	23
	209.9	3.5	2.5	2.8	209.9	3.5	2.5	2.8
20.0	.9350	.33	3.8	60	28.78	1.8	3.5	11
	14.81	.22	3.6	91	48.51	.90	3.3	22
	419.9	7.7	1.8	2.6	419.9	7.7	1.8	2.6

References

[Daub92] Daubechies, I.: Ten Lectures on Wavelets. Society for Industrial and Applied Mathematics, Philadelphia (1992)

[Mall98] Mallat, S.: A Wavelet Tour of Signal Processing. Academic Press, SanDiego London Boston New York Sydney Tokyo Toronto (1998)

Dynamical Systems and Control Theory

3-D Computerized ionospheric tomography with random field priors*

Orhan Arikan[1], Feza Arikan[2] and Cemil B. Erol[3]

[1] Department of Electrical and Electronics Engineering, Bilkent University, Bilkent, Ankara, Turkey oarikan@ee.bilkent.edu.tr
[2] Department of Electrical and Electronics Engineering, Hacettepe University, Beytepe, Ankara, Turkey arikan@hacettepe.edu.tr
[3] Tübitak - İltaren, Kavaklidere, Ankara, Turkey cemil.erol@iltaren.tubitak.gov.tr

Computerized Ionospheric Tomography (CIT) is a method to reconstruct ionospheric electron density image by computing Total Electron Content (TEC) values from the recorded GPS signals. Due to the multi-scale variability of the ionosphere and inherent biases and errors in the computation of TEC, CIT constitutes an underdetermined ill-posed inverse problem. In this study, CIT is performed by using a Bayesian approach with Gaussian random field priors. The 3-D mean and the covariance of the assumed Gaussian random field priors can either be obtained from ionospheric models such as IRI or they can be estimated by an iterative algorithm from the GPS measurements. Given sparse and non-uniform TEC measurements, the electron field is obtained from mean square estimation where the Gaussian random field structure provides regularization. Geographical and temporal variations of ionosphere can be observed by obtaining tomographic reconstructions of electron density distribution from Earth-based GPS stations for both quiet and disturbed days of ionosphere. 2-D slices that will be obtained from 3-D reconstructions can be compared with the model based reconstructions or with the available Global Ionospheric Maps from IGS centers.

1 Introduction

Computerized Ionospheric Tomography (CIT) is a method to reconstruct ionospheric electron density distribution by using the Global Positioning System (GPS) signals collected by the Earth-based receivers [HIK94]. In CIT, Total Electron Content (TEC) values and tomographic reconstruction algorithms are used together to obtain ionospheric electron density

* This work is supported by TÜBİTAK EEEAG grant 105E171.

Kenan Taş et al. (eds), Mathematical Methods in Engineering, 325–334.

distribution [HIK94, HP01]. TEC is defined as the number of electrons included in a cylinder with 1 m^2 cross-section [DH97]. TEC can be obtained from designated Global Positioning System (GPS) receiver measurements [CB98] or from ionospheric models such as International Reference Ionosphere (IRI) [IRI]. Ionosphere is a dispersive medium and exhibits temporal and spatial variation. Due to this multi-scale variability, the ionospheric electron density models are still under development and the GPS measurements are converted into TEC values with inherent biases and errors. Thus, CIT constitutes an underdetermined ill-posed inverse problem that requires advanced signal processing methods for solution under certain error bounds. Various approaches for the solution of the CIT include serial expansion of electron density into two dimensional basis functions, iterative algebraic reconstruction methods, neural network and statistical analysis methods [MMMT05, FS92, YR91, SN98, DK03, YAA05, YAAE05, AFL88]. In this study, CIT is performed by using a Bayesian framework with Gaussian random field priors [Van88]. The electron density distribution can be estimated by the minimum mean squared error algorithm given the prior distribution of the random field and the measurements. In ionospheric tomography application, the 3-D mean distribution and the covariance function can either be obtained from ionospheric models such as IRI or they can be estimated from the field itself by using an iterative algorithm. In this study, the random field theory is applied to the ionospheric electron density reconstruction problem for the first time and the field is estimated using the minimum mean squared error algorithm. Alternatives in the choice of the mean distribution and correlation functions are also investigated. Geographical and temporal variations of ionosphere can be observed by obtaining tomographic reconstructions of electron density distribution from Earth-based GPS stations for both quiet and disturbed days of ionosphere. 2-D slices that will be obtained from 3-D reconstructions can be compared with the model based reconstructions or with the available Global Ionospheric Maps (GIM) from IGS centers on the internet [IGS].

In Section 2, 3-D CIT model and set-up for CIT are provided. Section 3 includes the solution of the CIT problem in minimum mean squared error (MMSE) sense using random field priors in a Bayesian framework. The computation of the ionospheric measurement data from GPS recordings is given in Section 4. Section 5 is the discussion related to the application and implementation alternatives of the 3-D CIT solution introduced for the first time in this study.

2 3-D CIT model

In this section, 3-D CIT model will be discussed. Let $g(r, \theta, \phi, t)$ represent the ionospheric electron density profile in height, r, in angle measured from global zenith, θ, in angle measured from Greenwich longitude, ϕ, and time

t. The spherical angle θ corresponds to the angle measured from the Earth's geographical zenith (90°-latitude in northern hemisphere and 90°+latitude in southern hemisphere). Let us define a grid structure which expands the region of interest in ionosphere for a specific time instant t_s. Suppose the ionosphere is divided into N_r pixels in height, each pixel of length p_r. In θ, it is divided into N_θ pixels, each pixel of length p_θ. In ϕ, it is divided into N_ϕ pixels, each pixel of length p_ϕ. Any sample point in the region of interest at time instant t_s can be designated by $(r_{n_r}, \theta_{n_\theta}, \phi_{n_\phi}, t_s)$, where

$$r_{n_r} = r_i + (p_r/2)(2n_r - 1), \quad 1 \le n_r \le N_r \tag{1}$$

$$\theta_{n_\theta} = \theta_i + (p_\theta/2)(2n_\theta - 1), \quad 1 \le n_\theta \le N_\theta \tag{2}$$

$$\phi_{n_\phi} = \phi_i + (p_\phi/2)(2n_\phi - 1), \quad 1 \le n_\phi \le N_\phi \tag{3}$$

The r_i is the lower bound for the ionosphere in height, θ_i is the lower limit for the range in latitude, and ϕ_i is the lower limit for the range in longitude. Any sample of $g(r, \theta, \phi, t)$ at $(r_{n_r}, \theta_{n_\theta}, \phi_{n_\phi}, t_s)$ can be defined as

$$g(r_{n_r}, \theta_{n_\theta}, \phi_{n_\phi}, t_s) = g_s(n_r, n_\theta, n_\phi, t_s) = g_s(l, t_s) \tag{4}$$

where the lexicographical index

$$l = n_r + (n_\theta - 1)N_r + (n_\phi - 1)N_r N_\theta \tag{5}$$

is a short hand notation for three indices, namely, n_r, n_θ, n_ϕ. Let us define the electron density vector

$$\mathbf{g}(t_s) = [g_s(1, t_s) \ \dots \ g_s(l, t_s) \ \dots \ g_s(N_r N_\theta N_\phi, t_s)]^T_{1 \times N_r N_\theta N_\phi} \tag{6}$$

Let the number of measurements be denoted as $n_m = 1, \dots, N_m$, and the measurement at time instant, t_s, on the n_m^{th} projection to be $d_{n_m}(t_s)$ where

$$d_{n_m}(t_s) = \sum_{l=1}^{N_r N_\theta N_\phi} \beta_{n_m,l}(t_s) \ g_s(l, t_s) \tag{7}$$

and $\beta_{n_m,l}(t_s)$ is equal to 1 if the lth pixel is on the n_m^{th} projection; and 0 otherwise. Then, we can define a matrix $\mathbf{B}(n_m, l, t_s) = \beta_{n_m,l}(t_s)$ and the measurement vector

$$\mathbf{d}(t_s) = [d_1(t_s) \ \dots \ d_{n_m}(t_s) \ \dots \ d_{N_m}(t_s)]^T \tag{8}$$

as

$$\mathbf{d}_{N_m \times 1}(t_s) = \mathbf{B}_{N_m \times N_r N_\theta N_\phi}(t_s) \ \mathbf{g}_{N_r N_\theta N_\phi \times 1}(t_s). \tag{9}$$

Let N_u GPS receiving stations are located in the given geographical range of $[\theta_i, \theta_i + p_\theta N_\theta]$, $[\phi_i, \phi_i + p_\phi N_\phi]$ and the receiver coordinates are designated by (θ_u, ϕ_u), where $u = 1, \dots, N_u$. Suppose the number of measurements recorded by receiver u is $N_{u;m}$. Then, Eq. (7) will be modified to

$$d_{u;n_m}(t_s) = \sum_{l=1}^{N_r N_\theta N_\phi} \beta_{u;n_m,l}(t_s)\ g_s(l,t_s) \tag{10}$$

where $d_{u;n_m}(t_s)$ is the measurement from the n_m^{th} projection for the station u. The measurement vector for the uth receiver can be written as

$$\mathbf{d_u}(t_s) = [d_{u;1}(t_s)\ \ldots\ d_{u;n_m}(t_s)\ \ldots\ d_{u;N_m}(t_s)]^T \tag{11}$$

and for all the stations combined, the measurement vector in Eq. (9) can be modified to

$$\mathbf{d}(t_s) = [\mathbf{d_1}(t_s);\ldots;\mathbf{d_u}(t_s);\ldots;\mathbf{d_{N_u}}(t_s)]. \tag{12}$$

The matrix $\mathbf{B}$ in Eq. (9) should also be modified to $\mathbf{B}_u(u;n_m,l,t_s) = \beta_{u;n_m,l}(t_s)$. Then, for receiver u at time (t_s), the measurement equation will be given as

$$\mathbf{d_u}(t_s) = \mathbf{B}_u(t_s)\ \mathbf{g}(t_s). \tag{13}$$

For the following discussion, the time index (t_s) will be suppressed. In the next section, we will discuss the minimum mean square estimation of $\mathbf{g}(t_s)$ using Gaussian random field priors.

3 MMSE of electron density using random field priors

The electron density distribution $g(r,\theta,\phi,t)$ at time instant t_s can be represented as a vector given in Eq. (6). The prior Gaussian random field model of $\mathbf{g}$ can be expressed in the following form [Van88]:

$$\mathbf{g} \sim \mathcal{N}(\underline{\mu}_g, \mathbf{R}_{gg}) \tag{14}$$

where $\underline{\mu}_g = \mathcal{E}\{\mathbf{g}\}$ is the mean vector and $\mathbf{R}_{gg} = \mathcal{E}\{(\mathbf{g}-\underline{\mu}_g)(\mathbf{g}-\underline{\mu}_g)^T\}$ is the correlation matrix. $\mathcal{E}\{.\}$ represents the expectation operator and the superscript T denotes transpose. If there were no measurements, the best estimate for the electron density distribution would be its prior mean $\underline{\mu}_g$. However, with the available measurements for location u

$$\mathbf{d_u} = \mathbf{B}_u\ \mathbf{g} + \underline{\nu}_u \tag{15}$$

provides new information on the field $\mathbf{g}$. Here, $\underline{\nu}_u \sim \mathcal{N}(\mathbf{0}, \sigma_u^2 \mathbf{I})$ and represents the noise on the measurements. $\mathbf{I}$ is the identity matrix. We will assume that the noise between the stations u and v are uncorrelated (independent), i.e. $\mathcal{E}\{\underline{\nu}_u \underline{\nu}_v\} = 0$ if $u \neq v$. This model of measurements allow different noise variations for different locations. The joint probability density function (pdf) of the field and measurements is also Gaussian as follows [Van88]:

$$\begin{bmatrix}\mathbf{g}\\ \mathbf{d}\end{bmatrix} \sim \mathcal{N}\left(\begin{bmatrix}\underline{\mu}_g\\ \underline{\mu}_d\end{bmatrix}, \begin{bmatrix}\mathbf{R}_{gg}\ \mathbf{R}_{gd}\\ \mathbf{R}_{dg}\ \mathbf{R}_{dd}\end{bmatrix}\right) \tag{16}$$

where

$$\underline{\mu}_d = \mathcal{E}\{\mathbf{d}\} = \begin{bmatrix} \mathcal{E}\{\mathbf{d_1}\} \\ \vdots \\ \mathcal{E}\{\mathbf{d_{N_u}}\} \end{bmatrix} = \begin{bmatrix} \mathbf{B}_1\underline{\mu}_g \\ \vdots \\ \mathbf{B}_{N_u}\underline{\mu}_g \end{bmatrix} \tag{17}$$

and

$$\mathbf{R}_{gd} = \mathbf{R}_{dg}^T = \mathbf{R}_{gg}[\mathbf{B}_1^T \ \ldots \ \mathbf{B}_{N_u}^T] \tag{18}$$

and finally

$$\mathbf{R}_{dd} = \begin{bmatrix} \mathbf{B}_1\mathbf{R}_{gg}\mathbf{B}_1^T + \sigma_1^2\mathbf{I} \cdots & \mathbf{B}_1\mathbf{R}_{gg}\mathbf{B}_{N_u}^T \\ \vdots \quad \ddots & \vdots \\ \mathbf{B}_{N_u}\mathbf{R}_{gg}\mathbf{B}_1^T \quad \cdots & \mathbf{B}_{N_u}\mathbf{R}_{gg}\mathbf{B}_{N_u}^T + \sigma_{N_u}^2\mathbf{I} \end{bmatrix}. \tag{19}$$

Then, the minimum mean-squared estimate (MMSE) for the electron density field $\mathbf{g}$ given the measurement vector $\mathbf{g}$ is obtained by [Van88]

$$\mathbf{g}_{\text{MMSE}} = \mathcal{E}\{\mathbf{g}|\mathbf{d}\} = \underline{\mu}_g + \mathbf{R}_{gd}\mathbf{R}_{dd}^{-1}(\mathbf{d} - \underline{\mu}_d). \tag{20}$$

In the next section, the GPS measurement vector and possible means of obtaining $\underline{\mu}_g$, $\underline{\mu}_d$, and the correlation matrices will be discussed.

4 Computation of TEC from GPS recordings

The measurement vector for CIT in Eq. (13) corresponds to the TEC which is the number of electrons included in a cylinder with 1 m^2 cross-section. As it is widely discussed in the literature, TEC is the line integral of the electron density in the path joining the transmitter and the receiver. TEC measurements can be obtained by various means including digisondes, ionosondes, LEO satellites and GPS system. In recent years, GPS system is widely used in TEC computation due to world wide availability of recordings from internet with no additional cost to the users. The GPS constellation consists of a network of 24 satellites orbiting the Earth at 20,200 km [GPS]. The satellites are evenly distributed within 6 orbital planes inclined 55° with respect to the Earths equator and equally spaced at 60°. Each satellite has a period of 12 hours. GPS satellites transmit two simultaneous L-Band Pseudo Random (PRN) signals whose carrier frequencies are f_1 =1575.42 MHz and f_2 =1227.60 MHz, respectively. If the dispersive ionosphere were not on the path from satellite to receiver, the pseudo-range between the satellite to receiver, P, would be given as [GPS]:

$$P = (T - T_s)\, c \tag{21}$$

where T is the time the signal is received, T_s is the time the signal is transmitted from the satellite and c is the speed of light in vacuum, 3×10^8 m/sec.

Ionosphere at L-band presents an isotropic and dispersive plasma structure where the dielectric constant and refractive index is a function of frequency. Thus, each signal with different carrier frequencies will be refracted with different amounts. This refraction will be represented as a time delay for the transmitted signal since the speed of the electromagnetic wave in the ionosphere will be slower than that of free space. Therefore, Eq. (21) should be modified to include the group delay in the ionosphere, d_{ion}, and the group delay in the troposphere, d_{trop}. The index of refraction of the troposphere is not a function of frequency, so the delay of the electromagnetic wave in the troposphere will be the same for both L-band frequencies. Yet, the ionospheric delay will be different for the two signals with two different carrier frequencies due to frequency dispersion [GPS]. Thus, the pseudo-range for the two signals in meters can be expressed as:

$$P1 = P + d_{\text{ion1}} + d_{\text{trop}} - c\,(\varepsilon_{r1} + \varepsilon_{s1}) \quad (22)$$
$$P2 = P + d_{\text{ion2}} + d_{\text{trop}} - c\,(\varepsilon_{r2} + \varepsilon_{s2}) \quad (23)$$

where ε_r is the receiver bias and ε_s is the satellite bias, both in seconds. The subscripts 1 and 2 denote the respective values for f_1 and f_2. GPS receivers also record the phase of the received signals in two different frequencies and in meters as:

$$\lambda_1\varphi_1 = P + \lambda_1\varphi_{\text{ion1}} + \lambda_1\varphi_{\text{trop}} + c\,(\varepsilon_{r1} + \varepsilon_{s1}) + \lambda_1\,N1 \quad (24)$$
$$\lambda_2\varphi_2 = P + \lambda_2\varphi_{\text{ion2}} + \lambda_2\varphi_{\text{trop}} + c\,(\varepsilon_{r2} + \varepsilon_{s2}) + \lambda_2\,N2 \quad (25)$$

where λ_1 and λ_2 are the wavelengths of the electromagnetic wave corresponding to f_1 and f_2, respectively. φ_1 and φ_2 are the measured phase for f_1 and f_2 in cycles, respectively. φ_{ion1} and φ_{ion2} are the phase delays in the ionosphere for f_1 and f_2, and φ_{trop} is the phase delay in the troposphere. $N1$ and $N2$ are called the phase ambiguity corresponding to f_1 and f_2, respectively. The relationship between the group delay and phase delay in the ionosphere can be given as [GPS]:

$$d_{\text{ion}} = -\varphi_{\text{ion}}\frac{c}{f} \quad (26)$$
$$\approx A\,\frac{\text{STEC}}{f^2} \quad (27)$$

where the constant $A = 40.28\ \text{m}^3/\text{s}^2$. Slant TEC (STEC) corresponds to the number of free electrons in a cylinder of cross section $1\ \text{m}^2$ between the GPS satellite and the Earth-based receiver. If Eq. (26) is inserted into Eq. (22) to Eq. (25), the following will be obtained:

$$P4 = P2 - P1 = d_{\text{ion2}} - d_{\text{ion1}} - c\,(\varepsilon_{r2} - \varepsilon_{r1}) - c\,(\varepsilon_{s2} - \varepsilon_{s1}) \quad (28)$$
$$L4 = \lambda_1\varphi_1 - \lambda_2\varphi_2 = d_{\text{ion2}} - d_{\text{ion1}} - c(\varepsilon_{r2} - \varepsilon_{r1}) - c(\varepsilon_{s2} - \varepsilon_{s1}) + N \quad (29)$$

where $N = (\lambda_1 \ N1 - \lambda_2 \ N2)$. The difference $P4$ and $L4$ values are called the geometry free combination [GPS]. The difference of satellite and receiver biases for the two frequencies are called as Differential Code Bias (DCB) for satellite and receiver, respectively and denoted as

$$\mathrm{DCB_s} = (\varepsilon_{s2} - \varepsilon_{s1}), \tag{30}$$

$$\mathrm{DCB_r} = (\varepsilon_{r2} - \varepsilon_{r1}). \tag{31}$$

The differential code biases can be obtained from [CODE] for many GPS stations connected to the IGS network. If Eq. (27), Eq. (30) and Eq. (31) are inserted into Eq. (28) and Eq.(29), the following will be obtained:

$$P4 = P2 - P1 = A\left(\frac{f_1^2 - f_2^2}{f_1^2 f_2^2}\right) \mathrm{STEC} - c(\mathrm{DCB_r} + \mathrm{DCB_s}) \tag{32}$$

$$L4 = \lambda_1\varphi_1 - \lambda_2\varphi_2 = A\left(\frac{f_1^2 - f_2^2}{f_1^2 f_2^2}\right) \mathrm{STEC} - c(\mathrm{DCB_r} + \mathrm{DCB_s}) + N \tag{33}$$

STEC can be computed using the above equations as:

$$\mathrm{STEC_P} = \frac{1}{A}\left(\frac{f_1^2 f_2^2}{f_1^2 - f_2^2}\right)[P4 + c(\mathrm{DCB_r} + \mathrm{DCB_s})] \tag{34}$$

$$\mathrm{STEC_L} = \frac{1}{A}\left(\frac{f_1^2 f_2^2}{f_1^2 - f_2^2}\right)[L4 - N + c(\mathrm{DCB_r} + \mathrm{DCB_s})]\,. \tag{35}$$

$\mathrm{STEC_P}$ in Eq. (34) is called the absolute TEC. Although it is simple to compute, the measurements are very noisy and are usually corrupted by multipath especially for satellites under 60° of elevation angle [AEA03]. $\mathrm{STEC_L}$ in Eq. (35) is called the relative TEC. The measurements from Relative TEC are less noisy and are immune to multipath effects but it is usually very difficult to resolve the initial phase ambiguity N. Also, the receiver may lose synchronization with the satellite and this corresponds to a sudden phase difference in the measurements. This phenomena is commonly known as cycle slip [AEA03]. Thus, the cycle slips must be detected and corrected before the phase measurements to be included in the TEC computation. Some alternate methods combine the $P4$ and $L4$ recordings to compute STEC to obtain an unambiguous and low noise measurement. One of these methods is called the baseline method and the baseline for one satellite in one epoch is computed by [CB98]

$$B = \frac{1}{N_{me}} \sum_{n_{me}=1}^{N_{me}} [P4(n_{me}) - L4(n_{me})] \tag{36}$$

where N_{me} is the total number of recordings in a connected phase arc. Thus, within a connected phase arc, STEC can be computed using the above equation as

$$\mathrm{STEC_F} = \frac{1}{A}\left(\frac{f_1^2 f_2^2}{f_1^2 - f_2^2}\right)[(B + L4) + c(\mathrm{DCB_r} + \mathrm{DCB_s})]. \tag{37}$$

At each GPS receiver station, the pseudo range $P1, P2$ and phase values φ_1, φ_2 for the two frequencies are recorded in a Receiver INdependent EXchange (RINEX) file with a standard format. The measurements are usually recorded for every 30 s for all the satellites that are in view at that instant. Thus, the STEC values for each satellite at each epoch can be computed either using Eq. (34) or using Eq. (33) and Eq. (37). These measurements form the entries of the measurement vector in Eq. (11) and Eq. (15) for a receiver u.

In most studies in the literature, the slant TEC values are projected to the local zenith direction using a mapping function $M(\epsilon_m(n))$ and the vertical TEC (VTEC) is computed as

$$\text{VTEC} = \text{STEC}/M(\epsilon_m(n)) \tag{38}$$

where

$$M(\epsilon_m(n)) = \left[1 - \left(\frac{R\cos\epsilon_m(n)}{R+h}\right)^2\right]^{-1/2}. \tag{39}$$

In the above equation, $\epsilon_m(n)$ is the elevation angle in the local receiver coordinates for the mth satellite at the nth time sample; R is the effective mean earth radius; and h is the height of ionospheric pierce point for the thin shell model. In the following section, various application and implementation alternatives will be discussed.

5 Discussion

The STEC measurements computed from GPS recordings can be used in the MMSE solution presented in Section 3 for various receiver locations and for quiet and disturbed ionosphere. The slant TEC values in Eq. (34) or Eq. (37) are the entries of the data vector given in Eq. (13) for the receiver station u. In order to compute the matrices $\mathbf{B}_u$, the ionosphere should be segmented into $N_r N_\theta N_\phi$ pixels for the time instant, t_s corresponding to the measurement sample n. The indices l which represent the pixels that the ray connecting the satellite to the receiver pass through, should be determined and the β values in Eq. (7) corresponding to these indices l should be 1. In another implementation of the problem, the β coefficients can be assigned scaled values depending on the length of the ray segment within the lth pixel.

There are many issues in the implementation of Eq. (20). Some of these issues include the determination of $\underline{\mu}_g$, and the correlation matrix $\mathbf{R}_{gg}$. $\mathbf{R}_{gg}$ is necessary in computation of both $\mathbf{R}_{gd}$ and $\mathbf{R}_{dd}$ in Eq. (20). In one application, both $\underline{\mu}_g$ and $\mathbf{R}_{gg}$ can be determined from model ionospheres such as IRI [IRI] for quiet ionospheres and for night hours. Previous studies indicate that IRI provides reliable estimates of ionospheric electron density for mid-latitude receiver locations and for quiet night hours when compared with the measurements [AEA03]. Another alternative can be obtaining $\underline{\mu}_g$ and $\mathbf{R}_{gg}$

iteratively from the tomographic inversion itself. This method may be more appropriate for locations or times where IRI results not are in accordance with the measurement data. Another issue that needs to be addressed in the computation of Eq. (20) is the inversion of the matrix $\mathbf{R}_{dd}$. The size of $\mathbf{R}_{dd}$ can be quite limiting for standard matrix inversion and alternative inversion algorithms such as conjugate gradient methods can be considered.

The noise on the measurements can be estimated for each station separately and various alternatives in choosing different values on the reconstructions should be investigated. The number of receiving stations in a given region that is necessary for certain resolution constraints also needs to be investigated.

Due to the lack of a complete forward model, all the inversions related to the ionosphere are bound have certain biases and errors. Various centers of IGS produce Global Ionospheric Maps which are updated every two hours. 3-D reconstructions of the electron density matrix are not common and it is still under discussion in the literature. The method introduced in Section 3, can be applied to various locations and for both quiet and disturbed days and the reconstructions can be updated within the wide sense stationarity period of the ionosphere for that given region and time duration. Once the mean vector and the correlation matrices are determined a priori, the computational time even for the 3-D reconstruction is not a limiting factor and this method introduces a totally new approach and opens new horizons in CIT.

6 Conclusion

In this study, CIT is performed by using a Bayesian framework with Gaussian random field priors. The electron density distribution is estimated by the minimum mean squared error algorithm given the prior distribution of the random field and the measurements. In ionospheric tomography application the 3-D mean distribution and the covariance function can either be obtained from ionospheric models such as IRI or they can be estimated from the field itself by using an iterative algorithm. In this study, the random field theory is applied to the ionospheric electron density reconstruction problem for the first time and the field is estimated using the minimum mean squared error algorithm. Alternatives in the choice of the mean distribution and correlation functions are also discussed. The theory can be tried on various sections of ionosphere for both quiet and disturbed days and the 2-D slices of the 3-D reconstructions can be compared with Global Ionospheric Maps available from IGS centers. The advantages of the introduced method compared to the available 2-D reconstructions will be discussed in detail in future work. The 3-D CIT method introduced for the first time in this study is very promising for the automatic reconstruction algorithms that can be implemented on the internet and opens totally new horizons in CIT.

References

[HIK94] Hajj, G.A., Ibanez-Meier, R., Kursinski, E.R.: Imaging the Ionosphere with the Global Positioning System. International Journal of Imaging Systems and Technology, **5**, 174–184 (1994)

[HP01] Hocke, K., Pavelyev, A.G.: General aspect of GPS data use for atmospheric science. Adv. Space Res., **27(6-7)**, 1313–1320 (2001)

[DH97] Davies, K., Hartmann, G.K.: Studying the ionosphere with the Global Positioning System. Radio Sci., **32(4)**, 1695–1703 (1997)

[CB98] Calais, E., Bernard Minster, J.: GPS, earthquakes, the ionosphere, and the Space Shuttle. Phys. Earth Planet. Int., **105**, 167–181 (1998)

[IRI] http://nssdc.gsfc.nasa.gov./space/model/models/iri.html

[MMMT05] Ma, X.F., Maruyama, T., Ma, G., Takeda, T.: Three dimensional ionospheric tomography using observation data of GPS ground receivers and ionosonde by neural network. J. Geophys. Res.-Space Physics, **110-A05308**, 1–12 (2005)

[FS92] Fremouw, E.J., Secan, J.A.: Application of stochastic inverse theory to ionospheric tomography. Radio Sci., **27(5)**, 721–732 (1992)

[YR91] Yeh, K.C., Raymund, T.D.: Limitations of ionospheric imaging by ionospheric tomography. Radio Sci., **26(6)**, 1361–1380 (1991)

[SN98] Sutton, E., Na, H.: Time-varying reconstruction of the ionosphere. International Journal of Imaging Systems and Technology, **9**, 484–490 (1998)

[DK03] Dabas, R.S., Kersley, L.: Radio tomographic imaging as an aid to modelling of ionospheric electron density. Radio Sci., **38(3)-1035**, 1-1/1-20 (2003)

[YAA05] Yavuz, E., Arikan, F., Arikan, O.: A Hybrid reconstruction algorithm for computerized ionospheric tomography. In: Kurnaz, S., Ince, F., Onbasioglu, S., Basturk, S. (eds) Proceedings of 2nd International Conference on Recent Advances in Space Technologies(Istanbul,Turkey), IEEE 05EX1011, (2005)

[YAAE05] Yavuz, E., Arikan, F., Arikan, O., Erol, C.B.: Algorithms and basis functions in tomographic reconstruction of ionosphere electron density. In: Çetin, A.E., Tekalp, A.M. (eds) Proceedings of EUSIPCO'2005, 13th European Signal Processing Conference CD(Antalya, Turkey) (2005)

[AFL88] Austen, J.R., Franke, S.J., Liu, C.H.: Ionospheric imaging using computerized tomography. Radio Sci., **23(3)**, 299–307 (1988)

[Van88] Vanmarcke, E.: Random Fields. (3rd ed) The MIT Press, Cambridge, MA (1988)

[IGS] http://igs.ensg.ign.fr

[GPS] http://www.gmat.unsw.edu.au/snap/gps/gps_survey/

[CODE] ftp.unibe.ch/aiub/CODE2001

[AEA03] Arikan, F., Erol, C.B., Arikan, O.: Regularized estimation of vertical total electron content from Global Positioning System data. J. Geophys. Res.-Space Physics, **108(A12)**, 1469–1480 (2003)

Comparison of fuzzy and Volterra series nonlinear system modeling approaches

Musa H. Asyali[1] and Musa Alci[2]

[1] Dept. of Computer Engineering, Yasar University, Izmir, Turkey
musa.asyali@yasar.edu.tr
[2] Dept. of Electrical and Electronics Engineering, Ege University, Izmir, Turkey
musa.alci@ege.edu.tr

In this study, we investigated modeling performances of two popular nonlinear system identification methods, namely fuzzy modeling and Volterra series. In literature a general approach to nonlinear structure modeling does not exist, therefore both fuzzy models and Volterra series are interesting and widely used as they can approximate a large class of nonlinear functions. In fuzzy modeling, a dynamic system is modeled using a set of fuzzy membership functions and rules. The fuzzy model parameters are trained using optimization techniques. In Volterra series approach, the dynamic system is modeled using a set of kernel functions that represent the first and higher order convolutions. The kernel functions are typically estimated using an orthogonal expansion technique using a set of suitable basis functions such as Laguerre. We compared the modeling performance of these approaches on a hypothetical test system whose kernels or structure is known priori and observed that the Volterra modeling based on Laguerre basis expansion of kernels offers better performance.

1 Introduction

The use of the fundamental models for nonlinear system is not practical in most cases, as the models of realistic complexity may often involve hundreds of nonlinear differential equations. Therefore, empirical models such as the Volterra Series approach that are obtained from input-output data are commonly used in nonlinear system identification [WK03, AJ05]. Another empirical technique gaining popularity in the recent years is the fuzzy modeling approach [SZL95]. Both approaches are very important because they can approximate a large class of nonlinear systems. In this study, we aim at comparing modeling performances or efficiencies of these two modeling approaches on a test system. Such a comparison does not exist in literature. Therefore,

Kenan Taş et al. (eds), Mathematical Methods in Engineering, 335–345.

motivation behind is study is to address the question of which approach performs better. This will help modeling scientists select the right, i.e. better performing, approach.

We define modeling efficiency as the ability of a model to reduce the estimation error using as few parameters as possible. In this context, we used the minimum description length criterion to compare efficiencies of the two approaches.

In Section 2, we will briefly review Volterra series and the fuzzy modeling approaches. We will describe the test data that we used in our comparison studies and present the results in Section 3. Finally, in Section 4, we will discuss our results and make some concluding remarks.

2 Methods

2.1 Volterra series

Volterra has shown that a nonlinear time invariant system's output or response $y(t)$ to an input or stimulus $x(t)$ can be expressed by the following multi-convolution relation [WK03, AJ05, Mar93].

$$y(t) = C_{dc} + \sum_{i=1}^{\infty} \int_{0}^{\infty} \cdots \int_{0}^{\infty} x(t-\tau_1)\ldots x(t-\tau_i) H_i(\tau_1,\ldots,\tau_i) d\tau_1 \ldots d\tau_i \quad (1)$$

The sum given by (1) is known as the Volterra series. Here, $H_i(\tau_1,\ldots,\tau_i)$ denotes the system's ith order Volterra kernel which is associated with the system's ith order nonlinearity. The C_{dc} is a constant term that balances the means of the two sides. For a linear system C_{dc} is 0 and the remainder of the right-hand side of (1) reduces to the well-known convolution integral in which case $H_1(\tau_1)$ is called the impulse response.

As we will be using numerical techniques for kernel estimation, we need to translate equation (1) into discrete-time. We do this by assuming a default sampling period of 1 s and input-output data length of N points. If we denote input and output sequences by $x[n]$ and $y[n]$ $(n = 0, 1, \ldots, N-1)$ after discretization of (1) we obtain

$$y[n] = C_{dc} + \sum_{m=0}^{N_1-1} x[n-m]H_1[m] + \sum_{m_1=0}^{N_2-1}\sum_{m_2=0}^{N_2-1} H_2[m_1, m_2] \quad (2)$$

where, H_1 and H_2 denote the 1st and 2nd order kernels and N_1 and N_2 their lengths. Notice that, during the conversion to discrete-time, we included kernels up to 2nd order and assumed that the contribution of the higher order kernels are embedded in the error sequence $e[n]$.

Given a system with memory M, our sample input-output data must have a length greater than M, i.e. $N > M$, in order to capture the system dynamics. We need some prior knowledge in order to make a reasonable guess for the memory of the system. As for the selection of kernel lengths N_1 and N_2, these cannot be greater than the data length N, i.e. $N > N_1$ and $N > N_2$. Further, both N_1 and N_2 must be greater than M. From (2) we note that, H_1 and H_2 respectively has N_1 and $N_2(N_2 - 1)$ distinct values or parameters to be estimated, as such the number of values in the kernels increases exponentially with the order. Therefore, to keep the number of parameters at a reasonable level, one can select kernel lengths N_1 and N_2 such that $N > N_1 \geq N_2 \geq M$.

We note that even a 2nd order Volterra model is highly parameterized and this may cause the parameters to have high noise-sensitivity. One way to alleviate this difficulty associated with the estimation of Volterra kernels is to project the kernels onto a small number of orthogonal basis functions. The key issue here is to utilize basis functions that are morphologically similar to the kernels of the system under study. This enables accurate representation of kernels with a relatively small number of basis functions, which implies a reduction in the number of parameters to be estimated. Estimating a smaller number of parameters may improve the numerical condition of the estimation problem and produce coefficient estimates with less variance and hence a more reliable model. The series expansion utilizing discrete orthogonal Laguerre Basis Functions (LBF) has been used widely [WK03]. Discrete LBF can be defined/given more conveniently by their z-transform:

$$L_q[m] \overrightarrow{z - transform} L_q[z] = \sqrt{1 - \xi^2} \frac{z}{z - \xi} \left(\frac{1 - \xi z}{z - \xi} \right)^q \tag{3}$$

Here, ξ is the pole parameter ($0 < \xi < 1$) that determines how soon the LBF will die away and q is the order of the basis functions. As increases, the functions become more oscillatory and prolonged. Therefore, ξ must be chosen in accordance with memory of the system. Given the guessed value for the memory and the highest order of LBF to be used to represent the kernels, using (3) the most suitable value for ξ can be calculated easily. This is the approach employed typically in literature. However, knowing the memory of the system in advance accurately is rarely possible. Therefore we employ a different technique based on simplex optimization to select an optimal value for ξ in our kernel estimation studies. This issue is further explained in Section 3.2. For a more detailed treatment of the LBF, we refer the reader to [WK03, AJ05, Mar93].

We will now show that Volterra kernel estimation problem can be formulated as a multiple regression problem using LBF expansion and therefore solved using least squares estimation. We return to (2) where we employ two kernels to explain the nonlinear dynamics of a system. Even if the system under study may have higher order nonlinearities (kernels), we can still find out how well we can approximate the system's behavior with a 2nd order

Volterra model. This is why this approach is sometimes referred to as *truncated Volterra Series*. We first expand H_1 and H_2 using LBF L_q, q being the order, as

$$H_1[m] \cong \sum_{q=0}^{Q_1-1} C_q L_q[m], \quad m = 0, 1, \ldots, N_1 - 1 \tag{4}$$

$$H_2[m_1, m_2] \cong \sum_{q_1=0}^{Q_2-1} \sum_{q_2=0}^{Q_2-1} C_{q_1,q_2} L_{q_1}[m_1] L_{q_2}[m_2] \tag{5}$$

Here, C_q and C_{q_1,q_2} are coefficients or weights, and Q_1 and Q_2 are the number of basis functions that are used in the expansion of H_1 and H_2 respectively. Selection of proper values for Q_1 and Q_2 is of crucial importance. We will discuss this issue in Section 3.2. If we substitute (4) and (5) in (2) and define convolution of x with L_q as

$$v_q[n] = \sum_{m=0}^{N-1} x[n-m] L_q[m],$$

we can express $y[n]$ as

$$y[n] = \sum_{q=0}^{Q_1-1} C_q v_q[n] + \sum_{q_1=0}^{Q_2-1} \sum_{q_2=0}^{Q_2-1} C_{q_1,q_2} v_{q_1}[n] v_{q_2}[n] + e[n] \tag{6}$$

We should note here that the error terms in (6) and (2) are slightly different. The error term in (6) includes not only missing and/or ignored contribution of higher order kernels to the output but also the error introduced due the approximate kernel expansions (4) and (5) which are substituted in (2). By further defining column vectors corresponding to the output, convolution, and error sequences respectively as $\mathbf{y} = [y[0]...y[N-1]]^{\mathrm{T}}$, $\mathbf{v}_q = [\mathbf{v}_q[0]...\mathbf{v}_q[N-1]]^{\mathrm{T}}$, and $\mathbf{e} = [e[0]...e[N-1]]^{\mathrm{T}}$, we can put (6) into matrix form as

$$\mathbf{y} = \mathbf{VC} + \mathbf{e}, \tag{7}$$

where, $\mathbf{V} = [[1 \ldots 1]^{\mathrm{T}} | \mathbf{v}_0 \ \mathbf{v}_1 \ldots \mathbf{v}_{Q_2-1} | \mathbf{v}_{0,0} \ \mathbf{v}_{0,1} \ldots \mathbf{v}_{Q_2-1,Q_2-1}]$ is the $N \times P$ observation matrix formed by using $\mathbf{v}_q$'s and their element-wise multiplicative combinations

$$\mathbf{v}_{q_1,q_2} = \mathbf{v}_{q_1} \mathbf{v}_{q_2} \text{ and } \mathbf{C} = [C_{dc} | C_0 \ C_1 \ldots C_{Q_2-1} | C_{0,0} \ 2C_{0,1} \ldots C_{Q_2-1,Q_2-1}]^T$$

is the $P \times 1$ vector of coefficients. The column of 1's in $\mathbf{V}$ allows for the estimation of the constant term C_{dc}. Since $\mathbf{v}_{q_1,q_2} = \mathbf{v}_{q_2,q_1}$, we collected similar terms in the expansion of H_2 in (5) and doubled the corresponding coefficient in vector $\mathbf{C}$, hence $P = 1 + Q_1 + Q_2(Q_2 + 1)/2$.

The over-determined system of equations given in (7) can be solved conveniently for $\mathbf{C}$ using the least squares technique. Inclusion of the constant term in the regression assures that the error sequence will have zero mean.

Once the coefficients are estimated, they are substituted in the expansions (4) and (5) and the kernels are constructed. Extension of this technique to the estimation of higher order kernels is straightforward.

2.2 Fuzzy modeling

Fuzzy models are based on the concept of fuzzy logic, a notion which extends human decision making practices or heuristics into a formal system modeling and/or identification platform. Using fuzzy models one can formulate mapping from a given input to an output using the following elements: fuzzifier, inference engine, defuzzifier and rule base [Le90, LLW97]. Rule base consists of linguistic statements such as:

$$\text{If } x_1 = A_1^l \text{ and } x_2 = A_2^l \text{ and ... and } x_n = A_n^l \text{ , then } y = B^l \tag{8}$$

where $A_1^l, ..., A_n^l$, are the fuzzy sets represented by the input membership functions, B^l are the fuzzy sets represented by the output membership functions and $l = 1, 2, ..., M$ is the rule index.

Fuzzy models have been successfully applied in fields such as automatic control, expert systems, computer vision, and data clustering/classification. There are two types of fuzzy inference systems that are commonly used in the practice, Mamdani-type and Sugeno-type. Mamdani's model uses fuzzy sets in both antecedent and consequent parts of rules. Sugeno has shown that, it is also possible to use crisp functions as the output membership function rather than a distributed fuzzy set. This approach (also called Takagi, Sugeno and Kang model) enhances the efficiency of the defuzzification process as it requires less computation than the Mamdani method. A comprehensive survey of many other ways proposed to implement fuzzy rules and models can be found in [Le90]. Using singleton fuzzifier, product inference engine, center average defuzzifier, and Gaussian membership functions, the fuzzy model with respect to the given rule base is modeled in [Wan94, Wan97] as:

$$f(x, \bar{x}_i^l, \sigma_i^l, \bar{y}^l) = \frac{\sum_{l=1}^{M} \bar{y}^l \left[\prod_{i=1}^{n} \mu_{A_i^l}(x_i, \bar{x}_i^l, \sigma_i^l) \right]}{\sum_{l=1}^{M} \left[\prod_{i=1}^{n} \mu_{A_i^l}(x_i, \bar{x}_i^l, \sigma_i^l) \right]} \tag{9}$$

where M is the number of rules, n is the number of inputs,$\bar{y}^l$ parameters represent the center of output membership functions B^l , $\bar{x}_i^l$ parameters represent the center of input membership functions A^l , and σ_i^l parameters represent the input membership function widths. Considering the fact that fuzzy models are parametric, we can use optimization tools to calculate or train the system parameters. During the optimization, the following performance criterion is minimized:

$$E = \frac{1}{2} \sum_{n=0}^{N-1} (y[n] - \tilde{y}[n])^2 \tag{10}$$

Here, N is the length of input/output pairs and y and $\tilde{y}$ respectively denote the actual and estimated output values. In order to determine the parameters of the fuzzy model, fuzzy system is represented as a feed forward network. In our estimations, Levenberg-Marquardt algorithm with Fletcher strategy is used for tuning the parameters [KA05].

3 Experimental data and results

3.1 Experimental data

We have carried the modeling performance comparison study on synthetic data obtained using the following linear-nonlinear Wiener cascade (Fig. 1).

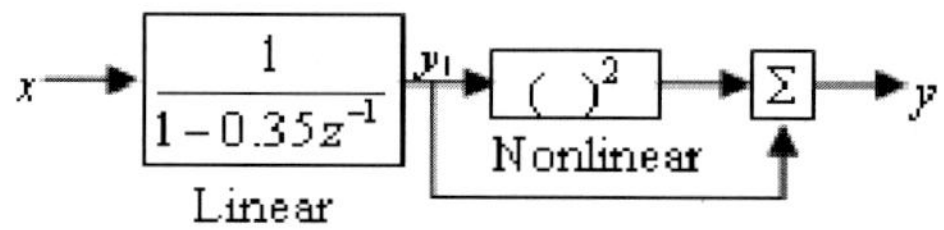

Fig. 1. Generation of synthetic test data.

This cascade corresponds to a 2nd order nonlinear system. Expressed algebraically, the relationship between input x and output y is: $y[n] = y_1[n] + y_1[n]^2$, where $y_1[n] = x[n-5] - 0.35y[n-1]$. According to this formulation, Volterra kernels of our test system are as follows

$$H_1[n] = \begin{cases} 0 & , n < 5 \\ (0.35)^{n-5}, & n \geq 5 \end{cases} \text{ and } H_2[n_1, n_2] = H_1[n_1]H_1[n_2].$$

The original 1st and 2nd order kernels of this system are shown in Fig.2. By driving this system with a 300 points long random sequence with zero mean and uniform density, i.e. $x \sim \text{uniform}(-0.5, 0.5)$, we have generated our test data (Fig. 3).

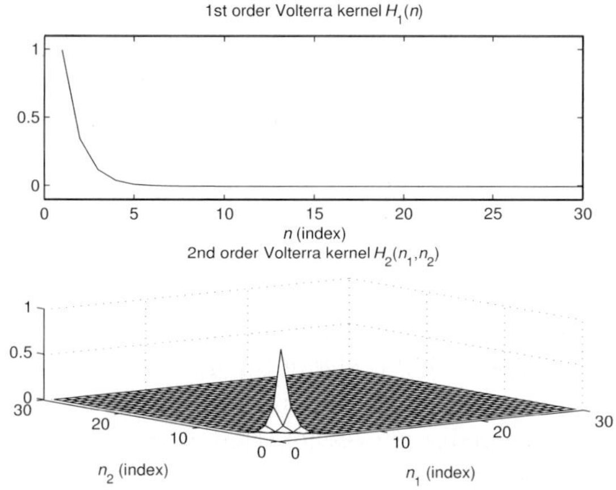

Fig. 2. 1st (upper panel) and 2nd (lower panel) order Volterra kernels of the test system shown in Fig.1.

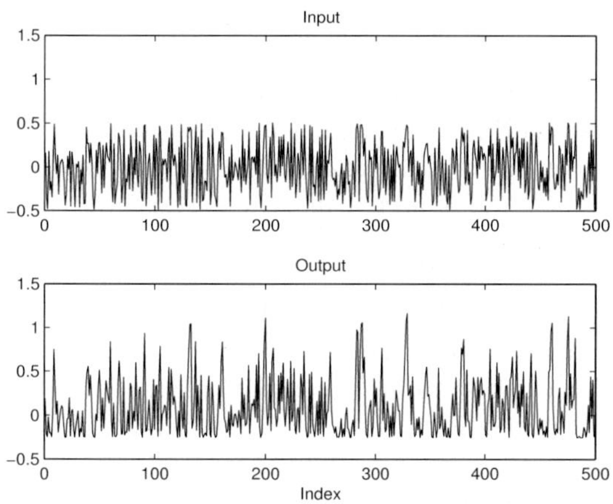

Fig. 3. Input-output data set used in the modeling performance comparison study.

3.2 LBE based Volterra kernel estimation on the test data

An important issue in Volterra kernel estimation is the selection and number of basis functions to be used in the expansions, namely Q_1 and Q_2 in (4) and (5). If too many basis functions are used or, equivalently, the number of model parameters P is large, we may be trying to explain even the noise in the data. This will diminish our model's ability to generalize and the numerical condition of the estimation problem will worsen, i.e., we may have coefficient estimates with large variances. Conversely, if P is too small, we may have a poor model with a large estimation error. The issue of determining an optimal value for the number of model parameters has been dealt with extensively in system identification literature and various criteria based on information theoretical results have been proposed. One of the most commonly used of such criteria is the Minimum Description Length (MDL) [BRY98, Ris84]:

$$N \log(\mathrm{SSE}) + P \log(N)$$

where N is the length of the data sequence, SSE is the sum of squared error, and $P = 1 + Q_1 + Q_2(Q_2 + 1)/2$ is the number of model parameters. We note that MDL involves two terms acting in opposite sense, the SSE that monotonically decreases and a penalty expression that increases with P. We followed the following simple strategy to determine optimum model parameters. We let both Q_1 and Q_2 vary from 1 to 20 with increments of 1 (total of 400 trials) and at each stop we carried out the least squares estimation given by (7). For any given value Q_1 and Q_2, we also to need select an optimal value for the pole parameter ξ of the basis functions. It is difficult to obtain the derivative of the error term in (7) with respect to ξ analytically. We therefore choose to utilize Nelder-Mead simplex (direct search) method [12] to find the optimal ξ that minimizes the SSE. Using a starting or initial value of 0.5, we established the optimal ξ value with a resolution of 10^{-4}. We then computed the MDL corresponding to this optimal model at each Q_1 and Q_2 stop. Finally we chose Q_1 and Q_2 pair that produces the smallest MDL and selected the corresponding model as the overall optimal 2nd order Volterra model. We report the parameters for the optimal Volterra modeling results for our test dataset in Table 1. The MSE reported in the table is the mean squared error, which is SSE/N.

We developed a set of routines in MATLAB™(The MathWorks, Inc. Natick, MA) to implement Volterra kernel estimation. Execution of the code on the test data took about 1 minute on a PC with 3 GHz Intel Pentium IV processor and 1 GB of memory, running under Microsoft Windows XP™.

3.3 Fuzzy modeling on the test data

We also implemented Sugeno-type fuzzy modeling approach, whose details are given in Section 2.2, in MATLAB. Table 2 presents fuzzy modeling results on the two data sets using two inputs ($x[n]$ and $y[n-1]$) and 4 rules, after 200 iterations. Execution time of the fuzzy modeling code on the test data test also took about 1 minute.

4 Discussion and conclusion

In this study, we compared modeling performances two popular nonlinear system identification and/or modeling schemes, namely Volterra series expan-

Table 1. LBE based Volterra Kernel Estimation Results. ($||H_1 - \tilde{H}_1||$ and $||H_2 - \tilde{H}_2||$ are Euclidean and Frobenius norm of the difference between the actual and estimated 1st and 2nd order kernels respectively.)

Optimal Q_1	5
Optimal Q_2	5
Optimal ξ	0.3505
Total number of model parameters	21
MSE	2.7724×10^{-33}
MDL	-3.4245×10^{4}
$\|\|H_1 - \tilde{H}_1\|\|$	2.3018×10^{-16}
$\|\|H_2 - \tilde{H}_2\|\|$	2.5349×10^{-16}

Table 2. Fuzzy Modeling Results. (MSF: Membership Function.)

Rule 1	
Input MSF centers	[−4.0096 − 0.8264]
Input MSF widths	[1.5304 0.7787]
Output MSF center	−0.2460
Rule 2	
Input MSF centers	[−0.0094 0.4661]
Input MSF widths	[0.4169 0.2663]
Output MSF center	−0.3050
Rule 3	
Input MSF centers	[0.3261 0.4984]
Input MSF widths	[0.3998 0.2630]
Output MSF center	1.0031
Rule 4	
Input MSF centers	[0.7675 0.5472]
Input MSF widths	[0.3274 0.2592]
Output MSF center	1.6532
Total number of model parameters	20
MSE	0.0251
MDL	-1.7147×10^{3}

sion and fuzzy modeling on a test system. There exist many different ways in both lines of modeling approach. That is, we can estimate the kernels in the Volterra series expansion using many different strategies and similarly we can devise many different fuzzy models. Due their popularity in their respective line of research in literature and our particular experience with these methods, we picked Laguerre basis function expansion based method for Volterra modeling and Sugeno-type approach (with singleton fuzzifier, product inference engine, center average defuzzifier, and Gaussian membership functions) for fuzzy modeling.

From the results presented in Tables 1 and 2, we observe that the Volterra series modeling approach based on Laguerre basis expansion (LBE) of kernels shows superior performance over the fuzzy modeling approach. While making this performance comparison we cannot rely on the mean squared error (MSE) figures only, as the number of model parameters for each approach may be different in general. Although we spent some effort to keep the number of model parameters about the same for both approaches, there is still a difference (21 for LBE versus 20 for fuzzy approach). We therefore resort to the use of minimum description length (MDL) criterion which takes into account both the error variance and the number of model parameters for modeling performance comparison. The MDL of the LBE approach is almost 20 times lower than the MDL of the fuzzy approach. This means that the LBE method is doing a much better job in terms of producing a lower estimation error with the same or fewer number of model parameters.

Furthermore, with the LBE approach, we can obtain the kernels of the system directly. For instance, the figures in the last two rows of Table 1 indicate that we have recovered the kernels of our test system with excellent accuracy, using the LBE approach. Whereas with the fuzzy modeling approach, we can only obtain information about the morphology (i.e., location and spread) of the membership functions utilized in the model. Although conversion of this information into kernels through Taylor series expansion could be possible [Sch80, CHMM98, Ste05], this is a difficult process in general. The advantage of the fuzzy model on the other hand is that the operation of the system or model can be understood more easily, as it is expressed in terms of some rules resembling human decision making process. However, in our case, this advantage of the fuzzy modeling approach is greatly overshadowed by the superior performance of the LBE approach.

As for the computational complexity, the implementation of both approaches took about 1 min on our computational platform whose details are given in Section 3.2, we therefore conclude that the computational complexities of both approaches are comparable.

The results we have obtained on our test system indicate that the LBE based Volterra series nonlinear dynamic system modeling is superior over the fuzzy modeling approach. However, a more general or stronger statement requires further studies involving many different nonlinear dynamic systems with different characteristics. In a future study, we will compare the performances of two approaches on a real modeling problem.

References

[WK03] Westwick, D.T., Kearney, R.E.: Identification of Nonlinear Physiological Systems. IEEE Biomedical Engineering Book Series, IEEE Press/Wiley, (2003)

[AJ05] Asyali, M.H., Juusola, M.: Use of the Meixner functions in estimation of Volterra kernels of nonlinear systems with delay. IEEE Transactions on Biomedical Engineering, **52(2)**, 229–237 (2005)

[SZL95] Sjoberg, J., Zhang, Q., Ljung, L., et al.: Nonlinear black-box modeling in system identification: a unified overview. Automatica, **31(12)**, 1691–1724 (1995)

[Mar93] Marmarelis, V.Z.: Identification of nonlinear biological system using Laguerre expansions of kernels. Ann. Biomed. Eng., **21(6)**, 573–589 (1993)

[Le90] Lee, C.C.: Fuzzy logic in control systems: fuzzy logic control parts I and II. IEEE Transactions on System, Man, and Cybernetics, **20(2)**, 404–435 (1990)

[LLW97] Leu, Y.G., Lee, T.T., Wang, W.Y.: Online tuning of fuzzy-neural network for adaptive control of nonlinear dynamical systems. IEEE Trans. on System, Man and Cybernetics, **27(6)**, 1034–1043 (1997)

[Wan94] Wang, L.X.: Adaptive Fuzzy Systems and Control: Design and Stability Analysis. PTR Prentice Hall Inc., Englewood Cliffs (NJ) (1994)

[Wan97] Wang, L.X.: A Course in Fuzzy Systems and Control. Prentice Hall Inc., Upper Saddle River (NJ) (1997)

[KA05] Karatepe, E., Alci, M.: A new approach to fuzzy-wavelet system modeling. International Journal of Approximate Reasoning, **40(3)**, 302–322 (2005)

[BRY98] Barron, A., Rissanen, J., Yu, B.: The minimum description length principle in coding and modeling. IEEE Trans. Information Theory, **44(6)**, 2743–2760 (1998)

[Ris84] Rissanen, J.: Universal coding, information, prediction, and estimation. IEEE Trans. Information Theory, **30**, 629–636 (1984)

[LRWW98] Lagarias, J.C., Reeds, J.A., Wright, M. H., Wright, P.E.: Convergence properties of the Nelder-Meads Simplex method in low dimensions. SIAM Journal of Optimization, **9**, 112–147 (1998)

[Sch80] Schetzen, M.: The Volterra and Wiener Theory of the Nonlinear Systems. Wiley and Sons, New York (1980)

[CHMM98] Chon, K.H., Holstein-Rathlou, N.-H., Marsh, D.J., Marmarelis, V.Z.: Comparative Nonlinear Modeling of Renal Autoregulation in Rats: Volterra Approach Versus Artificial Neural Networks. IEEE Transactions On Neural Networks, **9(3)**, 430–435 (1998)

[Ste05] Stegmayer, G.: Neural-based Identification for Nonlinear Dynamic Systems, CIMSA 2005- Proc. of IEEE International Conference on Computational Intelligence for Measurement Systems and Applications(Giardini Naxos, Italy), ISBN 0-7803-9026-1, (2005)

Novel hardware-oriented algorithms for TDOA positioning technique in cellular networks

E. Doukhnitch, M. Salamah and A. Sandouka

Department of Computer Engineering, Eastern Mediterranean University, Famagusta, N. Cyprus {evgueni.doukhnitch, muhammed.salamah, amjad.sandouka}@emu.edu.tr

Finding the location of an object is one of the important features of robotics applications, electronic warfare positioning, and the 3G/4G wireless communication systems. Many valuable location based services can be enabled by this new feature. Position estimation from Time Difference of Arrival (TDOA) measurements is one of the commonly used methods. This approach is based on intersections of hyperbolic curves defined by the time differences of arrival of signals received from different sources. The location is determined using standard complex computation methods that are usually implemented in software. In this paper, we propose new hardware-oriented algorithms that use only simple add and shift operations in the computation and therefore can be easily implemented in hardware.

1 Introduction

The location estimation of a mobile station (MS) in wireless cellular networks has gained considerable attention especially after the clear mandating requirement, stated by the US Federal Communications Commission (FCC), demanding service providers to generate PL (Position Location) estimates for Enhanced-911 (E-911) emergency calls.

Using measurement of signals, different methods can be applied to any cellular system to derive (estimate) the PL of the MS. These include angle of arrival (AOA), time of arrival (TOA), time difference of arrival (TDOA), and more [Lait01, Zha02, JAO99, CH94].

TDOA method is one of the most promising methods used. In a two dimensional (2-D) model where three base stations exist, two independent TDOA measurements are made with respect to the base station that receives the MS's signal first. Once the two measurements are obtained, they are converted into range (distance) difference measurements. Such conversion is simply done by multiplying every TDOA measurement by the velocity of the signal (speed of light). Each range difference defines a hyperbola, and thus with three base

Kenan Taş et al. (eds), Mathematical Methods in Engineering, 347–357.

stations two hyperbolas exit. The position of the base stations defines the foci of such hyperbolas. The intersection of the two hyperbolas defines the location of the mobile phone.

Algorithms solving the location of the mobile stations using TDOA method are usually software-oriented. In this paper, we propose new algorithms which can be implemented in hardware using, for example, a simple field programmable gate array (FPGA) chip. Only simple add, subtract, and shift operations are needed in the proposed algorithms.

2 The model used for the traditional algorithms

A general model for 2-D PL estimation has been adopted. Let (x,y) be the mobile's location, which will be estimated, and (X_i,Y_i) be the known location of BS_i, where i = 1,2,3. Referring to BS_1, two TDOA are obtained. The distance between the MS and the ith base station is given as:

$$R_i = \sqrt{(x - X_i)^2 + (y - Y_i)^2}. \tag{1}$$

The range difference between base stations with respect to BS_1 is:

$$\begin{aligned} R_{i,1} &= cd_{i,1} = R_i - R_1 \\ &= \sqrt{(x - X_i)^2 + (y - Y_i)^2} - \sqrt{(x - X_1)^2 + (y - Y_1)^2}, \end{aligned} \tag{2}$$

where c is the signal propagation speed, and $d_{i,1}$ is the estimated TDOA between the ith base station and the first base station.
Equation (2) defines the two nonlinear hyperbolic equations, for $i = 2$ and 3, whose solution gives the 2-D coordinates of the source. Rearranging the form of (2) into:

$$R_i^2 = (R_{i,1} + R_1)^2. \tag{3}$$

Equation (1) can now be written as:

$$R_{i,1}^2 + 2R_{i,1}R_1 + R_1^2 = X_i^2 + Y_i^2 - 2X_i x - 2Y_i y + x^2 + y^2. \tag{4}$$

Subtracting (1) at $i = 1$ from (4) yields to the general form of the hyperbolic equations, which can then be written as:

$$R_{i,1}^2 + 2R_{i,1}R_1 = X_i^2 + Y_i^2 - 2X_{i,1}x - 2Y_{i,1}y - X_1^2 - Y_1^2. \tag{5}$$

where $X_{i,1}$ and $Y_{i,1}$ are equal to X_i-X_1 and Y_i-Y_1 respectively. The set of equations in (5) are now linear with R_1, x, and y as the unknown variables.

Exploiting equation (5), Chan [CH94] developed his own methods to solve this set of equations, and thus determining the location of the mobile phone.

3 The proposed hardware algorithms

The algorithms which we proposed are also based on TDOA measurements and use the same source of information as the traditional ones. However, the way of finding the mobile's location is performed in a completely different way as illustrated in the flowchart of Figure 1.

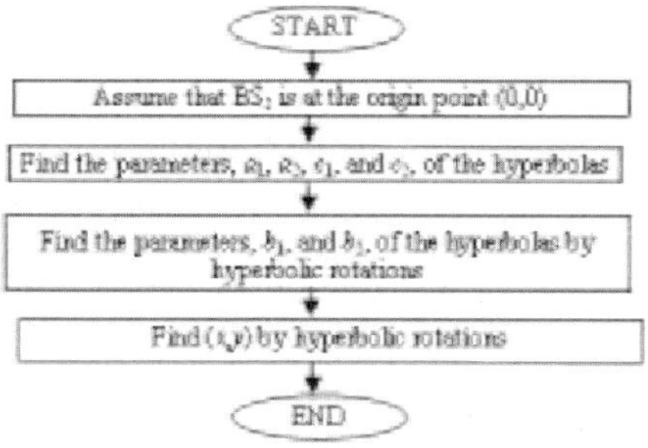

Fig. 1. The General Scheme for DSS algorithms

3.1 The DSS-1 algorithm

For linearly placed base stations, i.e. in highways, and establishing a local coordinate system, let the coordinates of BS_1, BS_2 and BS_3, be (X_1,0), (X_2,0) and (X_3,0) respectively, as shown in Figure 2. BS_1 is assumed to be the reference base station.

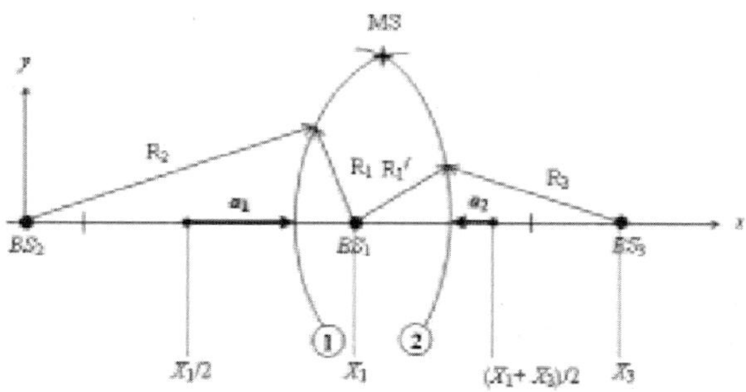

Fig. 2. TDOA Position Determination for Linearly Placed Base Stations

Although the figure shows vectors, for simplicity, scalars or magnitude representations are used. The parameters defining the two hyperbolas above can be written as:

$$a_1 = (R_2 - R_1)/2 = cd_{2,1}/2, \tag{6}$$

$$a_2 = (R_3 - R_1^{/})/2 = cd_{3,1}/2, \tag{7}$$

$$c_1 = X_1/2, \tag{8}$$

$$c_2 = (X_3 - X_1)/2, \tag{9}$$

$$b_1 = \sqrt{c_1^2 - a_1^2}, \tag{10}$$

$$b_2 = \sqrt{c_2^2 - a_2^2}. \tag{11}$$

For determining the location of the MS, we proposed a hardware oriented algorithm which we named as DSS. It works in two modes: static mode (DSS-1) and the dynamic mode (DDSS-1).In its static mode the idea is simple. Hyperbolic parallel rotation of vectors with a fixed step rotation angle $\sigma = \operatorname{arcsinh}(2^{-k})$ is used, and is done recursively, step by step. The value of k depends on the needed accuracy. From [Mull97] the hyperbolic rotation matrix M can be written as:

$$\mathrm{M} = \begin{bmatrix} \cosh\sigma \; \sinh\sigma \\ \sinh\sigma \; \cosh\sigma \end{bmatrix}. \tag{12}$$

Taking the *sinh* function as:

$$\sinh\sigma = 2^{-k}, \tag{13}$$

and knowing that:

$$\cosh^2\sigma - \sinh^2\sigma = 1, \tag{14}$$

cosh function can be written as:

$$\cosh\sigma = \sqrt{1 + \sinh^2\sigma} = \sqrt{1 + 2^{-2k}}. \tag{15}$$

It's clear that:

$$\sqrt{1 + 2^{-2k}} \le 1 + 2^{-2k-1}. \tag{16}$$

Hence,

$$\begin{aligned} 1 + 2^{-2k} &\le 1 + 2 \cdot 2^{-2k-1} + 2^{-4k-2} \\ &= 1 + 2^{-2k} + 2^{-4k-2}. \end{aligned} \tag{17}$$

For $k \ge 4$, one gets $2^{-4k-2} \le 2^{-18} < 10^{-5}$. So, for an accuracy of $\varepsilon = 10^{-5}$, cosh σ can be approximated as:

$$\cosh\sigma = 1 + 2^{-2k-1} = 1 + 2^{-(2k+1)} \tag{18}$$

Finally, recall that the parametric equations of a hyperbola are given as:

$$x = a\ cosh\ \sigma, \tag{19}$$

$$y = b\ sinh\ \sigma, \tag{20}$$

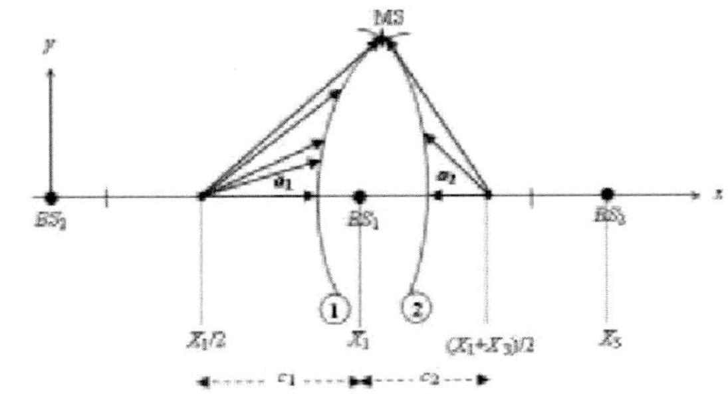

Fig. 3. Idea of Positioning for Linearly Placed Base Stations

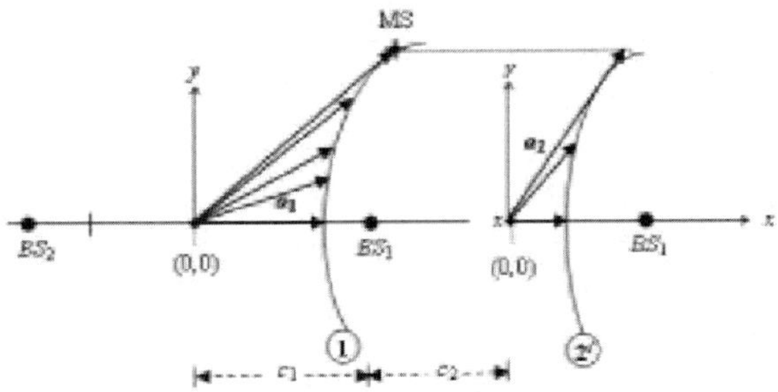

Fig. 4. Parallel Rotations of Vectors

with semi-major axis a parallel to the x-axis and semi-minor axis b parallel to the y-axis.

Figure 3 illustrated the idea of positioning the MS. Vectors a_1 and a_2 are hyperbolically rotated until their heads intersect with each other. With the coordinates being transferred to the origin points, an equivalent rotation is illustrated in Figure 4. Two stages of rotations are needed, for each of the vectors, to find the (x,y) coordinates of their heads after each step of rotation.

Before starting parallel rotations of the two vectors, a_1 and a_2, a check is made to determine the one with the smaller b value, since it needs more rotations. This is because of equations (10), (11), and the hyperbolic rotations given below where the y coordinates of the vectors' heads after rotation depend on b_1 and b_2. It will be rotated again until its y coordinate equals to the y coordinate of the second vector. Then, another parallel rotation can be used, and so on. Hence, if b_2 is assumed to be the larger, then the conditions of rotation will be as follows:

$While \quad (((c_1 + c_2) - (x_{1,i} + x_{2,i})) > \varepsilon)$
  $Rotate \quad a_2$
  $While \quad ((y'_{2,i} - y'_{1,i}) > \varepsilon)$
    $Rotate \quad a_1$
  $End \quad While$
$End \quad While$

Therefore, a_1 is recursively rotated and the new coordinates of its head after each rotation, using matrix (12), are as follows:

$$x_{1,i+1} = x_{1,i} + x_{1,i}u + y_{1,i}v, \tag{21}$$

$$y_{1,i+1} = y_{1,i} + y_{1,i}u + x_{1,i}v, \tag{22}$$

where $v = 2^{-k}$, $u = v^2/2 = 2^{-(2k+1)}$, and the initial values are $x_{1,0} = a_1$ and $y_{1,0} = 0$.

The resultant $x_{1,i+1} = a_1 \cosh\sigma$ can be used as the real x coordinate of the rotated vector a_1. However, the resultant $y_{1,i+1} = a_1 \sinh\sigma$ is not the real y coordinate of the rotated vector because of (20), where b_1 should be used to find the real y coordinate, and thus a second stage of rotation is needed and it's as follows:

$$x'_{1,i+1} = x'_{1,i} + x'_{1,i}u + y'_{1,i}v, \tag{23}$$

$$y'_{1,i+1} = y'_{1,i} + y'_{1,i}u + x'_{1,i}v. \tag{24}$$

The initial values are $x'_{1,0} = b_1$, and $y'_{1,0} = 0$. Now, the resultant $y'_{1,i+1} = b_1 \sinh\sigma$ can be used as the real y coordinate of the rotated vector a_1.

To calculate b_1 (10) again hyperbolic rotation is used. According to the CORDIC algorithm [Mull97] the above equations can be used in the rotation with a slight modification:

$$x_{1,i+1} = x_{1,i} + x_{1,i}u - y_{1,i}v, \tag{25}$$

$$y_{1,i+1} = y_{1,i} + y_{1,i}u - x_{1,i}v. \tag{26}$$

The initial values are $x_0 = c_1$, and $y_0 = a_1$. The stopping criterion is when y_i becomes nearly close to zero. The last iterated x_i is b_1. Note that more rotations are needed to approximate the value of b with larger a value.

So, it's clearly seen from the above equations that no trigonometric or complex calculations are used. Instead, simple add, subtract and shift operations are used which are the necessary requirements for simple hardware implementation.

To use the above 6 rotation equations for the second vector a_2, the mirror of that vector is taken around the y axis, and replacing x_1, y_1, x'_1, y'_1, a_1, c_1 and b_1 with x_2, y_2, x'_2, y'_2, a_2, c_2 and b_2 respectively. Therefore, the new coordinates of the mirrored a_2 vector are $(x_{2,i+1}, y'_{2,i+1})$.

The intersection point $(x_c,y_c) = (x_{1,i},y^{/}_{1,i})$, which is obtained from the last iterations of rotations. The calculated intersection point is obtained after coordinate transformation (Figure 4 to Figure 3). The real intersection point (x,y) can be written as $(x_{1,i} + X_1/2,\ y^{/}_{1,i})$. The intersection point gives the position of the mobile station.

3.2 The DDSS-1 algorithm

For the purpose of decreasing the number of rotations, we modified the DSS-1. Unlike the DSS-1 where the step angle is fixed, the step angle may vary from one rotation to an-other in DDSS-1 depending on distance criteria.

To approximate the unknown parameter of the first hyperbola b_1, where equations (25) and (26) define the rotations needed, the largest possible step angle is taken in every rotation. As the y coordinate of the vector decreases, i.e. approaches to zero, rotation with smaller step angle is performed. Approximation of v in each rotation is chosen as follows:

$$v = \sinh \sigma_i = 2^{-k}, \quad \text{if} \quad y_{1,i} \geq 2^{-k}. \tag{27}$$

The same method is used to approximate b_2.

Moreover, dynamic rotations of the two vectors a_1 and a_2 depend on the distance between their current heads' positions. If the distance is large, rotation is done with a bigger step angle, i.e. k is small. As the distance gets smaller, the two vectors are rotated with smaller step angles. To guarantee convergence and to minimize the error, the approximation of v in each iteration i is taken as follows [SDD04]:

$$v = \sinh \sigma_i = \begin{cases} 2^{-k-1}, & \text{if} \quad \Delta x_i \geq 2^{-k}, \\ 2^{-m-1}, & \text{if} \quad 2^{-m} \leq \Delta x_i < 2^{-m+1}, \quad \text{and} \quad k < m \leq n, \end{cases} \tag{28}$$

where $\Delta x_i = (c_1 + c_2) - (x_{1,i} + x_{2,i})$, $k = 4$ and $n = 11$.
Although v is only shown in (27) and (28), u should be changed accordingly.

After describing the particular case of linearly placed BSs, we present the hardware algorithm, in its static and dynamic modes, that can be used for arbitrarily placed base stations, which is the typical scenario in the infrastructure of a cellular system.

3.3 The DSS-2 algorithm

The idea of DSS-1 can be extended for arbitrarily placed base stations. Establishing a local coordinate system and having positions at (0,0),$(X_2$,0), and (X_3,Y_3), for BS_1, BS_2,and BS_3 respectively, the DSS-2 algorithm can be applied to find the position of the MS. As it's shown in Figure 5, BS_3 is assumed to lie on an angle of 60° above the horizontal axis, where the two base stations exist. Such angle is chosen, due to the hexagonal shape used for representing cellular networks.

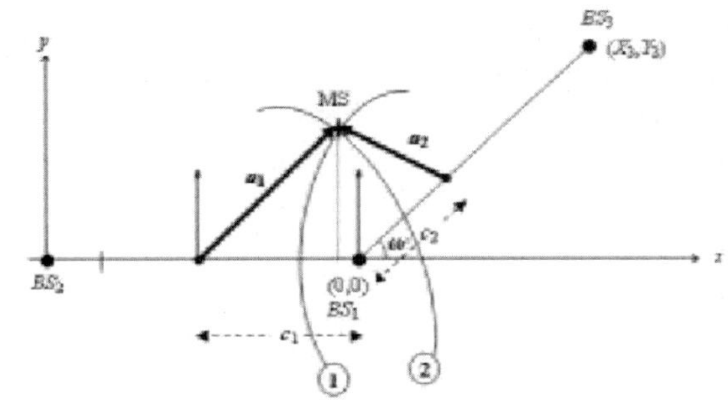

Fig. 5. Idea of Positioning for Arbitrarily Placed Base Stations

As in DSS-1, the same equations are used to describe the rotations needed for vector a_1, and the ones used to approximate b_1, and b_2. However, to rotate a_2 in iteration i and get its real coordinates $(x^{//}_{2,i}, y^{//}_{2,i})$ on the hyperbola, it should be first rotated as before using the hyperbolic equations (21 - 24) with $x_1, y_1, x^{/}_1, y^{/}_1, a_1, c_1$ and b_1 replaced by $x_2, y_2, x^{/}_2, y^{/}_2, a_2, c_2$ and b_2 respectively. Then, as seen in Figure 6, with local coordinates (0,0) at BS_2, the point with coordinates $(c_2 - x_{2,i}, y^{/}_{2,i})$ on the hyperbola should be circularly rotated by 60°. Equations (29) and (30) define the rotations needed to get the real point

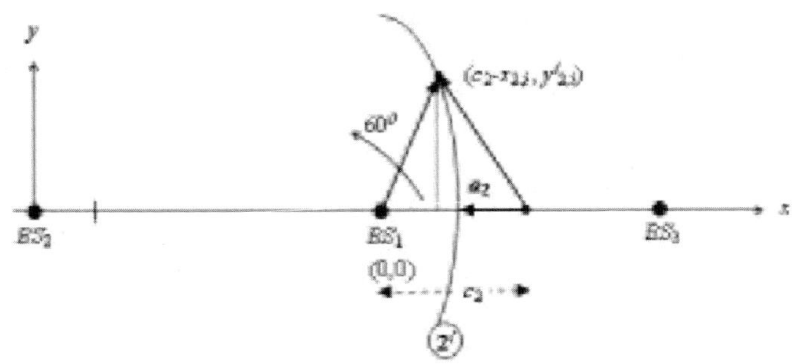

Fig. 6. Idea of Circular Rotation by 60°

on the second hyperbola.

$$x^{//}_{2,i} = (c_2 - x_{2,i})\cos(60^0) - y^{/}_{2,i}\sin(60^0), \tag{29}$$

$$y^{//}_{2,i} = y^{/}_{2,i}\cos(60^0) + (c_2 - x_{2,i})\sin(60^0). \tag{30}$$

From Figure 5 where the stopping criterion is shown, and assuming b_2 is the larger, the conditions of rotation can be written as:

$While \quad (((c_1 - (x_{1,i} - x_{2,i}^{//})) > \varepsilon))$
  $Rotate \quad a_2 \quad (Hyperbolic \quad rotation \quad followed \quad by \quad circular$
    $Rotation \quad by \quad 60°)$
  $While \quad ((y_{2,i}^{//} - y_{1,i}^{/}) > \varepsilon)$
    $Rotate \quad a_1$
  $End \quad While$
$End \quad While$

The intersection point can be obtained from the last iterations of rotations, and the real one can be calculated as before and written as $(x_{1,i} + X_1/2, y_{1,i}^{/})$.

It's noteworthy to mention that the multiplication by Sin(60°) can be implemented in hardware using a multiple operand adder [Mull97]. A multiplication time is reduced to an addition time. Also, multiplication by Cos(60°) = 0.5 can be implemented as a shift operation.

3.4 The DDSS-2 algorithm

The DDSS-2 (Dynamic DSS-2) is the dynamic version of DSS-2 where rotations of vectors are done in a dynamic fashion taking the maximum step angle that can be feasible to guarantee the convergence of the algorithm. The same criteria in (27) and (28) are used for the dynamic rotations of the vectors rotations along the hyperbolas, and for the approximation of the unknown parameters of the two hyperbolas.

3.5 Simulation results and analysis

Matlab 6.5 package was used in the analysis. To find the location of a mobile station in the coverage area of the three BSs, programs for both traditional and DSS algorithms were written. The experiments were repeated for many arbitrary positions of the mobile station, and results were taken for 95% confidence level.

For the purpose of comparing our algorithms with the traditional one, the average computational costs required to find the location of the handset, in each case, were calculated. Weights of the operations for 20 bits accuracy are taken as shown in Table 1 [Mull97]. Figure 7 shows the average computa-

Table 1. Weights of Operations

Operation	Addition	Subtraction	Shift	Multiplication	Division	Square root
Weight	1	1	1	40	40	100

tional cost needed for the DSS-1 versus sinh(σ) (which explicitly specifies the step rotation , and implicitly specifies the accuracy level). As it's seen, the computational cost increases as the step angle decreases (i.e. accuracy level increases). Figure 8 shows the error (in meters) in determining the mobile lo-

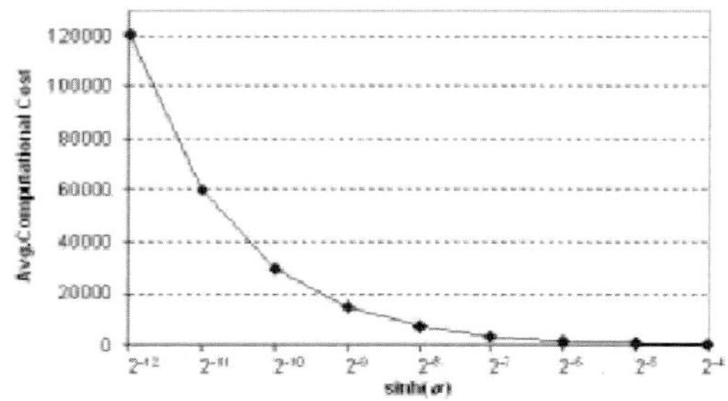

Fig. 7. Avg. Computational Cost Versus sinh(σ) for DSS-1 Algorithm

cation versus sinh(σ). The error decreases when the accuracy increases (step rotation angle decreases). For sin(σ) $= 2^{-7}$ a very good level of accuracy can be achieved. This corresponds to an accuracy of approximately 55m for base stations of radius 2.5km and thereby satisfying the E-911 standards. Figure

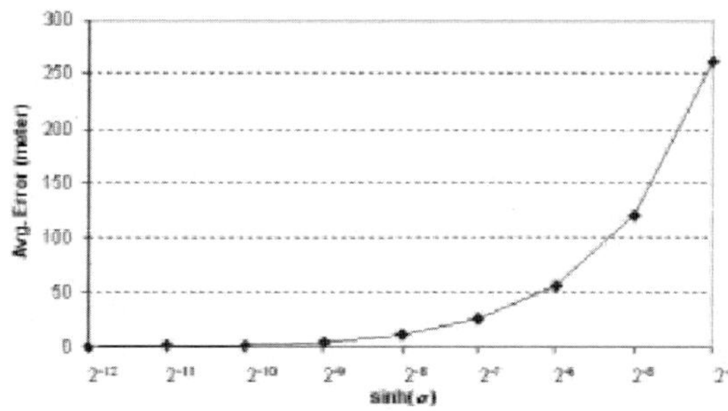

Fig. 8. Error of Estimating the Mobile's Location Versus sinh(σ) for DSS-1 Algorithm

9 shows the average computational cost versus the normalized Δx in (28) for our general algorithm in its dynamic mode, DDSS-2, and the traditional algorithm. The initial distance Δx between the two vectors' heads of the hyperbolas explicitly specifies the initial step rotation angle used, as discussed in section 3.2. As shown before, more number of rotations is needed to approximate the value of b as Δx decreases, i.e. a increases. Moreover, with the dynamic mode the accuracy increases, since rotation with a smaller step angle is performed. It's clearly seen that no rotations are needed for the traditional algorithm, it shows constant performance, and that DDSS-2 outperforms the traditional algorithm for most of the places where the handset can be located.

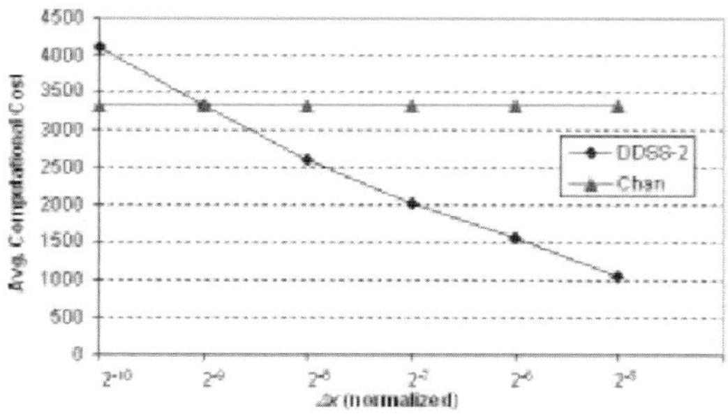

Fig. 9. Avg. Computational Cost Versus Δx for Chan's and DDSS-2 Algorithms

4 Conclusion

This paper presents new hardware oriented algorithms (DSS) based on TDOA measurements to determine the position of a mobile phone. Since all operations in our algorithms are simple add and shift operations, it can be easily implemented in hardware. As can be seen from the results a good level of accuracy can be obtained that satisfies the E-911 standards. The dynamic mode was proposed to decrease the number of operations needed and to get a better accuracy. As a future work, a direction of study could be to modify these algorithms for finding the position of a handset in 3-D space.

References

[Lait01] Laitinen, H. et al.: Cellular Location Technology. CELLO Project Technical Report, CELLO-WP2-VTT-D03- 007-Int, (2001)

[Zha02] Zhao, Y.: Standardization of Mobile Phone Positioning for 3G Systems. IEEE Communications Magazine, **40(4)**, 108-116 (2002)

[JAO99] Jami, I., Ali, M., Ormondroyd, R.F.: Comparison of Methods of Locating and Tracking Cellular Mobiles, Novel Methods of Location and Tracking of Cellular Mobiles and Their System Applications. Proc. of IEE Colloquium(London UK), 1-6 (1999)

[CH94] Chan, Y.T., Ho, K.C.: A simple and efficient estimator for hyperbolic location. IEEE Transactions on Signal Processing, **42(8)**, 1905-1915 (1994)

[Mull97] Muller, J.M.: Elementary Function Algorithms and Implementation. Birkhauser, Berlin (1997)

[SDD04] Salamah, M., Doukhnitch, E., Devrim, D.: A Fast Hardware-Oriented Algorithm for Cellular Mobiles Positioning. Lecture Notes on Computer Science LNCS 3280, Springer, Berlin (2004)

Unknown costs in a duopoly with differentiated products

Fernanda A. Ferreira[1,2], Flávio Ferreira[2] and Alberto A. Pinto[1]

[1] Faculdade de Ciências da Universidade do Porto, Rua do Campo Alegre, 687, 4169-007 Porto, Portugal aapinto@fc.up.pt

[2] ESEIG, Instituto Politécnico do Porto, Rua D. Sancho I, 981, 4480-876 Vila do Conde, Portugal {fernandaamelia, flavioferreira}@eseig.ipp.pt

We consider a duopoly model with unknown costs. The firms' aims are to maximize their profits by choosing the levels of their outputs. The chooses are made simultaneously by both firms.

In this paper, we suppose that each firm has two different technologies, and uses one of them following a probability distribution. The utilization of one or the other technology affects the unitary production cost. We show that this game has exactly one Bayesian Nash equilibrium. We analyze the advantages, for firms and for consumers, of using the technology with highest production cost *versus* the one with cheapest production cost. We also analyze the expected total quantity produced in each situation, which is of particular importance in the case that scanty natural resources are used in the production.

1 Introduction

Bayesian games are used to model situations in which there are players with privileged information, and where the payoff of each player depends upon this privileged information, besides to depend upon the actions of the payers. The case that we will study belongs to this class of games, since there are market conflicts in which each firm knows its production costs, but does not know the production costs of the other firm. We will consider a Cournot duopoly model, one of the classical models in the theory of duopoly(see [Cour1897]). Let E_1 and E_2 be two firms, each producing a differentiated product. Both firms simultaneously choose the quantity that must produce with the purpose to maximize their expected profit.

In §3, we consider an economic model in which we suppose that each firm has two different technologies, and uses one of them following a probability distribution. The utilization of one or the other technology affects the unitary production cost. We suppose that firm E_1's unitary production cost is c_A

Kenan Taş et al. (eds), Mathematical Methods in Engineering, 359–369.

with probability ϕ and c_B with probability $1-\phi$ (where $c_A > c_B$), and firm E_2's unitary production cost is c_H with probability θ and c_L with probability $1-\theta$ (where $c_H > c_L$). Both probability distributions of unitary production costs are common knowledge. In this work, we determine the quantities in the Bayesian Nash equilibrium for the above model, and we analyze the advantages, for firms and for consumers, of using the technology with highest production cost *versus* the one with cheapest production cost. We also analyze the expected total quantities produced in each situation which is of particular importance in the case that scanty natural resources are used in the production. In [Gib92], it is presented a simpler model for homogeneous goods and where just one of the firms has uncertainty costs.

We will start by describing, in §2, the standard Cournot model, that is, in which the unitary production costs are fixed.

2 The Cournot model

The standard Cournot model is a static game of complete information, that is, a one-shot game where the players' payoff functions are common knowledge. Cournot was at the forefront of the development of Game Theory. In [Cour1897], he proposed what now is known as standard Cournot model to describe a situation where a small number of firms compete in a homogeneous product market, simultaneously choosing output levels. We consider two firms, E_1 and E_2, each producing a differentiated product. The firms simultaneously choose output levels, respectively, $q_1 \geq 0$ and $q_2 \geq 0$. The representative consumer maximizes $U(q_1, q_2) - p_1 q_1 - p_2 q_2$, where p_i stands for the price of the good produced by the firm E_i, for $i \in \{1, 2\}$. The function U is defined by $U(q_1, q_2) = \alpha(q_1 + q_2) - (q_1^2 + 2\gamma q_1 q_2 + q_2^2)/2$, where $\alpha > 0$ and $0 \leq \gamma \leq 1$. The parameter γ expresses the degree of product differentiation (see [SV84]). This utility function gives rise to a linear demand structure. Inverse demands are given by

$$p_1 = \alpha - q_1 - \gamma q_2, \tag{1}$$

$$p_2 = \alpha - \gamma q_1 - q_2, \tag{2}$$

in the region of quantity space where prices are positive. We note that the two products are substitutes, and, since $\gamma \leq 1$, "cross effects" are dominated by "own effects" (see [Gal85]). Moreover, if $\gamma = 1$, then the goods are homogeneous, and if $\gamma = 0$, then the goods are independent. Assume that the total cost to firm E_i of producing quantity q_i is $C_i(q_i) = c q_i$. That is, there are no fixed costs and the marginal cost is constant at c, with $c < \alpha$. The payoff to firm E_i is given by the profit function

$$\pi_i(q_i, q_j) = q_i(\alpha - q_i - \gamma q_j - c).$$

Now, we are going to compute the Nash equilibrium of the Cournot game. If (q_1^*, q_2^*) is the Nash equilibrium, then q_i^* is the solution of

$$\max_{0 \le q_i \le \alpha} \pi_i(q_i, q_j^*),$$

for $i, j \in \{1, 2\}$ with $i \neq j$. Therefore, q_i^* solves

$$\max_{0 \le q_i \le \alpha} q_i(\alpha - q_i - \gamma q_j - c).$$

Hence, we get

$$\begin{cases} q_1^* = \frac{\alpha - c - \gamma q_2^*}{2} \\ q_2^* = \frac{\alpha - c - \gamma q_1^*}{2} \end{cases},$$

and, so, the Nash equilibrium is

$$\left(\frac{\alpha - c}{2 + \gamma}, \frac{\alpha - c}{2 + \gamma} \right).$$

Remark 1. The equilibrium price of the good produced by the firm E_i is

$$p_i^* = \frac{\alpha + \gamma}{2 + \gamma}.$$

Remark 2. Firm E_i's profit at equilibrium is

$$\pi_i^* = \left(\frac{\alpha - c}{2 + \gamma} \right)^2.$$

3 A Bayesian Cournot model

In this section, we consider a Cournot model with incomplete information. Recall that in a game of complete information the players' payoff functions are common knowledge. In a game of incomplete information, in contrast, at least one player is uncertain about another player's payoff function. These games are called Bayesian games.

Let us consider the same model as in the previous section, supposing now that each firm has two different technologies, and uses one of them following a probability distribution. The utilization of one or the other technology affects the unitary production cost. The following probability distributions of unitary production costs are common knowledge:

$$c_1 = \begin{cases} c_A, & \text{with probability } \phi \\ c_B, & \text{with probability } 1 - \phi \end{cases},$$

$$c_2 = \begin{cases} c_H, & \text{with probability } \theta \\ c_L, & \text{with probability } 1-\theta \end{cases} .$$

We suppose that $c_A > c_B$, $c_H > c_L$ and $c_A, c_B, c_H, c_L < a$. Moreover, we suppose that the highest unitary production cost of any firm is greater than the lowest unitary production cost of the other one, that is, $c_A > c_L$ and $c_H > c_B$. Otherwise, the conclusions will be obviously.

Firms' payoffs are given by

$$\pi_1(q_1(c_1), q_2(c_2)) = q_1(c_1)(\alpha - q_1(c_1) - \gamma q_2(c_2) - c_1),$$

$$\pi_2(q_1(c_1), q_2(c_2)) = q_2(c_2)(\alpha - \gamma q_1(c_1) - q_2(c_2) - c_2),$$

where c_i is firm E_i's unitary production cost, for $i \in \{1, 2\}$.

Firm E_1 should choose a quantity, $q_1^*(c_A)$ or $q_1^*(c_B)$, depending on its unitary production cost, to maximize its expected payoff; and firm E_2 should choose a quantity, $q_2^*(c_H)$ or $q_2^*(c_L)$, depending on its unitary production cost, to maximize its expected payoff.

Proposition 1. *Let $E(c_1) = \phi c_A + (1-\phi)c_B$ be the firm E_1's expected unitary production cost, and let $E(c_2) = \theta c_H + (1-\theta)c_L$ be the firm E_2's expected unitary production cost. For the Cournot model with differentiated goods and with uncertainty costs considered above, the Bayesian Nash equilibrium is*

$$((q_1^*(c_A), q_1^*(c_B)), (q_2^*(c_H), q_2^*(c_L))),$$

where

$$q_1^*(c_A) = \frac{2(2-\gamma)\alpha - (4-\gamma^2)c_A - \gamma^2 E(c_1) + 2\gamma E(c_2)}{2(4-\gamma^2)}, \tag{3}$$

$$q_1^*(c_B) = \frac{2(2-\gamma)\alpha - (4-\gamma^2)c_B - \gamma^2 E(c_1) + 2\gamma E(c_2)}{2(4-\gamma^2)}, \tag{4}$$

$$q_2^*(c_H) = \frac{2(2-\gamma)\alpha - (4-\gamma^2)c_H + 2\gamma E(c_1) - \gamma^2 E(c_2)}{2(4-\gamma^2)}, \tag{5}$$

$$q_2^*(c_L) = \frac{2(2-\gamma)\alpha - (4-\gamma^2)c_L + 2\gamma E(c_1) - \gamma^2 E(c_2)}{2(4-\gamma^2)}, \tag{6}$$

assuming $q_1 < (\alpha - c_H)/\gamma$ and $q_2 < (\alpha - c_A)/\gamma$.

Proof. If firm E_1's unitary production cost is high, $q_1^*(c_A)$ is the solution of

$$\max_{0 \le q_1 \le \alpha} (\theta(\alpha - q_1 - \gamma q_2(c_H) - c_A)q_1 + (1-\theta)(\alpha - q_1 - \gamma q_2(c_L) - c_A)q_1);$$

and if it is low, $q_1^*(c_B)$ is the solution of

$$\max_{0 \le q_1 \le \alpha} (\theta(\alpha - q_1 - \gamma q_2(c_H) - c_B)q_1 + (1-\theta)(\alpha - q_1 - \gamma q_2(c_L) - c_B)q_1).$$

If firm E_2's unitary production cost is high, $q_2^*(c_H)$ is the solution of

$$\max_{0 \le q_2 \le \alpha} (\phi(\alpha - \gamma q_1(c_A) - q_2 - c_H)q_2 + (1-\phi)(\alpha - \gamma q_1(c_B) - q_2 - c_H)q_2);$$

and if it is low, $q_2^*(c_L)$ is the solution of

$$\max_{0 \le q_2 \le \alpha} (\phi(\alpha - \gamma q_1(c_A) - q_2 - c_L)q_2 + (1-\phi)(\alpha - \gamma q_1(c_B) - q_2 - c_L)q_2).$$

Then,

$$\begin{aligned} q_1^*(c_A) &= \frac{\alpha - c_A - \gamma(\theta q_2^*(c_H) + (1-\theta)q_2^*(c_L))}{2}, \\ q_1^*(c_B) &= \frac{\alpha - c_B - \gamma(\theta q_2^*(c_H) + (1-\theta)q_2^*(c_L))}{2}, \\ q_2^*(c_H) &= \frac{\alpha - c_H - \gamma(\phi q_1^*(c_A) + (1-\phi)q_1^*(c_B))}{2}, \\ q_2^*(c_L) &= \frac{\alpha - c_L - \gamma(\phi q_1^*(c_A) + (1-\phi)q_1^*(c_B))}{2}, \end{aligned}$$

assuming $q_1 < (\alpha - c_H)/\gamma$ and $q_2 < (\alpha - c_A)/\gamma$. Therefore, we obtain equalities (3)-(6). □

Remark 3. From equalities (3) and (4), we get that the expected quantity produced by the firm E_1 is

$$\begin{aligned} E(q_1^*) &= q_1^*(c_A)\phi + q_1^*(c_B)(1-\phi) \\ &= \frac{(2-\gamma)\alpha - 2E(c_1) + \gamma E(c_2)}{4-\gamma^2}. \end{aligned} \tag{7}$$

From equalities (5) and (6), we get that the expected quantity produced by the firm E_2 is

$$\begin{aligned} E(q_2^*) &= q_2^*(c_H)\theta + q_2^*(c_L)(1-\theta) \\ &= \frac{(2-\gamma)\alpha + \gamma E(c_1) - 2E(c_2)}{4-\gamma^2}. \end{aligned} \tag{8}$$

In the case of firms producing independent goods, the expected quantity produced by each firm is minimum when the firm uses its most expensive technology, and it is maximum when the firm uses its cheapest technology (see Figure 1a). In the case of firms producing differentiated goods, the expected quantity produced by the firm E_1 is minimum when $\phi = 1$ and $\theta = 0$, and it is maximum when $\phi = 0$ and $\theta = 1$; the expected quantity produced by firm E_2 is minimum when $\phi = 0$ and $\theta = 1$, and it is maximum when $\phi = 1$ and $\theta = 0$ (see Figure 1b).

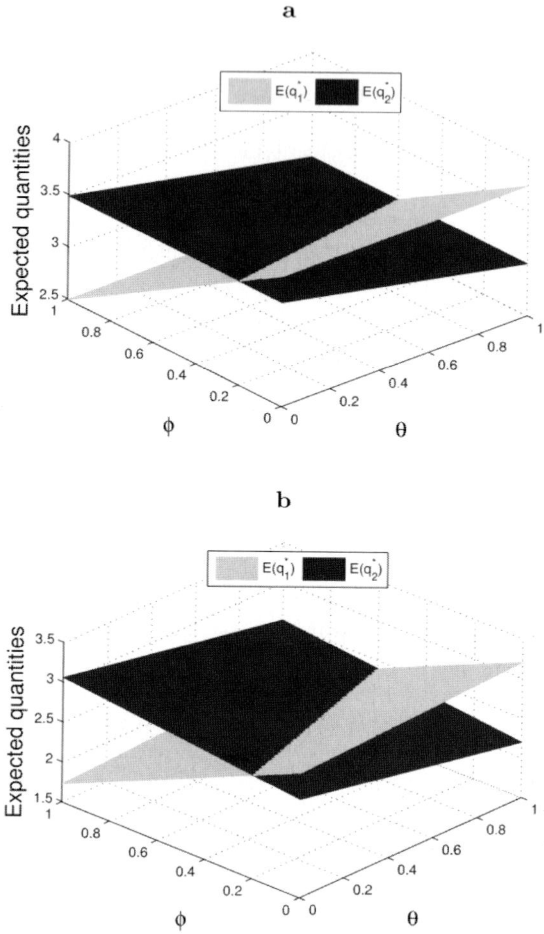

Fig. 1. Firms' expected quantities, $E(q_1^*)$ and $E(q_2^*)$, in the case of: (**a**) firms producing independent goods ($\gamma = 0$); and (**b**) firms producing differentiated goods with degree of differentiation $\gamma = 0.5$. Other parameters values: $\alpha = 10$, $c_A = 5$, $c_B = 2.5$, $c_H = 4$ and $c_L = 3$.

Remark 4. Since $E(Q^*) = E(q_1^*) + E(q_2^*)$, we obtain that the expected aggregate quantity produced is

$$E(Q^*) = \frac{2\alpha - E(c_1) - E(c_2)}{2 + \gamma}.$$

The changing of the expected aggregate quantity with respect to the probabilities ϕ and θ is illustrated in Figure 2a for the case of independent goods ($\gamma = 0$), and in Figure 2b for the case of an intermediate degree of differentiation of the goods ($\gamma = 0.5$), for some parameter region of the model. We see that the minimum of the expected aggregate quantity is attained when both firms use their most expensive technologies, and the maximum is attained when both firms use their cheapest technologies.

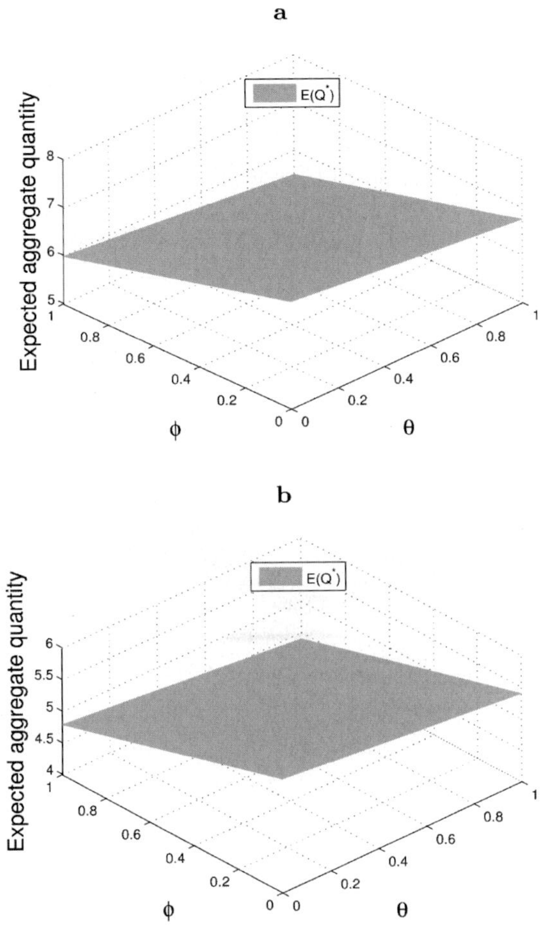

Fig. 2. Expected aggregate quantity, $E(Q^*)$, in the case of: (**a**) firms producing independent goods $\gamma = 0$; and (**b**) firms producing differentiated goods with degree of differentiation $\gamma = 0.5$. Other parameters values: $\alpha = 10$, $c_A = 5$, $c_B = 2.5$, $c_H = 4$ and $c_L = 3$.

Remark 5. From (1) and using (7) and (8), we get that the expected market price of the good produced by the firm E_1 is

$$\begin{aligned} E(p_1^*) &= \alpha - E(q_1^*) - \gamma E(q_2^*) \\ &= \frac{(2-\gamma)\alpha + (2-\gamma^2)E(c_1) + \gamma E(c_2)}{4-\gamma^2}, \end{aligned}$$

and from (2) and using (7) and (8), we get that the expected market price of the good produced by the firm E_2 is

$$E(p_2^*) = \alpha - \gamma E(q_1^*) - E(q_2^*)$$

$$= \frac{(2-\gamma)\alpha + \gamma E(c_1) + (2-\gamma^2)E(c_2)}{4-\gamma^2}.$$

In the case of firms producing independent goods, the expected market price of the good produced by each firm is minimum when the firm uses its cheapest technology, and it is maximum when the firm uses its most expensive technology (see Figure 3a). In the case of firms producing differentiated goods, the expected market price of the good produced by each firm is minimum when $\phi = \theta = 0$, and it is maximum when $\phi = \theta = 1$ (see Figure 3b).

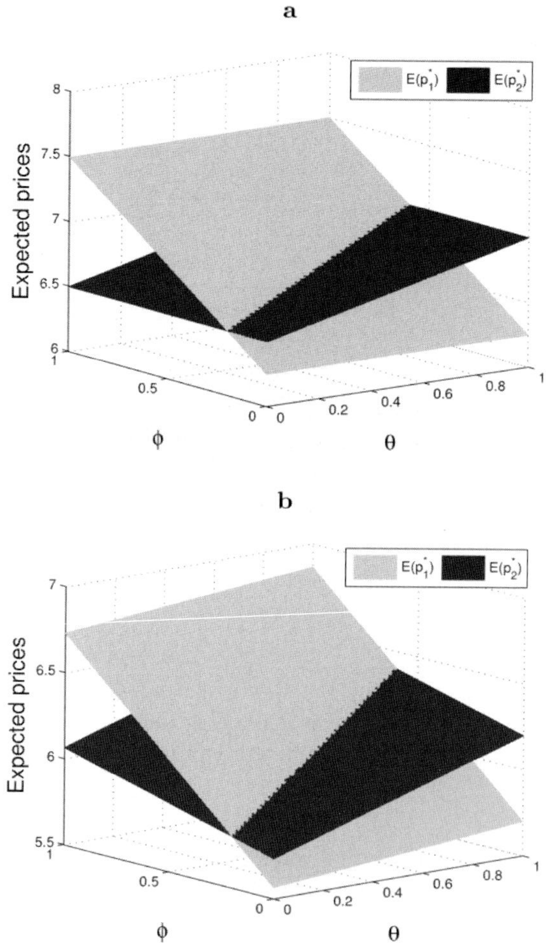

Fig. 3. Expected prices, $E(p_1^*)$ and $E(p_2^*)$, in the case of: (**a**) firms producing independent goods ($\gamma = 0$); and (**b**) firms producing differentiated goods with degree of differentiation $\gamma = 0.5$. Other parameters values: $\alpha = 10$, $c_A = 5$, $c_B = 2.5$, $c_H = 4$ and $c_L = 3$.

Remark 6. Firm E_1's expected profit is

$$\begin{aligned} E(\pi_1^*) = & \pi_1(q_1^*(c_A), q_2^*(c_H))\phi\theta + \\ & +\pi_1(q_1^*(c_A), q_2^*(c_L))\phi(1-\theta) + \\ & +\pi_1(q_1^*(c_B), q_2^*(c_H))(1-\phi)\theta + \\ & +\pi_1(q_1^*(c_B), q_2^*(c_L))(1-\phi)(1-\theta), \end{aligned}$$

and firm E_2's expected profit is

$$\begin{aligned} E(\pi_2^*) = & \pi_2(q_1^*(c_A), q_2^*(c_H))\phi\theta + \\ & +\pi_2(q_1^*(c_A), q_2^*(c_L))\phi(1-\theta) + \\ & +\pi_2(q_1^*(c_B), q_2^*(c_H))(1-\phi)\theta + \\ & +\pi_2(q_1^*(c_B), q_2^*(c_L))(1-\phi)(1-\theta). \end{aligned}$$

The effect of the probabilities ϕ and θ over the firms' expected profits is shown in Figure 4a for the case of independent goods ($\gamma = 0$), and in Figure 4b for the case of an intermediate degree of differentiation of the goods ($\gamma = 0.5$), for some parameter region of the model. In the case of firms producing differentiated goods, each firm profits more when it uses its cheapest technology and the other firm uses its more expensive technology (see Figure 4b).

a

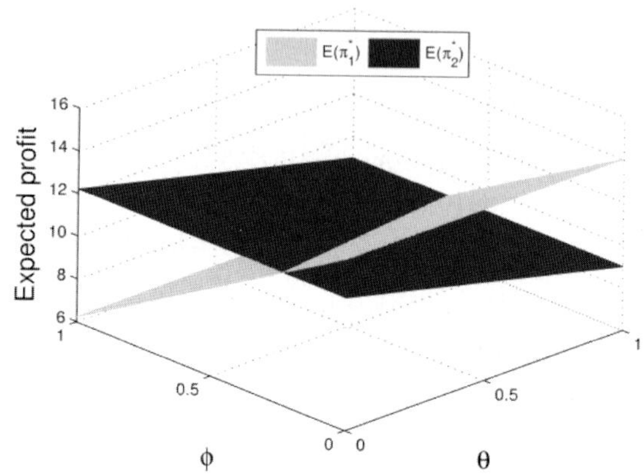

b

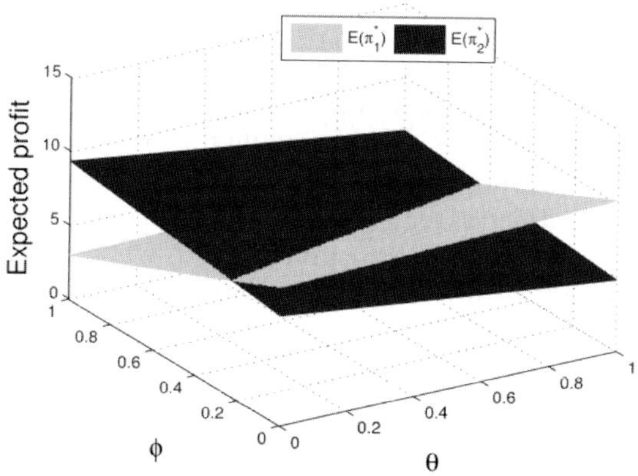

Fig. 4. Firms' expected profits, $E(\pi_1^*)$ and $E(\pi_2^*)$, in the case of: (**a**) firms producing independent goods ($\gamma = 0$); and (**b**) firms producing differentiated goods with degree of differentiation $\gamma = 0.5$. Other parameters values: $\alpha = 10$, $c_A = 5$, $c_B = 2.5$, $c_H = 4$ and $c_L = 3$.

Acknowledgments

We would like to thank Bruno Oliveira and Miguel Ferreira all the useful discussions. We thank the Programs POCTI and POSI by FCT and Ministério da Ciência, Tecnologia e do Ensino Superior, and Centro de Matemática da Universidade do Porto for their financial support. Fernanda Ferreira gratefully acknowledges financial support from ESEIG/IPP and from PRODEP III by FSE and EU. Flávio Ferreira also acknowledges financial support from ESEIG/IPP.

References

[Cour1897] Cournot, A.: Recherches sur les Principes Mathématiques de la Théorie des Richesses. Paris (1838). English edition: Bacon N., Researches into the Mathematical Principles of the Theory of Wealth. Macmillan, New York (1897)

[Gib92] Gibbons, R.: A Primer in Game Theory. Pearson Prentice Hall, Harlow (1992)

[SV84] Singh, N., Vives, X.: Price and quantity competition in a differentiated duopoly. RAND Journal of Economics, **15**, 546–554 (1984)

[Gal85] Gal-Or, E.: First mover and second mover advantages. Int. Econ. Rev, **26**, 649–653 (1985)

Bayesian price leadership

Fernanda A. Ferreira[1,2], Flávio Ferreira[2] and Alberto A. Pinto[1]

[1] Faculdade de Ciências da Universidade do Porto, Rua do Campo Alegre, 687, 4169-007 Porto, Portugal aapinto@fc.up.pt
[2] ESEIG, Instituto Politécnico do Porto, Rua D. Sancho I, 981, 4480-876 Vila do Conde, Portugal {fernandaamelia, flavioferreira}@eseig.ipp.pt

In this paper, we consider a linear price setting duopoly competition with differentiated goods and with unknown costs. The firms' aims are to choose the prices of their products according to the well-known concept of perfect Bayesian Nash equilibrium. There is a firm (F_1) that chooses first the price p_1 of its good; the other firm (F_2) observes p_1 and then chooses the price p_2 of its good.

We suppose that each firm has two different technologies, and uses one of them following a probability distribution. The utilization of one or the other technology affects the unitary production cost. We show that there is exactly one perfect Bayesian Nash equilibrium for this game. We analyze the advantages, for firms and for consumers, of using the technology with highest production cost *versus* the one with cheapest production cost.

1 Introduction

A game is a situation with any kind of interactions, and it has, by definition, participants who are called players. A player may be interpreted as an individual or as an organization making a "rational" decision. The players receive payoffs that depend on the combination of decisions just taken. There are games of complete information, in which the players' payoff functions are common knowledge and games of incomplete information (also called Bayesian games), in which at least one player is uncertain about another player's payoff function. In the first case, the usual solution concept is the Nash equilibrium: a decision combination is a Nash equilibrium when, if one player sticks rigidly to his decision in the combination, then the other player cannot increase his reward by selecting other than his decision in that combination. That is, each player's strategy must be a best response to the other player's strategies. A Nash equilibrium in a Bayesian dynamic game is called a perfect Bayesian Nash equilibrium. The model presented in §2 belongs to the first class, and the model considered in §3 belongs to the second one.

Kenan Taş et al. (eds), Mathematical Methods in Engineering, 371–379

In industrial organization, we find situations in which firms have to choose output levels or prices in a market. In the case of quantity competition, von Stackelberg [Sta34] proposed a dynamic model of duopoly in which a dominant (leader) firm moves first and a subordinate (follower) firm moves second. In this case, it is well-known that the leading firm has advantages over the follower (see [Gib92]).

In this paper, we consider a similar model, but instead of choosing the output levels, the firms choose the prices for their goods. The timing of the game is as follows: (i) The leading firm chooses a price $p_1 \geq 0$ for its good; (ii) The follower observes p_1, and then chooses a price $p_2 \geq 0$ for its good. Price competition is fundamentally different from quantity competition in that the leadership role is not now the most preferred one (see §2). We will study this model by considering that each firm has two different technologies, and uses one of them following a probability distribution. The utilization of one or the other technology affects the unitary production cost. We suppose that firm F_1's unitary production cost is c_A with probability ϕ and c_B with probability $1-\phi$ (where $c_A > c_B$), and firm F_2's unitary production cost is c_H with probability θ and c_L with probability $1-\theta$ (where $c_H > c_L$). Both probability distributions of unitary production costs are common knowledge. In this work, we determine the prices in the perfect Bayesian Nash equilibrium for the above model, and we show that, in contrast to the case with complete information, the leading firm may profit more than the follower (see §3). We also analyze the variations of the prices and the profits over the parameters of the probability distributions, for some different degrees of product differentiation.

Van Damme and Hurkens [DH04] studied a related question by considering that one firm has higher production cost, but in a game of complete information.

2 The model with complete information

We consider an economy with a monopolistic sector with two firms, F_1 and F_2. Firm F_i produces a differentiated product i at a constant marginal cost. We present the sequential-move model, with complete information, in which firms choose prices. The timing of the game is as follows: (i) Firm F_1 (leader) chooses a price $p_1 \geq 0$ for its good; (ii) firm F_2 (follower) observes p_1 and then chooses a price $p_2 \geq 0$ for its good.

The direct demands are

$$q_i = a - p_i + bp_j,$$

provided that the quantities q_i are positive, with $i, j \in \{1, 2\}$ and $i \neq j$, where $a > 0$ and $0 \leq b \leq 1$. Firm F_i's profit is given by

$$\begin{aligned}\pi_i(p_i, p_j) &= q_i(p_i - c)\\ &= (a - p_i + bp_j)(p_i - c),\end{aligned}$$

where $0 < c < a$ is the unitary production cost for both firms.

Using backwards-induction, we will first compute $p_2^*(p_1)$.

Firm F_2's reaction to an arbitrary price p_1 fixed by F_1, $p_2^*(p_1)$, is the solution of

$$\max_{p_2 \geq 0} \pi_2(p_1, p_2).$$

Then,

$$p_2^*(p_1) = \frac{a + c + bp_1}{2}.$$

Firm F_1 can anticipate $p_2^*(p_1)$. Thus,

$$\pi_1(p_1, p_2^*(p_1)) = (a - p_1 + bp_2^*(p_1))(p_1 - c) = \left(a - p_1 + b \cdot \frac{a + c + bp_1}{2}\right)(p_1 - c).$$

Hence, p_1^* is the solution of

$$\max_{p_1 \geq 0} \left(a - p_1 + b \cdot \frac{a + c + bp_1}{2}\right)(p_1 - c).$$

Then,

$$p_1^* = \frac{a(2 + b) + (2 + b - b^2)c}{2(2 - b^2)}$$

and

$$p_2^*(p_1^*) = \frac{a(4 + 2b - b^2) + (4 + 2b - b^2 - b^3)c}{4(2 - b^2)}.$$

So, the Nash equilibrium is

$$\left(\frac{a(2 + b) + (2 + b - b^2)c}{2(2 - b^2)}, \frac{a(4 + 2b - b^2) + (4 + 2b - b^2 - b^3)c}{4(2 - b^2)}\right).$$

Remark 1. The price of the good produced by the leading firm is higher than the price produced by the follower, unless the goods are independent ($b = 0$). In fact, we have that

$$p_1^* - p_2^* = \frac{b^2(a + c(b - 1))}{4(2 - b^2)} \geq 0.$$

Thus, if the goods are not independent, then the good produced by the firm F_1 has a higher price than the good produced by the firm F_2; if the goods are independent, then their prices are equal.

Remark 2. Firm F_1's profit at equilibrium is

$$\pi_1 = \frac{(a(2+b) + (b^2+b-2)c)^2}{8(2-b^2)};$$

and firm F_2's profit at equilibrium is

$$\pi_2 = \frac{\left(a(4+2b-b^2) + (-4+2b+3b^2-b^3)c\right)^2}{16(2-b^2)^2}.$$

Then, we get

$$\pi_1 - \pi_2 = -\frac{b^3(a+c(b-1))^2(3b+4)}{16(2-b^2)^2} \leq 0,$$

which means that the follower firm has advantages over the leader, unless the goods are independent ($b = 0$).

3 The model with incomplete information

In this section, we consider the model presented in the previous section, but now with incomplete information. In a game of complete information the players' payoff functions are common knowledge. In a game of incomplete information, in contrast, at least one player is uncertain about another player's payoff function. These games are called Bayesian games. We suppose that each firm has two different technologies, and uses one of them following some probability distribution. The utilization of one or the other technology affects the unitary production cost. The following probability distributions of the firms' production costs are common knowledge among both firms:

$$C_1(q_1) = \begin{cases} c_A\ q_1 & \text{with probability } \phi \\ c_B\ q_1 & \text{with probability } 1-\phi \end{cases},$$

$$C_2(q_2) = \begin{cases} c_H\ q_2 & \text{with probability } \theta \\ c_L\ q_2 & \text{with probability } 1-\theta \end{cases}.$$

We suppose that $c_A > c_B$, $c_H > c_L$ and $c_A, c_B, c_H, c_L < a$. Firms' profits are given by

$$\pi_1(p_1(c_1), p_2(c_2)) = (a - p_1(c_1) + bp_2(c_2))(p_1(c_1) - c_1),$$

$$\pi_2(p_1(c_1), p_2(c_2)) = (a - p_2(c_2) + bp_1(c_1))(p_2(c_2) - c_2),$$

where c_i is firm F_i's unitary production cost, for $i \in \{1, 2\}$.

Firm F_1 should choose a price for its good, $p_1^*(c_A)$ or $p_1^*(c_B)$, depending on its unitary production cost, to maximize its expected profit; and firm F_2 should choose a price for its good, $p_2^*(c_H)$ or $p_2^*(c_L)$, depending on its unitary production cost, to maximize its expected profit.

Proposition 1. *Let $E(c_2) = \theta c_H + (1-\theta)c_L$ be the firm F_2's expected unitary production cost. For the model with uncertainty costs considered above, the perfect Bayesian Nash equilibrium is*

$$((p_1^*(c_A), p_1^*(c_B)), (p_2^*(c_H|p_1^*(c_A)), p_2^*(c_H|p_1^*(c_B)), p_2^*(c_L|p_1^*(c_A)), p_2^*(c_L|p_1^*(c_B)))),$$

where

$$p_1^*(c_A) = \frac{a(2+b) + (2-b^2)c_A + bE(c_2)}{2(2-b^2)}, \tag{1}$$

$$p_1^*(c_B) = \frac{a(2+b) + (2-b^2)c_B + bE(c_2)}{2(2-b^2)}, \tag{2}$$

$$p_2^*(c_H|p_1^*(c_A)) = \frac{a(4+2b-b^2) + (2b-b^3)c_A + (4-2b^2)c_H + b^2E(c_2)}{4(2-b^2)}, \tag{3}$$

$$p_2^*(c_L|p_1^*(c_A)) = \frac{a(4+2b-b^2) + (2b-b^3)c_A + (4-2b^2)c_L + b^2E(c_2)}{4(2-b^2)}, \tag{4}$$

$$p_2^*(c_H|p_1^*(c_B)) = \frac{a(4+2b-b^2) + (2b-b^3)c_B + (4-2b^2)c_H + b^2E(c_2)}{4(2-b^2)}, \tag{5}$$

$$p_2^*(c_L|p_1^*(c_B)) = \frac{a(4+2b-b^2) + (2b-b^3)c_B + (4-2b^2)c_L + b^2E(c_2)}{4(2-b^2)}. \tag{6}$$

Proof. Using backwards-induction, we will first compute $p_2^*(p_1)$, by consider separately the cases where the production cost of firm F_1 is (i) c_A and (ii) c_B.

(i) Let us suppose that F_1 used the most expensive technology, i.e the price p_1 depends upon c_A, that we represent by $p_1(c_A)$. If firm F_2's unitary production cost is high, $p_2^*(c_H|p_1(c_A))$ is the solution of

$$\max_{p_2 \geq 0}(a - p_2 + bp_1^*(c_A))(p_2 - c_H);$$

and if it is low, $p_2^*(c_L|p_1(c_A))$ is the solution of

$$\max_{p_2 \geq 0}(a - p_2 + bp_1^*(c_A))(p_2 - c_L).$$

(ii) Let us suppose that F_1 used the cheapest technology, i.e the price p_1 depends upon c_B, that we represent by $p_1(c_B)$. If firm F_2's unitary production cost is high, then $p_2^*(c_H|p_1(c_B))$ is the solution of

$$\max_{p_2 \geq 0}(a - p_2 + bp_1^*(c_B))(p_2 - c_H);$$

and if it is low, $p_2^*(c_L|p_1(c_B))$ is the solution of

$$\max_{p_2 \geq 0}(a - p_2 + bp_1^*(c_B))(p_2 - c_L).$$

Then,

$$p_2^*(c_H|p_1(c_A)) = \frac{a + c_H + bp_1(c_A)}{2}, \tag{7}$$

$$p_2^*(c_L|p_1(c_A)) = \frac{a + c_L + bp_1(c_A)}{2}, \tag{8}$$

$$p_2^*(c_H|p_1(c_B)) = \frac{a + c_H + bp_1(c_B)}{2}, \tag{9}$$

$$p_2^*(c_L|p_1(c_B)) = \frac{a + c_L + bp_1(c_B)}{2}. \tag{10}$$

The firm F_1 can anticipate $p_2^*(p_1)$ and then use this value to compute p_1^*. If firm F_1's unitary production cost is high, $p_1^*(c_A)$ is the solution of

$$\max_{p_1 \geq 0}(\theta(a-p_1+bp_2^*(c_H|p_1(c_A)))(p_1-c_A)+(1-\theta)(a-p_1+bp_2^*(c_L|p_1(c_A)))(p_1-c_A));$$

and if it is low, $p_1^*(c_B)$ is the solution of

$$\max_{p_1 \geq 0}(\theta(a-p_1+bp_2^*(c_H|p_1(c_B)))(p_1-c_B)+(1-\theta)(a-p_1+bp_2^*(c_L|p_1(c_B)))(p_1-c_B)).$$

Then,

$$p_1^*(c_A) = \frac{a(2+b) + (2-b^2)c_A + b(\theta c_H + (1-\theta)c_L)}{2(2-b^2)} \tag{11}$$

and

$$p_1^*(c_B) = \frac{a(2+b) + (2-b^2)c_B + b(\theta c_H + (1-\theta)c_L)}{2(2-b^2)}. \tag{12}$$

Using (11) and (12) in (7), (8), (9) and (10), we get

$$p_2^*(c_H|p_1^*(c_A)) = \frac{a(4+2b-b^2) + (2b-b^3)c_A + (4-2b^2)c_H + b^2(\theta c_H + (1-\theta)c_L)}{4(2-b^2)},$$

$$p_2^*(c_L|p_1^*(c_A)) = \frac{a(4+2b-b^2) + (2b-b^3)c_A + (4-2b^2)c_L + b^2(\theta c_H + (1-\theta)c_L)}{4(2-b^2)},$$

$$p_2^*(c_H|p_1^*(c_B)) = \frac{a(4+2b-b^2) + (2b-b^3)c_B + (4-2b^2)c_H + b^2(\theta c_H + (1-\theta)c_L)}{4(2-b^2)},$$

$$p_2^*(c_L|p_1^*(c_B)) = \frac{a(4+2b-b^2) + (2b-b^3)c_B + (4-2b^2)c_L + b^2(\theta c_H + (1-\theta)c_L)}{4(2-b^2)}.$$

□

Remark 3. Let $E(c_1) = \phi c_A + (1-\phi)c_B$ be the firm F_1's expected unitary production cost, and let $E(c_2) = \theta c_H + (1-\theta)c_L$ be the firm F_2's expected unitary production cost. From equalities (1) and (2), we get that the expected price for the good produced by the firm F_1 is

$$\begin{aligned} E(p_1^*) &= p_1^*(c_A)\phi + p_1^*(c_B)(1-\phi) \\ &= \frac{a(2+b)+(2-b^2)E(c_1)+bE(c_2)}{2(2-b^2)}; \end{aligned}$$

From equalities (3)-(6), we get that the expected price for the good produced by the firm F_2 is

$$\begin{aligned} E(p_2^*) &= (p_2^*(c_H|p_1^*(c_A))\phi + p_2^*(c_H|p_1^*(c_B))(1-\phi))\,\theta + \\ &\quad + (p_2^*(c_L|p_1^*(c_A))\phi + p_2^*(c_L|p_1^*(c_B))(1-\phi))\,(1-\theta) \\ &= \frac{a(4+2b-b^2)+(2b-b^3)E(c_1)+(4-b^2)E(c_2)}{4(2-b^2)}. \end{aligned}$$

In Figure 1, for some particular values of the parameters ($a = 10$, $c_A = c_H = 5$, $c_B = c_L = 2$) and for independent ($b = 0$) and substitutable ($b = 0.9$) goods, we show how the probabilities ϕ and θ affect the prices of the goods produced by both firms. We see that, in the case of independent goods, the graphics of the expected prices are symmetric; Moreover, the probability ϕ does not affect the expected price of the good produced by the firm F_2 and the probability θ does not affect the expected price of the good produced by the firm F_1 (see Figure 1a). The expected prices are lower when both firms use their cheapest technologies with high probability (see Figure 1b).

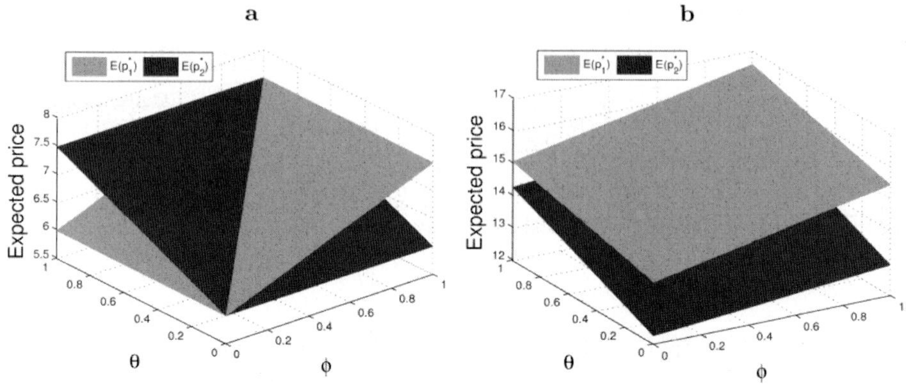

Fig. 1. Firms' expected prices, $E(p_1^*)$ and $E(p_2^*)$, (**a**) for firms producing independent goods, $b = 0$; and (**b**) for firms producing differentiated goods with degree of differentiation $b = 0.9$. Other parameteres values: $a = 10$, $c_A = c_H = 5$ and $c_B = c_L = 2$.

Remark 4. Firm F_1's expected profit is

$$\begin{aligned} E(\pi_1^*) = &\pi_1(p_1^*(c_A), p_2^*(c_H|p_1^*(c_A)))\phi\theta + \\ &+\pi_1(p_1^*(c_B), p_2^*(c_H|p_1^*(c_B)))(1-\phi)\theta + \\ &+\pi_1(p_1^*(c_A), p_2^*(c_L|p_1^*(c_A)))\phi(1-\theta) + \\ &+\pi_1(p_1^*(c_B), p_2^*(c_L|p_1^*(c_B)))(1-\phi)(1-\theta), \end{aligned}$$

and Firm F_2's expected profit is

$$\begin{aligned} E(\pi_2^*) = &\pi_2(p_1^*(c_A), p_2^*(c_H|p_1^*(c_A)))\phi\theta + \\ &+\pi_2(p_1^*(c_B), p_2^*(c_H|p_1^*(c_B)))(1-\phi)\theta + \\ &+\pi_2(p_1^*(c_A), p_2^*(c_L|p_1^*(c_A)))\phi(1-\theta) + \\ &+\pi_2(p_1^*(c_B), p_2^*(c_L|p_1^*(c_B)))(1-\phi)(1-\theta). \end{aligned}$$

In Figure 2, for some particular values of the parameters ($a = 10$, $c_A = c_H = 5$, $c_B = c_L = 2$) and for independent ($b = 0$) and substitutable ($b = 0.9$) goods, we show how the probabilities ϕ and θ affect the firms' expected profits. We see that, in the case of independent goods, the graphics of the expected profits are symmetric; Moreover, the probability ϕ does not affect the expected price of the good produced by the firm F_2 and the probability θ does not affect the expected price of the good produced by the firm F_1 (see Figure 2a). Each firm profits more when it uses its cheapest technology with high probability and the other firm uses its cheapest technology with low probability (see Figure 2b).

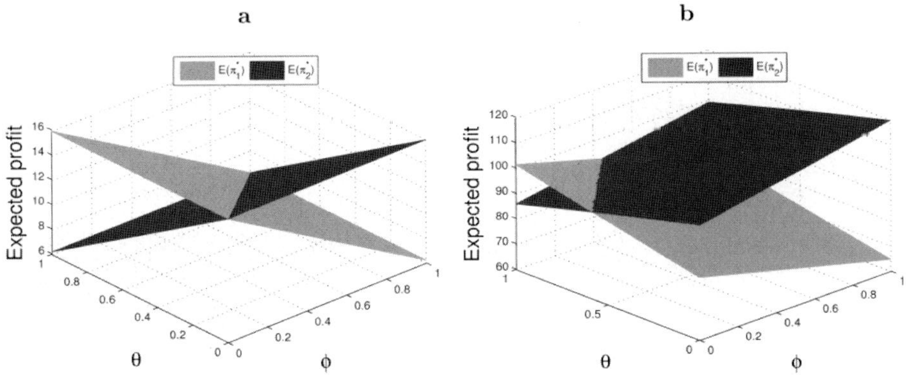

Fig. 2. Firms' expected profits, $E(\pi_1^*)$ and $E(\pi_2^*)$, (**a**) for firms producing independent goods, $b = 0$; and (**b**) for firms producing differentiated goods with degree of differentiation $b = 0.9$. Other parameteres values: $a = 10$, $c_A = c_H = 5$ and $c_B = c_L = 2$.

Acknowledgments

We would like to thank Bruno Oliveira and Miguel Ferreira all the useful discussions. We thank the Programs POCTI and POSI by FCT and Ministério da Ciência, Tecnologia e do Ensino Superior, and Centro de Matemática da Universidade do Porto for their financial support. Fernanda Ferreira gratefully acknowledges financial support from ESEIG/IPP and from PRODEP III by FSE and EU. Flávio Ferreira also acknowledges financial support from ESEIG/IPP.

References

[Sta34] von Stackelberg, H.: Marktform und Gleichgewicht. Julius Springer, Vienna (1934)

[Gib92] Gibbons, R.: A Primer in Game Theory. Pearson Prentice Hall, Harlow (1992)

[DH04] van Damme, E., Hurkens, S.: Endogenous price leadership. Games and Economic Behavior, **47**, 404–420 (2004)

Comparison of methodologies in river flow prediciton. The Paiva river case

Rui Gonçalves[1], Alberto Pinto[2] and Francisco Calheiros[1]

[1] Faculdade de Engenharia da Universidade do Porto, R. Dr. Roberto Frias, 4200-465 Porto {rjasg, xico}@fe.up.pt

[2] Centro de Matemática da Universidade do Porto, Rua Campo Alegre, 687, 4169 - 007 Porto aapinto@fe.up.pt

The aim of this work is to predict a future value of the daily mean discharge of the river Paiva. Several approaches are considered. Methods from Dynamical Systems and Stochastic Processes are applied. The Takens embedding shows an intermittent dynamical behaviour of the river Paiva where the laminar phase occurs in the absence of rainfall. The forcing of the system is non-deterministic and is due to the precipitation occurrence. Good predictability is found in the laminar regime.

1 Introduction

The analysis, modelling and prediction of a natural system like a River are of most importance for several reasons. Prevention of natural disasters and optimization of storage reservoirs for hydroelectric production among other reasons explain the influence of modelling and prediction in a natural system like a river. The river flow is a measurement of a complex system with many relevant variables: precipitation, inflow-runoff transformation, and the hydrogeological features of the river, etc. Some variables, such as rainfall, may change drastically in periods of time which are usually shorter than the period between observations. The effect of this sudden change is usually noticeable in the daily runoff values.

The relevant data for this work is the daily mean runoff of the river Paiva measured in Fragas da Torre from October 1946 to September 1999. We will discuss the predictability of a future value and we will show the difficulties in trying to do point prediction. We will also use different techniques coming from different areas as for example, Dynamical Systems and Stochastic Processes.

Several papers, [PR97, Siv00, CG03, PR96, LIRIL98], among others, concerning the application of chaos theory in hydrological data were published in the

Kenan Taş et al. (eds), Mathematical Methods in Engineering, 381–390.

last 15 years. The identification of chaos involves the use of nonlinear techniques for characterization and prediction in order to explore determinism in data. These techniques include correlation-dimension estimation, false nearest neighbours method, nonlinear prediction, Lyapunov exponent and entropy estimation among others. In some of the works, [PR97, PR96, JL94, LIRIL98] and [JG00] the authors were able to estimate the correlation-dimension for the reconstructed system. Here we also compute the correlation-dimension for the Paiva river runoff. The slopes of the correlation integral curve are computed for two different sets of data of Paiva runoff series, the original data set and for runoffs less than $20m^3/s$. The analysis led us to reason that for our data set an analogy may be established between the theoretical dynamical intermittence phenomena and the dynamics of the river Paiva. We also do a Singular Value Decomposition (SVD) analysis showing that the dynamics of runoffs are close to a segment line, table 2. Later we use the nearest neighbours method of prediction for one-step ahead prediction.

1.1 Data and preliminary analysis

The relevant data for this work consist of the time series of mean daily runoff of the Paiva river, measured at Fragas da Torre section, district of Beira-Alta. They are available for download in the *Instituto Nacional da Água* webpage[3]. The sample period runs from 1st of October of 1946 to 30th of September of 1999 for a total of 19358 observations (see chronogram of figure 1). The Paiva is considered to be one of the less polluted rivers in Europe. There are no dams at its basin which is still almost untouched by men. The Paiva is a small river when compared to the most economic relevant Iberian rivers like the Douro or Tejo. The Paiva is not an intermittent river in the sense that at the referred location and in the 53 years of observation the surface stream never dropped to zero. If we observe the series in more detail we can see that there is a huge difference in the flow between the summer and winter meaning that there are no glaciers or dams that may give water to the river in the summer (table 1, figure 1). This is typical for small rivers in southern Europe.

The monthly river flow descriptive statistics, (figure 2) shows structural differences specially between the summer and winter revealing the dominant regimes of each season. The spring and autumn months reveal a kind of mixed behaviour in terms of runoff statistics revealing a coexistence of the two regimes.

The sample autocorrelation function (figure 3) is characterized by the usual seasonality of this kind of data but also by an irregular behaviour specially for the winter where the correlation is positive.

[3] http://www.inag.pt

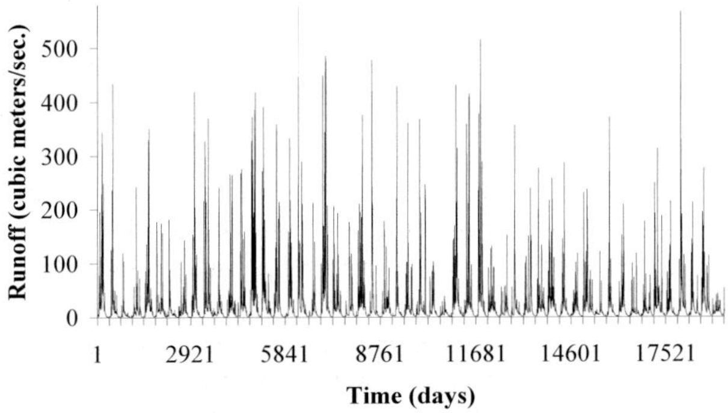

Fig. 1. Chronogram of the daily mean riverflow of Paiva measured at Fragas da Torre 1946-99.

Table 1. Descriptive statistics for the daily mean riverflow series of Paiva (1946-99) measured at Fragas da Torre.

Statistic	Value
Mean	20.73 m^3/s
Median	5.66m^3/s
Skewness	5.3
Kurtosis	45.98
Maximum	920.0 m^3/s
Minimum	0.06m^3/s

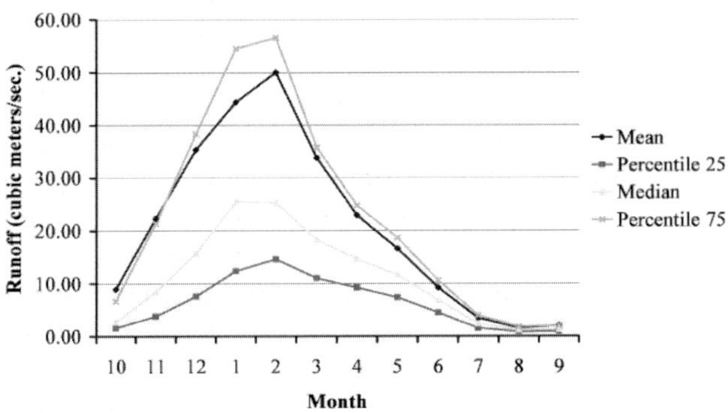

Fig. 2. Evolution of Monthly Paiva daily mean runoff time series statistics.

2 Correlation-dimension estimation

First we do a Takens reconstruction embedding[Tak80], using the Paiva river 1946-99 daily mean data for several dimensions 1-8 in an attempt to under-

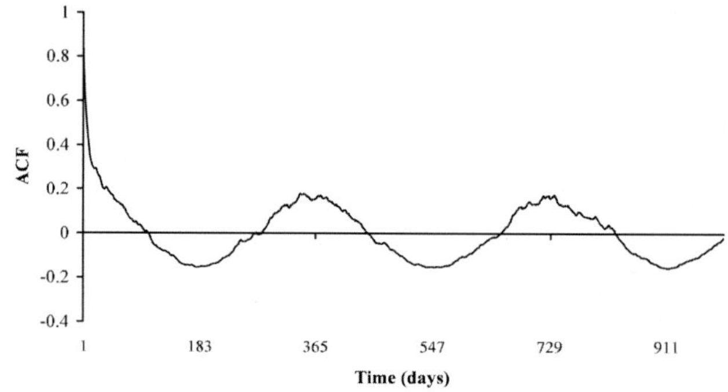

Fig. 3. Sample Autocorrelation Function of the daily mean runoff series (1946-99) of the Paiva river at Fragas da Torre.

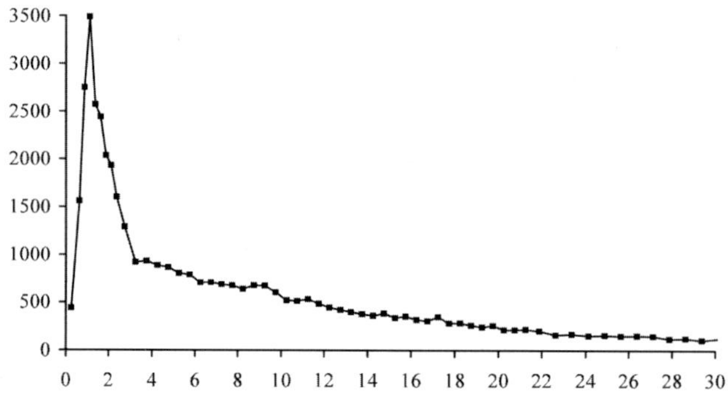

Fig. 4. Histogram of the mean daily runoff series of Paiva river.

stand deterministic and stochastic dynamics underneath the behaviour of the river runoff. In this direction we do a Correlation-Integral (CI) Analysis for all the data and then we consider only the runoffs less than $20m^3/s$ which represents about 75% of the data corresponding mainly to the periods without rain, figure 5. We realized that the CI integral slopes are close to 1 for runoffs less than $20m^3/s$. Afterwards we do a SVD analysis showing that the dynamics of runoffs of the laminar phase are close to a segment line. Furthermore, the majority of orbits stay even closer to a 3D hypervolume containing the segment line.

2.1 Correlation-integral analysis

The sample CI, $C_N^{(m)}(\varepsilon)$ of a reconstructed system is defined by,

$$C_N^{(m)}(\varepsilon) = \frac{2}{N(N-1)} \Theta \left\{ (i,j) : 1 \leq i < j \leq N, \|X_i - X_j\| < \varepsilon \right\} \qquad (1)$$

where $(X_t, X_{t+1}, \ldots, X_{t+m-1})$ is a reconstructed vector which elements are values of the time series, $\{X_t\}_{t=1}^{N}$, N is the number of data points of the series, Θ the Heaviside function, ε the neighborhood radius and m the embedding dimension of the reconstructed phase space. The sample CI is the relevant statistic proposed by [GP83] for the estimation of the correlation-dimension. Empirically the sample CI is the fraction of reconstruction vectors at a distance smaller than ε in the reconstructed phase space. The sum, (1), is computed for a set of distances, $\varepsilon_1, \ldots, \varepsilon_n$ evenly spaced on a logarithmic scale. A *scaling range* is said to exists if for such a range of values the sample correlation integral behaves like a power law or the same to say like an horizontal line on a log – log scale.

$$C_N^{(m)}(\varepsilon) \sim \alpha \varepsilon^{D_C}, \; \varepsilon \to 0, \; N \to \infty \tag{2}$$

For a constant α, $d(N, \varepsilon)$ is the slope of the CI curve for a certain range and D_C is then the estimate of the correlation dimension,

$$d(N, \varepsilon) = \frac{\partial \ln C_N^{(k)}(\varepsilon)}{\partial \ln \varepsilon} \quad \text{and} \quad D_C = \lim_{\varepsilon \to 0+} \lim_{N \to \infty} d(N, \varepsilon) \tag{3}$$

Hence the exponent can be estimated using linear regression of $\ln C_N^{(k)}(\varepsilon)$ over $\ln \varepsilon$.

On figure 5 three different behaviours for the correlation-integral curve can be noticed for different ranges of the radius, ε. For runoff values larger than $30m^3/s$ there is no scaling range and the slope continues to decrease towards the zero dimension range. For runoffs in the interval $[5 - 30m^3/s]$ there is a scaling range and the slope of the curves indicates a near 1 dimension which is the dimension of a line. These facts indicate the existence in the reconstructed phase-space of a one-dimensional manifold to which all the laminar phase orbits are close. It may be said that the orbits near this one-dimensional manifold constitute a ε-neighborhood.

2.2 Singular value decomposition analysis

The Singular Value Decomposition (SVD) is a technique used to find the Principal Components of a set of data. These Principal Components are linear combinations of random or statistical variables which have special properties in terms of variances. For example, the first principal component is the normalized linear combination with maximum variance. Transforming the original vector variable to the vector of principal components amounts to a rotation of coordinate axes to a new coordinate system made of normalized characteristic vectors of the covariance matrix. In the context of dynamical systems we take as vector variable the reconstruction vectors, $(X_t, X_{t+1}, \ldots, X_{t+m-1})$, where X_t is the daily mean runoff at day t. In this context the individuals

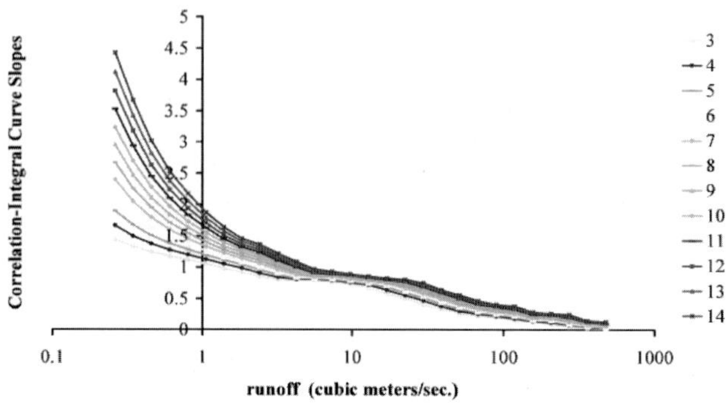

Fig. 5. Slopes of the sample correlation integral curve of the Paiva river data and for several embedding dimensions.

are the reconstruction vectors. From this point of view the data matrix is a homogeneous set and the metric to be used in the space of individuals is the Euclidean one.
Using the *SVD* one computes the principal directions of the data set and corresponding weights. This information is also relevant for the understanding of the correlation integral curve. Using the *SPAD* statistical package we have calculated the axes or principal factors for the covariance matrix of the daily mean runoff for different embedding dimensions which values are presented on table 2.

Table 2. Two largest Eigenvalues of the Covariance Matrix for the daily mean runoff series and percentage % of the total variance explained.

Dimension	1st Eigenvalue	%	2nd Eigenvalue	%
3	4018.45	86.66	424.67	9.16
4	5116.84	82.76	617.28	9.98
5	6147.33	79.51	821.15	10.62
6	6963.53	75.06	1024.27	11.04
7	7793.24	72.00	1367.12	12.63
8	8612.60	69.62	1699.14	13.74

For all considered dimensions the first two axes explain more than 80% of the total variance. The correlations between the principal components and the original variables are presented on the table 3. In the laminar phase there exists a principal component explaining more 90% of the variance. This is explained by a laminar dynamic close to a segment line, figure 6. According to the usual criteria to quantify the number of significant eigenvalues, (see [Sap90]) the reconstruction vectors (*individuals*) of the last two data sets are

Table 3. Correlation between the 1st Principal Component and the original variables for several embedding dimensions.

Dimension	3	4	5	6	7	8
X_t	0.92	0.88	0.85	0.83	0.81	0.78
X_{t+1}	0.96	0.94	0.91	0.89	0.86	0.84
X_{t+2}	0.92	0.94	0.93	0.91	0.89	0.87
X_{t+3}	-	0.88	0.91	0.91	0.90	0.88
X_{t+4}	-	-	0.85	0.87	0.88	0.87
X_{t+5}	-	-	-	0.77	0.82	0.84
X_{t+6}	-	-	-	-	0.77	0.81
X_{t+7}	-	-	-	-	-	0.76

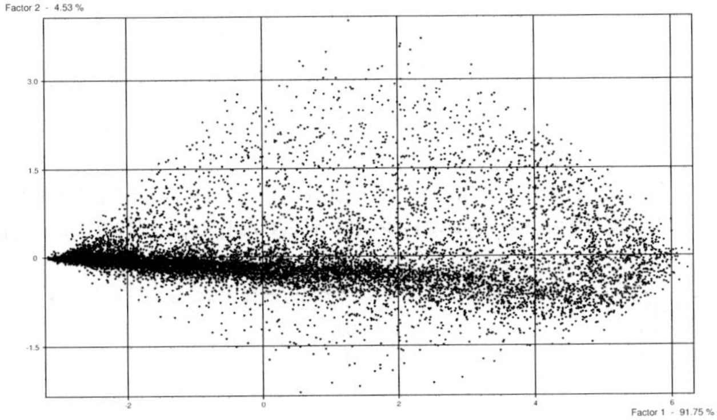

Fig. 6. Projection of *individuals* (reconstruction vectors) on the plane with principal axes.

almost one-dimensional. On figure 6 we present the *individuals* (reconstruction vectors) in the plane with the two principal components as axis.

2.3 Nonlinear prediction

Several authors used nonlinear prediction methods for river flow data locally in the phase space, [JL94, JLX02, PR96, PR01, LIRIL98] and [IS02] among others.

In this work we use a different version of the nearest neighbours method proposed by [KSc97] to predict the next day runoff for the years 1997/98 using the information of the historic series from 1946 to 1999.

Instead of using all the neighbours within a fixed radius we use the top ten closest neighbours. The prediction set is the phase space average of the neighbours images. Other authors, [LIRIL98] have used as predictors different local linear functions including the average of the closest neighbours images and concluded that the prediction results for this method were better than those obtained by other local functions. When the predictor is just the average of the Nearest neighbours then the method may be considered non-parametric. Nevertheless there are always the phase space parameters to be found following some criteria. Using such a predictor the prediction quality depends only on the reconstruction parameters which values reflect the number of neighbours and their distance from the prediction point. The good thing about these dependencies is that it is not expected for the prediction results to depend to much on the parameters values. There is no dynamical reason for that. On table 4 we present a summary of the usual fitting evaluation criteria.

Table 4. Mean Square Error for the hydrological year 1997/98 one step ahead prediction and for different embedding dimensions (m) and time delays, (τ).

τ/m	3	4	5	6	7	8	9	10	11	12	13	14	15
1	8.98	7.36	7.08	7.39	8.38	8.35	8.63	8.70	9.29	9.03	9.10	10.12	10.91
2	10.87	12.33	11.72	12.84	12.55	14.01	14.30	14.53	15.28	14.93	15.37	14.28	14.49
3	12.05	11.43	13.04	13.60	13.28	13.49	13.82	15.50	15.58	16.00	14.52	14.40	13.40
4	8.93	11.01	13.35	13.37	11.99	12.36	12.42	13.10	14.04	14.23	15.22	15.07	16.14
5	18.63	17.08	20.32	19.11	17.76	21.08	20.67	20.53	21.40	21.79	21.96	21.62	21.87
6	16.57	16.74	16.92	18.15	14.64	15.09	14.58	14.84	14.72	15.38	17.63	18.17	18.53
7	20.95	22.27	26.13	26.85	26.64	25.25	25.66	26.20	25.93	25.38	25.63	25.43	26.44
8	17.88	17.57	17.96	16.72	17.68	20.37	20.54	18.38	20.69	20.77	20.37	20.21	19.72
9	17.26	18.98	14.31	14.70	13.93	14.04	18.02	18.28	15.71	15.63	15.50	15.97	16.33
10	34.99	34.02	33.23	33.13	33.21	34.90	35.23	39.62	39.66	40.83	40.39	39.97	39.93

The best Mean Square Error[4] (MSE) result was found for an embedding dimension 5 and for time delay of one day, (see table 4). Nevertheless, the results for others dimensions were found to be of the same magnitude and therefore not significantly different.

Here we should mention the paper on river flow prediction, [IS02] where the authors also obtained results of the same magnitude for the *MSE* for different embedding dimensions.

Predicting locally in the phase space with linear functions can be a problem if the system has an intermittent behaviour. As we mention earlier in the laminar phase the flow of the Paiva river slowly converges to an equilibrium. This convergence is abruptly stopped if it starts raining at some point in time. If a rain event occurs then the flow increases and the system starts a much more erratic behaviour. A quality prediction comparison between the two incompatible dynamic regimes should be made. The phase space orbits to be included on the estimate calculation will depend on the choice for one

[4] The Mean Square Error of Prediction (MSE) is defined by MSE $= \frac{1}{n}\sum_{i=1}^{n}\left(X_t - \hat{X}_t\right)$. When several models are proposed for the same data the ultimate choice of one may depend on goodness of fit such as the MSE.

or the other regime. Increasing(decreasing) orbits for the river flow should be eliminated if the absence of rainfall (rainfall) regime is forecasted.

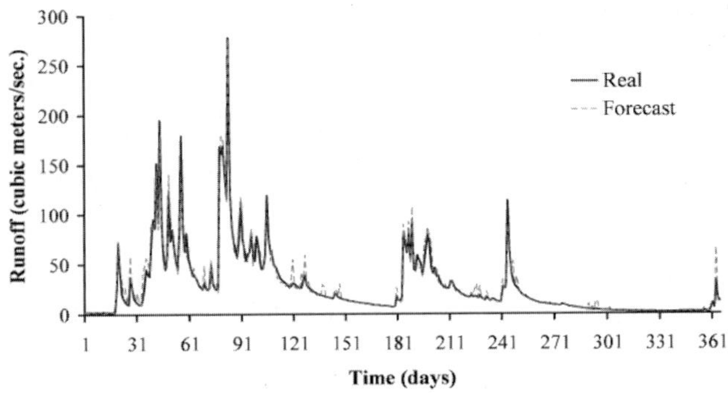

Fig. 7. One-step ahead prediction using 10 nearest neighbours.

The one-step ahead predictions for the Paiva river and for the hydrological year 1997/98 are presented on figure 7.

3 Conclusions

A Dynamical analysis of the Paiva river data was performed using Takens method of dynamical reconstruct, [Tak80]. Later we used the nearest neighbour's method of prediction for one-step ahead prediction. The results indicate a gain on quality prediction when one considers only the laminar phase. These differences are due to the action of rain that seems to be unpredictable when we are dealing with daily mean runoff data. The prediction results also reveal close MSE for different embedding dimensions of the phase space and the dimension 3 has been proven to be the best. We noted also that the principal component analysis of the reconstruction vectors confirmed the correlation curve analysis. This means that the information given by recent past runoffs which reflects non-observable variables does not improve predictions in a significant manner. This result is not due to a wrong tuning of the method but instead to a limitation of the data. We may conclude then that the most significant information for the next day mean runoff is the present day mean runoff.

References

[PR97] Porporato, L., Ridolfi, L.: Nonlinear analysis of a river flow time sequences. Water Resources Research, **33**, 1353–1367 (1997)

[Siv00] Sivakumar, B.: Chaos in hydrology: important issues and interpretations. J. of Hydrology, **227**, 1-20 (2000)

[CG03] Calheiros, F., Gonçalves, R.: Previsão em Hidrologia. In: Brito, P., Figueiredo, A., Sousa, F., Teles P. and Rosado, F., (eds.) Proceedings of the X Annual Congress of the Portuguese Statistical Society (SPE). 229–241 (2003)

[PR96] Porporato, L., Ridolfi, L.: Clues to the existence of deterministic chaos in river flow. Int. J. of Mod. Phys. B., **10**, 1821–1862 (1996)

[LIRIL98] Liu, Q., Islam, S., Rodriguez-Iturbe, I., Le, Y.: Phase-space analysis of daily streamflow: characterization and prediction. AWRA. **210**, 463–475 (1998)

[JL94] Jayawardena, A.W., Lai, F.: Analysis and prediction of chaos in rainfall and stream flow time series. Journal of Hydrology, **153**, 23–52 (1994)

[JG00] Jayawardena, A., Gurung, A.: Noise reduction and prediction of hydrometereological time series dynamical systems approach vs stochastic approach. Journal of Hydrology, **228**, 242–264 (2000)

[Tak80] Takens, F.: Detecting strange attractors in Turbulence. In: Rand D. A., Young L. (ed.) Lecture Notes in Math. **898**, Springer, Berlin (1980)

[GP83] Grassberger, P., Procaccia, I.: Measuring the strangeness of strange attractors. Physica 9D, **9**, 189–208 (1983)

[Sap90] Saporta, G.: Probabilités, Analyse des Donnes Et Statistique. Editions Technip, Paris (1990)

[JLX02] Jayawardena, A., Li, W., Xu, P.: Neighbourhood selection for local modelling and prediction of hydrological time series. Journal of Hydrology, **258**, 40–57 (2002)

[PR01] Porporato, A., Ridolfi, L.: Multivariate nonlinear prediction of river flows. Journal of Hydrology, **248**, 109–122 (2001)

[IS02] Islam, S., Sivakumar, B.: Characterization and prediction of runoff dynamics: a nonlinear dynamical view. Journal of American Water Resources Association, **25**, 179–190 (2002)

[KSc97] Kantz, H., Schreiber, T.: Nonlinear Time Series Analysis. Cambridge Univ. Press, Cambridge (1997)

A XY Spin Chain Models on Space Curves and Analogy with Kirchhoff Rods

Georgi G. Grahovski[1,2] and Rossen Dandoloff[2]

[1] Laboratoire de Physique Théorique et Modélisation, Université de Cergy-Pontoise, 2 avenue A. Chauvin, F-95302 Cergy-Pontoise Cedex, FRANCE
ggrahovs@ptm.u-cergy.fr rossen@ptm.u-cergy.fr

[2] Laboratory of Solitons, Coherence and Geometry, Institute for Nuclear Research and Nuclear Energy, Bulgarian Academy of Sciences,72 Tsarigradsko chaussée, 1784 Sofia, BULGARIA
grah@inrne.bas.bg

A XY Heisenberg spin chain model with two perpendicular spins par site is mapped onto a Kirchhoff thin elastic rod. It is shown that in the case of constant curvature the Euler–Lagrange equation leads to the static sine-Gordon equation. The case of a double-helical DNA-like configuration corresponds to two interacting Heisenberg spin chains and the corresponding Euler–Lagrange equation gives a system of coupled static sine-Gordon-type equations. The kink-antikink type and periodical static solutions for these models are derived. The soliton dynamics and the the nonlinear excitations of the systems are investigated. The interplay between curvature and nonlinear excitations is analyzed as well.

1 Introduction

The study of an elastic rods is a subject to increased interest especially in connection with the bio-mathematical models of proteins and of DNA [1, 2, 3, 4, 5, 6, 7, 8]. The main feature of a thin rod is a space curve (rod's axis) and the corresponding orthonormal frame with a tangent vector $\mathbf{t}$ to the axial curve [9, 10, 11, 12, 13]. The static energy of the elastic rod is related to the bending and twisting energies [14, 15]. It is tempting to map the elastic rod problem to a classical spin chain [16, 17, 18, 19] (in the continuum limit, where the normalized spin $\mathbf{S}$ is mapped onto the tangent vector $\mathbf{t}$). We will show however that the full mapping of the elastic rod onto a spin-chain model requires a system of two orthogonal spins [20, 21, 22].

The spin Hamiltonian for a Heisenberg spin chain is given by the following expression:

$$H = J_0 \sum_i \mathbf{S}_i(s) \cdot \mathbf{S}_{i+1}(s), \qquad \mathbf{S}_i^2 = \mathbf{S}_{i+1}^2 = 1. \tag{1}$$

In the continuum limit this Hamiltonian goes over to

$$H = J_0 \int_{-\infty}^{+\infty} \left(\frac{d\mathbf{S}(x)}{dx} \right)^2 dx. \tag{2}$$

Kenan Taş et al. (eds), Mathematical Methods in Engineering, 391–401.

In the case of XY spin chain the spin is given by the rotation angle $\theta(x)$: $\mathbf{S}(s) = (\cos\theta(x), \sin\theta(x))$. The Hamiltonian now reads $H = J_0 \int_{-\infty}^{+\infty} \left(\frac{d\theta}{dx}\right)^2 dx$.

This article is organized as follows: In section 2 are reviewed the static properties of the Kirchhoff equations for a thin elastic rod. It is shown that if a curvature is present the twist angle satisfies the static sine-Gordon equation. The mapping of the Kirchhoff model onto a XY spin chain is done in Section 3. In addition a XY spin model with two interacting Heisenberg chains for a system of two interacting Kirchhoff rods is proposed. In Section 4 the soliton-like solutions for the static sine-Gordon equations are briefly discussed.

2 Kirchhoff Model for Elastic Rods

The model introduced by Kirchhoff (1859) describes the shape and the dynamics of a thin elastic rod in equilibrium and is based on the analogy with the dynamics of a heavy spinning top (the Lagrange case). The shape is described by the *static Kirchhoff model* while the time evolution – by the *dynamical Kirchhoff model*. Here we will concentrate on the statics of thin elastic rods (here and below we shall call them Kirchhoff rods).

We consider a static space curve $\mathbf{R}(s) : \mathbb{R} \to \mathbb{R}^3$ as a smooth function mapping the arc-length interval $I \subset \mathbb{R}$ into the physical space $\mathbb{R}^3$. For every s we define the Frenet basis $(\mathbf{t}(s), \mathbf{n}(s), \mathbf{b}(s))$ to be the normal, binormal and the tangent vectors to the curve(s). The tangent vector is a unit vector given by $\mathbf{t} = \left(\frac{d\mathbf{R}}{ds}\right)$ and the curvature $\kappa(s)$ of the curve at the point s is then given by:

$$\kappa(s) := \left|\frac{d\mathbf{t}}{ds}\right| .$$

The triad $(\mathbf{t}(s), \mathbf{n}(s), \mathbf{b}(s))$ evolves along s according to the Frenet–Serret equations:

$$\frac{d\mathbf{t}}{ds} = \kappa\mathbf{n}(s) \qquad \frac{d\mathbf{n}}{ds} = -\kappa\mathbf{t}(s) + \tau\mathbf{b}(s) \qquad \frac{d\mathbf{b}}{ds} = -\tau\mathbf{n}(s), \tag{3}$$

where τ is the torsion of the curve $\mathbf{R}(s)$. If the curvature κ and the torsion τ are known for all s then the Frenet–Serret triad can be obtained as unique solution of (3). Next the space curve $\mathbf{R}(s)$ can be reconstructed by integrating the tangent vector $\mathbf{t}(s)$.

A thin rod can be modelled by a space curve $\mathbf{R}(s)$ joining the loci of the centroids of the cross sections together with the local basis $(\mathbf{d}_1(s), \mathbf{d}_2(s), \mathbf{d}_3(s))$ attached to the rod material. This local basis can be expressed through the Frenet–Seret triad as follows:

$$(\mathbf{d}_3(s), \mathbf{d}_2(s), \mathbf{d}_1(s)) = (\mathbf{t}(s), \mathbf{n}(s), \mathbf{b}(s)) \begin{pmatrix} 1 & 0 & 0 \\ 0 & \cos\phi & -\sin\phi \\ 0 & \sin\phi & \cos\phi \end{pmatrix},$$

where ϕ is the twist angle of the rod. The components of the derivatives of the local basis $(\mathbf{d}_3(s), \mathbf{d}_2(s), \mathbf{d}_1(s))$ with respect to s can be expressed by using the twist vector $\mathbf{k}(s) = \kappa_1\mathbf{d}_1 + \kappa_2\mathbf{d}_2 + \kappa_3\mathbf{d}_3$ as follows:

$$\frac{d\mathbf{d}_i}{ds} = \mathbf{k}(s) \times \mathbf{d}_i(s), \qquad i = 1, 2, 3.$$

The static Kirchhoff equations describe the shape of the rod under the effects of internal elastic stresses and boundary constraints, in the absence of external force fields. Let $\mathbf{F}(s)$ is the tension and $\mathbf{M}(s)$ is the torque of the rod. In the approximation of a linear theory (the Hook's law applies) the torque $\mathbf{M}$ is related to the twist vector $\mathbf{k}$ by $\mathbf{M}(s) = \mathcal{S} \cdot \mathbf{k}(s)$, where $\mathcal{S} = \mathrm{diag}\,(1, a, b)$. The constant a measures the asymmetry of the cross section and b is the scaled torsional stiffness. In particular for symmetric ($a = 1$) hyperelastic ($b = 1$) rods we have $\mathbf{M}(s) = \mathbf{k}(s)$. In the generic case the torque is

$$\mathbf{M}(s) = \kappa_1(s)\mathbf{d}_1(s) + a\kappa_2(s)\mathbf{d}_2(s) + b\kappa_3(s)\mathbf{d}_3(s) \tag{4}$$

and the elastic energy of the Kirchhoff rod is given by:

$$H = \frac{1}{2}\int_{s_1}^{s_2} \mathbf{M}(s) \cdot \mathbf{k}(s)\, ds = \frac{1}{2}\int_{s_1}^{s_2} (\kappa_1^2(s) + a\kappa_2^2(s) + b\kappa_3^2(s))\, ds \tag{5}$$

The conservation of the linear and angular momenta is provided by the static Kirchhoff equations:

$$\frac{d\mathbf{F}}{ds} = 0, \tag{6}$$

$$\frac{d\mathbf{M}}{ds} + \mathbf{d}_3(s) \times \mathbf{F}(s) = 0. \tag{7}$$

Here $\mathbf{F}(s)$ is the tension of the rod and the torque $\mathbf{M}(s)$ is given by (4) and the twist vector reads

$$\mathbf{k}(s) = (\kappa(s)\sin\phi, \kappa(s)\cos\phi, \tau + \phi_s), \tag{8}$$

where the twist angle ϕ is a function of the arc length parameter s: $\phi = \phi(s)$. The expression for the tension $\mathbf{F}(s)$ in the local basis $\mathbf{d}_1, \mathbf{d}_2, \mathbf{d}_3$:

$$\mathbf{F}(s) = F_1(s)\mathbf{d}_1(s) + F_2(s)\mathbf{d}_2(s) + F_3(s)\mathbf{d}_3(s)$$

reduces the Kirchhoff equations (6) and (7) to the following system of ODE's:

$$F_{1,s} + \kappa_2 F_3 - \kappa_3 F_2 = 0 \tag{9}$$

$$F_{2,s} + \kappa_3 F_1 - \kappa_1 F_3 = 0 \tag{10}$$

$$F_{3,s} + \kappa_1 F_2 - \kappa_2 F_1 = 0 \tag{11}$$

$$F_1 = -a\kappa_{2,s} + (b-1)\kappa_1\kappa_3 \tag{12}$$

$$F_2 = \kappa_{1,s} + (b-a)\kappa_2\kappa_3 \tag{13}$$

$$b\kappa_{3,s} + (a-1)\kappa_1\kappa_2 = 0. \tag{14}$$

Using the parameterization of the twist vector (8) from (14) for the case of constant curvature $\kappa(s) = \kappa_0$ we get the famous static (scalar) sine–Gordon equation:

$$\frac{d^2u}{ds^2} + \frac{(a-1)}{b}\kappa_0^2 \sin u(s) = 0, \qquad u(s) = 2\phi(s). \tag{15}$$

This second order differential equation is a completely integrable Hamiltonian system and allows so-called "soliton"-like solutions. It appears in a wide variety of physical problems for e.g. charge-density-wave materials, splay waves in membranes, magnetic flux in Josephson lines, torsion coupled pendula, propagation of crystal dislocations, Bloch wall motion in magnetic crystals, two-dimensional elementary particle models in the quantum field theory, etc.

For a symmetric rod, i.e. $a = 1$ equation (15) simplifies to $\frac{d^2u}{ds^2} = 0$, i.e. ϕ_s = const. This is the usual case widely discussed in the literature [23]. The cross-section of the symmetric rod ($a = 1$) has a continuous rotational symmetry around the central axis. Therefore the elastic energy density h does not depend on ϕ and from the variational principle it follows that h could be only a function of the derivatives ϕ_s. Therefore the only solution for a constant twist is ϕ_s = const.

There have been even attempts to generalize this result to the asymmetric case ($a \neq 1$) [24], i.e. to show that ϕ_s = const holds true for any Kirchhoff rod.

The asymmetric case has been overlooked for a long time. For a constant curvature and torsion along the center-line it represents another integrable case of the Kirchhoff equations for a thin elastic rod. This opens new possibilities for a more adequate modelling of bio-polymers and gives the phenomenological bases for the widely used DNA models. Here there is no more continuous rotational symmetry of the cross-section around the center axis and obviously h depends on ϕ as well. So the constant twist is no more a solution.

The solution of (15) is compatible with the full system of Kirchhoff equations (9)–(14) for constant curvature and torsion.

3 Space Curve Models of Spin Chains

In this section we consider two models of XY spin chains on space curves: a single space curve model and a double-helical (DNA-like) model. We will show that the Kirchhoff rod model from the previous section can be mapped into the XY Heisenberg spin chain model described below.

3.1 Single Kirchhoff Rod Model

Let us consider the following spin chain model with two perpendicular spins per site: the spin vectors $\mathbf{S}_1$ and $\mathbf{S}_2$ have different lengths and are given by

$$\mathbf{S}_1(s) = \mathbf{n}(s)\cos\phi + \mathbf{b}(s)\sin\phi, \qquad \mathbf{S}_1^2 = 1, \qquad \phi = \phi(s); \tag{16}$$

$$\mathbf{S}_2(s) = -\sqrt{\frac{b-a+1}{a+b-1}}\mathbf{n}(s)\sin\phi + \sqrt{\frac{b-a+1}{a+b-1}}\mathbf{b}(s)\cos\phi, \qquad \mathbf{S}_2^2 = \frac{b-a+1}{a+b-1}.$$

Here also both spin vectors are orthogonal: $\mathbf{S}_1(s)\cdot\mathbf{S}_2(s) = 0$ for every s and

$$\left|\frac{d\mathbf{S}_1}{ds}\right|^2 = (\tau + \phi_s)^2 + \kappa^2 \cos^2 \phi, \quad \left|\frac{d\mathbf{S}_2}{ds}\right|^2 = \frac{b-a+1}{a+b-1}\left[(\tau + \phi_s)^2 + \kappa^2 \sin^2 \phi\right].$$

This is an integrable system with Hamiltonian given by

$$H = J_0 \sum_i \left(\mathbf{S}_{1,i}(s) \cdot \mathbf{S}_{1,i+1}(s) + \mathbf{S}_{2,i}(s) \cdot \mathbf{S}_{2,i+1}(s)\right),$$

which in the continuum limit leads to:

$$\begin{aligned} H &= J_0 \int_{s_1}^{s_2} \left[\left(\frac{\partial \mathbf{S}_1}{\partial s}\right)^2 + \left(\frac{\partial \mathbf{S}_2}{\partial s}\right)^2\right] ds \\ &= J_1 \int_{s_1}^{s_2} \left(b(\tau + \phi_s)^2 + \kappa^2 (a-1) \sin^2 \phi\right) ds, \end{aligned} \tag{17}$$

where J_1 is the renormalized coupling constant, $s \in [s_1, s_2]$ and the subscript s means a derivative with respect to s. This Hamiltonian coincides with that one in (5) where $\mathbf{k}(s)$ has been replaced from (8). Thus the asymmetric Kirchhoff rod is mapped onto a two-spin chain. If the curvature is constant $\kappa(s) = \kappa_0$, then the Euler–Lagrange equation gives the (scalar) static sine–Gordon equation:

$$\frac{d^2\phi}{ds^2} + \kappa_0^2 \frac{1-a}{b} \sin\phi \cos\phi = 0. \tag{18}$$

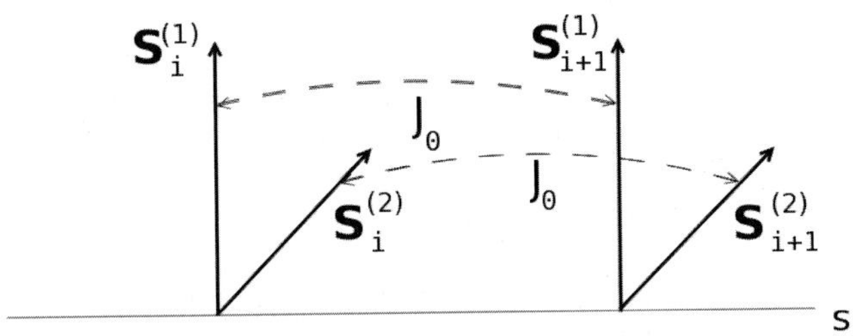

Fig. 1. A two-spin XY chain system with a coupling constant J_0 that corresponds to the Kirchhoff rod model.

When $a \to 1$ the Hamiltonian (17) simplifies to $\int_{s_1}^{s_2} (\tau + \phi_s)^2 \, ds$, so the Euler-Lagrange equation leads to $\phi_{ss} = 0$, or $\phi_s =$ const. Note that the case $a = 1$ corresponds to a symmetrical Kirchhoff rod. Such a rod-model is mapped onto a symmetrical spin chain system $(\mathbf{S}_1, \mathbf{S}_2)$ with $\mathbf{S}_1^2 = \mathbf{S}_2^2 = 1$. The mapping of the Kirchhoff symmetric rod needs a two-spin XY chain rather than a simple one-spin XY chain. ¿From (18) one can easily get that

$$\left(\frac{d\phi}{ds}\right)^2 + \kappa_0^2 \frac{1-a}{b} \sin^2 \phi(s) = 0,$$

which may be used for the calculating the corresponding static energy. For the asymmetric thin elastic rod the model Hamiltonian takes the form:

$$H = \int_{s_1}^{s_2} \left((\tau + \phi_s)^2 + \kappa_0^2 \frac{a-1}{b} \sin^2 \phi \right) ds \tag{19}$$

Below we shall discuss the properties of the static soliton solutions of the model (19) like the dynamics of nonlinear excitations and the kink-type solutions of (18) as well as its periodic soliton-like solutions (knoidal waves).

3.2 Double Helical Model

We specify the second model we will deal with: Consider a system of two helices with a common symmetry axis (Fig. 2). It is easy to see that at every point the Frenet basis (moving triads) attached to the two helices $(\mathbf{t}_i(s), \mathbf{n}_i(s), \mathbf{b}_i(s))$, $i = 1, 2$, where $\mathbf{t}_i(s)$, $\mathbf{n}_i(s)$ and $\mathbf{b}_i(s)$ are the tangent, the normal, and the binormal vectors to the curves respectively, are oriented in such a way that for every s the normal vectors to both helices are anti-parallel: $\mathbf{n}_1(s) = -\mathbf{n}_2(s)$. Each of the triads evolves along the curves according to the Frenet–Serret equations (3).

If $\mathbf{S}_i(s)$ are chosen to lie in the horizontal plane then the spin vectors have the form:

$$\begin{aligned} \mathbf{S}_1(s) &= \mathbf{n}(s)\sin\phi_1 + (\alpha\mathbf{t}(s) + \beta\mathbf{b}(s))\cos\phi_1, \quad ||\mathbf{S}_1||^2 = 1, \quad \phi_i = \phi(s), \\ \mathbf{S}_2(s) &= -\mathbf{n}(s)\sin\phi_2 + (\alpha\mathbf{t}(s) + \beta\mathbf{b}(s))\cos\phi_2, \quad \alpha^2 + \beta^2 = 1, \quad i = 1, 2; \end{aligned} \tag{20}$$

where α and β are constants. For this two-helical system we have an additional constraint $\alpha\kappa + \beta\tau = 0$ because $\alpha\mathbf{t}(s) + \beta\mathbf{b}(s)$ is to be horizontal (i.e. its z-component is zero).

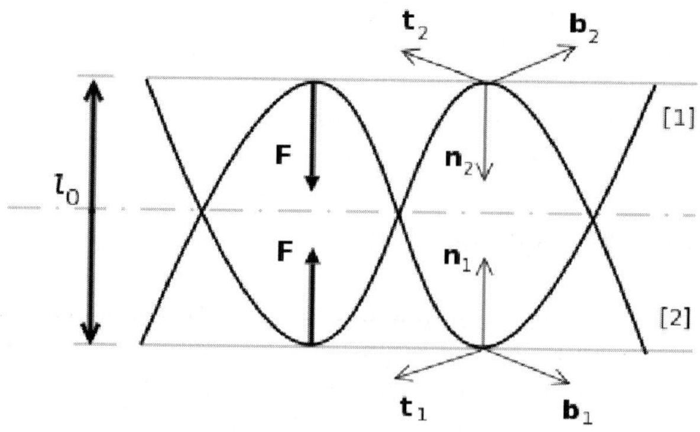

Fig. 2. Double-helical configuration with elastic interaction

We consider that the interaction of both spin-helices is of the usual form $J(a)\mathbf{S}_1(s)\cdot\mathbf{S}_2(s)$ (as it is shown of Figure 2) which adds the following expression (in the continuum limit) to the full Hamiltonian of the two single-helices system:

$$H_{\text{int}} = \int_{-\infty}^{\infty} J(l)\cos(\phi_1(s) - \phi_2(s))\,ds, \tag{21}$$

where l is the distance between the two chains. The "kinetic part" of the Hamiltonian becomes:

$$\begin{aligned} H_{\text{kin}} &= J_0 \int_{-\infty}^{\infty} \left[\left(\frac{\partial \mathbf{S}_1(s)}{\partial s} \right)^2 + \left(\frac{\partial \mathbf{S}_2(s)}{\partial s} \right)^2 \right] ds \\ &= J_0 \int_{-\infty}^{\infty} [\phi_{1,s} + (\alpha\kappa - \beta\tau)]^2\,ds + J_0 \int_{-\infty}^{\infty} [\phi_{2,s} + (\alpha\kappa - \beta\tau)]^2\,ds \end{aligned} \tag{22}$$

and the full Hamiltonian of the considered system reads: $H = H_{\text{kin}} + H_{\text{int}}$. The Euler-Lagrange equation gives the following system of coupled ODE's:

$$\begin{aligned} J_0 \frac{d^2\phi_1}{ds^2} - 2J(l)\sin(\phi_1 - \phi_2) &= 0, \\ J_0 \frac{d^2\phi_2}{ds^2} + 2J(l)\sin(\phi_1 - \phi_2) &= 0. \end{aligned} \tag{23}$$

This system to the best of authors knowledge is not integrable. However it has special reductions wick are integrable. A nontrivial example of such special solution is the case $\phi_2 = -\phi_1 = \phi$. Then (23) reduces to the (scalar) sine–Gordon equation:

$$J_0 \frac{d^2\phi}{ds2} + 2J(l)\sin\phi = 0. \tag{24}$$

In the generic case the relative angle $\Delta\phi = \phi_1 - \phi_2$ satisfies the same equation (24) (with twice bigger coupling constant $4J(l)$) and we also have an additional constraint: $\phi_{1,ss} + \phi_{2,ss} = 0$.

4 Soliton-like Solutions

The (scalar) static sine-Gordon equation has several types of soliton solutions. Here we will discuss briefly the nonlinear excitations generated by the kink and anti-kink and the periodic (soliton lattice) solutions for an appropriate boundary conditions.

The kink type solution of the static sine-Gordon equation (15), (18) is given by

$$\phi(s) = 2\arctan\left[\exp\left(\frac{1}{\kappa_0}\sqrt{\frac{b}{1-a}}s\right)\right], \tag{25}$$

and the corresponding static energy is

$$E_{\text{kink}} = 4\kappa_0\sqrt{\frac{1-a}{b}}\tanh\left(\kappa_0\sqrt{\frac{1-a}{b}}l\right) \tag{26}$$

The periodic (soliton lattice) solution of (15), (18) is

$$\phi(s) = 2\arccos\left[\text{sn}\left(\frac{\kappa_0}{k}\sqrt{\frac{1-a}{b}}s, k\right)\right] \tag{27}$$

with the periodicity $4\frac{\kappa_0}{k}\sqrt{\frac{1-a}{b}}K(k)$, where k is the modulus of the Jacobian elliptic function sn (sine amplitude), and $K(k)$ is the complete elliptic integral of the first kind. In the limit $k \to 1$ we have $K(k) \to \infty$ and the half-period tends to infinity as well, so we recover the single kink soliton solution (25).

The corresponding static energy per soliton of the soliton lattice is given by:

$$E_{\text{soliton}} = \frac{\kappa_0}{k}\sqrt{\frac{1-a}{b}}\left(E(k) - \frac{1}{3}(k')^2K(k)\right), \tag{28}$$

where $E(k)$ is the complete elliptic integral of second kind. In the single soliton limit ($k \to 1$) the lattice energy per soliton (28) reduces to eqn. (26).

5 Conclusions

We have shown that the single asymmetric elastic Kirchhoff rod model can be mapped onto a 2-spin XY Heisenberg chain and the spin vectors must have a different lengths. In this case the Euler-Lagrange equation for the spin chain Hamiltonian gives the static sine-Gordon equation. For the case of symmetric rods ($a = 1$) both spins have the same length. The symmetric ($a = 1$) and the asymmetric ($a \neq 1$) Kirchhoff rods have very different static properties. In general the family of thin elastic Kirchhoff rods falls into two groups: i) the group of symmetric rods ($a = 1$). Here the twist is constant along the rod and if the torsion is constant as well, the Kirchhoff equations are integrable and the curvature satisfies the non-linear Schrödinger equation [25]; ii) the group of asymmetric rods ($a \neq 1$). Here in general the twist and the curvature satisfy a coupled differential equations (for a constant torsion). In the special case where the curvature is constant the system of Kirchhoff equations is integrable again and the twist satisfies the sine-Gordon equation.

The new properties for the asymmetric rods discussed here appear under the presence of constant curvature of the rod. The twist and the twist energy are localized in the region where the curvature is present.

This new class of static properties of the thin Kirchhoff rod (especially the exactly solvable models that appear here) open a new direction where one may look for a better models of biopolymers and DNA.

The dynamics of such models, which is of interest for realistic biopolymers, should be investigated. Due to the Galilean invariance of the sine-Gordon equation (15) a special class of dynamical travelling wave type solutions can be obtained from the static ones by Galilean boost.

The general assumption that all thin rods exhibit constant twist should now be restricted to the class of symmetric thin rods only and to all straight rods as well. The class of asymmetric thin rods does not belong to this category. Here the twist is not constant and "interacts" with the curvature. In the case of constant curvature, the problem has exact solution. For non-constant curvature the case is more complex and should be of considerable interest e.g. for the problem of DNA supercoiling [26, 27, 28, 29, 30].

Acknowledgments

The work of GGG is supported by the Bulgarian National Scientific Foundation Young Scientists Scholarship for the project "Solitons, differential Geometry and Biophysical Models" (contract No. F-1867). The support by the National Science Foundation of Bulgaria, contract No. F-1410 is also acknowledged.

References

1. L. V. Yakushevich, *Nonlinear Physics of DNA*, Wiley-VCH, Weinheim (2004)
2. R. Dandoloff R. Balakrishnan, Quantum Effective Potential, Electron Transport an Conformons in Byopolimers, J. Phys. A: Math. Gen. **38**, 6121–6127 (2005).
3. M. Daniel and V. Vasumathi, Perturbes Soliton Excitations in Inhomogeneous DNA, Phys. Rev. E (In press).
4. A. F. Fonseca and M. A. M. de Aguiar,Near equilibrium dynamics of nonhomogeneous Kirchhoff filaments in viscous media, Phys. Rev. E **63** 016611 (2000)
5. I. Klapper, Biolgical Applications of the Dynamics of Twisted Elastic Rods, J. Comp. Phys. **125** 325–337 (1996)
6. W. R. Bauer, R. A. Lund, and J. H. White, Twist and writhe of a DNA loop containing intrinsic bends, Proc. Natl. Acad. Sci. USA **90** 833–837 (1993)

7. D. Bensimon, A. J. Simon, V. Croquette, and A. Bensimon, Stretching DNA with a receding meniscus: Experiments and models, Phys. Rev. Lett. **74** 4754–4757 (1995)
8. Ph. Cluzel, A. Lebrun, Ch. Heller, R. Lavery, J.-L. Viovy, D. Chatenay, and F. Caron, DNA: An Extensible Molecule, Science **271** 792–794 (1996)
   S. B. Smith, Y. Cui, and C. Bustamante, Overstretching B-DNA: The Elastic Response of Individual Double-Stranded and Single-Stranded DNA Molecules, ibid. **271** 795–799 (1996)
   T. R. Strick, J.-F. Allemand, D. Bensimon, and V. Croquette, The Elasticity of a Single Supercoiled DNA Molecule, ibid. **271** 1835–1837 (1996)
9. A. E H. Love, A Treatise on the Mathematical Theory od Elasticity, Dover Publications, New York (1944)
10. J. E. Marsden and Th. J. R. Hughes, Mathematical Foundations of Elasticity, Dover Publications, New York (1994)
11. A. Goriely and M. Tabor, Nonlinear dynamics of filaments I: Dynamical instabilities, Physica D **105** 20–44 (1997); II: Nonlinear analysis, ibid. **105** 45–61 (1997); III: Instabilities of helical rods, Proc. Roy. Soc. **453** 2583–2601 (1997); IV: Spontaneous looping of elastic rods, ibid. **455** 3183–3202 (1998)
12. M. Nizette and A. Goriely, Towards a classification of Euler-Kirchhoff filaments, J. Math. Phys. **40** 2830–2866 (1999)
13. A. Goriely and M. Nizette, Kovalevskaya rods and Kovalevskaya waves, Regul. Chaotic Dyn. **45** no. 1, 95–106 (2000)
14. B. D. Coleman, E. H. Dill, M. Lembo, Zh. Lu and I. Tobias, On the Dynamics of Rods in the Theory of Kirchhoff and Clebsch, Arch. Rational Mech. Anal. **121** 339–359 (1993)
15. S. S. Antman and T.-P.Liu, Travelling waves in hyperelastic rods, Quart. Appl. Math. **36** no. 4, 377–399 (1978/79)
16. M. Gaudin, *La fonction d'onde de Bethe*, Collection du Commissariat a' l'E'nergie Atomique: Se'rie Scientifique, Masson, Paris (1983)
17. K. M. Leung, Path integral approach to the statistical mechanics of solitons, Phys. Rev. B **26**, 226–244 (1983);
    Mechanical properties of double-sine-Gordon solitons and the application to anisotropic Heisenberg ferromagnetic chains, Phys. Rev. B **27**, 2877–2888 (1983).
18. M. Takahashi, Thermodynamics of One-Dimensional Solvable Models, Cambrigde University Press, Cambridge (2005)
19. H. B. Thacker, Exact Integrability in Quantum Field Theorry and Statistical Systems, Rev. Mod. Phys. **53** 253–285 (1981)
20. R. Dandoloff and A. Saxena, Interaction induced deformation of two coupled XY spin chains, J.Phys.:Condens.Matter **9**, L667–L670 (1997)
21. R. Dandoloff and A. Saxena, Nonlinear Sigma Model and the origin of geometric frustration on curved manifolds , Z. Phys. B **104** 661–668 (1997)
22. M. Lakshmanan and S. Rajasekar, Nonlinear Dynamics: Integrability, Chaos and Patterns (Advanced texts in Physics), Springer-Verlag, Berlin (2003)
23. L. D. Landau and E. M. Lifshitz, Theory of Elasticity (Course of Theoretical Physics, Vol 7), Pergamon Press, Oxford (1986)

24. T. McMillen and A. Goriely, Tendril Perversion in Intrinsically Curved Rods, J. Nonlin. Sci. **12**, 241–281 (2002)
25. R. Balakrishnan and R. Dandoloff, The Schrödinger equation as a moving curve, Phys.Lett. A **260**, 62–67 (1999)
26. N. G. Hunt and J. E. Hearst, Elastic model of DNA supercoiling in the infinite length limit, J. Chem. Phys. **12**, 9329–9336 (1991)
27. F. Tanaka and H. Takahashi, Elastic theory of supercoiled DNA, J. Chem. Phys. **83**, 6017–6026 (1985)
28. Y. Yang, I. Tobias, and W. K. Olson, Finite element analysis of DNA supercoiling, J. Chem. Phys. **98**, 1673–1686 (1993)
29. T. P. Westcott, I. Tobias, and W. K. Olson, Modeling self-contact forces in the elastic theory of DNA supercoiling, J. Chem. Phys. **107**, 3967–3980 (1997)
30. Y. Shi, A. E. Borivik and J. E. Hearst, Elastic rod model incorporating shear and extension, generalized Nonlinear Schrödinger equations, and novel closed-form solutions for supercoiled DNA, J. Chem. Phys. **103**, 3166–3183 (1995)

Approximate controllability of one-dimensional SDE driven by countably many Brownian motions

N.I.Mahmudov[1] and M.M. Matar[2]

[1] Mathematics Department, Eastern Mediterranean University, Gazimagusa, Mersin 10, Turkey nazim.mahmudov@emu.edu.tr
[2] Mathematics Department, Eastern Mediterranean University, Gazimagusa, Mersin 10, Turkey mohammed.matar@emu.edu.tr

We are given $(\Omega, \mathcal{F}, \mathbf{P})$ as a complete probability space with right continuous complete σ-algebra filtration $(\mathcal{F}_t)_{t\in[0,T]}$, generated by the infinite sequence of independent Brownian motions $(W^i)_{i\geq 1}$. Let, for every $t \in [0,T]$, $L_2(\Omega, \mathcal{F}_t, \mathbf{R})$ be the Hilbert space of all $\mathcal{F}_t$-measurable, and square-integrable variables in $\mathbf{R}$, and $L_2(\Omega, C([0,T], \mathbf{R}))$ be the space of all square integrable and a.e. continuous functions on $\mathbf{R}$ equipped with the norm $|X| = (\mathbf{E}\sup_{t\in[0,T]} |X(t)|^2)^{1/2}$. $L_2^{\mathcal{F}}([0,T], \mathbf{R})$ denotes the Hilbert space of all square-integrable and $\mathcal{F}_t$-adapted processes with values in $\mathbf{R}$. Define the sequence $\sigma(x) = (\sigma_i(x))_{i\geq 1}$, where for each $i \geq 1$, $\sigma_i(x) \in C([0,T], \mathbf{R})$ and that $\sigma(x) \in \ell_2$, i.e. $|\sigma(x)|^2 = \sum_{i=1}^{\infty} |\sigma_i(x)|^2 < \infty$. In this paper we study the approximate controllability of the one-dimensional semi-linear stochastic differential equation

$$\begin{cases} dX(t) = [AX(t) + Bu(t) + b(X(t))]\, dt + \displaystyle\sum_{i=1}^{\infty} \sigma_i(X(t)) dW^i(t) \\ X(0) = X_0,\ t \in [0,T], \end{cases}$$

where $A, B \in \mathbf{R}$, and $u \in L_2^{\mathcal{F}}([0,T], \mathbf{R})$ is a control. We obtain sufficient conditions for approximate controllability of the above system when coefficients b, and σ satisfy non-Lipschitz conditions.

1 Introduction

Controllability problems for linear and nonlinear stochastic systems are studied by many authors. Recently, Bashirov and Mahmudov [BM99] developed new sufficient conditions for different modes of controllability concepts in deterministic and stochastic linear systems. Mahmudov [DM02] investigated the

Kenan Taş et al. (eds), Mathematical Methods in Engineering, 403–413.

controllability of linear SDE (see also [MD00]). In spite of the deterministic controllability problems are widely introduced in linear and nonlinear sense, the controllability of the stochastic field still has wide area for researchers to develop, specially the nonlinear cases. One of such fields, when a sequence of Brownian motions defined on $\mathbf{R}$, is considered as the input of the system. Cao and He [GK03] prove the existence of the solution of such semi-linear differential system characterized with non-Lipschitz coefficients in a deterministic manner. In that paper, the idea is an extension of what Mao [Ma95] did in the case of backward differential system (in short BDS). However, the current paper is dealing with the controllability of a semi-linear stochastic differential system of non-Lipschitz coefficient (see the work of Dauer, Mahmudov, and Matar [DMM05] in the case of BDS) with a countably many Brownian motions as to be considered as input of one-dimensional system.

2 Preliminaries

We introduce, shortly, the idea of integration in term of a countably Brownian motions, and some definitions and assumptions concerning the problems. For this, in addition of the above spaces definitions, we define the space $\Im^2$ of all sequences $\varphi(t) \in \ell_2$ such that the processes $\varphi(t) \in L_2^{\mathcal{F}}(I, \mathbf{R})$, $I = [0, T]$. Throughout the sequel, we use the notation of the norm as $|\cdot|$.

As a beginning, it is convenient to recall some useful facts about stochastic integrals with respect to countably many Brownian motions (for the proof see Lemma (2.1) in [GK03]).

Lemma 1. *For any $\varphi \in \Im^2$, the martingale $M_t = \sum_{i=1}^{\infty} \int_0^t \varphi_i(s) dW^i(s)$, is a continuous martingale, for all $t \in I$.*

Proposition 1. *If H and G are in $\Im^2$, then so is $c_1 H + c_2 G$ for all c_1, and c_2 in $\mathbf{R}$. Moreover, $\sum_{i=1}^{\infty} \int_0^t (aH_i(s) + bG_i(s)) dW^i(s) = a \sum_{i=1}^{\infty} \int_0^t H_i(s) dW^i(s) + b \sum_{i=1}^{\infty} \int_0^t G_i(s) dW^i(s)$ is hold for every $t \in I$.*

In the above sense, one can deduce that, The Ito's formula, and B-D-G inequality are still valid.

Consider the following SDE

$$\begin{cases} dX(t) = [AX(t) + Bu(t) + b(X(t))]\, dt + \sum_{i=1}^{\infty} \sigma_i(X(t)) dW^i(t) \\ X_0 = X(0),\ t \in I, \end{cases} \tag{1}$$

where A and B are real numbers, $u \in L_2^{\mathcal{F}}(I, \mathbf{R})$, b and σ_i are in $C(I, \mathbf{R})$.

The following definitions are bases of what follows later.

Definition 1. *A stochastic process X is said to be a mild solution of eqn.(1) if $X \in L_2(\Omega, C(I, \mathbf{R}))$, and if for each $u \in L_2^{\mathcal{F}}(I, \mathbf{R})$, the process X satisfies the following integral equation:*

$$\begin{cases} X(t) = S(t)X_0 + \int_0^t S(t-s)[Bu(s) + b(X(s))]ds \\ \qquad + \sum_{i=1}^{\infty} \int_0^t S(t-s)\sigma_i(X(s))dW^i(s) \\ X(0) = X_0 \in L_2(\Omega, \mathcal{F}_0, \mathbf{R}). \end{cases} \tag{2}$$

where $S(t)$ =exp(At) is a real function.

Definition 2. *The system (1) is said to be approximately controllable on the interval I if $\overline{\mathcal{R}(T, X_0)} = L_2(\Omega, \mathcal{F}_T, \mathbf{R})$, where $\mathcal{R}(T, X_0) = \{X(T; X_0, u) : u \in L_2^{\mathcal{F}}(I, \mathbf{R})\}$.*

Remark 1. In general, approximate controllability is equivalent to convergence of function $\alpha(\alpha I + \Gamma_0^T)^{-1} : \mathbf{R} \to \mathbf{R}$ to zero as $\alpha \to 0^+$, where $\Gamma_t^T = \int_t^T S(T-s)BB^\star S^\star(T-s)ds = \frac{B^2}{2A}[S(2(T-t)) - 1]$, here $A \neq 0$ (see [MD00], Theorem 2).

We need some lemmas before proceeding to main results, so firstly, we give the following representation theorem on $\mathbf{R}$(see [Mah01, MD00]).

Lemma 2. *For any $h \in L_2(\Omega, \mathcal{F}_T, \mathbf{R})$, there exists $\varphi \in \Im^2$ such that*

$$h = \mathbf{E}h + \sum_{i=1}^{\infty} \int_0^T \varphi_i(s)dW^i(s). \tag{3}$$

Next lemma gives a formula for a control transferring the state X_0 to a neighborhood of an arbitrary state $h \in L_2(\Omega, \mathcal{F}_T, \mathbf{R})$.

Lemma 3. *The control $u \in L_2^{\mathcal{F}}(I, \mathbf{R})$ given by*

$$\begin{cases} u(t) = BS(T-t)(\alpha I + \Gamma_0^T)^{-1}(\mathbf{E}h - S(T)X_0) \\ \quad -BS(T-t)\int_0^t (\alpha I + \Gamma_s^T)^{-1} S(T-s)b(X(s))ds \\ \quad -BS(T-t)\sum_{i=1}^{\infty} \int_0^t (\alpha I + \Gamma_s^T)^{-1}[S(T-s)\sigma_i(X(s)) - \varphi_i(s)]dW^i(s), \end{cases} \tag{4}$$

transfers the system (2) from X_0 to X_T at time $t = T$, where

$$\begin{cases} X_T = h - \alpha(\alpha I + \Gamma_0^T)^{-1}(\mathbf{E}h - S(T)X_0) \\ \quad +\alpha \int_0^T (\alpha I + \Gamma_s^T)^{-1} S(T-s)b(X(s))ds \\ \quad +\alpha \sum_{i=1}^{\infty} \int_0^T (\alpha I + \Gamma_s^T)^{-1}[S(T-s)\sigma_i(X(s)) - \varphi_i(s)]dW^i(s). \end{cases} \tag{5}$$

Proof. By substituting eqn.(4) into eqn.(2) one can get

$$X(t) = S(t)X_0 + B^2 \int_0^t S(t-s)S(T-s)(\alpha I + \Gamma_0^T)^{-1}(\mathbf{E}h - S(T)X_0)ds$$

$$
\begin{aligned}
& -B^2 \int_0^t S(t-s)S(T-s)\int_0^s (\alpha I + \Gamma_r^T)^{-1} S(T-r) b(X(r)) dr ds \\
& -B^2 \int_0^t S(t-s)S(T-s) \\
& \times \sum_{i=1}^{\infty} \int_0^s (\alpha I + \Gamma_r^T)^{-1} [S(T-r)\sigma_i(X(r)) - \varphi_i(r)] dW^i(r) ds \\
& + \int_0^t S(t-s) b(X(s)) ds + \sum_{i=1}^{\infty} \int_0^t S(t-s)\sigma_i(X(s)) dW^i(s) \\
= {} & S(t)X_0 + \Gamma_0^t S(T-t)(\alpha I + \Gamma_0^T)^{-1}(\mathbf{E}h - S(T)X_0) \\
& - \int_0^t \Gamma_s^t S(T-t)(\alpha I + \Gamma_s^T)^{-1} S(T-s) b(X(s)) ds \qquad (6) \\
& - \int_0^t \Gamma_s^t S(T-t) \sum_{i=1}^{\infty} (\alpha I + \Gamma_s^T)^{-1} [S(T-s)\sigma_i(X(s)) + \varphi_i(s)] dW^i(s) \\
& + \int_0^t S(t-s) b(X(s)) ds + \sum_{i=1}^{\infty} \int_0^t S(t-s)\sigma_i(X(s)) dW^i(s).
\end{aligned}
$$

Now at $t = T$, and using the identity $\Gamma_0^T(\alpha I + \Gamma_0^T)^{-1} = I - \alpha(\alpha I + \Gamma_0^T)^{-1}$ one then can get eqn.(5).

The main assumption, which will be used in the main problems, is introduced as follows.
Assumption A

(A1) Assume ρ, and γ are concave non-decreasing continuous functions on $\mathbf{R}_+$ (the nonnegative real numbers) such that $\rho(0) = 0$, and $\gamma(0) = 0$.
(A2) For all x and y in $\mathbf{R}$,

$$
\begin{cases} |b(x) - b(y)| \leq \gamma(|x-y|) \\ |\sigma(x) - \sigma(y)| \leq \rho(|x-y|) \end{cases}.
$$

(A3) The integral $\int_{0+} \frac{x dx}{\rho^2(x) + \gamma^2(x)} = +\infty$.

The functions in Assumption A exist, we give the following:
Examples

(Ex1)$\gamma_1(x) = Kx,\ K > 0$.

(Ex2)$\gamma_2(x) = \begin{cases} 0, & x = 0, \\ Kx(\log \frac{1}{x})^{\alpha}, & 0 < x \leq \delta, \\ K\delta(\log \frac{1}{\delta})^{\alpha}, & x > \delta. \end{cases}$ where $\alpha \leq \frac{1}{2}$, $K > 0$ and $0 < \delta < 1$.

Before proceeding, we need the following lammas which are very useful for the proofs of desired results (for the proofs see [GK03]: lemmas and corollaries 3.2 to 3.8).

Lemma 4. *Under Assumption A, we have:*

(L1) b and σ are continuous functions on $\mathbf{R}$.
(L2) There exist some constant $K_2 > 0$ such that

$$\begin{cases} |b(x)|^2 \le K_2(1+\gamma^2(x)), \\ |\sigma(x)|^2 \le K_2(1+\rho^2(x)). \end{cases}$$

Hence $\sigma(x) \in \ell_2$ for all $x \in \mathbf{R}$.
(L3) There exist some constant $K_2 > 0$ such that $\rho^2(|x|) \le K_2(1+|x|^2)$ and $\gamma^2(|x|) \le K_2(1+|x|^2)$.
(L4) There exist some constant $K_2 > 0$ such that

$$|b(x)|^2 + |\sigma(x)|^2 \le K_2(1+|x|^2). \tag{7}$$

(L5) The function $g(x) = \rho^2(x^{\frac{1}{2}})$ is a concave non-decreasing continuous function on $\mathbf{R}_+$ with $g(0) = 0$.
(L6) The function $f(x) = \rho^2(x^{\frac{1}{2}})+\gamma^2(x^{\frac{1}{2}})$, $x \ge 0$ is a concave non-decreasing continuous function on $\mathbf{R}_+$ such that $f(0) = 0$ and $\int_{0+} \frac{1}{f(x)}dx = +\infty$.

Lemma 5. *Let $f(x)$ be a concave non-decreasing continuous function on R_+ such that $f(0) = 0$ and $\int_{0+} \frac{1}{f(x)}dx = \infty$. If Z is a non-negative continuous function on R_+ such that $Z_t \le \int_0^t f(Z_s)ds$ for all $t \ge 0$, then $Z = 0$.*

3 Existence and approximate controllability problems

Consider eqn.(2), we start with a construction of approximation by standard Picard's iteration. Let X^n be a sequence defined recursively by (see (6))

$$\begin{cases} X_t^0 = S(t)X_0, \\ X_t^n = X^n(t) = S(t)X_0 + \Gamma_0^t S^*(T-t)(\alpha I + \Gamma_0^T)^{-1}(\mathbf{E}h - S(T)X_0) \\ \quad - \int_0^t \left\{ \left(\Gamma_s^t S^*(T-t)(\alpha I + \Gamma_s^T)^{-1} S(T-s) - S(t-s) \right) b(X_s^{n-1}) \right\} ds \\ \quad - \sum_{i=1}^\infty \int_0^t \left\{ [\Gamma_s^t S^*(T-t)(\alpha I + \Gamma_s^T)^{-1} S(T-s) - S(t-s)] \sigma_i(X_s^{n-1}) \right. \\ \quad \left. - \Gamma_s^t S^*(T-t)(\alpha I + \Gamma_s^T)^{-1} \varphi_i(s) \right\} dW_s^i \right\}. \end{cases}$$

To simplify the previous iteration, denote $S_1^{\alpha,s}(t) = \Gamma_s^t S^*(T-t)(\alpha I + \Gamma_s^T)^{-1}$, $S_2^{\alpha,s}(t) = S_1^{\alpha,s}(t)S(T-s) - S(t-s)$, which are both bounded with respect to sup norm. Therefore, let $sup_{t\in I}|S_1^{\alpha,s}(t)| \le C_1$, $sup_{t\in I}|S_2^{\alpha,s}(t)| \le C_2$, and $sup_{t\in I}|S(t)| \le C_3$. Also, let $|S^\alpha(X^n)|^2 = \sum_{i=1}^\infty |S_i^\alpha(X^n)|^2 < \infty$, where $S_i^\alpha(s, X_s^n) = S_2^{\alpha,s}(s)\sigma_i(X_s^n) - S_1^{\alpha,s}(t)\varphi_i(s)$. Therefore, we can rewrite the previous iteration in the form

$$\begin{cases} X_t^0 = S(t)X_0, \\ X_t^n = S(t)X_0 + S_1^{\alpha,0}(t)(\mathbf{E}h - S(T)X_0) - \int_0^t S_2^\alpha(s) b(X_s^{n-1})ds \\ \quad - \sum_{i=1}^\infty \int_0^t S_i^\alpha(s, X_s^{n-1})dW_s^i. \end{cases} \tag{8}$$

The next lemma deals with the above Picard's iteration

Lemma 6. *Let b and σ satisfy eqn.(7). Then X^n is well defined, and it is continuous for $n \geq 1$. Moreover $S^\alpha(X^n) \in \Im^2$.*

Proof. *For all $n \geq 1$, and $0 \leq t \leq T$, then by Hôlder's inequality and using eqn.'s (7,8) we have*

$$\mathbf{E}|X_t^n|^2 \leq 4C_3^2\mathbf{E}|X_0|^2 + 8C_1^2((\mathbf{E}h)^2 + C_3^2\mathbf{E}|X_0|^2) + 4C_2^2 t\int_0^t \mathbf{E}|b(X_s^{n-1})|^2 ds$$
$$+8\int_0^t [C_2^2\mathbf{E}|\sigma(X_s^{n-1})|^2 + C_1^2|\varphi(s)|^2]ds$$
$$\leq C + C'(1 + \sup_{0\leq t\leq T} \mathbf{E}|X_t^{n-1}|^2,$$

where $0 < 8C_1^2(\mathbf{E}h)^2 + 4C_3^2(1 + 2C_1^2)|X_0|^2 + 8C_1^2 \int_0^T |\varphi(t)|^2dt \leq C$, $C' = 4K_2C_2^2T(T+2) > 0$, *and* $\sup_{0\leq t\leq T} \mathbf{E}|X_t^0|^2 = C_3^2|X_0|^2 < \infty$. *Hence* $\sup_{0\leq t\leq T} \mathbf{E}|X_t^n|^2 < \infty$, *for all $n \geq 1$. On the other hand, by lemma (1) and remark (1) one can get*

$$\int_0^t \mathbf{E}|S^\alpha(X_s^n)|^2 ds \leq \int_0^t (2C_2^2\mathbf{E}|\sigma(X_s^n)|^2 + 2C_1^2|\varphi(s)|^2)ds$$
$$\leq 2K_2TC_2^2(1 + \sup_{0\leq t\leq T} \mathbf{E}|X_t^n|^2) + 2C_1^2\int_0^T |\varphi(t)|^2dt < \infty.$$

Thus X^n is well defined, and it is a continuous for all $n \geq 1$, by lemma (1).

Lemma 7. *If Assumption A is satisfied. Then for any fixed $T > 0$, there exist constants $C_{2,T} > 0$ such that*

$$\begin{cases} \mathbf{E}\sup_{0\leq u\leq t} |X_u^n|^2 \leq C_{2,T} \\ \mathbf{E}\sup_{0\leq u\leq t} |X_u^n - X_u^m|^2 \leq C_{2,T} \end{cases} \tag{9}$$

for all n, $m \geq 1$, and $0 \leq t \leq T$.

Proof. *Let m be fixed, take $1 \leq n \leq m$ and $0 \leq t \leq T$. Then by Hôlder's and B-D-G inequalities and using eqn.(7), one can obtain*

$$\mathbf{E}\sup_{0\leq u\leq t} |X_u^n|^2 \leq 4C_3^2|X_0|^2 + 8C_1^2[|\mathbf{E}h|^2 + C_3^2|X_0|^2]$$
$$+4\mathbf{E}\sup_{0\leq u\leq t}\left|\sum_{i=1}^{\infty}\int_0^u S_i^\alpha(s, X_s^{n-1})dW_s^i.\right|^2$$
$$+4\mathbf{E}\sup_{0\leq u\leq t}\left|\int_0^u S_2^\alpha(s)b(X_s^{n-1})ds\right|^2$$
$$\leq 4C_3^2|X_0|^2 + 8C_1^2\left[|\mathbf{E}h|^2 + C_3^2|X_0|^2\right]$$

$$
\begin{aligned}
&+8\left[C_{2,2}\int_0^t \mathbf{E}\left|\sigma(X_s^{n-1})\right|^2 ds + C_{1,2}\int_0^t |\varphi(s)|^2 ds\right]\\
&+4tC_2^2\int_0^t \mathbf{E}\left|b(X_s^{n-1})\right|^2 ds\\
\le\ & 4C_3^2|X_0|^2 + 8C_1^2[|\mathbf{E}h|^2 + C_3^2|X_0|^2]\\
&+8C_{1,2}\int_0^T |\varphi(s)|^2 ds\\
&+4K_2(2C_{2,2}+C_2T\int_0^t (1+\mathbf{E}\left|X_s^{n-1}\right|^2)ds\\
\le\ & 4C_3^2|X_0|^2 + 8C_1^2[|\mathbf{E}h|^2 + C_3^2|X_0|^2] + 8C_{1,2}\int_0^T |\varphi(s)|^2 ds\\
&+4K_2T[2C_{2,2}+C_2^2T]\\
&+4K_2T[2C_{2,2}+C_2^2T]\int_0^t \sup_{0\le k\le m}\mathbf{E}\sup_{0\le u\le s}\left|X_u^k\right|^2 ds,
\end{aligned}
$$

where $C_{1,2}$, $C_{2,2}$ *are constants. Since* $\mathbf{E}\sup_{0\le u\le t}\left|X_u^0\right|^2 = C_3^2|X_0|^2$, *we have*

$$
\begin{aligned}
\sup_{0\le k\le m}\mathbf{E}\sup_{0\le u\le t}\left|X_u^k\right|^2 \le\ & 4C_3^2|X_0|^2 + 8C_1^2[|\mathbf{E}h|^2 + C_3^2|X_0|^2]\\
&+8C_{1,2}\int_0^T |\varphi(s)|^2 ds + 4K_2T[2C_{2,2}+C_2^2T]\\
&+4K_2[2C_{2,2}+C_2^2T]\int_0^t \sup_{0\le k\le m}\mathbf{E}\sup_{0\le u\le s}\left|X_u^k\right|^2 ds.
\end{aligned}
$$

By Gronwall-Bellman's lemma

$$
\begin{aligned}
\sup_{0\le k\le m}\mathbf{E}\sup_{0\le u\le t}\left|X_u^k\right|^2 \le\ & 4C_3^2|X_0|^2 + 8C_1^2[|\mathbf{E}h|^2 + C_3^2|X_0|^2]\\
&+8C_{1,2}\int_0^T |\varphi(s)|^2 ds\\
&+4K_2T[2C_{2,2}+C_2^2T]\\
&\quad\times\exp[4K_2[2C_{2,2}+C_2^2T]]t\\
\le\ & C_{2,T}
\end{aligned}
$$

where $C_{2,T} = [4C_3^2|X_0|^2 + 8C_1^2[|\mathbf{E}h|^2 + C_3^2\,|X_0|^2] + 8C_{1,2}\int_0^T |\varphi(s)|^2 ds +$ $4K_2T[2C_{2,2}+C_2^2T]]\exp[4K_2[2C_{2,2}+C_2^2T]]T > 0$, *for every arbitrary* $m \ge 1$. *Hence the first inequality is proved. The second inequality is obvious now.*

Lemma 8. *If Assumption A is satisfied. Then for any fixed* $T > 0$, *there exist constants* $C_{2,T} > 0$ *such that*

$$
\begin{cases}
\mathbf{E}\sup_{0\le u\le t}|X_u^n - X_u^m|^2 \le C_{2,T}\int_0^t f\left(\mathbf{E}\sup_{0\le u\le s}\left|X_u^{n-1}-X_u^{m-1}\right|^2\right) ds\\
\mathbf{E}|X_t^n - X_t^m|^2 \le C_{2,T}\int_0^t f\left(\mathbf{E}\left|X_s^{n-1}-X_s^{m-1}\right|^2\right) ds
\end{cases}
$$

for all $n,\ m \geq 1,\ 0 \leq t \leq T$, *where* $f(x) = \rho^2(x^{\frac{1}{2}}) + \gamma^2(x^{\frac{1}{2}})$, *for all* $x \geq 0$.

Proof. Let $n,\ m \geq 1$, *for every* $0 \leq u \leq t$, *we have*

$$X_u^n - X_u^m = \int_0^t S_2^\alpha(s)[b(X_s^{m-1}) - b(X_s^{n-1})]ds$$
$$+\sum_{i=1}^{\infty}\int_0^t S_2^\alpha(s)[\sigma(X_s^{m-1}) - \sigma(X_s^{n-1})]dW_s^i$$

Then

$$\mathbf{E}\sup_{0\leq u\leq t}|X_u^n - X_u^m|^2 \leq 2\mathbf{E}\sup_{0\leq u\leq t}\left|\int_0^u S_2^\alpha(s)[b(X_s^{n-1}) - b(X_s^{m-1})]ds\right|^2$$
$$+2\mathbf{E}\sup_{0\leq u\leq t}\left|\sum_{i=1}^{\infty}\int_0^u [S_2^\alpha(s)(\sigma(X_s^{n-1}) - \sigma(X_s^{m-1})]dW_s^i\right|^2.$$

Using now B-D-G inequality, and Assumption A we obtain

$$\mathbf{E}\sup_{0\leq u\leq t}|X_u^n - X_u^m|^2 \leq 2tC_2^2\int_0^t \mathbf{E}\left|b(X_s^{n-1}) - b(X_s^{m-1})\right|^2 ds$$
$$+2C_{2,2}\int_0^t \mathbf{E}\left|\sigma(X_s^{n-1}) - \sigma(X_s^{m-1})\right|^2 ds$$
$$\leq 2TC_2^2\int_0^t \mathbf{E}(\gamma^2(\left|X_s^{n-1} - X_s^{m-1}\right|))ds$$
$$+2C_{2,2}\int_0^t \mathbf{E}(\rho^2(\left|X_s^{n-1} - X_s^{m-1}\right|))ds.$$

Hence by lemma (4-L5), and Jensen's inequality we have

$$\mathbf{E}\sup_{0\leq u\leq t}|X_u^n - X_u^m|^2 \leq 2TC_2^2\int_0^t \gamma^2((\mathbf{E}\left|X_s^{n-1} - X_s^{m-1}\right|^2)^{\frac{1}{2}})ds$$
$$+2C_{2,2}\int_0^t \rho^2((\mathbf{E}\left|X_s^{n-1} - X_s^{m-1}\right|^2)^{\frac{1}{2}})ds$$
$$\leq C_{2,T}\int_0^t f\left(\mathbf{E}|X_s^{n-1} - X_s^{m-1}|^2\right)ds$$
$$\leq C_{2,T}\int_0^t f\left(\mathbf{E}\sup_{0\leq u\leq s}\left|X_u^{n-1} - X_u^{m-1}\right|^2\right)ds,$$

where $C_{2,T} = 2TC_2^2 \vee 2C_{2,2} > 0$, *and for any arbitrary* $n,\ m \geq 1$. *Applying the same procedure, one can obtain the second inequality.*

At least now we can prove the main results. At first, we prove the existence of a strong solution for (2) (see Kao and He [GK03], theorem 3.1).

Theorem 1. *Under Assumption A, the eqn.(2) has a unique solution.*

Proof. *Let $T > 0$ be given but fixed, and let $n,\ m \geq 1,\ 0 \leq u \leq t \leq T$. Then by lemma (8), we have*

$$\mathbf{E} \sup_{0 \leq u \leq t} |X_u^n - X_u^m|^2 \leq C_{2,T} \int_0^t f\left(\mathbf{E} \sup_{0 \leq u \leq s} |X_u^{n-1} - X_u^{m-1}|^2 \right) ds$$

where $f(x) = \rho^2(x^{\frac{1}{2}}) + \gamma^2(x^{\frac{1}{2}})$, for all $x \geq 0$. Let

$$Y_t = \limsup_{n,m \to \infty} \mathbf{E} \sup_{0 \leq u \leq t} |X_u^n - X_u^m|^2.$$

Then Y is nonnegative continuous function on I. Hence by lemma (7) and using Fatou's lemma, one can deduce that

$$Y_t \leq C_{2.T} \int_0^t f(Y_s) ds.$$

Now, lemmas (4-L6,8) imply that

$$Y_t = \limsup_{n,m \to \infty} \mathbf{E} \sup_{0 \leq u \leq t} |X_u^n - X_u^m|^2 = 0,$$

for all $0 \leq t \leq T$, i.e. the sequence X^n is a Cauchy sequence for any fixed $T > 0$. Denote this limit by X. The continuity of X follows by the continuity of X^n. As in the proof of lemma (8) one can get

$$\begin{aligned} \mathbf{E} \sup_{0 \leq u \leq t} |X_u^n - X_u|^2 &= \mathbf{E} \sup_{0 \leq u \leq t} \Bigg| \int_0^u S_2^\alpha(s)[b(X_s^{n-1}) - b(X_s]ds \\ &\quad + \sum_{i=1}^\infty \int_0^u [S_i^\alpha(s, X_s^{n-1}) - S_i^\alpha(s, X_s)] dW_s^i \Bigg|^2 \\ &\leq C_{2,T} \int_0^t f\left(\mathbf{E} \sup_{0 \leq u \leq s} \left|X_u^{n-1} - X_u\right|^2 \right) ds \to 0, \end{aligned}$$

as $n \to \infty$. So X_t satisfies eqn.(2). The uniqueness follows directly by the classical methods and by using lemma (8).

Next we prove the second main result in this approach, namely the approximate controllability of the semi-linear system (1). But firstly, due to remark (1), we need to show the approximate controllability of the corresponding linear system

$$\begin{cases} dX(t) = [AX(t) + Bu(t)]dt + \sum_{i=1}^\infty \sigma_i(t) dW^i(t) \\ X(0) = X_0, \ \ t \in I. \end{cases} \tag{10}$$

This can be done by the following:

Lemma 9. *The linear system (10) is approximate controllable.*

Proof. *If $A = 0$, then the case will be trivial. Hence let $A \neq 0$, then, by remark (1), one can get the inverse function of $(\alpha I + \Gamma_t^T)$ which is given by*

$$(\alpha I + \Gamma_t^T)^{-1} = T - \frac{1}{2A} ln|\frac{2A(t-\alpha)+B^2}{B^2}|.$$

This implies that $\alpha(\alpha I + \Gamma_0^T)^{-1}$ converges to 0 as $\alpha \to 0^+$, which ends the proof.

Theorem 2. *Let b and σ be bounded, and Assumption A be satisfied. Then the system (2) is approximate controllable.*

Proof. *Let X^α be a solution of (2). Then X^α at $t = T$ is given by (5). By lemma (2) and (5), this solution is continuous and is in $L_2(\Omega, C(I, \mathbf{R}))$. The only thing we have to prove the following*

$$\begin{aligned}
\mathbf{E}\left|X^\alpha(T) - h\right|^2 &\leq 6\left|\alpha(\alpha I + \Gamma_0^T)^{-1}\right|^2 \left(|\mathbf{E}h|^2 + |S(T)|^2 \ \mathbf{E}|X_0|^2\right) \\
&\quad + 3\mathbf{E}\left|\int_0^T \alpha(\alpha I + \Gamma_s^T)^{-1} S(T-s) b(X^\alpha(s)) ds\right|^2 \\
&\quad + 3\mathbf{E}\left|\sum_{i=1}^\infty \int_0^T \alpha(\alpha I + \Gamma_s^T)^{-1}[S(T-s)\sigma_i(X^\alpha(s)) - \varphi_i(s)]dW^i(s)\right|^2 \\
&\leq 6|\alpha(\alpha I + \Gamma_0^T)^{-1}|^2 \left[|\mathbf{E}h|^2 + C_3^2 \mathbf{E}\left|X_0\right|^2\right] \\
&\quad + 3C_3^2 \left(\int_0^T \left|\alpha(\alpha I + \Gamma_s^T)^{-1}\right|^2 ds\right)\left(\int_0^T \mathbf{E}\left|b(X^\alpha(s))\right|^2 ds\right) \\
&\quad + 6C_3^2 C_2 \left(\int_0^T \left|\alpha(\alpha I + \Gamma_s^T)^{-1}\right|^2 ds\right)\left(\int_0^T \mathbf{E}\left|\sigma(X^\alpha(s))\right|^2 ds\right) \\
&\quad + 6C_2 \left(\int_0^T \left|\alpha(\alpha I + \Gamma_s^T)^{-1}\right|^2 ds\right)\left(\int_0^T \left|\varphi(s)\right|^2 ds\right).
\end{aligned}$$

By Assumption A, and lemma (9), we deduce that $\mathbf{E}\left|X^\alpha(T) - h\right|^2 \to 0$ as $\alpha \to 0^+$.

Remark 2. If we assume that the functions γ, and ρ are written linearly through the origin then, Assumption A is still satisfied and the case will be reduced to Lipschitz one.

References

[BM99] Bashirov, A.E., Mahmudov, N.I.: On concepts of controllability of linear deterministic and stochastic systems. SIAM Journal: Control Optim., **37**, 1808–1821 (1999)

[DM02] Dauer, J.P., Mahmudov, N.I.: Approximate controllability of semilinear functional equations in Hilbert spaces. Journal of Mathematical Analysis and Applications, **273**, 310-327 (2002)

[MD00] Mahmudov, N.I., Denker, A.: On controllability of linear stochastic systems. International Journal of Control, **73**, 144-151 (2000)

[GK03] Guilan, C., Kai, H.: On a type of stochastic differential equations driven by countably many Brownian motions. Journal of Functional Analysis, **203**, 262–285 (2003)

[Ma95] Mao, X.R.: Adapted solutions of backward stochastic differential equations with non-Lipschitz coefficients. Stochastic Process. Appl., **58**, 281-292 (1995)

[DMM05] Dauer, J.P., Mahmudov, N.I., Matar, M.M.: Approximate controllability of backward stochastic evolution equations in Hilbert spaces. Journal of Mathematical Analysis and Applications(DOI information 10.1016/j.jmaa.2005.09.089), in press (2005)

[Mah01] Mahmudov, N.I.: Controllability of linear stochastic systems in Hilbert Spaces. Journal of mathematical Analysis and Applications, **259**, 64-82 (2001)

Synchronization between neuronal spiking activity and sub-threshold sinusoidal stimuli based on the FitzHugh-Nagumo model

Mahmut Ozer[1] and Muhammet Uzuntarla[2]

[1] Department of Electrical and Electronics Engineering, Engineering Faculty, Zonguldak Karaelmas University, Zonguldak, Turkey
mahmutozer2002@yahoo.com

[2] Department of Electrical and Electronics Engineering, Engineering Faculty, Zonguldak Karaelmas University, Zonguldak, Turkey muzuntarla@yahoo.com

The FitzHugh-Nagumo (FHN) model was proposed as a simplification of the neuronal model and provided insight into the more complex neuronal models. Recently, an analytical approach has been proposed for determining the response of a neuron or of the activity in a network of connected neurons based on the FHN model with Gaussian white noise current. In this study, we investigate the synchronization between neuronal spiking activity and subthreshold sinusoidal stimuli. For this purpose, we obtain the phase probability density of the spiking events for the sub-threshold stimuli. We show that the system exhibits the phase locking behaviour. We also show that the phase synchronization clusters the spiking activity on the positive phase of the subthreshold sinusoidal driving for smaller frequencies while it shifts the spiking activity towards the negative phase for larger frequencies.

1 Introduction

Ion channels constitute the fundamental elements for electrical signaling in nerve by providing conduction pathways for specific ions between intracellular and extracellular spaces. Voltage-gated ion channels are crucial for generating and propagating of action potentials, which are also called spikes. Hodgkin and Huxley (H-H) [HH52] proposed the first quantitative description of the voltage-dependent gating of the channels several decades ago. The H-H model describes dynamics of the membrane potential only within the limit of very large patch size, where conductance fluctuations are negligible. If membrane patch area comprises few channels, stochastic effects become more important on the neuronal dynamics. Finite populations of stochastic ion channels may give rise to random current fluctuations that can modify excitability, cause

Kenan Taş et al. (eds), Mathematical Methods in Engineering, 415–421.

spontaneous firing and result in variability of spike threshold, spike timing and interspike intervals [Rub95, CW96, SFS98]. One of the possible contributions of the stochastic ion channels is that the channel noise can provide improvements in the representations of weak signals in neuronal systems through a mechanism called stochastic resonance [BW95]. We examined the effect of the channel noise on the time-course of recovery from inactivation of sodium channels and showed that the channel noise provides both some amount of the non-inactivated channels and a smaller time-course for recovering from inactivation [OE05]. In a recent study, we investigated the effect of the subthreshold periodic current forcing on the regularity and synchronization of neuronal spiking activity by using a stochastic extension of the H-H model and showed that the intrinsic coherence resonance is independent of the forcing frequency for the very small patch size while it is dependent on the frequency for larger sizes [Oze06].

The neuronal systems and their dynamics have received considerable attention in recent years. The FitzHugh-Nagumo (FHN) model has deserved more attention since its simplicity. The FHN was proposed as a simplification of the H-H neuronal model and provided insight into the more complex neuronal models [Ftz61, NAY62]. Recently, an analytical approach has been proposed for determining the response of a neuron or of the activity in a network of connected neurons based on the FHN model with Gaussian white noise current [TR98]. In a recent study, Toral et.al [TMC03] characterized numerically the regime of anticipated synchronization in the coupled FHN model for neurons by considering two neurons, coupled in a master-slave configuration, subject to the same random external forcing. By constructing unidirectionally coupled FHN model neurons subject to a common random external forcing they demonstrated the anticipated synchronization, a regime in which the slave neuron fires the same train of pulses as the master neuron, but earlier in time [TMC03, CCT03]. In this study, we investigate the synchronization phenomenon between the neuronal spiking activity and subthreshold periodic forcing based on the FHN model. For this purpose, we obtain the phase probability density of the spiking events for the sub-threshold periodic stimuli.

2 The model

The FHN model consists of two variables. In the deterministic FHN model, the system does not include any noise source. By adding a noisy driving term and subthreshold periodic stimulus, we define our stochastic FHN model as follows:

$$\varepsilon \frac{dv}{dt} = v - \frac{1}{3}v^3 - \omega + I \tag{1}$$

$$\frac{dw}{dt} = \varepsilon(v + a - bw) + \xi(t) \tag{2}$$

where v and w represent the voltage and recovery variables, respectively. The first one is referred as fast variable due to its more rapid change compared to the second variable.ε is a small parameter separating time scales of the voltage and recovery variables. In the model the values of ε, a and b are selected as 0.2, 0.7 and 0.8, respectively. The model has two nullclines obtained directly from Eqs. (1-2). The second nullcline is controlled by the parameters a and b. In this study, the values of a and b are selected so that there is only one single equilibrium point at (v,w)=(-1.1994,-0.6243) while the other two roots of the intersection of the two nullclines result in a complex conjugate pair.I represents the subthreshold periodic stimulus:I=Asinωt where A and ω denote the amplitude and angular frequency, respectively. $\xi(t)$ is a zero mean, Gaussian white noise with the intensity D ; where $\langle\xi(t)\xi(t)\rangle = D\delta(t\text{-}t')$.

3 Synchronization phenomena

In order to investigate the synchronization between the spiking activity and subthreshold sinusoidal forcing, we obtain the phase probability density of the spiking events for the subthreshold periodic forcing. In the model the value of A is selected as 0.1. The simulation duration is taken as 1000 periods of the subthreshold periodic forcing. Each simulation is repeated 100 times. In each period, timing of the spikes is determined relative to forcing period. Then, by making necessary normalization the phase probability density of the spiking events,$P(\phi)$, is obtained. We calculated the phase probability density versus the phase,ϕ=wt , for six different noise intensities and four different frequencies of the forcing and plotted in Figure 1.

Figure 1 shows that the phase probability density becomes relatively flat for all frequencies as the intensity of the noise becomes stronger since the noise dominates over periodic forcing. With decreasing the intensity of the noise, a peak in $P(\phi)$ is getting more visible as shown in Figure 1. This means that the noise looses its strength and external forcing becomes dominant at smaller noise intensities, and therefore the timings of the spiking events concentrate on a specific phase of the stimulus, the phenomenon is called as a phase locking behaviour [RBH67]. On the other hand, for small frequencies, the membrane most frequently fires at the positive phase of the forcing and the phase lag between the maximum of $P(\phi)$ and the maximum of the forcing increases with increasing of the noise strength. This is due to the fact that increasing the noise helps neuron to cross the threshold more easily, consequently neuron fires more in advance before the maximum of the forcing is reached as indicated in [SGH03]. This effect is similar to the effect of anticipated synchronization between master and slave neurons [TMC03, CCT03].In this context, when the noise is large and the forcing frequency is small, the model can be considered as if it has a master-slave configuration. Therefore, the neuron exhibits an

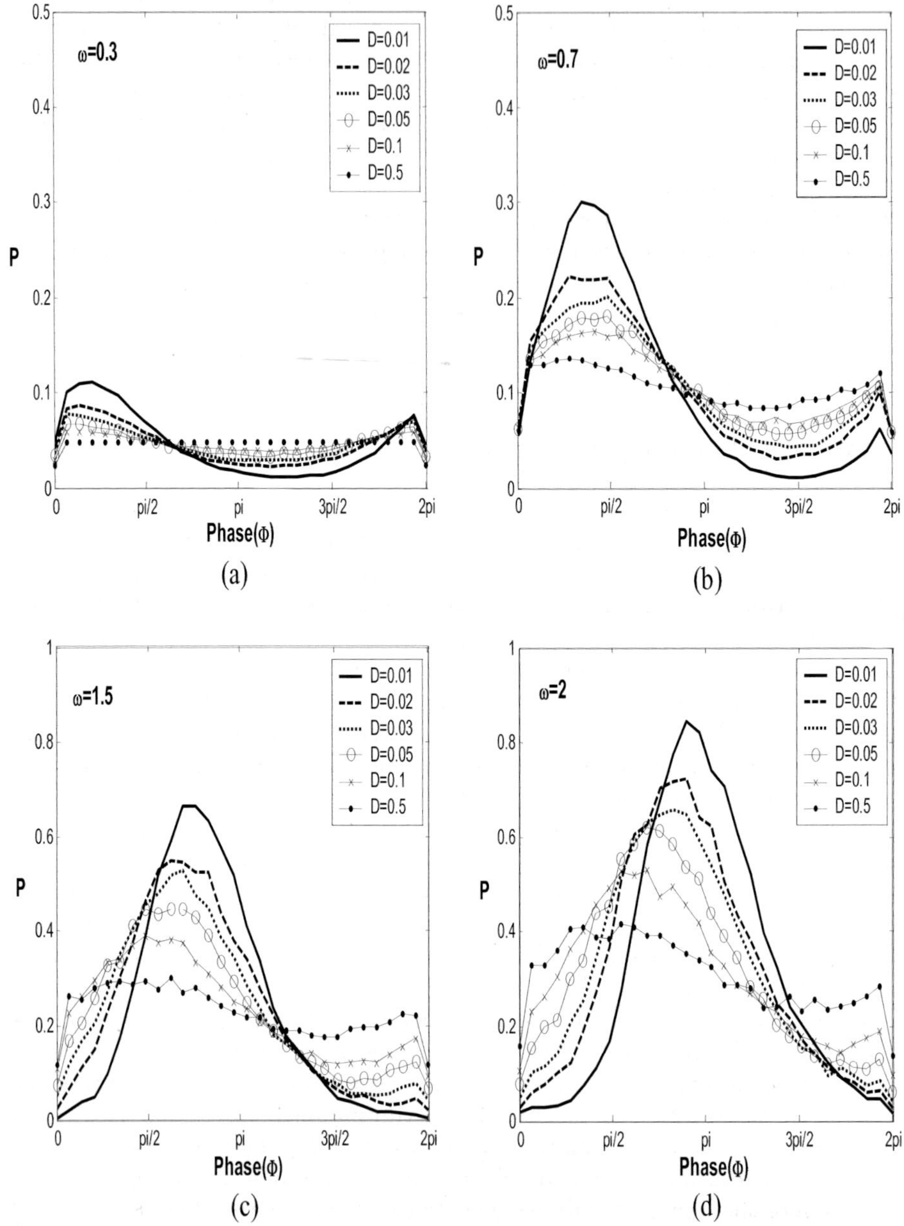

Fig. 1. The phase probability density of spiking events versus phase of the stimulus: (a) ω=0.3 ms^{-1} , (b) ω=0.7 ms^{-1} , (c) ω=1.5 ms^{-1} , and (d) ω=2 ms^{-1} for six different noise intensities.

anticipated synchronization by firing more frequently before maximum of the forcing, so predicting the response of the master neuron. Consequently, the phase lag between the maximum of the probability density and the maximum of the forcing occurs. We also observed that the phase probability density shifts towards negative phase of the forcing as the frequency is increased. For the smaller noise intensities, clustering of the spiking activities towards the negative phase is more visible. Although the shape of $P(\phi)$ covers all phases for larger noise intensities, it begins to have a normal distribution and its maximum shifts toward the negative phase as the forcing frequency is increased.

4 Conclusion

The periodic stimulation of nerve fibers or single neurons provides an effective tool to investigate the information processing capabilities and underlying mechanisms of the nervous system at cellular level [Hoo98, SER01]. Periodic current forcing has been used to characterize the intrinsic dynamics of squid giant axons [AMI84], subthreshold oscillations and resonance behavior [LY97, RBH03]. In this context, we investigated the synchronization between the neuronal spiking activity and subthreshold periodic forcing based on the FHN model by using the phase probability density of the spiking events. We show that the phase synchronization clusters the spiking activity on the positive phase of the forcing for smaller frequencies while it shifts the spiking activity towards the negative phase for larger frequencies. These observations are consistent with previously obtained findings for the stochastic H-H model relative to phase of the stimulus in [Oze06] and relative to period of the stimulus in [Ekm05], and they are also in agreement to some extent with the results in [TMC03]. Two different neuronal model result in similar findings. We also obtained same behaviour for the FHN model where a zero mean, Gaussian white noise source is added to sub-threshold periodic stimuli instead of adding into the recovery variable [OU06].

On the other hand, Figure 1 also shows that the phase probability density of the firing exhibits a nearly perfect sinusoidal modulation in response to low frequency sinusoidal stimulus while its shape becomes a normal distribution in response to the high frequency sinusoidal stimulus. This behaviour is more tractable for smaller noise intensities as shown in Figure 1. We also obtained similar behaviour for the stochastic H-H neuronal model in [Oze06]. Although the shape of the phase probability density covers all phases of the stimulus for smaller patch sizes in the stochastic H-H model in [Oze06] and for stronger noise intensities in the FHN model shown in Figure 1, it also begins to have a normal distribution as the frequency of the sinusoidal stimuli is increased. In this context, it was noted that the time dependent firing rate in response to relatively quiet low frequency sine waves shows a nearly perfect sinusoidal modulation around the spontaneous rate by illustrating the phase locking in Xenopus lateral line receptors [RVB98].

Synchronous activities across biological neuronal ensembles have been observed in many regions of brain by using multi-electrode recording techniques [EFS01, SS01, FSR02]. Synchrony is also considered as a mechanism for attentional selection, in which the responses of neurons in early processing stages that convey information to be selected are made prominent by increasing the degree of synchrony between these neurons [NHJ02]. Niebur [Nie02] indicated that the mechanism underlying the synchrony is common input from the structures controlling which stimuli are selected. Schreiber et al. [SEH04] provided experimental evidence that the frequency selectivity for periodic inputs does also extend to the non-periodic random stimuli in entorhinal cortex. Consequently, we think that our results may serve on the understanding of the synchronization between neuronal spiking activity and subthreshold periodic current forcing for a single neuron and provide insight into the synchronization of coupled neurons and the frequency-dependent information flow between neurons.

References

[HH52] Hodgkin, A.L., Huxley, A.F.: A quantitative description of membrane current and its ap-plication to conduction and excitation in nerve. J. Physiol. Lond., **117**, 500-544 (1952)

[Rub95] monograph Rubinstein, J.T.: Threshold fluctuations in N sodium channel model of the node of Ranvier. Biophys. J., **68**, 779-785 (1995)

[CW96] Chow, C.C., White, J.A.: Spontaneous action potentials due to channel fluctuations. Bio-phys. J., **71**, 3013-3021 (1996)

[SFS98] Schneidman, E., Freedman, B., Segev, I.: Ion channel stochasticity may be critical in deter-mining the reliability and precision of spike timing. Neural Comput., **10**, 1679-1703 (1998)

[BW95] Bezrukov, S.M., Vodyanov, I.: Noise induced enhancement of signal transduction across voltage-dependent ion channels. Nature, **378**, 362-364 (1995)

[OE05] M. Ozer and N. H. Ekmekci. Effect of channel noise on the time-course of recovery from in-activation of sodium channels. Physics Letters A, **338**, 150-154 (2005)

[Oze06] Ozer, M.: Frequency-dependent information coding in neurons with stochastic ion channels for subthreshold periodic forcing. Physics Letters A, **354**, 258-263 (2006)

[Ftz61] FitzHugh, R.A.: Impulses and physiological states in theoretical models of nerve membrane. Biophys. J., **1**, 445-466 (1961)

[NAY62] Nagumo, J., Arimoto, S., Yoshizawa, S.: An active pulse transmission line simulating nerve axon. Proc. Inst. Radio Eng., **50**, 2061-2070 (1962)

[TR98] Tuckwell, H.C., Rodriguez, R.: Analytical and simulation results for stochastic Fitzhugh-Nagumo neurons and neural networks. J. Comput. Neurosci., **5**, 91-113 (1998)

[TMC03] Toral, R., Masoller, C., Mirasso, C.R., Ciszak, M., Calvo, O.: Characterization of the anticipated synchronization regime in the coupled Fitzhugh-Nagumo model for neurons. Physica A, **325**, 192-198 (2003)

[CCT03] Ciszak, M., Calvo, O., Masoller, C., Mirasso, C.R., Toral, R.: Anticipating the response of ex-citable systems driven by random forcing. Phys. Rev. Lett., **90**, 2041021-2041024 (2003)

[RBH67] Rose, J.E., Brugge, J.F., Anderson, D.J., Hind, J.E.: Phase-locked response to low-frequency tones in single auditory nerve fibers of the squirrel monkey. J. Neurophysiol., **30**, 769-793 (1967)

[SGH03] Schmid, G., Goychuk, I., Hanggi, P.: Channel noise and synchronisation in excitable membranes. Physica A, **325**, 165-175 (2003)

[Hoo98] Hooper, S.L.: Transduction of temporal patterns by single neurons. Nature Neurosci., **1**, 720-726 (1998)

[SER01] Szcs, A., Elson, R.C., Rabinovich, M.I., Abarbanel, H.D.I., Selverston, A.I.: Nonlinear behavior of sinusoidally forced pyloric pacemaker neurons. J. Neurophysiol., **85**, 1623-1638 (2001)

[AMI84] Aihara, K., Matsumoto, G., Ikeyaga, Y.: Periodic and non-periodic responses of a periodi-cally forced Hodgkin-Huxley oscillator. J. Theor. Biol., **109**, 249-269 (1984)

[LY97] Lampl, I., Yarom, Y.: Subthreshold oscillations and resonant behavior: two manifestations of the same mechanism. Neuroscience, **78**, 325-341 (1997)

[RBH03] Richardson, M.J.E., Brunel, N., Hakim, V.: From subthreshold to firing-rate resonance. J. Neurophysiol., **89**, 2538-2554 (2003)

[Ekm05] Ekmekci, N.H.: Effect of ion channel fluctuations on neuronal dynamics. M.Sc. Thesis, Zonguldak Karaelmas University, Zonguldak (2005)

[OU06] Ozer, M., Uzuntarla, M.: Investigation of synchronization between neuronal spiking activity and subthreshold sinusoidal forcing. In: Jan, J., Kozumplik, J., Provaznik, I. (ed.) Analysis of Biomedical Signals and Images. Vutium Press, Brno (2006)

[RVB98] Rieke, F., Warland, D., van Stveninck, R.R., Bialek, W., Spikes: Exploring the Neural Code. The MIT Press, 30-31 (1998)

[EFS01] Engel, A.K., Fries, P., Singer, W.: Dynamic predictions: Oscillations and synchrony in top-down processing. Nature Rev. Neurosci., **2**, 704-716 (2001)

[SS01] Salinas, E., Sejnowski, T.J.: Correlated neuronal activity and the flow of neuronal information. Nature Rev. Neurosci., **2**, 539-550 (2001)

[FSR02] Fries, P., Schroder, J.H., Roelfsema, P.R., Singer, W., Engel, A.K: Oscillatory neuronal syn-chronization in primary visual cortex as a correlate of stimulus selection. J. Neurosci., **22**, 3739-3754 (2002)

[NHJ02] Niebur, E., Hsiao, S.S., Johnson, K.O.: Synchrony: a neuronal mechanism for attential selection. Curr. Opin. Neurobiol., **12**, 190-194 (2002)

[Nie02] Niebur, E.: Electrophysiological correlates of synchronous neural activity and atten-tion: a short review. Biosytems, **67**, 157-166 (2002)

[SEH04] Schreiber, S., Erchova, I., Heinemann, U., Herz, A.V.M.: Subthreshold resonance explains the frequency-dependent integration of periodic as well as random stimuli in the entorhinal cortex. J. Neurophysiol., 92, 408-415 (2004)

A characterization of the dynamics of Newton's derivative

Mehmet Özer[1,2,*], A. Valaristos[3], Yasar Polatoglu[1], Gürsel Hacibekiroglu[1], and Antanas Čenys[2,4] and A.N. Anagnostopoulos[3]

[1] Istanbul Kultur University, Atakoy Yerleskesi, Bakirkoy TR-34156, Istanbul, Turkey {m.ozer, y.polatoglu, g.hacibekiroglu}@iku.edu.tr

[2] Semiconductor Physics Institute, A.Gostauto 11 LT-01108 , Vilnius, Lithuania cenys@pfi.lt

[3] Aristotle University of Thessaloniki, GR-54124, Thessaloniki, Greece antonisv@hotmail.com, anagnost@physics.auth.gr

[4] Vilnius Technical University, Sauletekio 11, Vilnius, Lithuania

In the present report the dynamic behaviour of the one dimensional family of maps $F_{a,b,c}(x) = c[(1-a)x-b]^{\frac{1}{1-a}}$ is examined, for different ranges of the control parametres a, b and c. These maps are of special interest, since they are solutions of $N_f'(x) = a$, where N_f' is the Newton's method derivative. In literature only the case $N_f'(x) = 2$ has been completely examined. Simultaneously, they may be viewed as solutions of normal forms of second order homogeneous equations, $F''(x)+p(x)F(x) = 0$, with immense applications in mechanics and electronics. The reccurent form of these maps, $x_n = c[(1-a)x_{n-1} - b]^{\frac{1}{1-a}}$, after excessive iterations, shows an oscillatory behaviour with amplitudes undergoing the period doubling route to chaos. This behaviour was confirmed by calculating the corresponding Lyapunov exponents.

1 Introduction

The study of discrete dynamical systems expressed by maps [Wit91], through observation of their iterations has been in the full blaze of publicity since many years. Simultaneously, the 'long-term' behavior of actions of a (topological) group on a topological space, naturally denoted by a flow, has given rise to the study of the dynamics of differential equations, with numerous applications in physics, engineering, biology etc. Common denominator

* This work was supported by the PYTHAGORAS II project of the Greek Ministry of National Education and Religious Affairs and NATO ICS.EAP.CLG 981947. M.Özer acknowledges financial support from the Semiconductor Physics Institute, Vilnius, Lithuania (by the EC project PRAMA, contract Nr.G5MA-CT-2002-04014).

Kenan Taş et al. (eds), Mathematical Methods in Engineering, 423–433.

in all these efforts has been the discovery, investigation and interpretation of the chaotic behavior of such systems [Inc56, Bra93].

Newton's method for approximating the zeroes of a function, serves as a striking example of a dynamical system. Its recursion scheme, $N_f(x) = x_{n+1} = x_n - \frac{f(x_n)}{f'(x_n)}$, shows unpredictable behavior, depicted both in the complex and the real plane.

In this report, we analyze the dynamics of the map $f(x) = c[(1-a)x-b]^{\frac{1}{1-a}}$ for different values of the control parameters a, b and c. In the next two paragraphs we demonstrate two different approaches, which lead to the above map, indicating its importance in many applications.

2 Newton's method approach

For any real- or complex-valued function f, one can define the Newton derivative of f [OCPHAVA], as

$$N_f'(x) = \left(x - \frac{f(x)}{f'(x)}\right)' = \frac{f(x)f''(x)}{(f'(x))^2} \tag{1}$$

This is simply the first derivative of f by the Newton's method. In a previous work, we have examined the dynamic behavior of the maps satisfying $N_f'(x) = 2$, which are known to be fractional linear transformations, of the form $\frac{b}{x+a}$ [OPHVA05].

We now generalize the above results replacing 2 by a and we are studying

$$N_f'(x) = a \tag{2}$$

Simple calculations show that $N_f'(x) = a$ if and only if f is of the form

$$f(x) = c[(1-a)x - b]^{\frac{1}{1-a}} \tag{3}$$

with $a \neq 1$.

At this point, we incorporate in our discussion the idea of the Schwarzian derivative. Recall that the Schwarzian derivative of f is

$$S_f = \frac{f'''}{f'} - \frac{3}{2}\left(\frac{f''}{f'}\right)^2 \tag{4}$$

It has to be mentioned here that the existence of period doublings is allowed but not guaranteed by the possession of a negative Schwarzian derivative S_f. Using expression (4) and after some elementary calculations we obtain the explicit expression for S_f:

$$f'(x) = c[(1-a)x - b]^{\frac{1}{1-a}-1} \tag{5}$$

$$f''(x) = ca[(1-a)x-b]^{\frac{1}{1-a}-2} \tag{6}$$

$$f'''(x) = ca(2a-1)[(1-a)x-b]^{\frac{1}{1-a}-3} \tag{7}$$

Finally:

$$S_f = \frac{a(a-2)}{2[(1-a)x-b]^2} \tag{8}$$

Obviously, the control parameter a enters the map in an essential way. We restrict our attention to the case $1 < a < 2$, where $S_f < 0$.

3 Differential equations' approach

A general second order differential equation is an equation of the form

$$F(x, y, y', y'') = 0 \tag{9}$$

Extended theory has been developed to guarantee the existence and uniqueness of the solutions of (9). Many applications, including mechanical vibrations (undamped free, damped free, forced, damped forced) and electrical networks, have been studied by second order linear differential equations [CC02], [FFJ97]. In developing certain aspects of the theory of equations of the form $y'' + p(x)y' + q(x)y$, we often transform the dependent variable so that the equation takes the form $u'' + f(x)u = 0$, which is known as the normal form of the second order linear homogeneous equation [BP92, Inc56, Bra93].

On the other hand nonlinear differential equations (equivalently, systems of nonlinear differential equations) have been the milestone of research for many working groups worldwide. Nonlinearities enter the equations in many different ways (Van der Pol, Emden, Duffing, etc.), with interesting results [Dav60].

We consider the second order homogeneous differential equation

$$\frac{d^2 f(x)}{dx^2} + p(x)f(x) = 0 \tag{10}$$

where

$$p(x) = \frac{-a}{[(1-a)x-b]^2} \tag{11}$$

and a and b are real numbers ($a \neq 1$).

Its solution is

$$f(x) = c[(1-a)x-b]^{\frac{1}{1-a}} \tag{12}$$

where c is a real constant. This means that we are obtaining this way the same expression for $f(x)$ as in (3).

4 Exhibition of the dynamics - results

We examine the chaotic behavior of the family of maps

$$F_{a,b,c}(x) = c[(1-a)x - b]^{\frac{1}{1-a}} \tag{13}$$

for different values values of a, b and c, which serve as control parameters. We consider the recurrent form of (13):

$$x_n = c[(1-a)x_{n-1} - b]^{\frac{1}{1-a}} \tag{14}$$

For any point x_0 in the domain of the map, we study the orbit $\{x_n\}_{n=0}^{\infty}$, of it.

The parameters a, b, and c enter the map in an essential way. As these parameters are passing through some critical values, the sequence of iterates generated from the map also alters experiencing a transition from periodic to chaotic behavior.

To study these transitions in detail we have calculated numerically the bifurcation diagrams and the corresponding Lyapunov exponents for discrete values of b, c and a varying between 1 and 2. For this purpose we have used Mathcad [Hay93, DFran02, Kauf04] to calculate both the bifurcation diagrams and the Lyapunov exponents. Doing so, we took into account that the Lyapunov exponent can be found using the formula

$$\lambda = \lim_{N\to\infty} \frac{1}{N} \sum_{n=1}^{N} ln\left(\left|\frac{dx_{n+1}}{dx_n}\right|\right) \tag{15}$$

which in the case of our map becomes

$$\lambda = \frac{1}{N} \sum_{n=1}^{N} ln\left(\left|c[(1-a)x_n - b]^{\frac{1}{1-a}-1}\right|\right) \tag{16}$$

For the calculations of the Lyapunov exponent we have used instead of the general formula (15) the more specific (16). To avoid initial fluctuations we performed the averaging over the last 100 values of 10 000 iterations.

Setting the parameter c equal to 1, we are calculating numerically the bifurcations of (14) for $0.01 \le b \le 1$. The obtained results are complex numbers, confirming thus, the oscillatory behavior of the solutions. Plotting the magnitude (modulus) $|P_{k,i}|$ of the complex solutions versus a (denoted by p_i), with b as a parameter, the diagrams shown in Fig. 1 are obtained. From these diagrams, it can be concluded that for $0.01 \le b \le 0.4$, the modulus $|P_{k,i}|$ remains constant or increases slightly with a, except for a region to the left of the value $a = 2$, where it shows a chaotic behavior. The width of this chaotic region grows with increasing b, until the value $b = 0.4$ is obtained. For $0.4 \le b \le 1$, $|P_{k,i}|$ undergoes successive period doublings with increasing and new chaotic regions appear.

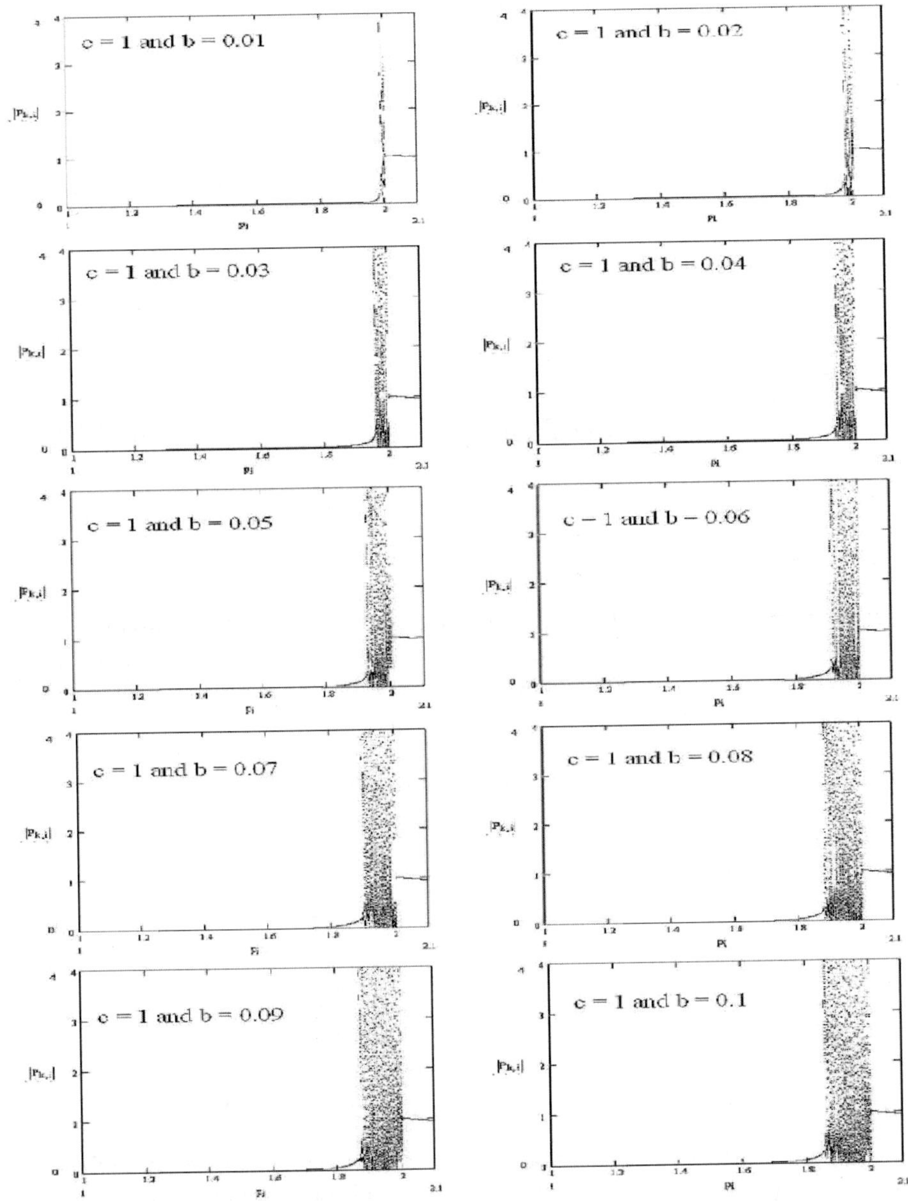

Fig. 1. The bifurcation diagram of the map for $c = 1$ and $0.01 \leq b \leq 0.1$

In a next step we are setting the parameter b equal to 1 and we are calculating numerically the bifurcations of (14) for $0.1 \leq c \leq 1$. The obtained results are again complex numbers, corresponding to an oscillatory behavior of the solutions. The diagrams shown in Fig. 2. are obtained by plotting $|P_{k,i}|$

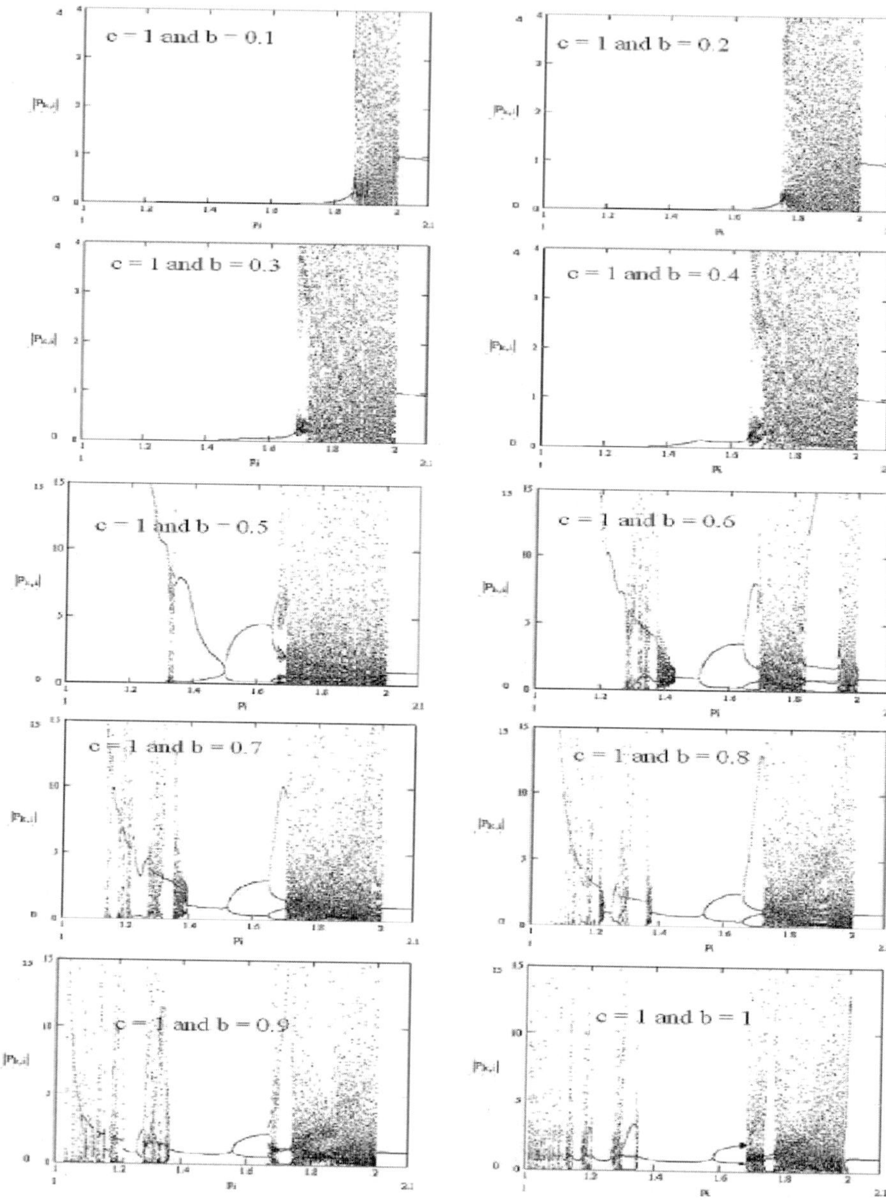

Fig. 1. (...continued) The bifurcation diagram of the map for $c = 1$ and $0.1 \leq b \leq 1$

of the complex solutions versus a (denoted by p_i), with c as a parameter. From these plots, it can be concluded that for $0.1 \leq c \leq 0.25$, modulus $|P_{k,i}|$ oscillates with a with a frequency inversely proportional to a. For $0.26 \leq c \leq 0.30$, this oscillatory behavior is interrupted by the appearance of a chaotic window

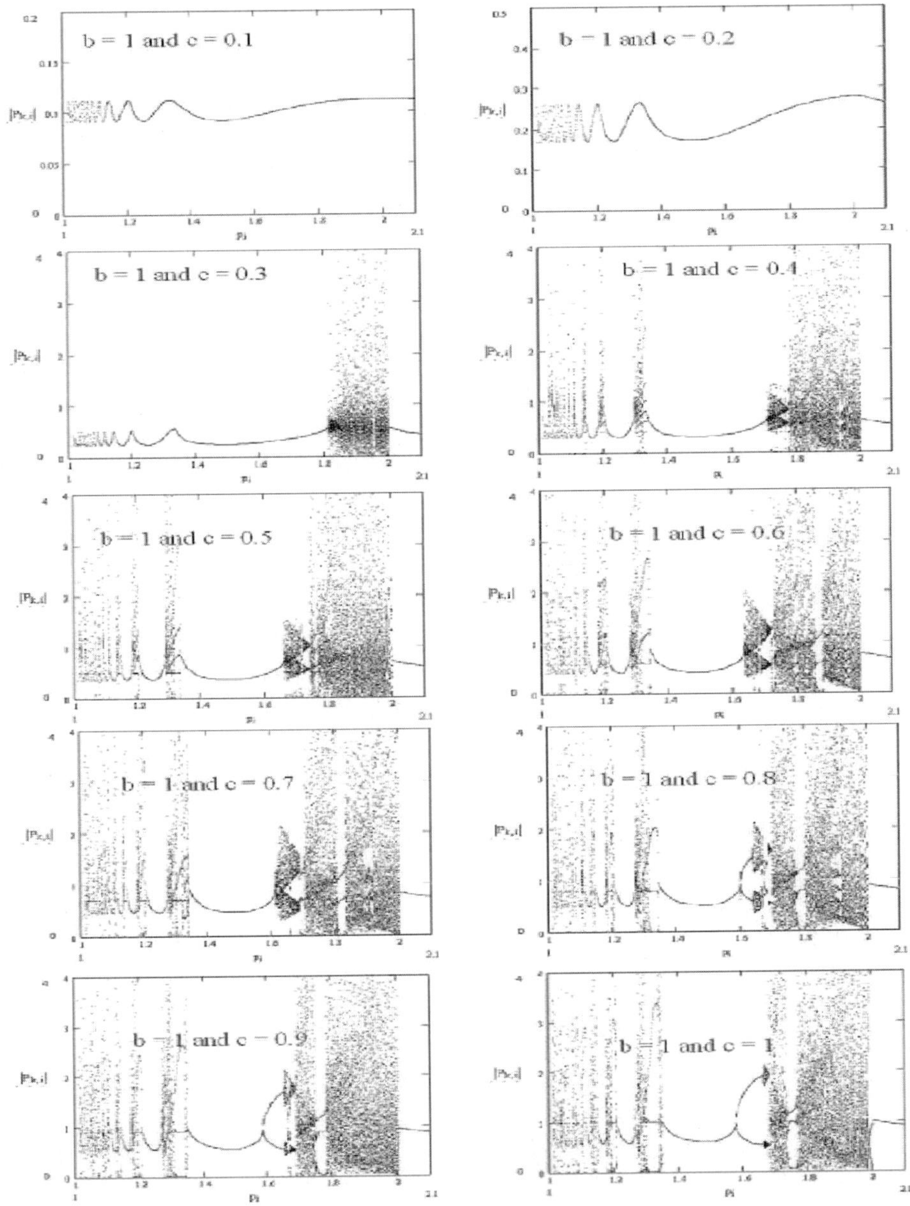

Fig. 2. The bifurcation diagram of the map for $b = 1$ and $0.1 \leq c \leq 1$

to the left of the value $a = 2$. The width of this chaotic window grows with increasing c, until the value $c = 0.30$ is obtained. For $0.30 \leq c \leq 1$, modulus $|P_{k,i}|$ undergoes successive period doublings and new chaotic windows appear. To visualize the behavior of the system, we plotted in Fig. 3. the real part

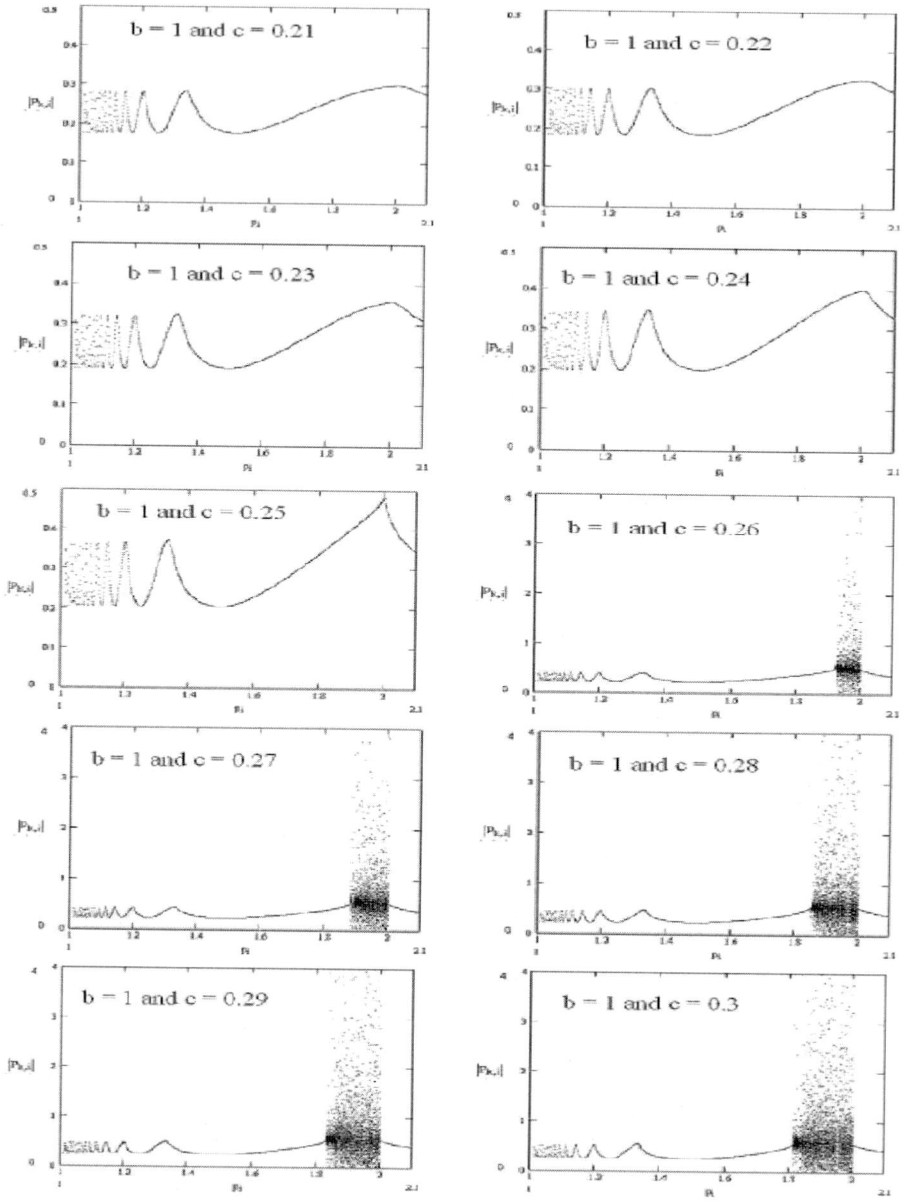

Fig. 2. (...continued) The bifurcation diagram of the map for $b = 1$ and $0.21 \le c \le 0.3$

$Re\,(P_{k,i})$ of the solutions versus their imaginary part $Im\,(P_{k,i})$. As expected for oscillations of constant amplitude, these plots are closed curves, interrupted at higher c's by a chaotic distribution of the $Re\,(P_{k,i})$-values around $Im\,(P_{k,i}) = 0$.

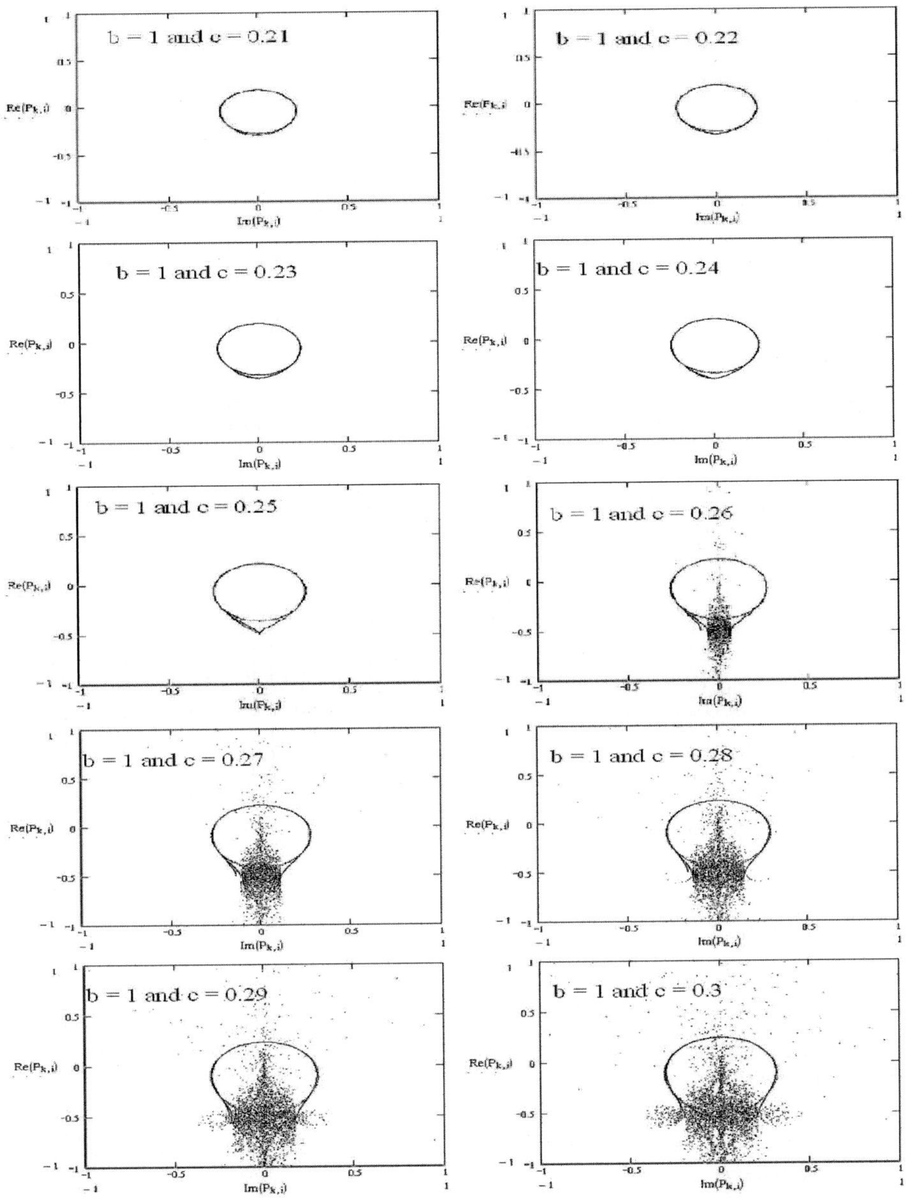

Fig. 3. $Re\,(P_{k,i})$ versus $Im\,(P_{k,i})$ of the solutions for $b = 1$, $0.21 \leq c \leq 0.30$

To confirm the true chaotic deterministic nature of the system in the windows of irregularity observed in the plots of Figs. 1.-3., we have calculated the corresponding Lyapunov exponents Λ_j of it. A representative calculation of Λ_j for the case $b = 1$ and $0.25 \leq c \leq 0.30$ is shown in Fig. 4. As expected, Λ_j

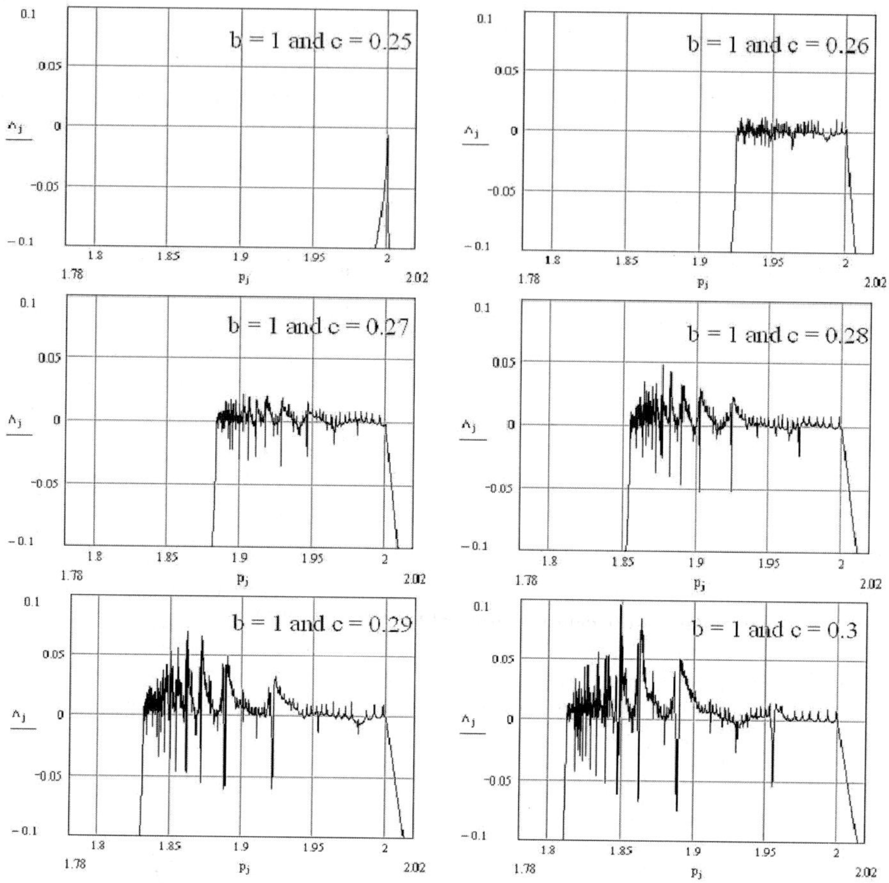

Fig. 4. Lyapunov exponents for $b = 1$ and $0.25 \leq c \leq 0.30$

remains negative in the case $b = 1$ and $c = 0.25$, where the orbit is periodic. In all other cases Λ_j is depicted in windows with positive values corresponding exactly to the chaotic windows of the plots of Fig. 2.

5 Comments

The dynamics of the map $F_{a,b,c}(x) = c[(1-a)x - b]^{\frac{1}{1-a}}$, are discussed and its chaotic behavior is realized for specific values of the control parameters. This family of maps is viewed as solution of a differential equation as well as solution of Newton's derivative equal to a constant a.

The differential equation (10) can be implemented by constructing a nonlinear electronic circuit, which will produce for $b = 1$ and $0.20 \leq c \leq 0.24$

a signal of constant modulus whose frequency decreases as a increases. Possible applications of this device are the sweeping of the frequency of signal generators. An interesting extension of the present study is to investigate the dynamics of the map resulting, after the incorporation relaxed Newton's derivative [BCR99, OPHVMAC06] and its connection to Briot-Bouquet differential equations.

References

[Wit91] Whittaker, J.V.: An Analytical Description of Some Simple Cases of Chaotic Behavior. American Mathematical Monthly, **98**, 489–504 (1991)

[Inc56] Ince, E.L.: Ordinary Differential Equations. Dover, New York (1956)

[Bra93] Braun, M.: Differential Equations and their Applications. Springer-Verlag, New York (1993)

[OCPHAVA] Özer, M., Čenys, A., Polatoglu, Y., Hacibekiroglu, G., Akat, E., Valaristos, A., Anagnostopoulos, A.N.: Bifurcations of Fibonacci Generating Functions. Chaos, Solitons and Fractals (accepted for publication)

[OPHVA05] Özer, M., Polatoglu, Y., Hacibekiroglu, G., Valaristos, A., Anagnostopoulos, A.N.: Some Results on Dynamics of Newton Diferential Equation. Journal of Naval Science and Engineering, **3(1)**, 23-38 (2005)

[FFJ97] Fulford, G., Forrester, P., Jones, A.: Modelling with Differential and Difference Equations. Cambridge University Press, New York (1997)

[CC02] Chapra, S.C., Canale, R.P.: Numerical Methods for Engineers. Mc Graw Hill Inc., New York (2002)

[BP92] Boyce, W.E., DiPrima, R.C.: Elementary Differential Equations and Boundary Value Problems. Wiley, New York (1992)

[Dav60] Davis, H.T.: Introduction to Nonlinear Differential and Integral Equations. Dover Publications, New York (1960)

[Hay93] Hayward, J.: Chaotic Iteration with MathCad. http://www.bham.ac.uk/ctimath/ reviews/aug93/mathcad.pdf (1993)

[DFran02] DiFranco, D.: Discovering Chaotic Iterations. http://www.mathcad.co.uk/ mcadlib/apps/chaotic.mcd (2002)

[Kauf04] Kaufman, M.: The Butterfly Effect. http://www.csuohio.edu/physics/kaufman/ Envphylab8.PDF (2004)

[BCR99] Billings, L., Curry, J.H., Robins, V.: Chaos in Relaxed Newton's Method: The Quadratic Case. Contemporary Mathematics, **252** (1999)

[OPHVMAC06] Özer, M., Polatoglu, Y., Hacibekiroglu, G., Valaristos, A., Miliou A.N., Anagnostopoulos, A.N., Čenys, A.: Dynamics on Relaxed Newton's Method Derivative. Proc. of I. Interdisciplinary Chaos Symposium on Chaos and Complex Systems(Istanbul, Turkey), to be appeared in Journal of Istanbul Kultur University (2006)

Dissipative solitons and nonlinear resonance dynamics in 2+1 dimensions

Oktay Pashaev

Izmir Institute of Technology, Izmir, Turkey oktaypashaev@iyte.edu.tr

We consider dissipative soliton (dissipaton) of the second member of SL(2,R) AKNS hierarchy in 1+1 dimension and show that it describes nonlinear doubled damped oscillator in 0+1 dimensions, where the velocity field plays the role of an effective damping. Combined with the third member of the hierarchy it give also rise to the real 2+1 dimensional solitons of KP-II and for KN hierarchy, to solitons of the MKP-II. By the Hirota bilinear form for both flows, we find new bilinear system and two soliton solution, showing resonance behaviour with creation of four virtual solitons. Our approach allows one to interpret the resonance soliton as a composite object of two dissipative solitons in 1+1 dimensions.

1 Noncompact symmetry of damped oscillator

It is well known that to set up the canonical formalism for dissipative system, a doubling of the degrees of freedom is required. It compliments the given dissipative system with its time-reversed image. In the case of a one-dimensional damped harmonic oscillator it leads to the doubled system

$$m\ddot{X} + \gamma\dot{X} + kX = 0 \tag{1}$$

$$m\ddot{Y} - \gamma\dot{Y} + kY = 0 \tag{2}$$

with time reversal symmetry, the total energy conservation and the Lagrangian

$$L = m\dot{X}\dot{Y} + \frac{\gamma}{2}(X\dot{Y} - \dot{X}Y) - kXY \tag{3}$$

The system possess the global $O(1,1)$ symmetry group $X \to Xe^{\alpha}, Y \to Ye^{-\alpha}$ and has realization as a motion in the hyperbolic plane [BGPV96]. Adding nonlinear coupling between these oscillators in a way to preserve this symmetry, we have the system

Kenan Taş et al. (eds), Mathematical Methods in Engineering, 435–444.

$$m\ddot{X} + \gamma\dot{X} + kX - m\frac{\Lambda}{4}X^2Y = 0 \tag{4}$$

$$m\ddot{Y} - \gamma\dot{Y} + kY - m\frac{\Lambda}{4}Y^2X = 0 \tag{5}$$

Despite of the nonlinear character, the last system admits the energy integral and an exact solution, which for the over-damping case $\Omega^2 = (1/m)(\gamma^2/4m - k) > 0$ is

$$X = \sqrt{\frac{8}{-\Lambda}}\Omega\frac{e^{-\frac{\gamma}{2m}t}}{\cosh \Omega t}, \qquad Y = \sqrt{\frac{8}{-\Lambda}}\Omega\frac{e^{\frac{\gamma}{2m}t}}{\cosh \Omega t} \tag{6}$$

2 Dissipatons and resonant NLS

The above nonlinear system (4), (5) is appear from $O(1,1)$ system

$$\begin{cases} e^+_t = e^+_{xx} + \frac{\Lambda}{4}e^+e^-e^+ \\ -e^-_t = e^-_{xx} + \frac{\Lambda}{4}e^+e^-e^- \end{cases} \tag{7}$$

for the traveling wave form of solution

$$e^+(x,t) = X(x - vt)e^{-\frac{k}{m}t}, \; e^-(x,t) = -Y(x - vt)e^{\frac{k}{m}t} \tag{8}$$

where moving frame coordinate $\tau = x - vt$ plays the role of effective time, while the velocity v has meaning of the damping coefficient $v = \gamma/m$. This system, the dissipative version of the Nonlinear Schrödinger equation, is called the Reaction-Diffusion (RD) system [MPS97]. It is connected with gauge theoretical formulation of 1+1 dimensional gravity[Jack84, MPS97], the constant curvature surfaces in pseudo-Euclidean space [MPS97] and the NLS soliton problem in the quantum potential (Resonant NLS) [PL02a, PL02b]

$$i\psi_t + \psi_{xx} + \frac{\Lambda}{4}|\psi|^2\psi = 2\frac{|\psi|_{xx}}{|\psi|}\psi \tag{9}$$

where $|\psi| = \sqrt{-e^+e^-}$, $\arg\psi = \frac{1}{2}\ln(-e^-/e^+)$. Dissipative analogs of solitons for this system are called dissipatons. They have reach resonance interaction phenomenology. The system is the first member of integrable hierarchy of equations with $O(1,1)$ symmetry which is generated by $SL(2,R)$ AKNS hierarchy.

3 KP-II and SL(2,R) AKNS hierarchy

The dissipative SL(2,R) AKNS hierarchy of evolution equations with times $t_0, t_1, t_2, ...t_N...$, for real functions $e^+(x,t_N), e^-(x,t_N)$,

$$\frac{1}{2}\sigma_3 \begin{pmatrix} e^+ \\ e^- \end{pmatrix}_{t_N} = \Re^{N+1} \begin{pmatrix} e^+ \\ e^- \end{pmatrix}, \tag{10}$$

where $N = 0, 1, 2, ...,$ $(\Lambda < 0)$, is generated by the recursion operator $\Re$

$$\Re = \begin{pmatrix} \partial_x - \frac{\Lambda}{4} e^+ \int^x e^- & -\frac{\Lambda}{4} e^+ \int^x e^+ \\ -\frac{\Lambda}{4} e^- \int^x e^- & \partial_x + \frac{\Lambda}{4} e^- \int^x e^+ \end{pmatrix}. \tag{11}$$

Then, the second member of the hierarchy is given by (7) while the third member appears as

$$\begin{cases} e^+_{t_2} = e^+_{xxx} + \frac{3\Lambda}{4} e^+ e^- e^+_x \\ e^-_{t_2} = e^-_{xxx} + \frac{3\Lambda}{4} e^+ e^- e^-_x \end{cases} \tag{12}$$

The AKNS hierarchy allows us to develop a method to find solution for (2+1) Kadomtsev-Petviashvili (KP-II)equation.

Theorem 1. *Let the functions* $e^+(x, y, t)$ *and* $e^-(x, y, t)$ *are solutions of equations (7) and (12) simultaneously. Then the function* $U(x, y, t) \equiv e^+ e^-$ *satisfies the Kadomtsev-Petviashivili (KP-II) equation*

$$(4U_t + \frac{3\Lambda}{4}(U^2)_x + U_{xxx})_x - 3U_{yy}. \tag{13}$$

3.1 Bilinear representation of KPII by AKNS flows

Using bilinear representations for systems (7) and (12) as in [PL02a, PL02b] and Theorem 1, we can find bilinear representation for KPII. We consider $G^\pm$ and F as real functions of three variables $G^{(\pm)} G^{(\pm)}(x, y, t)$, $F = F(x, y, t)$, and require for these functions to be a solution of corresponding bilinear systems for that equations simultaneously. Since the second equation in both systems is the same, it is sufficient to consider the next bilinear system

$$\begin{cases} (\pm D_y - D_x^2)(G^\pm \cdot F) = 0 \\ (D_t + D_x^3)(G^\pm \cdot F) = 0 \\ D_x^2(F \cdot F) = -2G^+ G^- \end{cases} \tag{14}$$

Then, according to Theorem 1, any solution of this system generates a solution of KPII. From the last equation we can derive U directly in terms of function F only

$$U = e^+ e^- = \frac{8}{-\Lambda} \frac{G^+ G^-}{F^2} = \frac{4}{\Lambda} \frac{D_x^2(F \cdot F)}{F^2} = \frac{8}{\Lambda} \frac{\partial^2}{\partial x^2} \ln F \tag{15}$$

Simplest solution of this system

$$G^\pm = \pm e^{\eta_1^\pm}, \;\; F = 1 + \frac{e^{(\eta_1^+ + \eta_1^-)}}{(k_1^+ + k_1^-)^2}, \tag{16}$$

where $\eta_1^\pm = k_1^\pm x \pm (k_1^\pm)^2 y - (k_1^\pm)^3 t + \eta_1^{\pm(0)}$, defines one-soliton solution of KPII according to (42)

$$U = \frac{2(k_1^+ + k_1^-)^2}{\Lambda \cosh^2 \frac{1}{2}[(k_1^+ + k_1^-)x + (k_1^{+2} - k_1^{-2})y - (k_1^{+3} + k_1^{-3})t + \gamma]}, \tag{17}$$

where $\gamma = -\ln(k_1^+ + k_1^-)^2 + \eta_1^{+(0)} + \eta_1^{-(0)}$. This soliton is a planar wave wall travelling in an arbitrary direction and called the line soliton.

3.2 Two soliton solution

Continuing Hirota's perturbation we find two soliton solution in the form

$$G^\pm = \pm(e^{\eta_1^\pm} + e^{\eta_2^\pm} + \alpha_1^\pm e^{\eta_1^+ + \eta_1^- + \eta_2^\pm} + \alpha_2^\pm e^{\eta_2^+ + \eta_2^- + \eta_1^\pm}), \tag{18}$$

$$F = 1 + \frac{e^{\eta_1^+ + \eta_1^-}}{(k_{11}^{+-})^2} + \frac{e^{\eta_1^+ + \eta_2^-}}{(k_{12}^{+-})^2} + \frac{e^{\eta_2^+ + \eta_1^-}}{(k_{21}^{+-})^2} + \frac{e^{\eta_2^+ + \eta_2^-}}{(k_{22}^{+-})^2} + \beta e^{\eta_1^+ + \eta_1^- + \eta_2^+ + \eta_2^-}, \tag{19}$$

where $\eta_i^\pm = k_i^\pm x \pm (k_i^\pm)^2 y - (k_i^\pm)^3 t + \eta_i^{\pm(0)}$, $k_{ij}^{ab} = k_i^a + k_j^b$, $(i,j = 1,2)$, $(a,b = +-)$,

$$\alpha_1^\pm = \frac{(k_1^\pm - k_2^\pm)^2}{(k_{11}^{+-} k_{21}^{\pm\mp})^2}, \quad \alpha_2^\pm = \frac{(k_1^\pm - k_2^\pm)^2}{(k_{22}^{+-} k_{12}^{\pm\mp})^2}, \quad \beta = \frac{(k_1^+ - k_2^+)^2 (k_1^- - k_2^-)^2}{(k_{11}^{+-} k_{12}^{+-} k_{21}^{+-} k_{22}^{+-})^2}.$$

Then, it provides two-soliton solution of KPII according to (42).

3.3 Degenerate four-soliton solution

However for KPII another bilinear form in terms of function F only is known [Hir71]

$$(D_x D_t + D_x^4 + D_y^2)(F \cdot F) = 0 \tag{20}$$

Thus, it is natural to compare soliton solutions of our bilinear equations (14) with the ones given by this equation. To solve equation (20) we consider $F = 1 + \varepsilon F_1 + \varepsilon^2 F_2 + \ldots$. The solution $F_1 = e^{\eta_1}$, where $\eta_1 = k_1 x + \Omega_1 y + \omega_1 t + \eta_1^0$, and dispersion $k_1\omega_1 + k_1^4 + \Omega_1^2 = 0$ with $F_n = 0$, (n = 2,3,...), under identification $k_1 = k_1^+ + k_1^-$, $\Omega_1 = \sqrt{3}(k_1^{+2} - k_1^{-2})$, $\omega_1 = -4(k_1^{+3} + k_1^{-3})$, and rescaling $4t \to t$, $\sqrt{3}y \to y$, determines one soliton solution of KPII (13). We realize that it coincides with our one soliton solution (17). But two soliton solution of equation (20) does not correspond to our two-soliton solution (18),(19). Appearance of four different terms $e^{\eta_i^\pm + \eta_k^\pm}$ in equation (19), suggests that our two-soliton solution should correspond to some degenerate case of four soliton solution of (20). To construct four soliton solution first we find following solutions of bilinear equations (20), $F_1 = e^{\eta_1}$, $F_2 = e^{\eta_2}$, $F_4 = e^{\eta_3}$, where $\eta_i = k_i x + \Omega_i y + \omega_i t + \eta_i^0, i = 1, 2, 3$, dispersion

$$k_i \omega_i + k_i^4 + \Omega_i^2 = 0 \tag{21}$$

and

$$F_3 = \alpha_{12} e^{\eta_1+\eta_2},\ F_5 = \alpha_{13} e^{\eta_1+\eta_3},\ F_6 = \alpha_{23} e^{\eta_1+\eta_3}, \tag{22}$$

where

$$\alpha_{ij} - \frac{(k_i - k_j)(\omega_i - \omega_j) + (k_i - k_j)^4 + (\Omega_i - \Omega_j)^2}{(k_i + k_j)(\omega_i + \omega_j) + (k_i + k_j)^4 + (\Omega_i + \Omega_j)^2}, (i, j = 1, 2, 3) \tag{23}$$

Then we parameterize our solution in the form

$$\begin{array}{l} k_1 = k_1^+ + k_1^-, \omega_1 = -4(k_1^{+3} + k_1^{-3}),\ \Omega_1 = \sqrt{3}(k_1^{+2} - k_1^{-2}), \\ k_2 = k_2^+ + k_2^-, \omega_2 = -4(k_2^{+3} + k_2^{-3}),\ \Omega_2 = \sqrt{3}(k_2^{+2} - k_2^{-2}), \\ k_3 = k_1^+ + k_2^-, \omega_3 = -4(k_1^{+3} + k_2^{-3}),\ \Omega_3 = \sqrt{3}(k_1^{+2} - k_2^{-2}) \\ k_4 = k_2^+ + k_1^-, \omega_4 = -4(k_2^{+3} + k_1^{-3}),\ \Omega_4 = \sqrt{3}(k_2^{+2} + k_1^{-2}), \end{array} \tag{24}$$

satisfying dispersion relations (21). Substituting these parameterizations to above solutions we find that $\alpha_{13} = 0 \Rightarrow F_5 = 0$, $\alpha_{23} = 0 \Rightarrow F_6 = 0$. Continuing Hirota's perturbation with solution $F_7 = e^{\eta_4}$, where $\eta_4 = k_4 x + \Omega_4 y + \omega_4 t + \eta_4^0$, we find that $F_8 = \alpha_{14} e^{\eta_1+\eta_4}$, where

$$\alpha_{14} - \frac{(k_1 - k_4)(\omega_1 - \omega_4) + (k_1 - k_4)^4 + (\Omega_1 - \Omega_4)^2}{(k_1 + k_4)(\omega_1 + \omega_4) + (k_1 + k_4)^4 + (\Omega_1 + \Omega_4)^2} \tag{25}$$

and after the parameterizations given above (24) it also vanishes $\alpha_{14} = 0 \Rightarrow F_8 = 0$. The next solution $F_9 = \alpha_{24} e^{\eta_2+\eta_4}$, where

$$\alpha_{24} = -\frac{(k_2 - k_4)(\omega_2 - \omega_4) + (k_2 - k_4)^4 + (\Omega_2 - \Omega_4)^2}{(k_2 + k_4)(\omega_2 + \omega_4) + (k_2 + k_4)^4 + (\Omega_2 + \Omega_4)^2}, \tag{26}$$

also is zero $\alpha_{24} = 0 \Rightarrow F_9 = 0$. Then we have $F_{10} = 0$, and $F_{11} = \alpha_{34} e^{\eta_3+\eta_4}$, where

$$\alpha_{34} - \frac{(k_3 - k_4)(\omega_3 - \omega_4) + (k_3 - k_4)^4 + (\Omega_3 - \Omega_4)^2}{(k_3 + k_4)(\omega_3 + \omega_4) + (k_3 + k_4)^4 + (\Omega_3 + \Omega_4)^2}. \tag{27}$$

When it is checked for higher order terms we find that $F_{12} = F_{13} = ... = 0$. Thus, we have degenerate four-soliton solution of equations (20) in the form

$$F = 1 + e^{\eta_1} + e^{\eta_2} + e^{\eta_3} + e^{\eta_4} + \alpha_{12} e^{\eta_1+\eta_2} + \alpha_{34} e^{\eta_3+\eta_4} \tag{28}$$

Comparing this solution with the one in (19) and taking into account that according parameterizations (24), $\eta_1 + \eta_2 = \eta_3 + \eta_4$, we see that they coincide. The above consideration shows that our two-soliton solution of KP-II corresponds to the degenerate four soliton solution in the canonical Hirota form (20). Moreover, it allows us to find new four virtual soliton resonance for KPII.

3.4 Resonance interaction of planar solitons

Choosing different values of parameters for our two soliton solution we find resonance character of soliton's interaction. For the next choice of parameters $k_1^+ = 2, k_1^- = 1, k_2^+ = 1.5, k_2^- = 0.5$, and vanishing value of the position shift constants, we obtained two soliton solution moving in the plane with constant velocity, with creation of four, so called virtual solitons.

The resonance character of our planar soliton interactions is related with resonance nature of dissipatons in 1+1 AKNS hierarchy. It has been reported also in several systems, but the four virtual soliton resonance does not seem to have been done for KPII [IR00] prior to our work. Recently we realized that resonance solitons for KPII have been constructed independently also by Biondini and Kodama [BK03, K04] using Sato's theory. Then, the comparision shows that our bilinear constraint plays the similar role as the Toda lattice in their paper.

4 MKP-II resonance solitons

4.1 MKP-II and Kaup-Newell hierarchy

The KN hierarchy for functions $q(x, t_N), r(x, t_N)$ has the form [Yan00]

$$\begin{pmatrix} q \\ r \end{pmatrix}_{t_N} = JL^N \begin{pmatrix} q \\ r \end{pmatrix} \tag{29}$$

where the operator

$$J = \begin{pmatrix} 0 & \partial_x \\ \partial_x & 0 \end{pmatrix} \tag{30}$$

is the first symplectic form, while

$$L = \frac{1}{2} \begin{pmatrix} -\partial_x - r\int^x q\partial_x & -r\int^x r\partial_x \\ -q\int^x q\partial_x & \partial_x - q\int^x r\partial_x \end{pmatrix} \tag{31}$$

is the recursion operator of the hierarchy. For the SL(2,R) case of KN hierarchy we have real time variables t_2, t_3 which we denote as $y \equiv t_2/2$, and $t \equiv -t_3/4$. In this case functions q and r are real, and we denote them as $e^+ \equiv q, \quad e^- \equiv -r$. Then, as the second member we have the DRD system [LLP00]

$$\begin{cases} e_y^+ = e_{xx}^+ - (e^+e^-e^+)_x \\ e_y^- = -e_{xx}^- - (e^+e^-e^-)_x \end{cases} \tag{32}$$

and for the third one

$$\begin{cases} e_t^+ = e_{xxx}^+ - 3(e^+e^-e_x^+)_x + \frac{3}{2}((e^+e^-)^2e^+)_x \\ e_t^- = e_{xxx}^- + 3(e^+e^-e_x^-)_x + \frac{3}{2}((e^+e^-)^2e^-)_x \end{cases} \tag{33}$$

Now we consider the pair of functions of three variables $e^+(x, y, t)$ and $e^-(x, y, t)$ satisfying the systems (32) and (33).

Theorem 2. . *Let the functions* $e^+(x,y,t)$ *and* $e^-(x,y,t)$, *are solutions of the systems (32) and (33) simultaneously. Then, the function* $U(x,y,t) \equiv e^+e^-$ *satisfies the modified Kadomtsev-Petviashvili equation (MKP-II)*

$$(-4U_t + U_{xxx} - \frac{3}{2}U^2U_x - 3U_x\partial_x^{-1}U_y)_x = -3U_{yy} \tag{34}$$

4.2 Bilinear form for the second and third flows

Now we will construct bilinear representation for systems (32) and (33) to find solutions of MKP-II according to our Theorem 2. In our paper [LLP00] we applied the Hirota bilinear method to integrate RDR (32). Now we will apply the same method as in the first section to (33) and MKP-II. To have the standard Hirota substitution, following [Lee89, LLP00] we first rewrite the systems (32) and (33) in terms of new functions Q^+, Q^-:

$$e^+ = e^{+\int^x Q^+Q^-}Q^+, \quad e^- = e^{-\int^x Q^+Q^-}Q^-, \tag{35}$$

and as result we have the systems

$$Q_y^{\pm} = \pm Q_{xx}^{\pm} + Q^{\pm}Q^{\pm}Q_x^{\mp} \mp \frac{1}{2}(Q^+Q^-)^2Q^{\pm}, \tag{36}$$

and

$$Q_t^{\pm} = Q_{xxx}^{\pm} \pm 3Q_x^+Q_x^-Q^{\pm} - \frac{3}{2}(Q^+Q^-)^2Q_x^{\pm}, \tag{37}$$

Then, due to the fact that $Q^+Q^- = e^+e^- = U$ the systems (36), (37) provide also solution of MKP-II which we can formulate as below

Theorem 3. *Let the functions* $Q^+(x,y,t)$ *and* $Q^-(x,y,t)$, *are solutions of the systems (36) and (37) simultaneously. Then, the function* $U(x,y,t) \equiv Q^+Q^-$ *satisfies the modified Kadomtsev-Petviashvili equation (MKP-II) (34).*

To solve the systems (36) and (37) we introduce four real functions g^+, g^-, f^+, f^- according to the formulas

$$Q^+ = \frac{g^+}{f^+}, \quad Q^- = \frac{g^-}{f^-}, \tag{38}$$

or using (35) and (38) for the original variables e^+ and e^- we have the following substitution

$$e^+ = \frac{g^+f^+}{(f^-)^2}, \quad e^- = \frac{g^-f^-}{(f^+)^2}. \tag{39}$$

Comparing bilinear forms for equations (36),(37), for simultaneous solution of both equations we have the next system

$$\begin{cases} (D_y \mp D_x^2)(g^\pm \cdot f^\pm) = 0, \\ (D_t - D_x^3)(g^\pm \cdot f^\pm) = 0, \\ D_x^2(f^+ \cdot f^-) + \frac{1}{2}D_x(g^+ \cdot g^-) = 0, \\ D_x(f^+ \cdot f^-) - \frac{1}{2}g^+g^- = 0 \end{cases} \tag{40}$$

From the last equation we have

$$U = e^+e^- = Q^+Q^- = \frac{g^+g^-}{f^+f^-} = 2\frac{D_x(f^+ \cdot f^-)}{f^+f^-} = 2\frac{f_x^+f^- - f^+f_x^-}{f^+f^-} \tag{41}$$

which provides solution of MKP-II by the following formula

$$U = 2(\ln \frac{f^+}{f^-})_x \tag{42}$$

4.3 Resonance solitons of MKP-II

Now we consider a solution of the system (40), giving 2+1 dimensional solution of MKP-II. For one-soliton solution we have

$$g^\pm = e^{\eta_1^\pm}, \quad f^\pm = 1 + e^{\phi_{11}^\pm} e^{\eta_1^+ + \eta_1^-}, \quad e^{\phi_{11}^\pm} = \pm \frac{k_1^\mp}{2(k_1^+ + k_1^-)^2}, \tag{43}$$

where, $\eta_1^\pm = k_1^\pm x \pm (k_1^\pm)^2 y + (k^\pm)^3 t + \eta_0^\pm$. The regularity condition requires $k_1^+ \leq 0$, $k_1^- \geq 0$. Then we have

$$U(x,y,t) = \frac{2k^2}{\sqrt{p^2 - k^2}\cosh k(x - py + \frac{k^2+3p^2}{4}t - a_0) + p}, \tag{44}$$

where $k = k_1^+ + k_1^-$, $p = k_1^- - k_1^+ > 0$, and bounded from the below parameter $p^2 > k^2$ is positive $p > 0$. The geometrical meaning of this parameter is $p^{-1} = \tan\alpha$, where α is the slope of the soliton line. Due to the condition $p > 0$, the direction of this line is restricted between $0 < \alpha < \pi/2$. (This is the space analog of the chirality property of dissipaton in 1+1 dimensions for DNLS [LLP00], when it propagates only in one direction.) The velocity of soliton is two dimensional vector $\mathbf{v} = (\omega, -\omega/p)$, where $\omega = (k^2 + 3p^2)/4$, directed at angle γ to the soliton line, where $\cos\gamma = 1 - 1/p^2$. When $p = 1$, the velocity of soliton is orthogonal to the soliton line.

For two soliton solution we have

$$g^\pm = e^{\eta_1^\pm} + e^{\eta_2^\pm} + \alpha_1^\pm e^{\eta_2^+ + \eta_2^- + \eta_1^\pm} + \alpha_2^\pm e^{\eta_1^+ + \eta_1^- + \eta_2^\pm}, \tag{45}$$

$$f^\pm = 1 + \sum_{i,j=1}^{2} e^{\phi_{ij}^\pm} e^{\eta_i^+ + \eta_j^-} + \beta^\pm e^{\eta_1^+ + \eta_1^- + \eta_2^+ + \eta_2^-}, \tag{46}$$

where $\eta_i^\pm = k_i^\pm x \pm (k_i^\pm)^2 y + (k^\pm)^3 t + \eta_{i0}^\pm$, $k_{ij}^{nm} \equiv (k_i^n + k_j^m)$ and

$$\alpha_1^{\pm} = \pm\frac{1}{2}\frac{k_2^{\mp}(k_1^{\pm}-k_2^{\pm})^2}{(k_{22}^{+-})^2(k_{12}^{\pm\mp})^2}, \quad \alpha_2^{\pm} = \pm\frac{1}{2}\frac{k_1^{\mp}(k_1^{\pm}-k_2^{\pm})^2}{(k_{11}^{+-})^2(k_{21}^{\pm\mp})^2}, \tag{47}$$

$$\beta^{\pm} = \frac{(k_1^{+}-k_2^{+})^2(k_1^{-}-k_2^{-})^2}{4(k_{11}^{+-}k_{12}^{+-}k_{21}^{+-}k_{22}^{+-})^2}k_1^{\mp}k_2^{\mp}, \tag{48}$$

$$e^{\phi_{ii}^{\pm}} = \pm\frac{k_i^{\mp}}{2(k_{ii}^{+-})^2}, \quad e^{\phi_{ij}^{+}} = \frac{k_j^{-}}{2(k_{ij}^{+-})^2}, \quad e^{\phi_{ij}^{-}} = -\frac{k_i^{+}}{2(k_{ij}^{+-})^2}. \tag{49}$$

The regularity conditions now are the same as for one soliton $k_i^+ \leq 0$, $k_i^- \geq 0$. Then this solution describes a collision of two solitons propagating in plane and at some value of parameters creating the resonance states.

5 Conclusions

In the present paper we have constructed virtual soliton resonance solutions for 2+1 dimensional KP-II and MKP-II in terms of dissipative solitons of 1+1 dimensional equations as the Reaction-Diffusion equation and Derivative Reaction-Diffusion equation and their higher members of SL(2,R) AKNS and Kaup-Newell hierarchies. We also have established the relation with 0+1 dimensional nonlinear doubled damped oscillator model, where the velocity of dissipaton plays the role of effective damping.

The idea to use couple of equations from the AKNS hierarchy to generate a solution of KP, and the KN hierarchy to generate a solution of MKP, can be applied also to multidimensional integrable sytems with zero curvature structure having form of the Chern-Simons gauge theory. Our three dimensional zero curvature representation of KP-II gives then flat non-Abelian connection for $SL(2, R)$ and corresponds to a sector of three dimensional gravity theory.

This work was supported partially by Izmir Institute of Technology (Grant No: 2005-IYTE-13), Izmir, Turkey.

References

[BGPV96] Blasone, M., Graziano, E., Pashaev, O.K., Vitiello, G.: Dissipation and topologically massive gauge theories in the pseudo-Euclidean plane. Annals of Physics, **252(1)**, 115–132 (1996)

[Jack84] Jackiw R.: In: S. Christensen (ed.) Quantum Theory of Gravity. Adam Hilger, Bristol (1984); Teitelboim C.: In: S. Christensen (ed.) Quantum Theory of Gravity. Adam Hilger, Bristol (1984)

[MPS97] Martina, L., Pashaev, O.K., Soliani, G.: Integrable dissipative structures in the gauge theory of gravity. Class. Quantum Grav., **14(12)**, 3179-3186 (1997); Phys. Rev. D, **58**, 084025 (1998)

[PL02a] Pashaev, O.K., Lee, J.H.: Resonance solitons as black holes in Madelung fluid. Mod. Phys. Lett. A, **17(24)**, 1601-1619 (2002)

[PL02b] Pashaev, O.K., Lee, J.H.: Black holes and solitons of the quantized dispersionless NLS and DNLS equations. ANZIAM Journal, **44**, 73-81 (2002)

[Hir71] Hirota, R.: Exact Solution of the Korteweg-de Vries Equation for Multiple Collisions of Solitons. Phys. Rev. Lett., **27**, 1192-1194 (1971); In: Bullough, R.K., Caudrey, P.J. (eds) Solitons. Springer, New York (1980)

[IR00] Infeld, E., Rowlands, G.: Nonlinear Waves, Solitons and Chaos. Cambridge University Press, Cambridge (2000)

[BK03] Biondini, G., Kodama, Y.: On a family of solutions of the Kadomtsev-Petviashvili equation which also satisfy the Toda lattice hierarchy. J. Phys. A: Math. Gen., **36**, 10519-10536 (2003)

[K04] Kodama, Y.: The Young diagrams and N-soliton solutions of the KP equation. arXiv: nlin.SI/0406033, (2004)

[Yan00] Yan, Z.: Liouville integrable N-Hamiltonian structures, involutive solutions and separation of variables associated with Kaup-Newell hierarchy. Chaos, Solitons and Fractals, **14**, 45-56 (2000)

[LLP00] Lee, J.H., Lin, C.K., Pashaev, O.K.: Equivalence relation and Bilinear representation for derivative NLS type equations. In 20 years after NEEDS'79, World Sci. Pub., Singapore (2000)

[Lee89] Lee J.H.: Global solvability of the derivative nonlinear Schrodinger equation. Transactions of the AMS, **314(1)**, 107-118, (1989)

Implementation of floating point arithmetics using an FPGA

Suhap Sahin, Adnan Kavak, Yasar Becerikli and H. Engin Demiray

Department of Computer Engineering, University of Kocaeli, Izmit, 41040, Turkey
{suhapsahin, akavak, becer, hedemiray}@kou.edu.tr

Floating point operations, which find their applications in vast areas such as many mathematical optimization methods, digital signal and image processing algorithms, and Artificial Neural Networks (ANNs), require much area and time for ordinary implementation on Field Programmable Gate Arrays (FPGAs). However, meaningful floating point arithmetic implementation on FPGAs is quite difficult with low level design specifications due to mapping difficulties and the complexity of floating point arithmetic. Design and implementation of floating point arithmetic and mapping of this into an FPGA become easier with the emergence of new generation FPGAs and development of high level languages such as VHDL tools. This paper presents the implementation methodologies of various floating point arithmetic operations such as addition, subtraction, multiplication, and division using 32-bit IEEE 754 floating point format. The implementation is performed using Xilinxs Spartan 3 FPGAs. The algorithms and implementation steps used for different operations are discussed in detail. As an example, an ANN application is presented using these algorithms.

1 Introduction

With the introduction of field programmable gate arrays (FPGAs), it is feasible to provide custom hardware for application specific computation design. The changes in designs in FPGAs can be accomplished within a few hours, and thus result in significant savings in cost and design cycle. FPGAs offer speed comparable to dedicated and fixed hardware systems for parallel algorithm acceleration [ZG99]. Floating point implementation on FPGAs is a challenging problem because floating point numbers require more fields than fixed point numbers and availability of physical resources on FPGAs (memory, gates, etc.) is limited. The floating point implementations on FPGAs require bit-width variation as a means to control precision. Several researchers [LMMSU98, LJC96] have implemented floating point adders and multipliers

Kenan Taş et al. (eds), Mathematical Methods in Engineering, 445–453.

on FPGAs, which meet IEEE 754 floating point format. Most commercial floating point implementations provide units that comply with the IEEE 754 standard [Nall01]. A fixed point design is larger and 40 percent slower than a corresponding floating point design with the same dynamic range [Ash01]. Here, we study implementation of various floating point arithmetic operations such as addition, subtraction, multiplication, and division on Xilinxs Spartan 3 FPGAs. The 32-bit IEEE 754 floating point format is used in implementations.

Artificial Neural Networks(ANNs) can solve great variety of problems in areas of pattern recognition, image processing and medical diagnostic [PZS93]. Implementation of ANNs falls into two categories: Software implementation and hardware implementation. ANNs are implemented in software, and are trained and simulated on general-purpose sequential computers for emulating a wide range of neural networks models. Software implementations offer flexibility. However hardware implementations are essential for applicability [RFP93]. Specific-purpose fixed hardware implementations, i.e. VLSI, are dedicated to a specific ANN model. VLSI implementations of ANNs provide high speed in real time applications and compactness. However, they lack flexibility for structural modification and are prohibitively costly. In this paper, as an example study, we are interested in building a different class of hardware environment for ANNs, i.e. FPGA-based reconfigurable computing environment. We explore that how efficiently 32 bit floating-point numeric representation is used in FPGA based implementation of ANNs. Spartan-IIIE series FPGAs are used in the implementation. A VHDL library referred to *fp_lib* is designed for using ANN's on FPGAs.

As an example of demonstration of floating point arithmetics, an artificial neural network (ANN) implementation is presented. The concept of ANNs is emerged from the principles of brain that are adapted to digital computers [Hay99]. Each neuron in ANNs takes some information as an input from another neuron or from an external input. This information is propagated as an output that are computed as weighted sum of inputs and applied as non-linear function.

2 Floating point algorithms

2.1 Floating point addition and subtraction

Our floating point addition and subtraction algorithm shown in Figure 1 is analogous to what is done in most conventional processor. Let F_1 and F_2 represent the two single precision floating numbers, F_{sum} is the sum of these two numbers and F_{minus} is $F_1 - F_2$. Since floating point format uses a signed-magnitude representation, the equation $F_{minus} = F_1 - F_2$ can be rewritten as; $F_{minus} = F_1 + (-F_2)$. Hence, this section describes the addition algorithm only. Subtraction is a variation of addition in which the sign bit of F_2 is

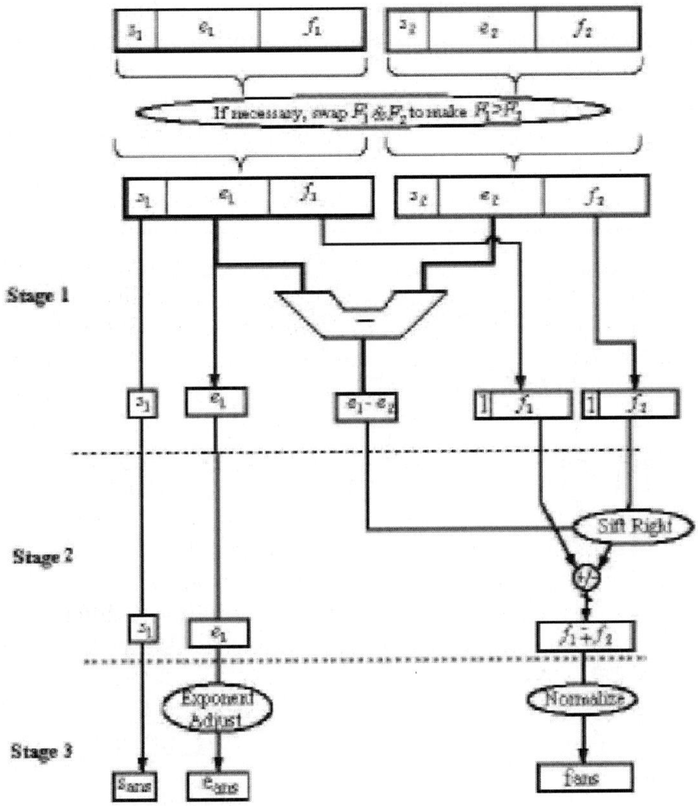

Fig. 1. 32-Bit floating point addition and subtraction algorithm on an FPGA.

inverted. The addition/subtraction operation is performed in three stages. The notation s_i, e_i and f_i are the sign field, fraction field and the exponent field of the floating point number F_i. Let the inputs are $F_1 = (s_1, f_1, e1)$ and $F_2 = (s_2, f_2, e2)$, where s_i denotes information bits, e_i denotes exponent bits, and f_i denotes fraction bits of floating point number F_i. The result is $F_{ans} = (s_{ans}, f_{ans}, eans) = F_1 + F_2$ or $F_1 + (-F_2)$. The algorithm is described as follows:

Step 1: If the absolute value of F_1 is less than the absolute value of F_2 then swap F_1 and F_2. Subtract e_1 from e_2 to calculate number of positions to shift f_2 to the right. Add the leading bit , $(1.f_1)$ and $(1.f_2)$.

Step 2: Shift $1.f_1$ to the right with $(e_1 - e_2)$ bits. If s_1 equals s_2 add $(1.f_1)$ to $(1.f_2)$, else subtract $(1.f_1)$ from $(1.f_2)$. Set the sign and exponent of the final result, F_{ans}, to the sign and exponent of f_1 value which is greater one.

Step 3: Shift the f_{ans} to the left until leading bit is a one and calculate number of shift bits. Adjust the e_{ans} by subtract number of bits from e_1.

2.2 Floating point multiplication

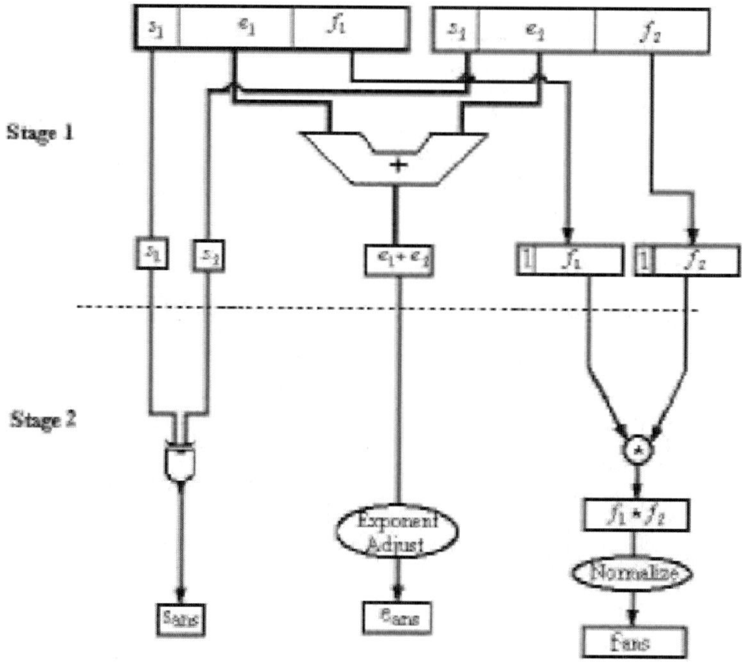

Fig. 2. 32-Bit floating point multiplication algorithm on an FPGA.

Floating point multiplication is similar to integer multiplication. Therefore, floating point multiplication is simpler than floating point addition. As in the architecture of the floating point adder, the floating point multiplier unit is a three-stage pipeline that produces a result on every clock cycle. The block diagram for floating point multiplier is shown in Figure 2. The algorithm for multiplication is performed in two stages. For simplicity, the algorithm does not check any special cases such as negative zero, illegal number, and so on F_i.

Requirement: $F_1 = (s_1, f_1, e_1)$, $F_2 = (s_2, f_2, e_2)$.

Result: $F_{ans} = (s_{ans}, f_{ans}, eans) = F_1 * F_2$

Step 1: The exponents, e_1 and e_2 are added and the result is stored e_{ans}. Add the leading bit $(1.f_1)$ and $(1.f_2)$.

Step 2: The fractions $(1.f_1)$ and $(1.f_2)$ are multiplied and the 23 bit to 45 bit of the result are f_{ans}. Calculate the sign bit (s_{ans}) by using the XOR of the sign bit of the two operands.

2.3 Floating point division

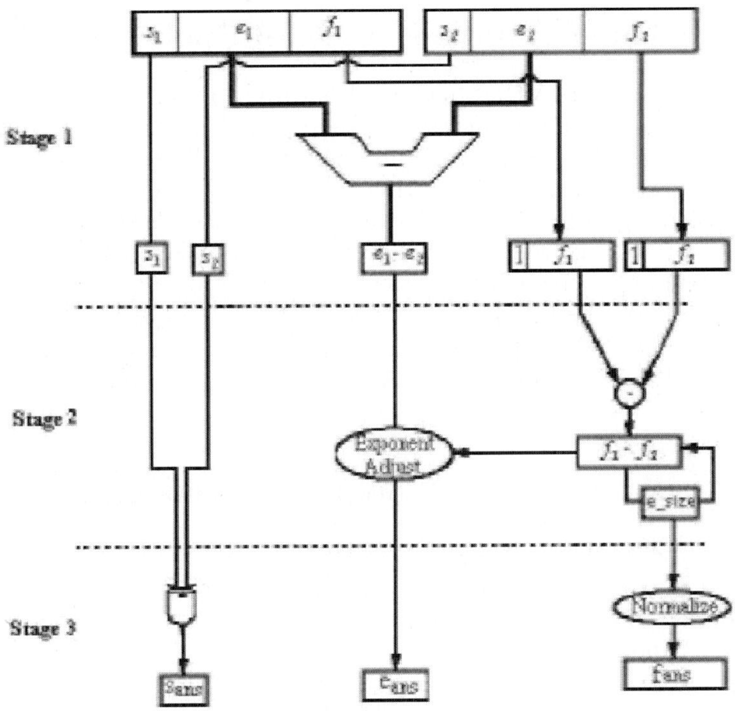

Fig. 3. 32-Bit floating point division algorithm on an FPGA.

Let F_1 and F_2 represent the two single precision floating numbers, F_{ans} is the division of these two numbers. As in the architecture of the floating point divider, the floating point divider unit is two-stage pipeline that produces a result on every clock cycle. The block diagram for floating point divider is shown in Figure 3. The algorithm for division is performed in two stages. For simplicity, the algorithm does not check any special cases such as negative zero, illegal number and so on F_i.

Requirement: $F_1 = (s_1, f_1, e_1)$, $F_2 = (s_2, f_2, e_2)$

Result: $F_{ans} = (s_{ans}, f_{ans}, eans) = F_1/F_2$

Step 1: The exponents, and are subtracted. Add the leading bit, $(1.f_1)$ and $(1.f_2)$.

Step 2: The fractions $(1.f_1)$ and $(1.f_2)$ are subtracted iteratively until e-size. In each iteration of subtraction, if the minuend is greater than subtrahend, digit value of the number is normalized to 1 otherwise 0. Moreover, a value for adjusting e values is obtained during each iteration.

Step 3: Shift the f_{ans} to the left until leading bit is a one and calculate number of shift bits. Calculate the sign bit s_{ans} by using the XOR of the sign bit of two operands.

3 FPGA implementations of floating point algorithms

This section presents VHDL code and hardware implementation results for the algorithms previously described. Digilentic demo board is used for the implementation. The board has Xilinx SpartanII [Xil93] 2s200epq208-6 and 50 MHz clock frequency. SpartanII chip has 2352 slices and 14 blocks RAM. The division algorithm design previously described requires much more area than the effective area of SpartanII. A VHDL library that it is referred as *fp_lib* was designed for hardware implementation on FPGAs. The *fp_lib* has two separate algorithms shown in Table 1. Implementation results for addition/subtraction and multiplication algorithms are summarized in Table 2 and Table 1, respectively, which show the architecture complexity of these algorithms.

Table 1. Summary of custom arithmetic VHDL algorithm.

HDL-Design	Description
fp_lib	IEEE 32-bit single precision floating point library
fp_mul	IEEE 32-bit single precision floating point pipelined parallel multiplier
fp_add	IEEE 32-bit single precision floating point pipelined parallel adder/subtractor

Table 2. Implementation results for floating point addition/subtraction on an FPGA.

Selected Device: 2s200epq208-6	
Number of Slices:	387 out of 2352 16%
Number of Slices Flip Flops:	106 out of 4704 2%
Number of 4 input LUTs:	903 out of 4704 15%
Number of bonded IOBs:	103 out of 146 70%

Table 3. Implementation results for floating point multiplication on an FPGA.

Selected Device: 2s200epq208-6	
Number of Slices:	326 out of 2352 13%
Number of Slices Flip Flops:	65 out of 4704 1%
Number of 4 input LUTs:	642 out of 4704 13%
Number of bonded IOBs:	103 out of 146 70%

4 Artificial Neural Network (ANN) implementation on the FPGA

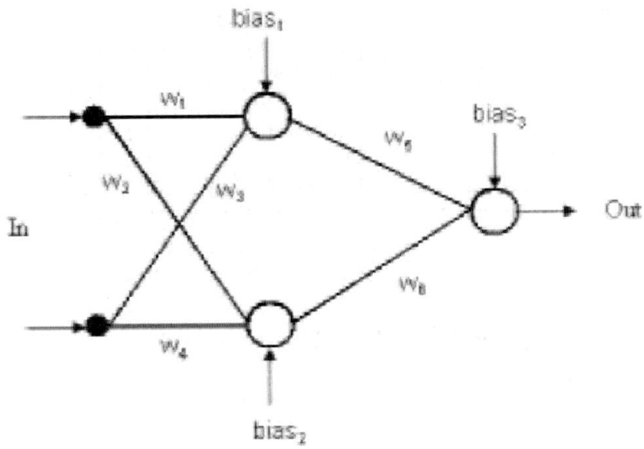

Fig. 4. An example of a three layer multi layer perceptron ANN.

An artificial neural network (ANN) that we consider here consists of an input layer, one hidden layer and an output layer as depicted in Figure 4. Sigmoid function is used as the activation function. In the fully parallel ANN's architecture, number of multipliers per neuron equals to number of connections to this neuron and number of the full adders equals to number of connections to the previous layer minus one. For example in 2-4-1 network, output neuron has 4 multipliers and 3 adders. A VHDL library designed for floating point addition *fp_add* and floating point multiplication *fp_mul* is used in the ANN application. A VHDL library that it is referred to as *fp_lib* is designed for implementing the ANN on the FPGA.

The weight coefficients used in Figure 4 are set as follows: $w1 = -9.4345$, $w2 = -13.595$,$w3 = -14.154$, $w4 = -24.822$, $w5 = 24.149$, $w6 = -37.329$. We used 16-bit floating point numbers to realize the ANN on Spartan-3 FPGA. Figure 5 shows entities realized on the FPGA. As seen in this Figure, the

ANN architecture consists of two inputs, single output, and a clock signal. This architecture contains a bolme.vhd entity to realize division operations, an xor.vhd entity that provides general control, and a paket.vhd entity. Floating point addition and multiplication operations in the ANN are called as functions in the xor.vhd entity. The cells in the ANN are called by the function that is inside the xor.vhd entity. The results for the digital XOR function and its FPGA based implementation using ANN are given in Table 4.

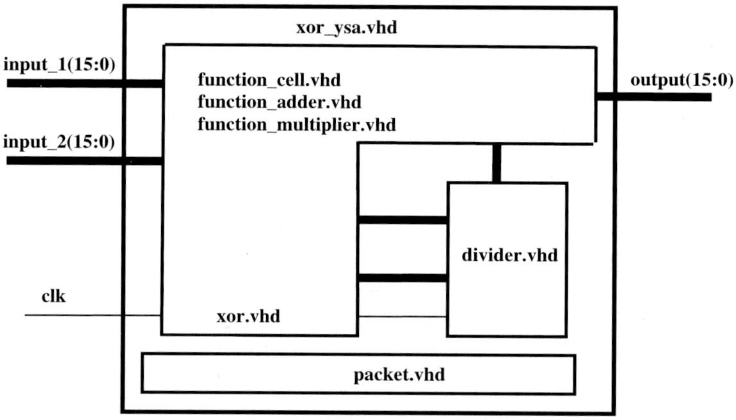

Fig. 5. Implementation architecture of the ANN entity on the FPGA.

Table 4. The results for the FPGA implementation of the ANN.

inputs	XOR(out)	FPGA based ANN(out)	ERROR
0,0	0.0	0.0596	0.0596
1,0	1.0	0.951	0.0488
0,1	1.0	0.949	0.051
1,1	0.0	0.0423	0.0423

5 Conclusions

We have presented some floating point arithmetic algorithms (addition, subtraction, multiplication, and division) that are suited for implementing on FPGAs. The implementations are performed on Xilinxs Spartan3 family FPGA. The implementation steps used for these operations are discussed in detail.

As an example of using these algorithms, implementation of an ANN on the FPGA is also presented. This study shows that FPGAs are versatile devices for implementing many different applications since FPGAs allow the hardware design via its configuration on software control. The improvement of circuitry design is just a matter of modification, debugging and downloading the new configuration code in a short time.

References

[ZG99] Zhu, J., Gunther, B.K.: Towards an FPGA based reconfigurable computing environment for neural network implementations. Proc. IEEE the 9th International Conference on Artificial Neural Networks (ICANN'99), IEE Conference Proceedings,**470**, 661–667 (1999)

[LMMSU98] Ligon, W. B., McMillan, S., Mpnn, G., Stivers, F., Underwood, K.D.: A re-evaluation of the practicality of floating point operations on FPGAs. Proc. IEEE Symposium on Field-Programmable Custom Computing Machines(Napa, CA), 206215 (1998)

[LJC96] Louca, L., Johnson, W.H., Cook, T.A.: Implementation of IEEE single precision floating point addition and multiplication on FPGAs. Proc. IEEE Workshop on FPGAs for Custom Computing Machines(Napa, CA), 107–116 (1996)

[Nall01] IEEE754 Floating Point Core. Nallatech Inc., http://www.nallatech.com/products/ip/, (2001)

[Ash01] Ashenden, P.J.: VHDL standards. IEEE Design Test of Computers, **18(6)**, 122–123 (2001)

[PZS93] DF Poliac, M., Zanetti, J., Salerno, D.: Performance mesuraments of seismocardiogram interpretation using neural networks. IEEE Computer in Cardiology, 573–576 (1993)

[RFP93] Rucket, U., Funke, A., Pintaske, C.: Accelerator board for neural associative memories. Neurocomputing, **5(1)**, 3949 (1993)

[Hay99] Haykin, S.: Neural Networks: A Comprehensive Foundation. Prentice Hall , New Jersey, (1999)

[Xil93] Xilinx Inc.: The Programmable Logic Data Book. San Jose, California (1993)

A method for the recovery of the electric field vibration inside vertical inhomogeneous anisotropic dielectrics

Valery Yakhno and Ali Sevimlican

Dokuz Eylul University, Izmir, Turkey
{valery.yakhno, ali.sevimlican}@deu.edu.tr

In the paper a new method of finding the time-dependent electric field in a layered inhomogeneous uniaxial anisotropic dielectric is suggested. This method is related to an initial value problem solving for finding the electric field. The permittivity $\mathcal{E}$ is a diagonal matrix and the components of $\mathcal{E}$ are smooth functions of the variable x_3 only. The density of the electric current is the source of the electric waves. The Fourier transform of electric current density with respect to x_1 and x_2 variables is assumed to be a continuous function. The suggested method consists of finding the Fourier image with respect to x_1 and x_2 variables of the electric field. The problem of finding the Fourier image of the electric field is reduced to an operator integral equation. This operator integral equation is solved by successive approximations method. After that the time-dependent electric field is found by the inverse Fourier transform.

1 Introduction

Electromagnetic wave propagation inside of anisotropic materials has attracted a great deal of interests [RWD94, Yak05, LBM00, MTK02, OZ93]. The propagation of electric waves in non-dispersive uniaxial anisotropic dielectrics can be described by the following relations [Yak05]

$$\mathcal{A}\frac{\partial^2 \mathbf{E}}{\partial t^2} + curl_x curl_x \mathbf{E} = \mathbf{f}(x,t), \quad x \in R^3, \quad t \in R, \tag{1}$$

$$\mathbf{E}|_{t=0} = 0, \; \left.\frac{\partial \mathbf{E}}{\partial t}\right|_{t=0} = 0, \tag{2}$$

where $\mathbf{f}(x,t) = -\mu\partial\mathbf{j}(x,t)/\partial t$, $\mu > 0$ is the constant (the magnetic permeability), $\mathbf{j}(x,t) = (j_1(x,t), j_1(x,t), j_1(x,t))$ is the density of electric current; $\mathcal{A} = \mu\mathcal{E}$, $\mathcal{E}$ (permittivity) is the diagonal matrix of the form $\mathcal{E} = diag(\varepsilon_{11}, \varepsilon_{11}, \varepsilon_{33})$. The main problem of the present paper is the Initial Value Problem (IVP)

Kenan Taş et al. (eds), Mathematical Methods in Engineering, 455–465.

for finding electric field $\mathbf{E}$ satisfying (1), (2) if the vector function $\mathbf{f}(x,t)$ and the matrix $\mathcal{A}$ are given. In our paper we suppose that the Fourier transform of the vector function $\mathbf{f}$ with respect to variables x_1, x_2 has components which are continuous relative to all variables simultaneously. We assume also that elements of the matrix $\mathcal{E} = diag(\varepsilon_{11}, \varepsilon_{11}, \varepsilon_{33})$ are twice continuously differentiable functions depending on x_3 variable only and such that $\mu\varepsilon_{jj}(x_3) = a_j^2(x_3) > 0$ for $x_3 \in R$, $j = 1, 3$. The main result of this paper is a new method for the recovery of the electric field $\mathbf{E}$ in the vertical inhomogeneous uniaxial anisotropic dielectrics by solving (1), (2).
Let $\tilde{\mathbf{E}}(\nu, x_3, t)$ be defined by

$$\tilde{\mathbf{E}}(\nu, x_3, t) = \mathcal{F}_{x_1x_2}[\mathbf{E}](\nu, x_3, t), \;\; \nu = (\nu_1, \nu_2) \in R^2,$$

where $\mathcal{F}_{x_1x_2}$ is the Fourier transform with respect to x_1, x_2, i.e.

$$\mathcal{F}_{x_1x_2}[\mathbf{E}](\nu, x_3, t) = \int_{-\infty}^{\infty}\int_{-\infty}^{\infty} \mathbf{E}(x,t)e^{i(\nu_1x_1+\nu_2x_2)}dx_1dx_2, \quad i^2 = -1.$$

Applying the operator $\mathcal{F}_{x_1x_2}$ to (1), (2) and using the properties of the Fourier transform we can write the problem (1), (2) in terms of the Fourier image $\tilde{\mathbf{E}}(\nu, x_3, t)$ as follows

$$\frac{\partial^2 \tilde{E}_j}{\partial t^2} - \frac{1}{a_1^2(x_3)}\frac{\partial^2 \tilde{E}_j}{\partial x_3^2} = \frac{1}{a_1^2(x_3)}\big[-\nu_k^2\tilde{E}_j + \nu_j\nu_k\tilde{E}_k + i\nu_j\frac{\partial \tilde{E}_3}{\partial x_3} + \tilde{f}_j\big], \quad (3)$$

$$\frac{\partial^2 \tilde{E}_3}{\partial t^2} + \frac{(\nu_1^2+\nu_2^2)}{a_3^2(x_3)}\tilde{E}_3 = \frac{1}{a_3^2(x_3)}\Big[i\nu_1\frac{\partial \tilde{E}_1}{\partial x_3} + i\nu_2\frac{\partial \tilde{E}_2}{\partial x_3} + \tilde{f}_3\Big], \quad (4)$$

$$\tilde{\mathbf{E}}|_{t=0} = 0, \quad \frac{\partial \tilde{\mathbf{E}}}{\partial t}|_{t=0} = 0, \quad (5)$$

where $j = 1, 2$; k is different from j and runs values $1, 2$.

2 Reduction of IVP (3) - (5) to a vector integral equation

The aim of this section is to show that IVP (3) - (5) is reduced to an integral equations system.
Let us consider the following transformation

$$y = \tau(x_3), \;\; \tau(x_3) = \int_0^{x_3} a_1(\xi)d\xi. \quad (6)$$

We note that the function $y = \tau(x_3)$ has the inverse $x_3 = \tau^{-1}(y)$. Let us denote

$$\tilde{U}_l(\nu, y, t) = \tilde{E}_l(\nu, x_3, t)|_{x_3=\tau^{-1}(y)}, \; l = 1, 2, 3, \tag{7}$$

then

$$\frac{\partial \tilde{E}_m}{\partial x_3}(\nu, x_3, t)|_{x_3=\tau^{-1}(y)} = a_1(\tau^{-1}(y))\frac{\partial \tilde{U}_m}{\partial y}(\nu, y, t), \; m = 1, 2, 3. \tag{8}$$

Equation (3), (4) may be written in the form

$$\frac{\partial^2 \tilde{U}_j}{\partial t^2} - \frac{\partial^2 \tilde{U}_j}{\partial y^2} = A(y)\frac{\partial \tilde{U}_j}{\partial y} - \nu_k^2 B^2(y)\tilde{U}_j + \nu_j \nu_k B^2(y)\tilde{U}_k$$

$$+i\nu_j B(y)\frac{\partial \tilde{U}_3}{\partial y} + g_j(\nu, y, t), \; j = 1, 2; \; k \neq j, \; k = 1, 2; \tag{9}$$

$$\frac{\partial^2 \tilde{U}_3}{\partial t^2} + (\nu_1^2 + \nu_2^2)C(y)\tilde{U}_3 = C(y)\Big[a_1(\tau^{-1}(y))\Big(i\nu_1\frac{\partial \tilde{U}_1}{\partial y} + i\nu_2\frac{\partial \tilde{U}_2}{\partial y}\Big) + \tilde{f}_3(\nu, y, t)\Big], \tag{10}$$

where

$$A(y) = \frac{a_1'(x_3)}{a_1^2(x_3)}\Big|_{x_3=\tau^{-1}(y)}, \quad B(y) = \frac{1}{a_1(x_3)}\Big|_{x_3=\tau^{-1}(y)}, \tag{11}$$

$$C(y) = \frac{1}{a_3^2(x_3)}\Big|_{x_3=\tau^{-1}(y)}, \quad g_j(\nu, y, t) = \frac{\tilde{f}_j(\nu, x_3, t)}{a_1^2(x_3)}\Big|_{x_3=\tau^{-1}(y)}. \tag{12}$$

We seek a solution of (9), (10) in the following form

$$\tilde{U}_l(\nu, y, t) = S(y)\tilde{W}_l(\nu, y, t), \; l = 1, 2, 3, \tag{13}$$

where the function $S(y)$ is defined by

$$S(y) = exp(-\frac{1}{2}\int_0^y A(\xi)d\xi). \tag{14}$$

Substituting (13) into (9) we find

$$\frac{\partial^2 \tilde{W}_j}{\partial t^2} - \frac{\partial^2 \tilde{W}_j}{\partial y^2} = \Big[q(y) - \nu_k^2 B^2(y)\Big]\tilde{W}_j + \nu_j \nu_k B^2(y)\tilde{W}_k$$

$$+i\nu_j B(y)\Big[-\frac{1}{2}A(y)\tilde{W}_3 + \frac{\partial \tilde{W}_3}{\partial y}\Big] + \frac{g_j(\nu, y, t)}{S(y)},$$

$$j = 1,2;\ k \neq j,\ k = 1,2; q(y) = -\frac{1}{2}A'(y) - \frac{1}{4}A^2(y). \tag{15}$$

Using D'Alambert formula we can show that equation (15) with zero initial data is equivalent to the following integral equation

$$\tilde{W}_j(\nu, y, t) = \frac{1}{2}\int_0^t \int_{y-(t-\tau)}^{y+(t-\tau)} \Big\{\Big[q(\xi) - \nu_k^2 B^2(\xi)\Big]\tilde{W}_j(\nu, \xi, \tau)$$

$$+\nu_j\nu_k B^2(\xi)\tilde{W}_k(\nu, \xi, \tau) + i\nu_j B(\xi)\Big[-\frac{1}{2}A(\xi)\tilde{W}_3(\nu, \xi, \tau) + \frac{\partial \tilde{W}_3}{\partial \xi}(\nu, \xi, \tau)\Big]$$

$$+\frac{g_j(\nu, \xi, \tau)}{S(\xi)}\Big\}d\xi d\tau, \quad j = 1,2;\ k \neq j,\ k = 1,2. \tag{16}$$

Using the formula

$$B(y)\frac{\partial \tilde{W}_3}{\partial y}(\nu, y, t) = \frac{\partial}{\partial y}\Big(B(y)\tilde{W}_3(\nu, y, t)\Big) - B'(y)\tilde{W}_3(\nu, y, t)$$

equation (16) may be written as follows

$$\tilde{W}_j(\nu, y, t) = \frac{1}{2}\int_0^t \int_{y-(t-\tau)}^{y+(t-\tau)} \Big\{\Big[q(\xi) - \nu_k^2 B^2(\xi)\Big]\tilde{W}_j(\nu, \xi, \tau)$$

$$+\nu_j\nu_k B^2(\xi)\tilde{W}_k(\nu, \xi, \tau) + \frac{i\nu_j}{2}B(\xi)\Big[-\frac{1}{2}A(\xi) + B'(\xi)\Big]\tilde{W}_3(\nu, \xi, \tau)\Big\}d\xi d\tau$$

$$+\frac{i\nu_j}{2}\int_0^t \Big[B(y + (t-\tau))\tilde{W}_3(\nu, y + (t-\tau), \tau)$$

$$-B(y - (t-\tau))\tilde{W}_3(\nu, y - (t-\tau), \tau)\Big]d\tau$$

$$+\frac{1}{2}\int_0^t \int_{y-(t-\tau)}^{y+(t-\tau)} \frac{\tilde{f}_j(\nu, \tau^{-1}(\xi), \tau)}{a_1^2(\tau^{-1}(\xi))S(\xi)}d\xi d\tau, \quad j = 1,2; k = 1,2; j \neq k. \tag{17}$$

After changing a variable in the second integral, the equation (17) has the form

$$\tilde{W}_j(\nu, y, t) = \frac{1}{2}\int_0^t \int_{y-(t-\tau)}^{y+(t-\tau)} \Big\{\Big[q(\xi) - \nu_k^2 B^2(\xi)\Big]\tilde{W}_j(\nu, \xi, \tau)$$

$$+\nu_j\nu_k B^2(\xi)\tilde{W}_k(\nu, \xi, \tau) + \frac{i\nu_j}{2}B(\xi)\Big[-\frac{1}{2}A(\xi) + B'(\xi)\Big]\tilde{W}_3(\nu, \xi, \tau)\Big\}d\xi d\tau$$

$$+\frac{i\nu_j}{2}\Big\{\int_y^{y+t} B(\eta)\tilde{W}_3(\nu,\eta,y+t-\eta)d\eta$$

$$-\int_{y-t}^{y} B(\mu)\tilde{W}_3(\nu,\mu,-y+t+\mu)d\mu\Big\}$$

$$+\frac{1}{2}\int_0^t\int_{y-(t-\tau)}^{y+(t-\tau)}\frac{\tilde{f}_j(\nu,\tau^{-1}(\xi),\tau)}{a_1^2(\tau^{-1}(\xi))S(\xi)}d\xi d\tau,\ j=1,2;\ k=1,2;\ j\neq k. \tag{18}$$

Differentiating (18) with respect to y we get equations the left hand sides of which contain $\dfrac{\partial\tilde{W}_j}{\partial y}$, $j=1,2$. These are the following equations

$$\frac{\partial\tilde{W}_j}{\partial y}(\nu,y,t)=\frac{1}{2}\int_0^t\Big\{\Big[q(\xi)-\nu_k^2B^2(\xi)\Big]\tilde{W}_j(\nu,\xi,\tau)$$

$$+\nu_j\nu_kB^2(\xi)\tilde{W}_k(\nu,\xi,\tau)+\frac{i\nu_j}{2}B(\xi)\Big[-\frac{1}{2}A(\xi)+B'(\xi)\Big]\tilde{W}_3(\nu,\xi,\tau)$$

$$+B(\xi)\frac{\partial\tilde{W}_3}{\partial t}(\nu,\xi,\tau)\Big\}\Big|_{\xi=y-(t-\tau)}^{\xi=y+(t-\tau)}d\tau-\frac{B(y)}{a_3^2(\tau^{-1}(y))}\int_0^t\Big[i\nu_1a_1(\tau^{-1}(y))\Big(S'(y)$$

$$\times\tilde{W}_1(\nu,y,\tau)+S(y)\frac{\partial\tilde{W}_1}{\partial y}(\nu,y,\tau)\Big)+i\nu_2a_1(\tau^{-1}(y))\Big(S'(y)\tilde{W}_2(\nu,y,\tau)$$

$$+S(y)\frac{\partial\tilde{W}_2}{\partial y}(\nu,y,\tau)\Big)\Big]\frac{\sin\big(d(\nu,y)(t-\tau)\big)}{d(\nu,y)}d\tau+G_j(\nu,y,t), \tag{19}$$

where

$$G_j(\nu,y,t)=\frac{1}{2}\int_0^t\Big\{\frac{\tilde{f}_j(\nu,\tau^{-1}(\xi),\tau)}{a_1^2(\tau^{-1}(\xi))S(\xi)}\Big\}\Big|_{\xi=y-(t-\tau)}^{\xi=y+(t-\tau)}d\tau$$

$$-\frac{B(y)}{a_3^2(\tau^{-1}(y))}\int_0^t\tilde{f}_3(\nu,\tau^{-1}(y),\tau)\frac{\sin\big(d(\nu,y)(t-\tau)\big)}{d(\nu,y)}d\tau,$$

$$j=1,2;\ k\neq j,\ k=1,2.$$

The notation $\Big\{...\Big\}\Big|_{\xi=y-(t-\tau)}^{\xi=y+(t-\tau)}$ means the difference of the expression which is inside bracket for $\xi=y+(t-\tau)$ and $\xi=y+(t-\tau)$.

Integrating the equation (10) twice with respect to t with zero initial data and using the equation (13) we find

$$\tilde{W}_3(\nu, y, t) = C(y) \int_0^t \Big[i\nu_1 a_1(\tau^{-1}(y)) \Big(\frac{-A(y)}{2} \tilde{W}_1(\nu, y, \tau)$$

$$+ \frac{\partial \tilde{W}_1}{\partial y}(\nu, y, \tau) \Big) + i\nu_2 a_1(\tau^{-1}(y)) \Big(\frac{-A(y)}{2} \tilde{W}_2(\nu, y, \tau)$$

$$+ \frac{\partial \tilde{W}_2}{\partial y}(\nu, y, \tau) \Big) + \frac{\tilde{f}_3(\nu, \tau^{-1}(y), \tau)}{S(y)} \Big] \times \frac{\sin\big(d(\nu, y)(t-\tau)\big)}{d(\nu, y)} d\tau, \tag{20}$$

where

$$d(\nu, y) = \frac{\sqrt{\nu_1^2 + \nu_2^2}}{a_3(\tau^{-1}(y))}.$$

Differentiating (20) with respect to t we find a relation containing $\frac{\partial \tilde{W}_3}{\partial t}$ in the left-hand side:

$$\frac{\partial \tilde{W}_3}{\partial t}(\nu, y, t) = C(y) \int_0^t \Big[i\nu_1 a_1(\tau^{-1}(y)) \Big(\frac{-A(y)}{2} \tilde{W}_1(\nu, y, \tau)$$

$$+ \frac{\partial \tilde{W}_1}{\partial y}(\nu, y, \tau) \Big) + i\nu_2 a_1(\tau^{-1}(y)) \Big(\frac{-A(y)}{2} \tilde{W}_2(\nu, y, \tau)$$

$$+ \frac{\partial \tilde{W}_2}{\partial y}(\nu, y, \tau) \Big) + \frac{\tilde{f}_3(\nu, \tau^{-1}(y), \tau)}{S(y)} \Big] \cos\big(d(\nu, y)(t-\tau)\big) d\tau. \tag{21}$$

Equations (16), (19), (20), (21) represent system of integral equations with respect to unknowns $\tilde{W}_j$, $\frac{\partial \tilde{W}_j}{\partial y}$, $j = 1, 2$; $\tilde{W}_3$, $\frac{\partial \tilde{W}_3}{\partial t}$. This system can be written in the form

$$\mathbf{V}(\nu, y, t) = \mathbf{G}(\nu, y, t) + \int_0^t (\mathbf{K}\mathbf{V})(\nu, y, t, \tau) d\tau, \tag{22}$$

where $\mathbf{V} = (V_1, V_2, V_3, V_4, V_5, V_6)$ is unknown vector-function whose components are

$$V_1 = \tilde{W}_1, V_2 = \tilde{W}_2, V_3 = \tilde{W}_3, V_4 = \frac{\partial \tilde{W}_1}{\partial y}, V_5 = \frac{\partial \tilde{W}_2}{\partial y}, V_6 = \frac{\partial \tilde{W}_3}{\partial t}; \tag{23}$$

$\mathbf{G} = (G_1, G_2, G_3, G_4, G_5, G_6)$ is the given vector-function whose components are defined by

$$G_j(\nu,y,t)=\frac{1}{2}\int_0^t\int_{y-(t-\tau)}^{y+(t-\tau)}\frac{\tilde{f}_j(\nu,\tau^{-1}(\xi),\tau)}{a_1^2(\tau^{-1}(\xi))S(\xi)}d\xi d\tau,\quad j=1,2, \tag{24}$$

$$G_3(\nu,y,t)=\frac{C(y)}{S(y)}\int_0^t\tilde{f}_3(\nu,\tau^{-1}(y),\tau)\frac{\sin\big(d(\nu,y)(t-\tau)\big)}{d(\nu,y)}d\tau, \tag{25}$$

$$G_{3+j}(\nu,y,t)=\frac{1}{2}\int_0^t\Big\{\frac{\tilde{f}_j(\nu,\tau^{-1}(\xi),\tau)}{a_1^2(\tau^{-1}(\xi))S(\xi)}\Big\}\Big|_{\xi=y-(t-\tau)}^{\xi=y+(t-\tau)}d\tau$$

$$+\frac{B(y)}{a_3^2(\tau^{-1}(y))}\int_0^t\tilde{f}_3(\nu,\tau^{-1}(y),\tau)\frac{\sin\big(d(\nu,y)(t-\tau)\big)}{d(\nu,y)}d\tau, j=1,2; \tag{26}$$

$$G_6(\nu,y,t)=\frac{C(y)}{S(y)}\int_0^t\tilde{f}_3(\nu,\tau^{-1}(y),\tau)\cos\big(d(\nu,y)(t-\tau)\big)d\tau. \tag{27}$$

The components of the vector-operator $\mathbf{K}=(\mathcal{K}_1,\mathcal{K}_2,\mathcal{K}_3,\mathcal{K}_4,\mathcal{K}_5,\mathcal{K}_6)$ are defined by

$$(\mathcal{K}_j\mathbf{V})(\nu,y,t,\tau)=\frac{1}{2}\int_{y-(t-\tau)}^{y+(t-\tau)}\Big\{\Big[q(\xi)-\nu_k^2B^2(\xi)\Big]V_j(\nu,\xi,\tau)$$

$$+\nu_j\nu_kB^2(\xi)V_k(\nu,\xi,\tau)+\frac{i\nu_j}{2}B(\xi)\Big[-\frac{1}{2}A(\xi)+B'(\xi)\Big]\tilde{V}_3(\nu,\xi,\tau)\Big\}d\xi$$

$$+\Big\{\frac{i\nu_j}{2}B(\xi)V_3(\nu,\xi,\tau)\Big\}\Big|_{\xi=y-(t-\tau)}^{\xi=y+(t-\tau)},\ j=1,2;\ k\neq j,\ k=1,2, \tag{28}$$

$$(\mathcal{K}_3\mathbf{V})(\nu,y,t,\tau)=\frac{1}{a_3^2(\tau^{-1}(y))}\Big\{i\nu_1a_1(\tau^{-1}(y))\Big[\frac{-A(y)}{2}V_1(\nu,y,\tau)$$

$$+V_4(\nu,y,\tau)\Big]+i\nu_2a_1(\tau^{-1}(y))\Big[\frac{-A(y)}{2}V_2(\nu,y,\tau)$$

$$+V_5(\nu,y,\tau)\Big]\Big\}\frac{\sin\big(d(\nu,y)(t-\tau)\big)}{d(\nu,y)}, \tag{29}$$

$$(\mathcal{K}_{3+j}\mathbf{V})(\nu,y,t,\tau)=\frac{1}{2}\Big\{\Big[q(\xi)-\nu_k^2B^2(\xi)\Big]V_j(\nu,\xi,\tau)$$

$$+\nu_j\nu_k B^2(\xi)V_k(\nu,\xi,\tau) + B(\xi)V_6(\nu,\xi,\tau) + \frac{i\nu_j}{2}B(\xi)\Big[-\frac{1}{2}A(\xi)$$

$$+B'(\xi)\Big]V_3(\nu,\xi,\tau)\Big\}\Big|_{\xi=y-(t-\tau)}^{\xi=y+(t-\tau)} - \frac{B(y)}{a_3^2(\tau^{-1}(y))}\Big[i\nu_1 a_1(\tau^{-1}(y))\Big(\frac{-A(y)}{2}$$

$$\times V_1(\nu,y,\tau) + V_4(\nu,y,\tau)\Big) + i\nu_2 a_1(\tau^{-1}(y))\Big(\frac{-A(y)}{2}V_2(\nu,y,\tau)$$

$$+V_5(\nu,y,\tau)\Big)\Big]\frac{\sin\big(d(\nu,y)(t-\tau)\big)}{d(\nu,y)}, j=1,2;\ k\neq j,\ k=1,2, \tag{30}$$

$$(\mathcal{K}_6\mathbf{V})(\nu,y,t,\tau) = \frac{1}{a_3(\tau^{-1}(y))}\Big\{i\nu_1 a_1(\tau^{-1}(y))\Big[\frac{-A(y)}{2}V_1(\nu,y,\tau)$$

$$+V_4(\nu,y,\tau)\Big] + i\nu_2 a_1(\tau^{-1}(y))\Big[\frac{-A(y)}{2}V_2(\nu,y,\tau)$$

$$+V_5(\nu,y,\tau)\Big]\Big\}\cos\big(d(\nu,y)(t-\tau)\big). \tag{31}$$

As a result of above reasoning we conclude that the following propositions take place.

Proposition 1. *Under above notations and assumptions the Initial Value Problem (3)-(5) is equivalent to the operator integral equation (22).*

Proposition 2. *Let T be a fixed positive number, $\Delta(T) = \{(y,t)|\ 0\leq t\leq T-|y|\}$, components of $\boldsymbol{G} = (G_1, G_2, ..., G_6)$ be defined by (24)-(27). Then under above assumptions $G_j(\nu,y,t)$, $1\leq j\leq 6$ are continuous functions for $\nu\in R^2, (y,t)\in\Delta(T)$.*

Proposition 3. *Let components of the vector operator $\boldsymbol{K} = (\mathcal{K}_1, \mathcal{K}_2, ..., \mathcal{K}_6)$ be defined by (28)-(31). Then under above assumptions the expression $\int_0^t \big(\mathcal{K}_j\,\boldsymbol{V}\big)(\nu,y,t,\tau)d\tau$ is a continuous function for any $\leq j\leq 6$ and any vector function $\boldsymbol{V}(\nu,y,t)$ with continuous components for $\nu\in R^2, (y,t)\in\Delta(T)$.*

Proposition 4. *Let $\boldsymbol{K}$ be the operator defined by (28)-(31). Then under above assumptions the following inequalities are satisfied*

$$\Big|\int_0^t \Big(\mathcal{K}_j\,\boldsymbol{V}\Big)(\nu,y,t,\tau)d\tau\Big| \leq M\int_0^t \|\,\boldsymbol{V}\|(\nu,\tau)d\tau,\ j=1,2,...6;$$

where

$$\|\,\boldsymbol{V}\|(\nu,\tau) = \max_{1\leq j\leq 6}\ \max_{\xi\in[-(T-\tau),(T-\tau)]}|V_j(\nu,\xi,\tau)|$$

and M is a positive number depending on T, ν such that $M(\nu,T) = \mathrm{O}(|\nu|^2)$ as $|\nu|\to\infty$.

3 Solving the vector integral equation by successive approximations

Let T be a fixed positive number. Let us consider now the vector integral equation (22) for $t \in [0, T]$. For solving this equation we apply the following successive approximations

$$\begin{aligned} \mathbf{V}^{(0)}(\nu, y, t) &= \mathbf{G}(\nu, y, t), \\ \mathbf{V}^{(n)}(\nu, y, t) &= \int_0^t (\mathbf{K}\mathbf{V}^{(n-1)})(\nu, y, t, \tau) d\tau, \ n = 1, 2, ... \end{aligned} \tag{32}$$

The goal of this section is to show that the series $\sum_{n=0}^{\infty} \mathbf{V}^{(n)}(\nu, y, t)$ is uniformly convergent to a vector function $\mathbf{V}(\nu, y, t)$ with continuous components for $\nu \in R^2$, $(y, t) \in \Delta(T)$ and this vector function $\mathbf{V}(\nu, y, t)$ is a solution of (22). Indeed, we find from (32) and Propositions 2 – 4 that $\mathbf{V}^{(n)}(\nu, y, t)$, $n = 0, 1, 2...$ are vector functions with continuous components for $\nu \in R^2$, $(y, t) \in \Delta(T)$ and

$$|V_j^{(n)}(\nu, y, t)| \leq M \int_0^t \|\mathbf{V}^{(n-1)}\|(\nu, \tau) d\tau, \ t \in [0, T] \tag{33}$$

It follows from (33) that

$$|V_j^{(n)}(\nu, y, t)| \leq \frac{(Mt)^n}{n!} \|\mathbf{G}\|(\nu, t), n = 0, 1, 2.. \tag{34}$$

The uniform convergence of $\sum_{n=0}^{\infty} V_j^{(n)}(\nu, y, t)$ to a continuous function $V_j(\nu, y, t)$ follows from inequality (34) and the first Weierstrass theorem [Apo67] . Let us consider the vector function $\mathbf{V}(\nu, y, t) = (V_j(\nu, y, t), 1 \leq j \leq 6$. We show below that this vector function $\mathbf{V}(\nu, y, t)$ is a solution of (22). Summing the equation (32) with respect to n from 1 to N we have

$$\sum_{n=1}^{N} \mathbf{V}^{(n)}(\nu, y, t) = \sum_{n=0}^{N-1} \int_0^t (K\mathbf{V}^{(n)})(\nu, y, t, \tau) d\tau, \tag{35}$$

where

$$\sum_{n=1}^{N} \mathbf{V}^{(n)}(\nu, y, t) = \Big(\sum_{n=1}^{N} V_1^{(n)}(\nu, y, t), ..., \sum_{n=1}^{N} V_6^{(n)}(\nu, y, t) \Big)$$

adding both sides $\mathbf{G}(\nu, y, t)$ of the equation (35) we get

$$\sum_{n=0}^{N} \mathbf{V}^{(n)}(\nu, y, t) = \mathbf{G}(\nu, y, t) + \int_0^t \sum_{n=0}^{N-1} (K\mathbf{V}^{(n)})(\nu, y, t, \tau) d\tau, \tag{36}$$

approaching N the infinity and using the second Weierstrass theorem [Apo67] we find that $\mathbf{V}(\nu, y, t)$ satisfies (22) for $\nu \in R^2$, $(y,t) \in \Delta(T)$, $t \in [0,T]$. This means that $\mathbf{V}(\nu, y, t)$ is a solution of (22) for $\nu \in R^2$, $(y,t) \in \Delta(T)$.

4 Finding the electric field

Let T be a given number, $\nu \in R^2, (y,t) \in \Delta(T)$; $\mathbf{V}(\nu, y, t) = \Big(V_1(\nu,y,t), V_2(\nu,y,t), V_3(\nu,y,t), V_4(\nu,y,t), V_5(\nu,y,t), V_6(\nu,y,t)\Big)$ be the solution of (22) found for $\nu \in R^2$, $(y,t) \in \Delta(T)$ by the method of successive approximations described in the section 3. Using the formulae (7), (8), (14) and (23) we find the Fourier transform images of the electric field components $\tilde{E}_l(\nu, x_3, t)$, $l = 1,2,3$ and $\frac{\partial \tilde{E}_j}{\partial x_3}(\nu, x_3, t)$, $j = 1,2$; $\frac{\partial \tilde{E}_3}{\partial t}(\nu, x_3, t)$ by the following relations

$$\tilde{E}_l(\nu, x_3, t) = S(\tau(x_3)) V_l(\nu, \tau(x_3), t), \quad l = 1,2,3, \tag{37}$$

$$\begin{aligned} \frac{\partial \tilde{E}_j}{\partial x_3}(\nu, x_3, t) = a_1(x_3)\Big[&S'(\tau(x_3))V_j(\nu, \tau(x_3), t) \\ &+ S(\tau(x_3))V_{j+3}(\nu, \tau(x_3), t)\Big], \quad j = 1,2; \end{aligned} \tag{38}$$

$$\frac{\partial \tilde{E}_3}{\partial t}(\nu, x_3, t) = S(\tau(x_3)) V_6(\nu, \tau(x_3), t). \tag{39}$$

Applying the inverse Fourier transform $\mathcal{F}_\nu^{-1}$ to formulae (37)-(39) we find the electric field components $E_l(x,t)$, $l = 1,2,3$ and their derivatives $\frac{\partial E_j}{\partial x_3}(x,t)$, $j = 1,2$; $\frac{\partial E_3}{\partial t}(x,t)$ for $(x_1,x_2) \in R^2$, $(x_3,t) \in \tilde{\Delta}(T)$. Here

$$\tilde{\Delta}(T) = \{(x_3,t)|\ 0 \le t \le T - |\tau(x_3)|\},$$

$$\mathcal{F}_\nu^{-1}[\tilde{E}_l](x,t) = \frac{1}{(2\pi)^2}\int_{-\infty}^{\infty}\int_{-\infty}^{\infty} \tilde{E}_l(\nu, x_3, t) e^{i(\nu_1 x_1 + \nu_2 x_2)} d\nu_1 d\nu_2.$$

5 Conclusion

In the paper a new method of finding the time-dependent electric field in a layered inhomogeneous uniaxial anisotropic dielectric is described. The density of the electric current is the source of electric waves. The Fourier transform of the electric current density with respect to x_1 and x_2 variables is assumed to be a continuous function. The suggested method consists of finding the Fourier image with respect to x_1 and x_2 variables of the electric field. The problem of finding the Fourier image of the electric field is reduced to an operator integral equation. This operator integral equation is solved by successive approximations method. After that the time-dependent electric field is found by the inverse Fourier transform.

6 Acknowledgment

This work is supported by Dokuz Eylul University, Turkey, under a research grant number 03.KB.FEN.049 .

References

[RWD94] Ramo, S., Whinnery, J.R., Duzer, T.: Fields and Waves in Communication Electronics. John Wiley and Sons, New York (1994)

[Yak05] Yakhno, V.G.: Constructing Green's function for the time-dependent Maxwell system in anisotropic dielectrics. Journal of Physics A : Mathematical and General, **38**, 2277– 2287 (2005)

[LBM00] Losada, V., Boix, R.R., Horno, M.: Full-wave analysis of circular microstrip resanators in multilayered media containing uniaxial anisotropic dielectrics, magnetized ferrites, and chrial materials. IEEE Trans. Microwave Theory Tech., **48**, 1057–1064 (2000)

[MTK02] Moss, C.D., Teixeria, F.L., Kong, J.A.: Analysis and compensation of numerical dispersion in the FDTD methods for layered, anisotropic media. IEEE Trans. Antennas Propagat., **50**, 1174–1184, (2002)

[OZ93] Olyslager, F., Zutter, D.D.: Rigorous boundray integral equation solution for general isotropic and uniaxial anisotropic dielectric waveguides in multilayered media including losses, gain and leakeage. IEEE Trans. Microwave Theory Tech., **41**, 1385–1392 (1993)

[Apo67] Apostol, T.M.: Calculus: Volume I. Blaisdell Publishing Company, Massachusetts (1967)

Author index